About the Cover Design

*The cover design is copied from drawings of the
Panamanian golden frog (Atelopus various zeteki)
found on pottery belonging to pre-Columbian Coclé
Indian culture. This bright golden frog occurs only in
the Valle de Antón, an area approximately two miles in
diameter. It is one of the attractions which has made the
valley a popular resort and is modeled on trinkets sold
in the shops of Panama. Unfortunately, the golden frog
is endangered because of collecting by tourists and
exploiters of its beauty.*

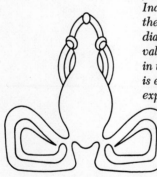

HERPETOLOGY

KENNETH R. PORTER

Department of Biological Sciences
The University of Denver

W. B. SAUNDERS COMPANY · PHILADELPHIA · LONDON · TORONTO

W. B. Saunders Company: West Washington Square
Philadelphia, Pa.   19105

12 Dyott Street
London, WC1A 1DB

833 Oxford Street
Toronto 18, Ontario

Herpetology

ISBN 0-7216-7295-7

Print No.: 9   8   7   6   5   4   3   2

This book is dedicated
to all researchers
whose publications provided its contents.

PREFACE

I wrote this book because many of my colleagues and I have felt a need for a modern and general reference on major aspects of the biology of amphibians and reptiles. In assigning texts for our courses, those of us who teach herpetology have had to choose between publications which are detailed treatments of particular subjects, field guides which are designed largely for the identification of species in a given region, and books which are abbreviated or restricted in scope. It is my hope that this book will be appropriate as a general text and will also be useful as a reference for all those who have an interest in amphibians and reptiles.

In choosing the contents for this book, I have been swayed by my opinion that students of recent years have been taught a considerable amount about molecular and cellular biology but have been shortchanged in their education regarding the structure and functioning of organisms as intact individuals, members of populations, and members of biotic communities. Consequently, I have included some basic material which might otherwise have been assumed to be in the background of the reader. I hope that readers with more general backgrounds will not find the amount of basic material excessive. Throughout the book, I have tried to document major points with citations to pertinent literature, and I hope that all readers will find the list of references at the end of each chapter useful when more detailed knowledge of the subject is desired.

I have attempted in this book to encompass the classes Amphibia and Reptilia in their entirety rather than to restrict the discussion to extant or regional forms. The inclusion of paleontological material is, perhaps, a departure from the expected contents of a herpetological text, but I view some knowledge of extinct forms as fundamental to an understanding of living forms. As is noted repeatedly in the book, many aspects of amphibian and reptilian biology have been studied in only a few species. However, an attempt has been made to discuss subjects in terms of principles which should be applicable to amphibians and reptiles everywhere. Subjects which are particularly in need of research have been noted.

Since much remains to be learned regarding the phylogenetic relationships of most groups, I have attempted neither to revise existing taxonomic schemes nor to classify amphibians and reptiles below the family or subfamily level. General characteristics of families are given, but the reader will certainly want to supplement this volume with appropriate field guides or taxonomic keys to species in his region of concern.

In recent years, the volume of herpetological literature has become so extensive that no one book can be expected to encompass more than a portion of it. In fact, one of the greatest difficulties in writing a general reference is that of becoming aware of all the pertinent information which has appeared in a great diversity of publications. I am indebted to many persons who have called to my attention various references. I have been exceptionally fortunate in having the entire first draft of *Herpetology* thoroughly read and criticized by Richard G. Zweifel and William E. Duellman. Their many suggestions and leads to the literature have greatly improved the contents of the book, and I am very grateful for the laborious tasks which they performed. I also wish to acknowledge the artistic work of Wendy F. Porter, who did most of the preparation of original illustrations. I am indebted to many authors and publishers who granted permission to reproduce illustrations already in the literature.

Finally, I would appreciate having corrections and other suggestions for the improvement of *Herpetology* forwarded to me.

KENNETH R. PORTER

CONTENTS

CHAPTER 1

INTRODUCTION

THE RELATION OF HERPETOLOGY TO OTHER FIELDS OF BIOLOGY

Herpetology includes all aspects of biology in which attention is focused on amphibians and reptiles. Thus, herpetology, in contrast to mammalogy and ornithology, which are studies of the class Mammalia and the class Aves respectively, is a field of study devoted to two distinct classes of animals just as ichthyology is the study of several classes of fishlike animals. Herpetological studies may be made from the functional approaches of genetics, physiology, ecology, behavior, and so forth, or from molecular, cellular, organismal, populational, and such levels of biological organization. The only restriction on herpetological research is that it must involve amphibians, reptiles, or both. Consideration will now be given to the taxonomic position and characteristics of amphibians and reptiles.

All amphibians and reptiles are contained in the phylum Chordata, whose members may be differentiated from those of other animal phyla by a notochord, a dorsal hollow nerve tube, and gill slits (or traces of these three features) at some stage in development. All of them are also members of the subphylum Vertebrata, which includes all chordates possessing an enlarged anterior end to the nerve cord which forms the brain, a cranium surrounding the brain, a segmented spinal column composed of vertebrae enclosing the nerve cord, and sense organs in the head. Because of the development of the cranium, this subphylum is sometimes referred to as the subphylum Craniata.

Birds, mammals, reptiles, and amphibians are grouped in the superclass Tetrapoda, the tetrapods, whereas the various fishlike vertebrates are contained in the superclass Pisces. Tetrapods are those vertebrates in which the paired appendages, if present, are typically in the form of limbs and usually end in five digits. In addition, tetrapods exhibit several features associated with terrestrial life: the outer layer of their skin is cornified, lungs are typically present in the adult form, the skeleton is ossified for the most part, a sternum is usually present, the number of bones in the cranium is reduced in comparison to that of the fish, and there is a reduction in the visceral skeleton.

The superclass Tetrapoda contains four distinct classes of vertebrates: class Aves, class Mammalia, class Reptilia, and class Amphibia. The amphibians and reptiles are distinguishable from the other tetrapods because their skin is devoid of hair and feathers; they possess a sinus venosus which is incorporated into the right auricle of birds and mammals; there are five pairs of pharyngeal pouches in the embryo, whereas four pairs appear in birds and mammals; they depend primarily on ex-

1

ternal rather than internal sources of heat in regulating their body temperature; and, with the exception of crocodiles and alligators which have a perfect four-chambered heart like the other tetrapods, amphibians and reptiles lack complete separation of arterial and venous blood because of an incomplete partition dividing the two halves of the ventricle.

THE HISTORY OF HERPETOLOGY

As was true of biology in general, herpetology initially consisted of collecting, naming, and describing new forms; many of the early herpetologists were museum workers. The first truly systematic arrangement of animals, including amphibians and reptiles, was published in 1693 by John Ray (1628–1705), an English naturalist, and titled *Synopsis Methodica Animalium Quadrupedum et Serpentini Generis*. Ray, the first to use the term species in the biological sense, was also the first to group amphibians and reptiles together because of their similarity in structure of the heart.

The great Swedish naturalist Carolus Linnaeus (1707–1778) appears to have had little interest in the reptiles and amphibians that he knew, for in the summary of his 10th *Systema Naturae*, published in 1758, he stated:

These foul and loathsome animals are distinguished by a heart with a single ventricle and a single auricle, doubtful lungs and a double penis.

Most amphibia are abhorrent because of their cold body, pale colour, cartilaginous skeleton, filthy skin, fierce aspect, calculating eye, offensive smell, harsh voice, squalid habitation, and terrible venom; and so their Creator has not exerted his powers [to create] many of them.

However, the 10th *Systema Naturae* made an important contribution to herpetology, just as to other fields of biology, for it introduced the concept of a binomial nomenclature together with priority concepts in nomenclature. It became the foundation for classification of all forms of life.

DEVELOPMENTS IN EUROPE

From the time of Linnaeus through the 19th century, herpetology was primarily descriptive; this was the period when vast numbers of amphibians and reptiles were first named and described. Some workers, however, began to concern themselves with the physiology, ecology, anatomy, and general natural history of these animals. This was when France was near its peak in the scientific community, and several French naturalists played important roles in herpetology. Georges L. L. de Buffon (1707–1788) provided much of the stimulation for other Frenchmen in the area of biology. He was a naturalist who also studied law and the physical sciences, especially mathematics. When he was 25 years old, he inherited a fortune from his mother; from then on he devoted his life to regular scientific labor. At first he was concerned primarily with mathematics, physics, and agriculture, but from 1739 until his death he concentrated on biology. He published, but did not author, several editions of *Histoire Naturelle* with various volumes written by different authors, including Bernard Lacépède and François M. Daudin. A total of 44 volumes were included in the last edition of the *Histoire Naturelle*, the last eight dealing in part with reptiles (the others covered plants, minerals, and other kinds of animals). Buffon's major contribution to herpetology was the stimulation he provided other French naturalists.

Bernard Lacépède (1756–1825) studied under Buffon and continued the *Histoire Naturelle* series under the title *Histoire des Quadrupèdes Ovipares et des Serpents*, published in 1789. In 1798, Lacépède published the first volume of *Histoire Naturelle des Poissons;* the fifth volume appeared in 1803. From then until he died, his participation in politics prevented him from making any further contributions of importance to science.

François M. Daudin (1774–1804) was another French naturalist who studied both birds and reptiles. In 1802, he published *Histoire Naturelle des Reptiles,* which complemented Buffon's *Histoire*

Naturelle. In the same year, he published *Histoire Naturelle des Quadrupèdes Ovipares*.

Between 1834 and 1854, the famous French anatomist André Marie Constant Duméril (1764–1860) and Gabriel Bibron (1806–1848) published a 10-volume series plus atlas titled *Erpétologie Générale ou Histoire Naturelle Complète des Reptiles;* Auguste H. A. Duméril (1812–1870), son of the senior author, aided in the preparation of volumes 7 and 9 after the death of Bibron. The *Erpétologie Générale* included a comprehensive account of the structure, physiology, and systematics of amphibians and reptiles; a general account of each of the orders recognized (amphibians were classified as the order Batrachia and crocodilians were grouped with lizards); descriptions of 121 species of turtles, 468 species of "lizards," 586 species of snakes, and 218 species of "batrachians;" and an atlas of 120 colored plates. This work was a major summary of the field of herpetology, including all information on reptiles and amphibians known at the time, and still remains a publication of major importance.

Gabriel Bibron was strictly a zoologist, one of the first to confine himself to this division of biology. He worked at the Musée National d'Histoire Naturelle at Paris with the Dumérils and made many of the field trips to Italy, England, and Holland which provided some of the information incorporated into the *Erpétologie Générale*. Bibron also published *Les Reptiles de Cuba* in 1840, one of the first pieces of herpetological literature dealing with the New World fauna.

Another associate of the Dumérils was Marie-Firmin Bocourt, who joined the Paris museum staff in 1834 as a 15-year-old preparator and became both a competent herpetological illustrator and field collector. Bocourt made an expedition to Siam in 1861–1862 and in 1864 was placed in charge of the French scientific mission to Mexico and Central America that was associated with the attempt to establish a French empire in Mexico under Maximilian. After the downfall of Maximilian, Bocourt went to Guatemala and other areas of Central America to collect. Returning to Paris in 1867, his collections were described and illustrated in *Mission Scientifique au Mexique et dans l'Amérique Centrale; Études sur les Reptiles et les Batraciens,* a 17-volume publication produced between 1870 and 1909 and coauthored by Auguste Duméril, Paul Brocchi, F. Mocquard, and Bocourt; Brocchi wrote the three-volume amphibian section, Duméril, Bocourt, and Mocquard, the reptile section. The accompanying atlas included 101 plates, 95 by Bocourt and the last six by Fernand Angel, another Musée National herpetologist. This work included illustrations of many type specimens; it is a basic reference for Central American herpetology.

Particularly through publication of the *Erpétologie Générale,* the French were the recognized leaders in the field of herpetology during the early 19th century. However, major herpetological contributions were also being made by a number of other Europeans. Hermann Schlegel (1804–1884), a Dutch naturalist and Director of the Museo Imperial at Leyden, published *Essai Sur la Physionomie des Serpents* (1837). The only general compendium of data on the morphology of amphibians and reptiles was compiled by Christian Karl Hoffmann (1841–1903) of the University of Leyden, and published in Bronn's *Klassen und Ordnungen des Thierreichs* between 1873 and 1890.

In Berlin, Wilhelm Carl Hartwig Peters (1815–1883), Professor of Zoology at the University of Berlin and Director of the Zoological Museum, described many new species of amphibians and reptiles from throughout the world. Having personally made an expedition to Africa, Peters also authored three of the five volumes of *Reise nach Mossambique,* including that on "amphibians" (this included both amphibians and reptiles since he followed the Linnaean classification); the African reports were published between 1852 and 1882.

Franz Steindachner (1834–1919), a prominent ichthyologist as well as herpetologist, founded an Austrian school of herpetologists at Vienna. Steindachner, who joined the staff of the Naturhistorisches Museum in 1860, published many papers on amphibians and reptiles. These included descriptions and notes on collections he made in Africa, Brazil, the Galápagos Islands, and southwestern Asia.

The Russian Imperial Academy of Sciences and the Museum of Natural History in St. Petersburg became prominent in herpetology in the latter part of the 19th century. Alexander Strauch (1832–1893), associated with the Zoological Museum of the Imperial Academy of Sciences from 1861 to 1892 and its director for the last 13 of these years, published herpetological papers in the *Memoirs* and *Bulletin* of this academy which included revisions of the descriptions and classifications of the crocodilians, turtles, and viperine snakes of the world (all in German). A. M. Nikolsky, who succeeded Strauch in Russian herpetology, wrote *Faune de la Russie,* comprehensive accounts of the herpetological fauna of Russia published between 1915 and 1918. Yakov Vladimirovitch Bedryagha (Jacques de Bedriaga) made extensive studies of the amphibians and reptiles of the Mediterranean region, Europe, and Mongolia; among his publications is an account of the amphibians of Europe, *Die lurch fauna Europas,* published between 1889 and 1897.

Major 19th century Italian herpetological publications include the second volume of C. L. Bonaparte's *Iconografia della Fauna Italica* (1832–1841), in which 53 colored plates illustrated the amphibians and reptiles of Italy, and *Iconographie Générale des Ophidiens* (1860–1881), started by George Jan (1791–1866) when he was Director of the Museum of Natural History in Milan. This publication was finished after Jan's death by Ferdinand Sordelli; it was an attempt to illustrate the snakes of the world but was reduced in scope by the refusal of the British Museum to lend its specimens to the illustrators.

By the late 19th Century, leadership in the field of herpetology resided with the British Museum in London. John Edward Gray (1800–1875), Keeper of Zoology, was a prolific publisher of papers on amphibians and reptiles and initiated publication of the museum catalogues. More importantly, in 1857 he selected a German named Albert C. L. G. Günther (1830–1914) to be his assistant in the divisions of ichthyology and herpetology. Günther, who later became Gray's successor as Keeper of Zoology, was an outstanding ichthyologist and a superb herpetologist. Among his important publications are the 1858 museum catalogues of Batrachia Salientia and of colubrine snakes, *Reptiles of British India* (1864), and volume 7, *Reptilia and Batrachia,* of *Biologia Centrali-Americana* (1885–1902). In 1867, Günther examined a tuatara and, to the great surprise of other herpetologists, concluded that it was not a lizard; his subsequent studies showed that the tuatara's skeleton differs from those of all other living reptiles and that, among other features, the upper jaw is beaklike and exactly as in the extinct reptiles Sir Richard Owen had named Rhynchocephalus in 1842. Günther used this name for a separate new order of reptiles, the order Rhynchocephalia. His discovery of the "living fossil" aroused much interest throughout the world, and a considerable volume of publications appeared on this animal during the following years. What is often said to be Günther's greatest contribution to herpetology, however, was his choice of the Belgian George Albert Boulenger as his successor at the British Museum.

Boulenger (1858–1937) studied at the University of Brussels and while a student assisted in the identification of specimens in the Belgian National Museum, also in Brussels. He published 20 papers on his work there and because of their quality was invited by Günther in 1880 to become Curator of Reptiles at the British Museum, one of the most prestigious positions in herpetology anywhere. Boulenger immediately began working on revisions of the museum catalogues. Following the suggestions made by Cope in his "Sketch of the Primary Groups of Batrachia Salientia" (1865), Boulenger completely revised the classification of amphibians in his *Catalogue of the Batrachia Gradienta s. Caudata and Batrachia Apoda in the Collection of the British Museum* and

Catalogue of the Batrachia Salientia s. Ecaudata in the Collection of the British Musem. Both of these catalogues appeared in 1882 and were followed by his three-volume *Catalogue of the Lizards in the British Museum* (1885–1887), the one-volume *Catalogue of the Chelonians, Rhynchocephalians, and Crocodiles in the British Museum* (1889), and the three-volume *Catalogue of the Snakes in the British Museum* (1893–1896). These nine volumes not only included important contributions to the classification of each group but represented a summary of the amphibians and reptiles of the world as known in 1896, since they each include descriptions of species in addition to those in the British Museum. What is amazing is that, during the same period of time that Boulenger was working on these catalogues (1881–1896), he published 279 herpetological papers in scientific journals and a volume on the amphibians and reptiles of British India (Schmidt, 1955)! Other Boulenger publications of note include *Tailless Batrachians of Europe* (1896–1897), *Les Batraciens et principalement ceux d'Europe* (1910), "Reptilia and Batrachia" in *Vertebrate Fauna of the Malay Peninsula* (1912), *Snakes of Europe* (1913), and *Monograph of the Lacertidae* (1920–1921). Boulenger published a total of more than 875 papers in scientific journals, of which 618 were on amphibians and reptiles and 257 on fish (Schmidt, 1955). After 40 years at the British Museum, Boulenger retired in 1920 to return to Belgium, where he pursued an old interest in wild roses.

THE NORTH AMERICAN CONTRIBUTION

During the early part of the 19th century, many exploration parties were traversing the North American continent; these generally included geologists and doctors or, occasionally, naturalists, who made collections of organisms encountered. Consequently, the early American herpetologists were primarily concerned with North American faunas, whereas the Europeans had begun to make worldwide studies. Thomas Say (1787–1843), son of the "fighting Quaker" of the Revolutionary War, Benjamin Say, was one of the first American naturalists to publish on amphibians and reptiles. Say was active in the founding of the Academy of Natural Sciences in Philadelphia and, in 1812, became its first secretary. In 1817 he became a member of the McClure expedition to Georgia and upon returning from that was recruited for the Long Yellowstone Expedition; he also accompanied Long on the 1823 expedition. Between these two assignments, Say was appointed Curator of the American Philosophical Society and occupied the Chair of Natural History at the University of Pennsylvania (1822–1828). Being the zoologist on Long's expeditions, Say described most of the collections which resulted from them. Richard Harlan (1796–1843) also worked at the Philadelphia Academy of Natural Sciences and published on both living and fossil forms. His most notable work, *Fauna Americana,* appeared in 1825.

John Edward Holbrook (1794–1871) was a practicing physician, herpetologist, and ichthyologist. After receiving his doctorate in medicine from the University of Pennsylvania in 1818, Holbrook spent the next four years in travel and graduate study in Europe where he established lifelong friendships with both the Dumérils and Bibron. These Frenchmen drew Holbrook's attention to reptiles. When he returned to the United States to settle at Charleston, South Carolina as a practicing physician, he began preparation of what he planned to be a monograph on the reptiles and batrachians of the United States; two years after arriving there, he helped establish the Medical College of South Carolina and was himself chosen as Professor of Anatomy, a position he held for thirty years in addition to retaining his medical practice. Being financially able, Holbrook hired the Italian artist J. Sera to make colored figures from living specimens of all the American reptiles he could obtain. These plates plus an accompanying text were bound together in the sequence in which they were completed and titled *North American Her-*

petology; the first volume appeared in 1836 and two more volumes were completed in 1838. Realizing that the organization of these three volumes was both inconvenient and unscientific, Holbrook in 1842 published five quarto volumes under the same name with the plates and text arranged in a systematic sequence. This 1842 publication immediately gained the respect of the Europeans, and Holbrook became regarded as the leading American zoologist of his day. Subsequent to publication of *North American Herpetology,* Holbrook shifted his interests to fish and became a recognized ichthyologist.

Spencer Fullerton Baird (1823–1887) had a very broad background in the sciences and languages; he simultaneously held chairs of chemistry and natural history at Dickinson College in Pennsylvania. In 1850 he went to the Smithsonian Institution as an assistant secretary and became the father of herpetological work at that institute. One of his duties as Keeper of Cabinet was to make collections for the Smithsonian Museum, and, through a vast network of private and governmental agencies plus personal contacts, Baird kept a steady stream of material flowing into the Smithsonian from practically all over the world. Much to his credit, this museum is now one of the great museums of the world. Although primarily an ornithologist, Baird played an important role in the development of American herpetology. Aside from his many descriptions and reports on collections received, he served as a major stimulus for many of the younger American zoologists of his day, just as had Buffon in France about a century earlier. One of Baird's more extensive publications in herpetology was *Catalogue of North American Reptiles,* Part I (snakes), which he co-authored with Charles Girard.

Charles Frédéric Girard (1822–1895) was educated at Neuchâtel, Switzerland, where he was a pupil and assistant of Louis Agassiz. When Agassiz came to the United States in 1847 he brought Girard with him. In 1850, when Baird was made an assistant secretary of the Smithsonian Institution, he invited Girard to join him and assist in the establishment of the United States National Museum, this being ac-

complished in 1857. Fish and reptiles were Girard's chief interest, but he also studied insects and other invertebrates. Girard frequently co-authored publications with Baird, although he generally did most of the work himself. For example, "Herpetology" of the Wilkes exploring expedition (1858) was written entirely by Girard but co-authored with Baird. Girard became a naturalized citizen of the United States in 1854 and in 1856 obtained a medical doctorate from Georgetown College. While visiting in Europe in 1861, Girard received the Cuvier Prize from the Institute of France. The troubles leading to the Civil War began while Girard was in Paris, and he returned to the Confederacy to supply the Southerners with drugs and surgical instruments. After the war ended, he returned to Paris and entered a medical career in which he stayed until his death. Girard is probably best known in American herpetology for his descriptions of specimens collected by exploration and survey parties in western North America.

Edward Drinker Cope (1840–1897) was the outstanding American herpetologist of the 19th century. His formal education was at Friends' School at Westtown, Pennsylvania, supplemented by one year at the University of Pennsylvania and some private tutoring. In 1859, he went to Washington to study reptiles under Baird at the Smithsonian Institution and while there published his first paper in that year's *Proceedings of the Academy of Natural Sciences of Philadelphia,* titled "On the Primary Division of the Salamandridae, with a Description of Two New Species." Returning to Philadelphia, Cope worked under Joseph Leidy at the University of Pennsylvania and spent considerable time each day at the Academy. He was elected to the Academy when only 21 years old, in 1861, and became a curator at the age of 25. When he was only 22, Cope was recognized as one of the country's leading anatomists. In 1864, he accepted the Chair of Comparative Zoology and Botany at Haverford College but resigned from this in 1867 because of ill health. He then spent much of his time describing fossils, including those of dinosaurs, collected by the Hayden Survey in Wyoming and other explorations; much of this work was done together with Leidy. Cope ran into financial trouble through bad investments and, after losing much of his fortune, was forced to become Professor of Geology and Mineralogy at the University of Pennsylvania in 1889. In 1895, after Leidy's death, he assumed the Chair of Zoology and Comparative Anatomy, holding this position until his death two years later. Cope was an extremely influential force in anatomy, publishing as many as 50 papers a year—approximately 1395 in all. His major herpetological publications include *The Batrachia of North America* (1889), *Classification of the Ophidia* (1895), and *The Crocodilians, Lizards, and Snakes of North America*, which was published in 1900, three years after he died. He published on Asian, South American, and African species as well as those from the United States. In addition, from 1878 until his death he was owner and senior editor of the *American Naturalist*, a major scientific journal. Clearly, herpetology in the United States had its roots in the Academy of Natural Sciences of Philadelphia and originated at an early time in this nation's history.

The United States National Museum, established through the efforts of Baird and Girard in 1857, came into prominence in herpetology when Leonhard Stejneger (1851–1943) was appointed Curator of Herpetology in 1889. Stejneger was born and educated in Norway and originally studied medicine because of an interest in botany and zoology. Later, he changed his mind about a medical career and graduated from law school in 1855 in order to go into business. After failing in his business career, he emigrated to the United States in 1881, determined to be a professional zoologist. Going directly to the Smithsonian Institution, he was given employment by Baird at the National Museum. After eight years, during which he confined himself almost entirely to studies of birds, Stejneger was made Curator of the Division of Herpetology (1889) and subsequently concentrated on the systematics of amphibians and reptiles. He soon established himself as an authority on the subject, especially emphasizing detailed descriptions of type specimens and type localities (the modern concept of species as variable populations had not yet been introduced at that time). Among

his important publications are *The Poisonous Snakes of North America* (1895), *The Herpetology of Porto Rico* (1904), and *Herpetology of Japan* (1907). Thomas Barbour (1884–1946), who worked as Director of the Harvard University Museum of Comparative Zoology, was a close friend of Stejneger's, and together they initiated the *Check List of North American Amphibians and Reptiles*. The first edition of this publication appeared in 1917, and they later published four additional editions. The sixth edition of this volume was authored by the late Karl P. Schmidt and was published in 1953.

LEADING CENTERS, HERPETOLOGISTS, AND TRENDS IN MODERN HERPETOLOGY

The modern field of herpetology consists primarily of investigations of behavior, morphology and karyology, ecology, speciation, and physiology, in addition to the description and naming of taxa still being discovered. The "new systematics" includes attempts to apply numerical taxonomy. Advances in herpetology have, of course, frequently been possible only because of the support given to research by a wide variety of university, museum, governmental, and private agencies, but they are also due to the influence of pioneering workers who generated an interest in amphibians and reptiles in their students and broadened the spectrum of herpetological research.

IN THE UNITED STATES

The University of Michigan has employed several outstanding herpetologists. Alexander G. Ruthven (1882–1971) made many important contributions while he was associated with the Museum of Zoology. In 1908, he published *Variations and Genetic Relationships of the Garter Snakes* (Bulletin 61 of the United States National Museum), which includes the first attempt at biostatistics. His studies of variation, geographic distribution, and relationships of reptiles gave much support to the now generally accepted rule that directly related forms on any line of descent ordinarily occupy neighboring environments. From the Museum of Zoology, of which he was curator from 1906 until 1913, Ruthven rose through administrative ranks to become President of the University of Michigan from 1929 until 1951. As both a research worker and an administrator, Ruthven provided the stimulus which led to the development of an active group of herpetologists at Michigan. Frank N. Blanchard (1888–1937), formerly in the Department of Zoology, was one of the first herpetologists to develop techniques for studying live animals in the field and was especially responsible for the development of marking techniques. Blanchard concerned himself primarily with studies of snakes and salamanders. Another influential herpetologist who was associated with the University of Michigan is Helen T. Gaige, who was Curator of Herpetology at the Museum of Zoology and was instrumental in organizing and developing the American Society of Ichthyologists and Herpetologists. Many of the leading herpetologists of today were students under these people at Michigan; its strong position in herpetology has been maintained by such people as Norman Hartweg, L. C. Stuart, and Charles F. Walker. Carl Gans, a leading reptilian morphologist who stresses the relationship between structure and function and is an editor of the *Biology Of The Reptilia* series (1969–), has recently joined the University of Michigan; and there has been an influx of other herpetologists with diversified interests.

A number of herpetologists have worked in the northeastern United States. Mary C. Dickerson (1866–1923) was the first Curator of Amphibians and Reptiles at the American Museum of Natural History in New York City. She authored *The Frog Book* (1906), a general reference on anuran amphibians, and initiated publication of *Natural History,* the first significant museum magazine. She also influenced the careers of a number of herpetologists who were her assistants, including Charles L. Camp, Emmett Reid Dunn, G. Kingsley Noble, and Karl P. Schmidt. Each of these went on to gain recognition of his own: Camp published many papers on the anatomy and classifications of reptiles and amphibians, including "Classification of

the Lizards" (1923). Since leaving the American Museum in 1917, he has been associated with the Museum of Paleontology at the University of California, Berkeley. Dunn (1894–1956), as a Professor at Haverford College, completed much of the classic work on salamanders. His publications include *Salamanders Of The Family Plethodontidae* (1926) and *American Caecilians* (1942). Noble (1894–1940) was Mary Dickerson's successor as Curator of Herpetology at the American Museum of Natural History and another expert on amphibians. His dissertation, *The Phylogeny of the Salientia,* was completed at Columbia under the direction of W. K. Gregory and upon publication in 1922 was recognized as a major herpetological contribution. In 1931, Noble published *The Biology Of The Amphibia,* still recognized as a classic and being reprinted.

Karl P. Schmidt (1890–1957) spent six years at the American Museum of Natural History, working under both Dickerson and Noble. In 1922, he assumed directorship of the newly-established Division of Amphibians and Reptiles at the Field Museum of Natural History, now the Chicago Natural History Museum, becoming Curator of Zoology in 1941 and Emeritus Curator of Zoology in 1955. During this time, he was generally regarded as the dean of American herpetologists; he built the museum into one of the foremost herpetological departments of the world. In addition to the sixth edition of *A Check List of North American Amphibians and Reptiles,* Schmidt co-authored *Field Book of Snakes of the United States and Canada* with D. D. Davis (1941) and *Living Reptiles of the World* with R. F. Inger (1957); in addition, he had very broad interests in biology and co-authored *Ecological Animal Geography* (1951) and *Principles of Animal Ecology* (1949). He authored an extensive list of publications on amphibians, reptiles, and zoogeography and served as herpetological editor for *Copeia* (journal of the American Society of Ichthyologists and Herpetologists) from 1937 to 1949. Associated with Schmidt at the Chicago Museum was Clifford H. Pope, reptilian biologist with many publications, including *Snakes Alive and How They Live* (1937), *The Turtles of North America* (1939), *The Reptile World* (1956) and *The Giant Snakes* (1962).

Following Noble's death in 1940, Charles M. Bogert became Assistant Curator in Charge of the Department of Herpetology at the American Museum; he subsequently advanced to become curator and chairman of the department. Bogert continued Noble's pattern of combining experimental and anatomical techniques in systematic studies and has an extensive list of publications dealing with the thermal regulation of reptiles and amphibians, behavior, distribution, taxonomy, and morphology of reptiles and amphibians, and the ecology and evolution of reptiles. James A. Oliver, author of *The Natural History of North American Amphibians and Reptiles* (1955) and *Snakes In Fact and Fiction* (1963), is former Director of the Bronx Zoo and recent Director of the American Museum. Edwin H. Colbert, who has been Curator of Fossil Reptiles at the American Museum, has also written several excellent books and many shorter publications on reptiles and their evolution. Two of his more recent books are *Dinosaurs: Their Discovery and Their World* (1962) and *The Age of Reptiles* (1965). Bogert's successor at the American Museum is Richard G. Zweifel, who became Curator of the Department of Herpetology in 1965 and is now Chairman and Curator. John A. Moore, formerly on the faculty at Columbia University and now at the University of California at Riverside, is a research associate of the American Museum. Moore has done extensive work on the common leopard frog *Rana pipiens* and has championed the detailed study of populations and geographical variation within species. Herndon G. Dowling, noted for his studies of reptiles and developer of the Herpetological Information Search Systems, is another Research Associate of the American Museum, as are Archie F. Carr, Jr., and Roger Conant, both of whom will be discussed later. These people form the nucleus of a group of herpetologists which make the American Museum of Natural History very strong in the field of herpetology.

Other outstanding New York herpetologists include Albert Hazen Wright (1879–1970), formerly at Cornell University, who co-authored with Anna Allen Wright *Handbook of Frogs and Toads of the United States and Canada* (1949) and *Handbook of Snakes of the United States and Canada* (1957). Sherman C. Bishop, while Professor of Vertebrate Zoology at the University of Rochester, authored *Handbook of Salamanders* (1947).

Louis Agassiz (1807–1873) was largely responsible for the development of the Museum of Comparative Zoology at Harvard into a great research museum and personally accumulated very large collections of amphibians and reptiles for it. This museum, like the American Museum, has had a succession of leading herpetologists associated with it, including Samuel Garmen (1834–1927) and Thomas Barbour. Alfred Sherwood Romer, Director Emeritus of the Museum of Comparative Zoology, is the leading vertebrate paleontologist of our time and author of innumerable publications which deal with the classification and anatomy of both living and fossil forms. Among his major publications important to herpetology are *Osteology of the Reptiles* (1956), *The Vertebrate Story* (4th edition, 1959), *Vertebrate Paleontology* (3rd edition, 1966), and *Notes and Comments on Vertebrate Paleontology* (1968). Arthur Loveridge, who worked under Barbour, has done extensive field work in both Africa and Australia and has contributed much to the systematics of amphibians and reptiles from these two continents. Ernest E. Williams, Professor and Curator of Reptiles and Amphibians, has contributed much to our understanding of the taxonomy, paleontology, and morphology of reptiles.

Following Stejneger's era at the United States National Museum, the Division of Herpetology was headed by Doris M. Cochran (1898–1968), who first went there as an aide in 1919 and then spent literally her entire professional life with that division, attaining the positions of curator in charge and research curator. She was the author of 11 books and monographs plus numerous shorter papers that include fundamental publications on the herpetology of the West Indies, Malaysia, and South America. Her *Living Amphibians of the World* (1961) has already been translated 13 times into eight languages. Her latest publications, *The New Field Book Of Reptiles and Amphibians* (1970) and *Frogs of Colombia* (1970), both co-authored by C. J. Goin, appeared after her death. The Division of Herpetology of the United States National Museum is now headed by James A. Peters, Curator and

Supervisor of Reptiles and Amphibians, whose interests are in the biology of reptiles and the zoogeography of Latin America. Peters is the author of *Dictionary of Herpetology* (1964), which contains definitions of terms used in anatomy, physiology, and systematics of amphibians and reptiles.

There is a strong nucleus of herpetologists at the University of Florida and the Florida State Museum. Walter Auffenberg is an authority on vertebrate paleontology with an interest in the behavior of amphibians and reptiles. Archie F. Carr is the world's leading authority on turtles and his studies of the ecology and migration of sea turtles have provided much of the currently available information on these poorly known reptiles. Carr has written both popular and scientific publications on tropical ecology and is the author of *Handbook of Turtles: The Turtles of the United States, Canada, and Baja California* (1952), the volume in the Life Nature Library titled *The Reptiles* (with the editors of Life, 1963), and *The Turtle: A Natural History* (1968), as well as numerous other publications. Coleman J. Goin has made many valuable contributions to the literature on the taxonomy, ecology, and genetics of amphibians. As mentioned, he is co-author of *The New Field Book of Reptiles and Amphibians* and *Frogs of Colombia;* and with Olive B. Goin he co-authored *Introduction to Herpetology* (1962).

The University of Kansas developed into a leading herpetological center through the influence of Edward H. Taylor and the Museum of Natural History. Taylor and his students made many of the early collections of amphibians and reptiles from Mexico, Central America, the Philippines, and Southeast Asia; these resulted in tremendous numbers of descriptive papers. One of Taylor's outstanding students is Hobart M. Smith, who, after spending many years at the University of Illinois, is now at the University of Colorado. Smith and Taylor have provided nearly all of the catalogues and taxonomic keys to the amphibians and reptiles of Mexico and Central America, and both have been very prolific publishers. Among Taylor's major works is *The Caecilians of the World* (1968), published after his retirement. Hobart Smith is author of *Handbook of Lizards: Lizards of the United States and Canada* (1946). Taylor's successor at the Museum of Natural History is William E. Duellman, who was a student at the University of Michigan. Largely through the efforts of Duellman and Henry S. Fitch, a very active group of herpetologists has located at the University of Kansas. Duellman is the world's leading authority on tree frogs and has an extensive list of publications on the systematics, morphology, distribution, and ecology of New World amphibians and reptiles; his *The Hylid Frogs of Middle America* was published in 1970. Fitch, who obtained his doctoral degree from the University of California, is an outstanding population ecologist and has made detailed studies of several species of reptiles.

California has had a number of eminent herpetologists. The Museum of Vertebrate Zoology of the University of California at Berkeley was first directed by Joseph C. Grinnell, whose work on amphibians and reptiles of California supplemented that of Charles L. Camp. The museum's present Curator of Herpetology is Robert C. Stebbins, who, in addition to developing several leading young herpetologists, has made important contributions to the literature of herpetology. His *Amphibians of Western North America* (1951) was the first general herpetological handbook to give recognition to the biological concept of a species as a populational entity. In addition to numerous shorter publications, Stebbins also authored the popular *Amphibians and Reptiles of Western North America* (1954) and the Peterson Field Guide volume *A Field Guide to Western Reptiles and Amphibians* (1966).

Many important contributions to herpetology have been made at the California Academy of Sciences at San Francisco. John van Denbergh (1872–1924) made many of the fundamental studies of reptiles, and his publications include the two-volume *Reptiles of Western North America* (1922). Largely through his efforts, the Academy made many of the initial collections in Baja, California and the Galápagos Islands. The truly outstanding herpetologist in California, however, was an amateur who had such high professional standards that his work was unsurpassed: Laurence Monroe Klauber (1883–1968). Klauber was employed by the San Diego Gas and Electric Company, initially as a salesman and ultimately as Chairman of the Board of Directors, and had a burning interest in lizards and snakes. He had a long association with the San Diego Zoological Garden, serving as Consulting Curator of Reptiles from 1922, as a member of the board of trustees from 1943, and as President from 1949 to 1951. He was also Consulting Curator of Herpetology and a member of the board of directors of the San Diego Natural History Museum from 1927 until his death. He had a private collection of some 35,000 specimens of amphibians and reptiles (now in the San Diego Natural History Museum) and a herpetological library containing nearly 1500 volumes and some 19,000 reprints and pamphlets, one of the finest in the United States (now also in the San Diego Natural History Museum). Having a good background in mathematics, Klauber applied statistical methods to taxonomic and other herpetological problems; this approach was emphasized in his studies of variation and growth within rattlesnake populations. Klauber's greatest contribution to herpetological literature is his two-volume *Rattlesnakes: Their Habits, Life Histories, and Influence on Mankind* (1956), one of the most complete and well-organized herpetological monographs ever written.

In the southcentral United States, the work of A. P. Blair at the University of Tulsa and W. F. Blair at the University of Texas in Austin on isolating mechanisms of amphibians and their importance in speciation has been most influential in the development of evolutionary theory and in the understanding of mating behavior. W. F. Blair is author of *The Rusty Lizard: A Population Study* (1960), in addition to many other publications. The late Arthur N. Bragg (1897–1968), for many years associated with the University of Oklahoma, was widely known for his studies of anuran life histories. These were published in more than 200 papers and a book, *Gnomes of the Night: The Spadefoot Toads* (1965). Tulane University became prominent in herpetology during the tenure of Fred R. Cagle (1915–1968), who had a very broad interest in biology but is probably best known for his studies of amphibian and turtle populations. E. Peter Volpe, also at Tulane, is a former student of John A. Moore's and has done outstanding research on the em-

bryological development, genetics, and hybridization of anuran amphibians.

OTHER CENTERS OF HERPETOLOGY

Although the United States has led the world in herpetology for many years, there are numerous centers of interest in this subject in other nations. In South America, herpetology has received the greatest emphasis in Argentina and in Brazil. Among Argentine herpetologists, Miguel and Kati Fernandez are noted for their descriptions of the life histories of a variety of anuran amphibians; Joseph M. Cei is a leading authority on the systematics of anurans, particularly of the family Leptodactylidae, and is one of several researchers utilizing biochemical approaches to systematics. In Brazil, Bertha Lutz of Rio de Janeiro and C. A. Bokermann of Sao Paulo have made important contributions to the literature of anuran amphibians, while Afranio do Amaral, who has been Director of the Instituto Butantan, has worked primarily on lizards and snakes.

An interest in Australian herpetology was initiated by J. R. Kinghorn at the Australian Museum in Sydney, and has been perpetuated through the work of a growing group of herpetologists: L. Harrison, M. J. Littlejohn, A. A. Martin, V. A. Harris, H. R. Bustard, and Eric Worrell. Harrison, Littlejohn, and Martin have worked primarily on amphibians of the Australian region, while Harris, Bustard, and Worrell have concentrated on reptiles. Vernon Harris was formerly at Ibadan University in Nigeria and is well known for his natural history studies of the rainbow lizard (*Agama agama*); these are described in *The Anatomy of the Rainbow Lizard Agama agama* (L.) (1963) and *The Life of the Rainbow Lizard* (1964). Worrell is author of *Reptiles of Australia* (1963) and *Australian Snakes, Crocodiles, Tortoises, Turtles, Lizards* (1967). Richard Sharell has published a well-illustrated volume on the amphibians and reptiles of New Zealand, titled *The Tuatara, Lizards, and Frogs of New Zealand* (1966).

In the Orient, C. C. Liu, author of *The Amphibians of West China* (1950), represents a Chinese center of herpetology, while a number of herpetologists are active in Japan. Among the latter are Hisaski Iwasawa at Niigata University and Mitsuru Kuramoto of Fukuota University, both of whom are publishing on the ecology, physiology, and systematics of anuran amphibians.

Germany has a very strong representation of herpetologists. The Senckenberg Museum at Frankfort first became prominent in herpetology through Oskar Boettger (1833–1910), author of the amphibian and reptile volume for the third edition of Brehm's *Tierleben* (1892). Boettger was succeeded by Robert Mertens, one of the truly great herpetologists of the world. Mertens has personally made expeditions to the East Indies, West Africa, the West Indies, and Central America. He is a leading authority on monitor lizards (family Varanidae) and has been very much interested in insular lizards and speciation processes in the Mediterranean. He co-authored the first two editions of *Die Amphibien und Reptilien Europas* (1928 and 1940) with Lorenz Müller and the 1960 edition with Heinz Wermuth; these are check lists patterned after those of North America. Mertens' *The World of Amphibians and Reptiles* (translated by H. W. Parker, 1960) is one of the most popular and best-illustrated general accounts in herpetology. Wermuth formerly worked at the Zoologisches Museum in Berlin and now is at the Staatliches Museum für Naturkunde in Stuttgart; his replacement in herpetology at Berlin is Günther Peters. Still another prominent German herpetologist is Walter Hellmich, Chief Keeper of the Zoological Collection in Munich, who studied amphibians and reptiles in South America and is author of *Reptiles and Amphibians of Europe* (translated from *Die Lurche Und Kriechtierre Europas* and first published in English in 1962).

Herpetology at the British Museum has been well represented in recent years by H. W. Parker (1891–1968), who replaced Boulenger and is author of *A Monograph of the Frogs of the Family Microhylidae* (1934), and the late Malcolm A. Smith (1875–1958), former Court Physician to the Court of Siam and a very able research associate of the museum. Dr. Smith developed his interest in herpetology while in Bangkok and combined eminence as a naturalist with a successful medical career. From Bangkok, he was able to go on zoological expeditions to the interior of Siam, Indochina, Hainan, and the Malay Arch-

ipelago; he even found time to study material in American museums. He was primarily a systematist but later became interested in the life histories and behavior of British amphibians and reptiles. His major publications are *Monograph of the Sea Snakes* (1926); *The Fauna of British India, Ceylon, and Burma, Including the Whole of the Indo-Chinese Sub-region: Reptilia and Amphibia* (three volumes which do not include amphibians, 1931–1943); and *The British Amphibians and Reptiles*, first published in 1951 with a third edition in 1964. Dr. Smith was largely responsible for the founding of the British Herpetological Society. Another British herpetologist of note is Angus d'A. Bellairs, associated with the University of London and Honorary Herpetologist to the Zoological Society of London. Bellairs has specialized in the anatomy of reptiles and has an extensive list of publications. He has authored *Reptiles* (2nd edition, 1968) and the two-volume *The Life of Reptiles* (1970). He is also co-author of *The World of Reptiles* (with R. Carrington, 1966).

A Russian center of herpetology has been represented by such people as A. G. Bannikov, S. A. Tchernov, A. M. Nikol'skii, and P. V. Terentjev. Bannikov has studied the population dynamics of anuran amphibians. Nikol'skii is author of *Fauna of Russia and Adjacent Countries*, which includes amphibians and reptiles (first published in 1915–1916 and translated in 1963–1964). Terentjev (1903–1970) and Tchernov co-authored *Key to Amphibians and Reptiles* (1965), a handbook to Russian species; and Terentjev also published a general herpetology text titled *Herpetology: A Manual On Amphibians and Reptiles* (1965), with an emphasis on systematics and geographical variations, the latter being one of his principal areas of interest.

The South Africans, headed by C. G. S. de Villiers, have done extensive research on the cranial morphology of amphibians. V. F. Fitzsimmons has provided a systematic survey of the lizards and snakes of southern Africa in his *The Lizards of South Africa* (1943) and *Snakes of Southern Africa* (1962).

M. Gabe and H. Saint Girons represent, together with others, a modern French center of herpetology. Using both light microscopy and electron microscopy, they have contributed much to our understanding of the histology and microscopic anatomy of reptiles. Gabe and Saint Girons authored *Contribution à l'histologie de Sphenodon punctatus Gray* (1964), an atlas of the histology of various organs as seen with the light microscope. Saint Girons has also made important studies of reptilian endocrinology.

It is impossible to mention here all the valuable contributions of the world's ever-growing list of herpetologists. However, the reader will gain an insight into the activities of many researchers by referring to the references listed at the end of each chapter.

MAJOR HERPETOLOGICAL SOCIETIES AND PUBLICATIONS

The oldest society promoting herpetology in the United States is The American Society of Ichthyologists and Herpetologists, founded in 1913. This society publishes *Copeia*, a quarterly journal containing original research papers, notes, reviews, and comments on both herpetological and ichthyological subjects. The American Society of Ichthyologists and Herpetologists, through its Herpetological Catalogue Committee, initiated publication of *Catalogue of American Amphibians and Reptiles*. This catalogue is a series of loose-leaf accounts, each authored by an expert in the field. Each account gives, in one to four pages, a wealth of information about a particular taxon, including common and scientific names (with a synonymy), a definition, a description, information of fossil occurrence, a detailed map of geographic distribution, and a comprehensive introduction to the literature. One hundred and twenty-five accounts were published from 1963 through March, 1972. Most of the accounts concern species occurring in the United States and Canada.

The Herpetologists' League is an international organization established in 1936 and dedicated entirely to furthering knowledge of the biology of amphibians and reptiles. The Herpetologists' League publishes *Herpetologica*, another quarterly publication, and also co-sponsors, together with the Society for the Study of Amphibians and Reptiles, *Herpetological Review*.

Herpetological Review is published bimonthly and is a newsletter on people, institutions, programs, and current events of interest to herpetologists. It includes "Current Herpetological Titles," provided by the Herpetological Information Search Systems at the American Museum of Natural History to keep readers abreast of literature in their field.

The Ohio Herpetological Society was founded in 1958 and published five volumes (1958–1966) of the *Journal of the Ohio Herpetological Society*. Following rapid growth in membership and expenses, the organization became the Society for the Study of Amphibians and Reptiles (1967) and now publishes the *Journal of Herpetology*, issued semimonthly and international in scope; Volume I of the *Journal of Herpetology* appeared in 1968. The Society also makes various important books and papers on amphibians and reptiles available through facsimile reprints. It has published the *Catalogue of American Amphibians and Reptiles* since 1971.

The British Herpetological Society was founded in 1947 to promote the study, care, and conservation of reptiles and amphibians, particularly in Britain and Europe. This society semiannually publishes *The British Journal of Herpetology*.

Research papers on the biology of amphibians and reptiles appear in any of nearly 300 journals. Some of these publications, such as the German *Aquarien und Terrarien*, are devoted partly or entirely to herpetology. The majority, however, represent functional approaches to biology and include papers on amphibians and reptiles together with studies of other kinds of organisms. Thus, the difficulty of staying current in herpetological fields, as in other areas of science, is increasing each year. Fortunately, various abstracting services provide both summaries and cross-indices which allow one to locate articles on specific subjects in a great number of biological journals. Herpetological literature is abstracted in both the *Zoological Record*, published by the Zoological Society of London, and *Biological Abstracts*, published by the BioSciences Information Service in Philadelphia. The time lag before current literature appears is about three years for *Zoological Record* and six months to a year for *Biological Abstracts*. The "Current Herpetological Titles" listing put out by the Herpetolog-

Leonhard Stejneger

Karl P. Schmidt

Doris M. Cochran

Laurence M. Klauber

(Photos of Leonhard Stejneger and Laurence M. Klauber from Schmidt, K. P.: Herpetology. *In* California Academy of Sciences: *A Century of Progress in the Natural Sciences, 1853–1953.* Photo of Karl P. Schmidt from D. D. Davis, *in* Copeia, October, 1959; courtesy of Field Museum of Natural History. Photo of Doris M. Cochran from Coleman J. Goin, *in* Herpetologica *8(4),* 1952.)

A. M. C. Duméril   Gabriel Bibron

G. A. Boulenger   E. D. Cope

(Photos from Schmidt, K. P.: Herpetology. In California Academy of Sciences: A Century of Progress in the Natural Sciences, 1853–1953.)

ical Information Search Systems in *Herpetological Review* is also useful for keeping up with the literature: this is a listing of herpetological papers published during a specified period of time and deals only with Recent orders of amphibians and reptiles. Herpetological Information Search Systems also offers consultation services and special bibliographies on amphibians and reptiles.

References

Bellairs, A. d'A. 1970. The Life of Reptiles. Volume I. Universe Natural History Series. New York, Universe Books. Pp. 1–16.

Black, J. H. and C. C. Carpenter. 1969. Arthur Norris Bragg, 18 December 1897–27 August 1968. Copeia *1969*:419–420.

Brattstrom, B. H. 1968. Laurence M. Klauber, 1883–1968. Herpetologica *24*:271–272.

Davis, D. D. 1959. Karl Patterson Schmidt, 1890–1957. Copeia *1959*:189–192.

Goin, C. J. and O. B. Goin. 1962. Introduction to Herpetology. San Francisco, W. H. Freeman. Pp. 1–13.

Grandison, A. G. C. 1969. H. W. Parker. Copeia *1969*:416–417.

Schmidt, K. P. 1955. Herpetology. *In:* Kessel, E. L. (ed.): A Century of Progress in the Natural Sciences – 1853–1953. San Francisco, California Academy of Sciences. Pp. 591–627.

Shaw, C. E. 1969. Laurence Monroe Klauber, 1883–1968. Copeia *1969*:417–419.

Smith, H. M. 1968. Doris M. Cochran, 1898–1968. Herpetologica *24*:268–270.

Smith, M. 1964. The British Amphibians and Reptiles. The New Naturalist (Series). London, Collins. Pp. 1–13.

CHAPTER 2

STRUCTURAL AND FUNCTIONAL CHARACTERISTICS OF LIVING AMPHIBIANS

The class Amphibia is in many ways transitional between the classes of fishes and reptiles and, in its entirety, is rather difficult to characterize morphologically. The most primitive of the extinct labyrinthodont amphibians were very similar to the rhipidistian crossopterygian fishes from which they evolved. The extinct anthracosaurian amphibians, which gave rise to the class Reptilia, form a series which progressively becomes more reptilian so that it is almost impossible to separate advanced amphibian fossils from those of primitive reptiles. The obvious difference between the primitive amphibians and the fishes lies in the development of limbs in the former; the real distinction between amphibians and reptiles involves the structure of the egg and the development of embryonic membranes. Amphibian eggs lack an outer protective shell and are of the anamniotic type; none of the inner embryonic membranes (the amnion, chorion, and allantois) develops (Fig. 2–1). Anamniotic eggs must be deposited in moist environments or they soon be-

Figure 2–1 An oviducal egg (left) and a comparable egg after fertilization with swollen jelly membrane (right). (From Balinsky, B. I.: *An Introduction to Embryology.* 3rd Ed. W. B. Saunders Co., 1970.)

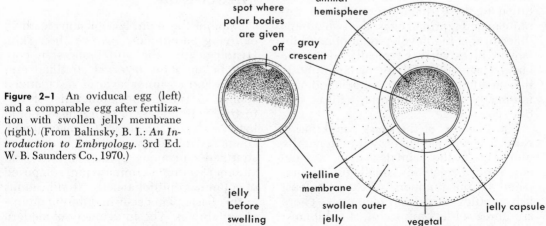

unpigmented spot where polar bodies are given off

animal hemisphere

gray crescent

jelly before swelling

vitelline membrane

swollen outer jelly

jelly capsule

vegetal hemisphere

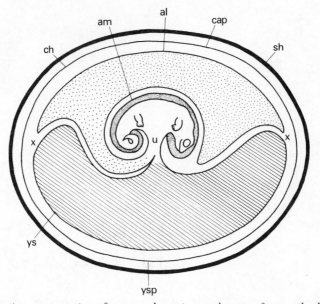

Figure 2–2 Diagrammatic representation of extra-embryonic membranes of a reptile during the later stages of embryonic life. Actually, there may be one or more shell membranes between the shell and the chorion, while the inner surface of the chorion and the outer surfaces of the amnion, allantois, and yolk sac are each covered with a layer of extra-embryonic mesoderm. Owing to fusion of these adjacent mesodermal layers, an amnio-chorion and a chorio-allantois are formed. The chorion and yolk sac may also fuse in places, and in viviparous forms where the shell is reduced, a placental area may be established at either the chorio-allantoic or chorion-yolk-sac sites, or at both. In some lizards (and perhaps in reptiles in general), the edges of the allantois acquire a more complex relationship with the yolk sac than is shown, and appear to fuse with it at the regions marked *x*. *al*, allantois, outer layer (allantoic cavity stippled). *am*, amnion (amniotic cavity darkly stippled). *cap*, site of chorio-allantoic placenta in viviparous forms. *ch*, chorion. *sh*, egg-shell. *u*, umbilical cord. *ys*, yolk sac (yolk shown by slanting lines). *ysp*, site of yolk-sac placenta in viviparous forms. (From Bellairs, A.: *The Life of Reptiles.* Universe Books, 1970.)

come desiccated. Reptilian eggs, on the other hand, possess an outer protective shell and are amniotic; all three embryonic membranes form early in development (Fig. 2–2). Amniotic eggs are relatively resistant to desiccation and represent one of the most important evolutionary advancements in vertebrate history. In addition to these differences between amphibian and reptilian eggs, the two groups differ in that the development of amphibians involves what is typically an aquatic larval stage followed by metamorphosis into an adult form. Reptilian development is directly to the adult form, with no distinct larval stage.

The majority of the class Amphibia is extinct; living members, while generally numerous in the habitats where they occur, form a relatively small group compared to the numbers of extant teleost fish, reptiles, birds, and mammals. There are three orders of living amphibians; these include salamanders, caecilians, and frogs. They are the products of evolu-

tionary innovations and selection over a period of about 400 million years and, consequently, are much modified from their Devonian ancestors and relatively easy to distinguish from other vertebrates.

INTEGUMENT

Some of the more conspicuous features of living amphibians involve their skin. Histologically, the vertebrate skin consists of an outer layer of stratified epidermis and an inner layer of vascularized spongy dermis. The epidermis of terrestrial vertebrates, in turn, consists of at least two layers: an outer stratum corneum, composed of compacted and keratinized stratified squamous epithelial cells, and an inner stratum germinativum, composed of columnar epithelial cells. Modifications of this basic plan occur in different groups of vertebrates. The integument of modern amphibians performs several functions, probably more than in any other group of

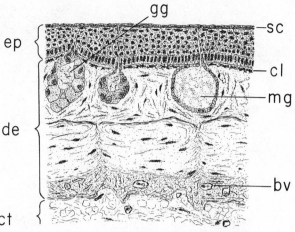

Figure 2-3 Representative section through the skin of an amphibian with mucous glands and granular glands. *bv*, blood vessel. *cl*, chromatophore layers. *ct*, subepithelial connective tissue. *de*, dermis. *ep*, epidermis. *gg*, granular gland. *mg*, mucous gland. *sc*, stratum corneum. (After Torrey, T. W.: *Morphogenesis of the Vertebrates.* 2nd Ed. John Wiley and Sons, 1967; also after Andrew, W.: *Textbook of Comparative Histology.*)

vertebrates, and the evolution of these multiple roles has made it distinctive from those of other contemporary vertebrates (Fig. 2–3).

Several integumentary features of living amphibians are related to respiration. This process, through which the blood releases dissolved carbon dioxide to the environment and gains dissolved oxygen, requires only a highly vascularized moist surface which is in contact with the environment (air or water) and is permeable enough to allow diffusion of these gases. In vertebrates, the respiratory surfaces are typically gills in aquatic forms and alveolar surfaces in the lungs in terrestrial species. Respiration in modern amphibians, however, is not confined to these two kinds of organs. Depending upon the species, there may also be cutaneous respiration through the integument or buccopharyngeal respiration through the lining of the oral cavity and pharynx. A combination of two or three respiratory surfaces is commonly used, and the lungs of adults, when present, are simple in structure and not partitioned to any great degree. Probably all amphibians, larvae and metamorphosed adults, utilize cutaneous respiration to some degree. The terrestrial salamanders in the family Plethodontidae lack lungs and depend entirely upon cutaneous and buccopharyngeal respiration.

EPIDERMIS

Smooth, thin skins undoubtedly facilitate cutaneous respiration, and the amphibian epidermis, particularly the stratum corneum layer, is relatively poorly de-

veloped compared to that of reptiles. Epidermal scales, a prominent characteristic of reptiles, are completely absent in amphibians. Among living amphibians, there tends to be an inverse correlation between the thickness of the stratum corneum and the degree to which cutaneous respiration is depended upon (Fig. 2–4). However, some amphibians with reduced lungs have blood capillaries which penetrate a relatively thick stratum corneum and lie near the surface of the skin. The adult male hairy frog, *Astylosternus robustus* (family Ranidae), of Cameroun has a growth of fine glandular filaments resembling hairs on its sides and hind legs (Fig. 2–5). These filaments are actually short vascular papillae and presumably function in increasing the respiratory surface of the animal and help to compensate for its reduced lungs. Hellbenders, in the family Cryptobranchidae, have moderately well-developed lungs but also have an efficient cutaneous respiratory system. In these primarily aquatic salamanders, there are highly vascularized folds of skin along the body which apparently function in a gill-like manner. Each fold has a number of capillary diverticulae which penetrate the relatively thick epidermis almost to the surface (Fig. 2–4B). Aquatic caecilians in the family Typhlonectidae exhibit a comparable degree of vascularization of their epidermis.

At least the superficial layer of the stratum corneum is molted in all metamorphosed amphibians at regular intervals of a few days to a month or more. This layer of the epidermis is held to the underlying epidermal layers by a number of flask-

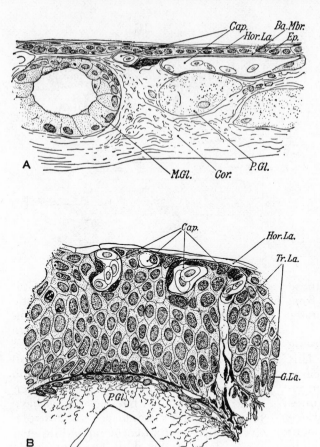

Figure 2–4 Cutaneous respiration of amphibians is made more efficient by thinning of the epidermis over the superficial cutaneous capillaries, as in the plethodontid salamander *Desmognathus quadramaculatus* (A), and by the penetration of capillaries through the epidermis to a position very near the surface, as in the hellbender *Cryptobranchus alleganiensis* (B). *Ba.Mbr.*, basal membrane; *Cap.*, capillary; *Ep.*, epidermis; *G.La.*, germinal layer; *Hor.La.*, horny layer; *M.Gl.*, mucous gland; *P.Gl.*, poison gland; *Tr.La.*, transitional layer of epidermis; *Cor.*, corium. (From Noble: *The Biology of the Amphibia.*)

shaped cells. The action of these cells is under hormonal control of the thyroid and pituitary glands, and at the time of molting all flask cells simultaneously release their secretion and withdraw from the stratum layer. Thus, molting occurs on all surfaces of the body at one time; in some forms (true frogs in the genus *Rana*, family Ranidae) the stratum corneum comes off in pieces while in others (true toads in the genus *Bufo*, family Bufonidae) it is pulled off in one piece by the mouth and forelegs. Many amphibians swallow the old corneum as it is being pulled off, the swallowing process probably facilitating its removal, and shed skins are rarely found. That the molting process is under the control of the thyroid and pituitary glands may be easily demonstrated, for either hypophysectomy or thyroidectomy results in the piling-up of cornified epidermal layers and the development of a thick stratum corneum.

Local thickenings of the epidermis fre-quently occur naturally in amphibians and produce a variety of adaptive structures, including warts. True claws, consisting of a dorsal layer of compressed stratum corneum forming compact horn and a ventral horn sole of less compact stratum corneum, are characteristic of reptiles, birds, and mammals; they do not occur in amphibians. However, various salamanders in the families Hynobiidae, Ambystomatidae, and Plethodontidae, which live in swift mountain streams, have digits adapted for clinging to the substrate which are tipped with thickened and partly horny epidermal caps; those of *Onychodactylus japonicus* (family Hynobiidae) larvae have evolved into sharp claws which superficially resemble those of lizards but lack the ventral horn sole. Pointed clawlike caps also occur on the digit tips of sirens (family Sirenidae). Sirens occur in swamps and marshes and completely lack hind limbs; thus, the "claws" on their forefeet are probably an adaptation allowing the ani-

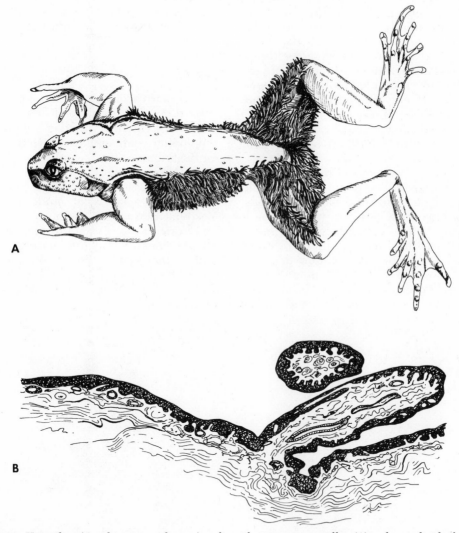

Figure 2–5 Hairy frog (*Astylosternus robustus*) male with cutaneous papillae (*A*) and an individual vascular cutaneous papilla of such a male hairy frog (*B*). (*A*, after Noble: *The Biology of the Amphibia; B*, from Andrew: *Textbook of Comparative Histology.*)

mal to hold onto submerged vegetation or other substrates. Asian frogs in the genus *Staurois* (family Ranidae) breed in mountain streams, and their tadpoles possess an abdominal suction disc which is covered with horny epidermal tubercles and is used to cling to rocks in swift currents.

Black, pointed, clawlike caps also occur on the inner three toes of frogs in the aquatic genus *Xenopus* (family Pipidae) (Fig. 2–6A), apparently as an adaptation for digging in the mud for insects (Cochran, 1961). Mexican burrowing toads (*Rhinophrynus dorsalis*, family Rhinophrynidae), spadefoot toads (family Pelobatidae), many true toads (family Bufonidae), and some leptodactylids (family Leptodactylidae) are fossorial (burrowing) forms and have cornified "spades" on their hind feet for digging.

Epidermal accessory breeding structures which facilitate clasping the female are found on males of a variety of frogs. Males of many species in the genus *Leptodactylus* (family Leptodactylidae) and a few other frogs have spines on their thumbs, sometimes in addition to a terminal horny cap (Fig. 2–6B). Some *Leptodactylus* males have heavy, horny protuberances on their chests, and males

of some salamanders, including the black-spotted newt (*Notophthalmus meridionalis*, family Salamandridae), develop transverse horny ridges on the under-surface of their thighs as well as horny tips to their toes during the breeding season. Horny chest tubercles also occur on the males of other salamanders.

Many amphibian larvae have horny teeth which function as rasping organs and are used in feeding; these "teeth" are produced through modification of epidermal cells (Fig. 2–7).

DERMIS

The dermis of amphibian skin is basically the same as that of fish except that it is more highly vascularized in forms employing cutaneous respiration. The earliest amphibians were covered with dermal scales like those of their fish ancestors, but these have been completely lost in the majority of modern species. Vestigial dermal scales, however, do occur, buried in the costal fold skin of caecilians. Generally lacking both epidermal and dermal scales, the amphibian skin is naked for the most part, and in this way, too, it is distinctive from that of any other vertebrate class.

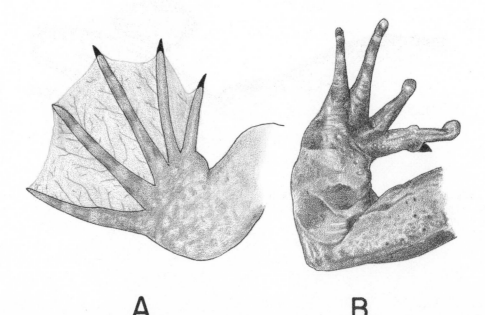

A   B

Figure 2–6 Dorsal view of left hind foot of clawed frog (*Xenopus laevis*) showing clawlike caps on inner three toes (*A*), and dorsal view of left forefoot of leptodactylid (*Leptodactylus pentadactylus*) showing spine on thumb (*B*).

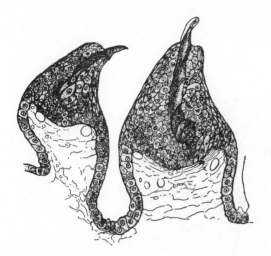

Figure 2–7 Horny larval teeth of a spadefoot toad (*Scaphiopus holbrooki*). (From Noble: *The Biology of the Amphibia*.)

Some frogs, however, have secondary deposits of bone (osteoderms) in the skin of the head. Others have co-ossified skin which is characterized by bone deposition continuous between, and in, the dense connective tissue of the dermal layers of the integument; this obliterates the deeper dermal layer and unites the epidermis and superficial dermis with the underlying bone (Trueb, 1966). In *Brachycephalus* (family Brachycephalidae), dermal plates develop in the back and fuse to the neural spines of the second to seventh vertebrae to form a broad shield.

The functional significance of co-ossification is not clear, but in tree frogs (family Hylidae) with casqued heads (expanded and thickened dermal roofing bones and jaw bones), the co-ossified condition is associated with frogs which practice phragmosis, utilizing their heads to plug the entrance to burrows or crevices. Trueb (1970) suggests that co-ossification in casque-headed phragmotic tree frogs probably reduces the danger of desiccation in arid environments. She has observed that the dermal and epidermal layers of skin which overlie membrane bone are poorly vascularized and have reduced numbers of glands and that the significance of the skin as a sensory and protective organ is diminished because of the extensive development of bone. Thus, although the skin covering the top of the skull undoubtedly becomes desiccated during phragmosis, it probably has few, if any, detrimental effects on the animal.

GLANDS

A principal evolutionary advance of the amphibian skin over that of fishes is the presence of large numbers of multicellular alveolar and, in some cases, tubular glands. Alveolar glands, epidermal in origin but located in the dermis (see Figure 2–3), occur as two types, both of which are found in all living amphibians. The first type includes large numbers of relatively small mucous glands which are distributed over the body surface and secrete a clear watery to viscous mucus. These secretions keep the skin moist in terrestrial situations and lubricate it when in water. Keeping the skin moist is essential if it is to function as a respiratory surface, and the secretion of mucus may also be important in the regulation of body temperature through evaporative cooling of the surface. In species of newts and frogs where it is toxic, the secretion may provide some protection from predators.

The second kind of alveolar gland includes the granular glands, which produce milky secretions that are much more toxic than mucus. Granular glands, so named because they have a granular appearance when stained with plasma dyes, serve primarily as a protective device against predators. They are often concentrated in strategic areas, as in the parotoid glands behind the heads of toads and some frogs (Fig. 2–8) or the glandular ridges along the backs of salamanders. Granular gland secretions contain alkaloid substances which resemble digitalis in action. These venoms may be sufficiently potent to kill large vertebrates by increasing the tonicity of the heart, weakening respiration, and causing general muscular paralysis and nausea.

Modified mucous glands performing special functions occur in a variety of amphibians. Nuptial pads on the thumbs (Fig. 2–9) or chests of certain male frogs contain glands which become active during the breeding season and secrete a particularly sticky mucus that assists the male in clasping the female during oviposition and fertilization. It is of interest that the unusually well-preserved remains

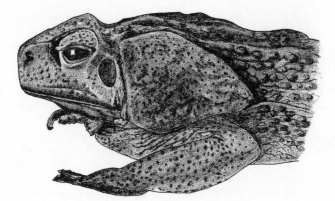

Figure 2–8   Marine toad (*Bufo marinus*) with exceptionally large triangular parotoid glands behind the head.

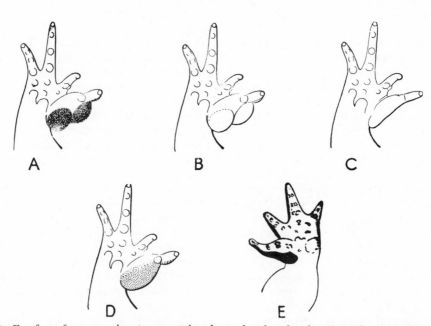

Figure 2–9   Forefeet of anurans showing nuptial pads on thumbs of males. *A.* Male of common frog (*Rana temporaria*) with nuptial pad during breeding season. *B.* Male of common frog when breeding season is over. *C.* Female of common frog. *D.* Male of edible frog (*Rana esculenta*) in breeding season. *E.* Male of common toad (*Bufo bufo*) in breeding season. (From Smith, M. A.: *The British Amphibians and Reptiles.* 3rd Ed. Collins, 1964.)

of *Eodiscoglossus santonjae* Vill (family Discoglossidae), from the Upper Jurassic of northern Spain, include a manus of five digits with a clearly revealed pigmented nuptial pad completely enclosing the first digit and apparently supported on one side by the second digit; the third digit also bears a similar integumentary imprint on its distal portion (Hecht, 1963). The first functional digit (thumb) on the manus of modern anurans represents the second digit primitively, and in some of the more primitive species, such as the tailed frog (*Ascaphus truei*, family Ascaphidae), the pad is alongside the thumb and over the vestiges of the first digit; in some species of true frogs (family Ranidae) and true toads (family Bufonidae), nuptial pads occur on the medial surfaces of the first, second, and third functional digits. Friction pad glands on the digits of tree frogs (family Hylidae) secrete a similarly sticky mucus which aids them in clinging to smooth surfaces.

Olfaction plays a major role in species recognition in salamanders, and qualities of the skin secretions of females are apparently recognized by males of the same species. Hedonic glands, differentiated mucous glands on the chin of male newts and plethodontid salamanders, secrete a substance which has a quiescent effect on the female and makes her amenable to courtship.

Tubular and sometimes branching glands around the nasal orifices of many amphibians serve the function of flushing dirt and water from the nostrils; these are especially well-developed in the nasal grooves of the plethodontid salamanders. Although not derived from mucous glands, mention must also be made of the hatching gland cells appearing on the snout of amphibian larvae shortly before hatching; these produce a secretion which digests the egg capsule and frees the young animals.

PIGMENTATION

The integument of amphibians is primarily responsible for their coloration, although diffuse pigments may also occur in deeper tissue and show through translucent areas of the skin. The external coloration of amphibians provides protection from predators and protection from injurious rays of the sun and functions in the absorption and dissipation of heat. Amphibian skin contains both scattered pigment granules in the epidermis and special pigment cells, called chromatophores, located in the dermis. Chromatophores also occur in fish and reptiles, but are not found in birds or mammals. Three major kinds of chromatophores occur in amphibians: melanophores, with black or brown pigment; lipophores, with drops or granules of red (erythrophores) or yellow (xanthophores) carotenoids; and guanophores, which contain crystals (leucophores) or platelets (iridophores) of guanine or guaninelike purines. The positioning of the chromatophores is such that they are stratified: the lipophores lie directly under the epidermis, the guanophores are found under the lipophores, and the melanophores are beneath the guanophores.

The prevailing color of an amphibian may result from the action of pigments absorbing some light rays and reflecting others (chemical color), interference phenomena in the outer layers of the skin (physical or structural color), or a combination of both chemical color and physical color. Dark spots are due to clustering of melanophores and golden flecks result from aggregations of yellow xanthophores.

The brilliant green which is characteristic of many frogs is often due to a combination of chemical and physical color. The epidermis of green skin is generally translucent, and, when light strikes it, the longer wavelengths penetrate to the melanophores where they are absorbed. Diffraction, affecting primarily the shorter wavelengths, occurs as light strikes the guanine crystals of the guanophores. This results in the reflection of blue through green wavelengths from this layer of chromatophores. As these reflected rays pass back through the outer layer of lipophores, absorption of the blue, indigo, and violet wavelengths occurs; only the green color passes through. Thus, green frogs usually appear green because this color includes the only wavelengths which escape absorption. Iridescent blue frogs have been found from time to time in populations of such ranid species as the leopard frog (*Rana pipiens*) and the green frog (*Rana clamitans*), which are normally green dorsally. The blue variant results from the same mechanism as the normal green color, except that the carotenoids are absent and all of the shorter wavelengths of blue, indigo, violet, and green are reflected from the guanophores out through the epidermis to the observer. Some amphibians exhibit a bright red color; this is produced by erythrophores plus a general reduction in melanophores, whose absence eliminates the absorption of the long wavelengths of red. When xanthophores are present and melanophores absent, a bright yellow color results. A more complete failure to develop pigments results in albinism; this has been reported in a variety of amphibians.

Although many frogs owe their green colors to the arrangement of chromatophores, as described above, Jones (1967) found that a number of Neotropical frogs in the families Hylidae, Centrolenidae, and Pseudidae have green pigmentation in bone, soft tissues or plasma. He identified this pigment as the bile pigment biliverdin, a green excretory product. The chlorotic condition was found to vary within a family, genus, species, or population. Jones observed that the green pigmentation appears at metamorphosis of hylid tadpoles and concluded that it results from the degradation of larval hemoglobin; in paradoxical frogs (*Pseudis paradoxa,* family Pseudidae), however, the green pigment appears prior to metamorphosis and may result from a different cause from that in hylids. The green pigmentation may disappear shortly after metamorphosis or continue into the adult stages. Among adult frogs, Jones found that chlorosis appears to develop at the end of the breeding season and that erythrocytes of frogs with the greatest concentrations of biliverdin appear immature or

hemolyze rapidly. Thus, he concluded that an immature or malfunctioning liver and an increase of biliverdin due to increased hemolysis could account for chlorosis at metamorphosis or following the breeding period; the hemolytic condition may be caused by high temperatures. Chlorotic species, except for the aquatic pseudids, are primarily arboreal; their high concentrations of biliverdin may be the result of selection for the protective coloration which it provides.

Many amphibians have the ability to change their color rapidly. In general, all such color changes are the result of changes in the distribution of pigment within chromatophores, changes in the relative positions of the different kinds of chromatophores, or a combination of both. Darkening of the skin results from the dispersion of melanin pigment into the pseudopods of melanophores extending into spaces between guanophores and lipophores; lightening of the skin results when this same pigment is withdrawn from the various processes into the bodies of the melanophores (Fig. 2–10). Changes in color from green to yellow occur from the combined effect of melanin concentration in the bodies of the melanophores and amoeboid movement of lipophores until they lie between, or even below, the guanophores (Fig. 2–11). These changes result in a higher intensity of reflected yellow rays from the xanthophores than of blue rays from the guanophores, giving a yellow appearance to the skin.

The control mechanism for color changes in amphibians differs from that of teleost fish. In teleosts, chromatophores are controlled primarily by postganglionic sympathetic nerve fibers whose direct stimulation causes contraction or expansion

Figure 2–10 Melanophores of common frog (*Rana temporaria*) with melanin pigment dispersed into pseudopods (*A*) and concentrated in cell bodies (*B*). (From Noble: *The Biology of the Amphibia.*)

A

B

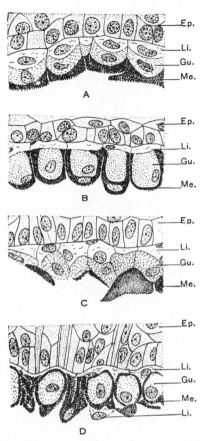

Figure 2-11 Diagrammatic section of the skin of a tree frog during color change. *A*. Bright green. The lipophores are arranged over the guanophores and the melanophores are partly expanded. *B*. Dark green. The guanophores are cylindrical and are nearly surrounded by the melanophores. *C*. Lemon yellow. Lipophores and guanophores are irregularly arranged and the melanophores are greatly contracted. *D*. Gray. The lipophores are greatly flattened and some are squeezed between the guanophores. The latter are completely surrounded by the melanophores. *Ep.*, epidermis; *Gu.*, guanophores; *Li.*, lipophores; *Me.*, melanophores. (From Noble: *The Biology of the Amphibia.*)

of cytoplasmic pigment within the cells and, therefore, lightening or darkening of the skin. Although amphibian color changes are to some degree under nervous control, they are induced primarily by the neurohumor intermedine, which is produced by the intermediate lobe of the pituitary gland; high blood levels of intermedine cause darkening of the amphibian, and its disappearance from the blood induces blanching. Because the control of their chromatophores is primarily hormonal, rather than nervous, amphibians are unable to change color as rapidly as

teleost fish. The biological significance of amphibian coloration and color changes will be discussed in a later chapter.

SKELETAL SYSTEM

SKULL

In comparison with that of the teleost fish, the modern amphibian skull is simple and contains far fewer bones; it is also much flatter. Much of the upper surface of the head, especially in anurans, lacks a bony covering, and parts of the primitive chondrocranium are unossified, degenerate, or lost. The only dermal bones generally left in the roof of anuran skulls are paired premaxillaries, septomaxillaries, maxillaries, nasals, quadratojugals, frontals, parietals, and squamosals; in some species, including the hylid *Hyla septentrionalis* (Trueb, 1966), an unpaired dermal sphenethmoid is also present (Fig. 2-12). The frontals and parietals of each side are fused to one another to form paired frontoparietals, and in some, such as clawed frogs (*Xenopus*, family Pipidae), the two frontoparietals may fuse with one another. Salamanders tend to have more primitive skulls than those of anurans, and in more primitive species paired lacrimals and prefrontals are still retained in addition to the dermal roof elements mentioned above; the quadratojugal only appears temporarily as a separate element during the ontogeny of salamanders. Postfrontals and jugals are present in the Apoda but have been lost in the other two orders of amphibians. As noted by Noble (1931), the elimination of elements from the roof of the skull has been correlated with an increase in the size of amphibian eyes, whereas the reduction of elements in the temporal region has been associated with increased freedom of the temporal muscles.

The following dermal bones have been lost from the posterior skull of salamanders and anurans: basioccipital, supraoccipital, postparietal, intertemporal, supratemporal, and tabular. Two or three centers of ossification generally appear in cartilaginous elements in the posterior part of the braincase. Ossification of the exoccipitals generally occurs, and each of these usually extends into the posterior wall of

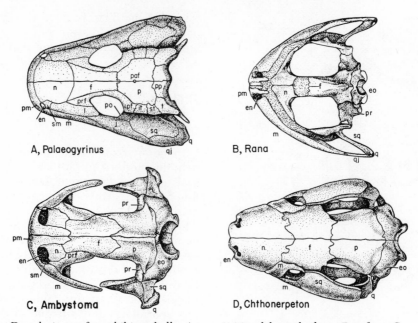

Figure 2–12   Dorsal views of amphibian skulls. *A*, a primitive labyrinthodont; *B*, a frog; *C*, a salamander; *D*, an apodan. *en*, external naris; *eo*, exoccipital; *f*, frontal (fused with parietal in frogs); *it*, intertemporal; *j*, jugal; *l*, lacrimal; *m*, maxilla; *n*, nasal; *p*, parietal; *paf*, parietal foramen; *pf*, postfrontal; *pm*, premaxilla; *po*, postorbital; *pp*, postparietal; *pr*, prootic; *prf*, prefrontal; *q*, quadrate; *qj*, quadratojugal; *sm*, septomaxilla; *sq*, squamosal; *st*, supratemporal; *t*, tabular. (From Romer, A. S.: *The Vertebrate Body*. 4th Ed. W. B. Saunders Co., 1970.)

the otic (ear) capsule. The anterior wall of the otic capsule becomes ossified in most anurans and in some salamanders, forming the prootic. Finally, a separate center of ossification, the opisthotic, appears in the posterior wall of the otic capsule in the salamander family Proteidae (*Necturus* and *Proteus*). The opisthotic is missing in all other amphibians; in lungless salamanders (family Plethodontidae), neither prootic, opisthotic, nor exoccipitals form separate ossifications, even in the larvae (Noble, 1931). The exoccipitals and prootics commonly fuse synosteotically in anurans, and in some members of the family Pipidae the exoccipital-prootic combination on one side may fuse with that on the other side, forming a single element around the posterior surface of the skull. In contrast to reptiles, which have a single occipital condyle, amphibians possess two; each exoccipital has a condyle which articulates with the atlas of the vertebral column. The interorbital walls of the braincase of most anurans and salamanders ossify, forming a sphenethmoid element on each side. The endochondral ethmoid may also ossify and fuse with the sphenethmoid.

Throughout the evolutionary history of Amphibia there has been, in addition to alterations of the posterior skull, a large number of changes in the ventral elements of the skull. There is much variation among the palates of modern forms (Fig. 2–13). Because amphibians have no secondary palate, the internal nares open through the anterior roof of the mouth. The ventral side of the braincase is covered by the parasphenoid, a dermal bone, on either side of which are the large orbits; the parasphenoid is very broad in caecilians. Anterior to the parasphenoid is usually a pair of prevomers, also dermal in origin; prevomers are absent in some anurans, and in others, such as the firebellied toads (*Bombina*, family Discoglossidae) and clawed frogs (*Xenopus*, family Pipidae), they fuse to form a single element. In salamanders, the prevomers and palatines fuse, and in the family Salamandridae a tooth-bearing process of the prevomer-palatine combination grows posteriorly on each side lateral to the parasphenoid; in the family Plethodontidae the same processes grow along the ventral surface of the parasphenoid.

Most anurans and primitive salamanders

in the family Salamandridae have skulls which are similar in appearance ventrally, having short prevomers, long maxillae, and quadratojugals. In advanced salamanders such as those of the family Plethodontidae the prevomers grow posteriorly, the maxillaries are short and fail to reach the quadrate, and the quadratojugals are missing; tailed frogs (*Ascaphus truei*, family Ascaphidae), some spadefoot toads (*Scaphiopus*, family Pelobatidae), and a few other anurans also lack quadratojugals. With the increase in size of the eyes there has been an increase in the size of the interpterygoid vacuities and a corresponding reduction in the width of the pterygoid elements of the modern amphibians. Epipterygoids are not present in amphibians. The attachment of the upper jaw to the cranium is normally autostylic in modern amphibians, but a movable basipterygoid articulation is present among salamanders in the family Hynobiidae and at least metamorphic stages of primitive frogs (Eaton, 1959).

The lower jaw in elasmobranch fishes consists of a pair of Meckel's cartilages which are covered with dermal bones in other gnathostomes. As one progresses upward through the vertebrates, there is an evolutionary trend in the lower jaw toward not only an ossification of Meckel's cartilages but also a reduction in the number of dermal bones; the amphibian lower jaw is intermediate in this trend. Incomplete ossification of Meckel's cartilages occurs in anuran amphibians, producing a small mentomeckelian bone, and the articular portion of the jaw remains cartilaginous. In salamanders, more or less complete ossification of the cartilage occurs to produce the articular bone. In contrast to primitive amphibians that had numerous dermal bones in the lower jaw, modern amphibians have a maximum of three bones in addition to the ossified Meckel's cartilage. A large dentary is always present and this forms the major portion of the jaw. Posterior to the dentary there may be a small splenial (considered to be the cornoid by some authors), and medial to the dentary there may be a relatively large angular (considered by some workers to be the prearticular).

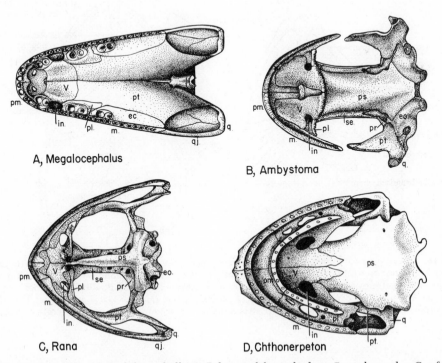

Figure 2–13 Ventral views of amphibian skulls. *A*, Paleozoic labyrinthodont; *B*, a salamander; *C*, a frog; *D*, an apodan. *ec*, ectopterygoid; *eo*, exoccipital; *in*, internal naris; *m*, maxilla; *pl*, palatine; *pm*, premaxilla; *pr*, prootic; *ps*, parasphenoid; *pt*, pterygoid; *q*, quadrate; *qj*, quadratojugal; *se*, sphenethmoid; *sq*, squamosal; *v*, vomer. (From Romer, A. S.: *The Vertebrate Body.* 4th Ed. W. B. Saunders Co., 1970.)

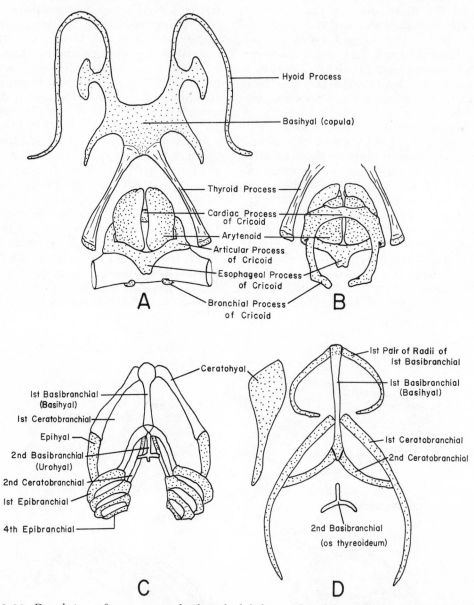

Figure 2–14 Dorsal views of representative hyobranchial skeletons of amphibians. *A*, an anuran (*Leptodactylus ocellatus*); *B*, ventral view of laryngeal skeleton shown in *A*; *C*, a perennibranch salamander (*Siren lacertina*); *D*, a metamorphosed terrestrial salamander (*Chioglossa lusitanica*). (*A* and *B*, after Trewavas: The hyoid and larynx of the Anura; *C*, after Noble: *The Biology of the Amphibia*; *D*, After Özeti and Wake: The morphology and evolution of the tongue and associated structures in salamanders and newts [family Salamandridae].)

HYOBRANCHIAL SKELETON

The hyobranchial skeleton consists of the hyoid apparatus and the cartilages of the larynx. The hyoid apparatus is a plate of cartilage or bone situated in the throat with projecting processes extending to the otic region. It functions as a support for the tongue and larynx, provides for muscle attachment, and in amphibians is important in both feeding and buccal respiration. The larynx is a chamber located at the anterior end of the windpipe whose walls are supported by cartilages derived from the branchial arches. The larynx is functionally important in respiration and, in anurans, in breeding behavior. Having extreme evolutionary plasticity, the hyobranchial skeleton varies considerably among different kinds of amphibians (Fig. 2–14).

In salamanders belonging to the family Hynobiidae, the hyoid arches are continuous with the basihyal (first basibranchial or copula) and are elongated; in other salamanders, the epihyals (lateral portions of the hyoid arches) are free from the basihyal, and the basihyal, which may be carried out of the mouth when the tongue is protruded, may support one or two pairs of processes (cornua). In addition, the second basibranchial (urohyal) of many salamanders separates from the rest of the apparatus at metamorphosis and becomes ossified as the *os thyreoideum*. Permanently aquatic salamanders such as sirens (*Siren*, family Sirenidae) have an adult hyobranchial skeleton which is essentially like that of larvae, with four pairs of epibranchials and two ceratohyals (Fig. 2–14C). Members of the family Hynobiidae retain two pairs of epibranchials after metamorphosis, while other salamanders retain only one pair plus, occasionally, rudiments of a second pair.

In the hyobranchial apparatus of adult caecilians, the hyoid arch and first branchial arch fuse; in addition to this structure, two or three separate branchial arches are present.

The hyobranchial skeleton of anurans consists of a cartilaginous plate supporting three or four pairs of processes (Fig. 2–14A, B). In most species the hyoid arches are continuous with the basihyal; these are the most anterior pair of processes and are long and slender, extending posteriorly to attach to the skull. The thyroid processes are the most posterior pair and are generally well ossified; they support the larynx. There is a tremendous amount of variation among different species of anurans in regard to this skeleton. Trewavas (1933) has described the skeleton and muscles of the hyolaryngeal apparatus for 60 species of phaneroglossan (tongued) anurans and the reader is referred to this monumental work for details of variability. The loss of a definitive tongue in the family Pipidae has involved the loss of the hyoid and the development of a boxlike hyobranchial apparatus (Ridgewood, 1898).

VERTEBRAL COLUMN, STERNUM, AND RIBS

The vertebrae of modern amphibians exhibit the beginning of a trend in modifications associated with adaptations to terrestrial life. Primitively, both the posterior and anterior faces of the vertebral centra were concave (amphicoelous centra), and vertebrae were unable to articulate directly with one another. The evolutionary trend in tetrapod vertebrae has included a reduction in the number of embryonic components; a general loss of the hemal arch; deviations from amphicoelous centra which allow each vertebra to articulate directly with those in front and behind it; strengthening of the intervertebral joints by paired processes from each centrum; and regional specialization of vertebrae correlated with differentiation of the body into cervical, thoracic, lumbar, sacral, and caudal portions. Primitively, the primary element of the central region of the vertebrae was the notochord, and the bony elements (two sets per neural arch) were ineffective structures lying alongside (hypocentra) or dorsolateral to (pleurocentra) the notochordal sheath. During the evolution of tetrapods, the pleurocentra and hypocentra increased in thickness and functionally replaced the notochord so that, except in neotenic amphibians which retain larval features, the notochord is retained only, if at all, as a cushioning element between a series of bony segments. Primitively, each vertebra had two functional centra (hypocentrum and pleurocentrum); where, through evolution, there has resulted only a single functional centrum, it may be the development of either the hypocentrum

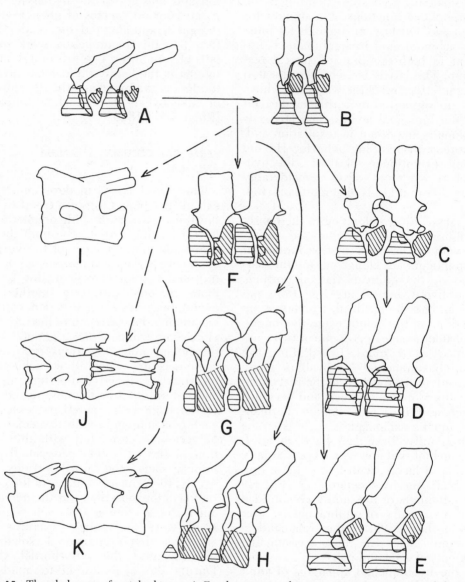

Figure 2–15 The phylogeny of vertebral types. *A, Eusthenopteron*, the crossopterygian stage; *B, Ichthyostega*, the ichthyostegid stage; *C, Eryops*, a rhachitome; *D, Mastodonsaurus*, a stereospondyl; *E, Lyriocephalus*, a trematosaurid; *F, Archeria*, an embolomere; *G, Seymouria*, a seymouriamorph; *H, Sphenodon*, the primitive amniote stage; *I, Cardiocephalus*, a lepospondyl; *J*, a salamander; *K*, a frog. Horizontal lines indicate intercentra, diagonal lines indicate pleurocentra. (Mainly after Williams: Gadow's arcualia and the development of tetrapod vertebrae.)

or the pleurocentrum. Williams (1959) has presented evidence that the hypocentrum became the functional centrum in only one branch of tetrapods, the Paleozoic Amphibia termed Rhachitomi and Stereospondyli which died out in the late Triassic, and that in all living tetrapods, including the three orders of amphibians, the pleurocentrum is the functional element of the centrum (Fig. 2–15). At any rate, the degree to which each embryonic element contributes to the adult vertebra, the presence or absence of a cartilaginous stage of development of centra, and the shape of the centra are used to classify vertebrae and have had an important bearing on the taxonomy of vertebrates.

Living amphibians fall into two categories in regard to the development of the centra. In salamanders and caecilians, there is direct deposition of bone around the notochord, and for this reason, in the past, these were lumped together in the subclass Lepospondyli. The centra of anurans, on the other hand, develop initially as blocks of cartilage and are subsequently replaced by bone (Fig. 2–16). Consequently, in older classifications anurans were placed in the subclass Apsidospondyli. As will be discussed in the next chapter, the splitting of modern amphibians into two subclasses is no longer generally considered realistic.

The shape of the centrum varies taxonomically among living amphibians (Figs. 2–16, 2–17). The vertebrae of most salamanders are opisthocoelous – the centrum is concave on its posterior face and convex on its anterior face. However, some of the permanently aquatic salamanders, such as the mudpuppy (*Necturus*, family Proteidae), have amphicoelous ver-

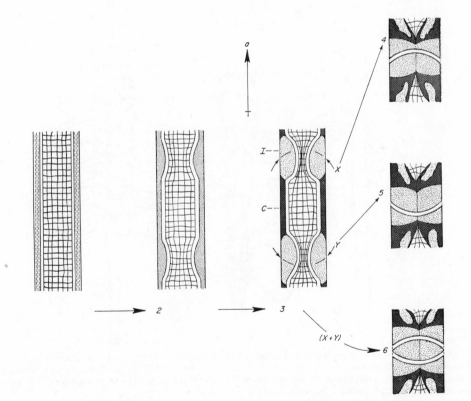

Figure 2–16 Diagrammatic representation of vertebral development in Anura and illustration of different shapes of centra. *1*, early larval stage; *2*, 22 mm tadpole of common frog (*Rana temporaria*); *3*, common frog two years of age; *4* and *5*, respectively, opisthocoelous and procoelous vertebrae of common frog three to four years old; *6*, "free disc" condition of pelobatid *Megophrys major*, four years of age. X and Y represent invading connective tissue arcs; (X + Y) represents invasion of a single intervertebral body by two connective tissue arcs. C, ossified cylinder of centrum; I, cartilaginous intervertebral discs. (From *Introduction to Herpetology*, Second Edition, by Coleman J. Goin and Olive B. Goin. W. H. Freeman and Company. Copyright © 1971.)

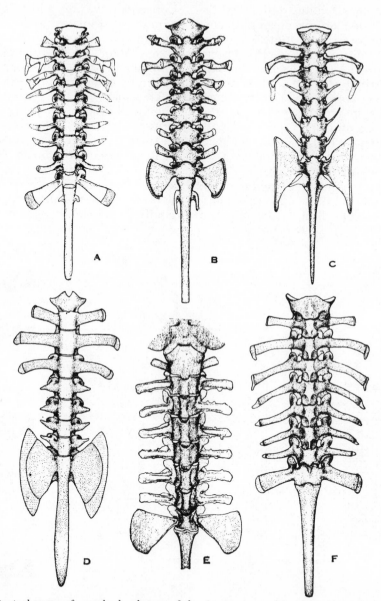

Figure 2–17 Principal types of vertebral columns of the Anura. *A*, amphicoelous (*Ascaphus truei*); *B*, opisthocoelous (*Alytes obstetricans*); *C*, opisthocoelous with fused coccyx (*Xenopus tropicalis*); *D*, anomocoelous (*Scaphiopus couchi*); *E*, procoelous (*Atelopus varius*); *F*, diplasiocoelous (*Rana virgatipes*). All are viewed from the ventral side. (From Noble: *The Biology of the Amphibia.*)

tebrae with persistent notochordal elements. Most anurans have procoelous vertebrae, the reverse of opisthocoelous, in which the concave surface faces anteriorly and the posterior face is convex. Three families of anurans (Discoglossidae, Pipidae, Rhinophrynidae), however, have opisthocoelous vertebrae; four families (Ranidae, Rhacophoridae, Hyperoliidae, and Microhylidae) are diplasiocoelous, a condition in which the presacral vertebra is amphicoelous and the other vertebrae are procoelous; the family Ascaphidae has amphicoelous vertebrae with free intervertebral discs; the vertebrae of the pelobatids (family Pelobatidae) are variable, being either procoelous or amphicoelous with free intervertebral discs. Caecilians have amphicoelous vertebrae and persistent intervertebral discs.

The intervertebral joints of amphibians are reinforced by a pair of prezygapophyses and a pair of postzygapophyses, processes arising from dorsal connecting pieces between successive neural arches during development (Fig. 2–18); these processes later break apart to form the zygapophysial joint. As in tetrapods in general, the articulation of these processes is such that the prezygapophyses, extending anteriorly from the vertebra, lie below the postzygapophyses that extend posteriorly from the vertebra ahead. In addition, some salamanders have hemapophyses, ventral arch elements which, if present, lie directly under the zygapophyses; lateral processes of the ventral arches, parapophyses, may also be present. Caecilians characteristically have "infrazygapophyses" which extend forward alongside the intervertebral region and the posterior part of the next anterior centrum; these apparently are ventral arch elements (Williams, 1959).

In fishes, the body and vertebral column are differentiated into trunk and tail regions only; to these, in amphibians, are added a short cervical (neck) region and a sacral region. Amphibians have a single cervical vertebra, the atlas. This articulates with the skull by means of two concavities on the anterior face of the centrum into which fit the two occipital condyles of the skull. The posterior surface of the atlas is similar to those of the remaining vertebrae (convex or concave, depending upon the amphibian). Anurans and salamanders possess in addition to the atlas a single sacral vertebra which articulates with the pelvic girdle. Between the atlas and the sacral vertebra are trunk vertebrae that are relatively uniform in salamanders and moderately differentiated in anurans. Posterior to the sacral vertebra of salamanders are the caudal vertebrae of the tail, which are unique among tetrapods in having a hemal arch that encloses the caudal artery and vein. The caudal vertebrae of anurans are fused together to form the coccyx (urostyle), an unseg-

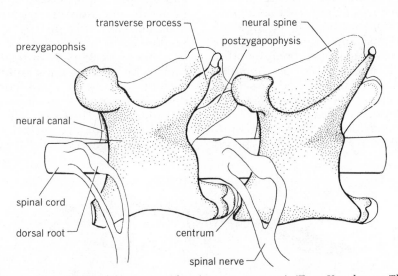

Figure 2–18   Two trunk vertebrae of the common frog (*Rana temporaria*). (From *Vertebrates: Their Structure and Life*, by W. B. Yapp. © 1965 by Oxford University Press, Inc. Reprinted by permission.)

mented bony rod which articulates with, or is fused to, the posterior face of the sacral vertebra. With the exception of the atlas, the vertebrae of caecilians are all nearly identical in shape.

The number of vertebrae varies greatly among living amphibians. The caecilians have up to 250. Salamanders have between 30 and 100, and there may be an increase in the number of vertebrae during an individual's life, particularly in the caudal region (Noble, 1931). The number of functional vertebrae in anurans is greatly reduced in adaptation to their jumping type of locomotion; the usual number is nine and there may be as few as six. The ascaphids are distinguishable from all other Recent families of anurans in having 10 vertebrae.

An obvious characteristic of most teleost fishes is the presence of numerous thin bony ribs articulating with trunk and caudal vertebrae; some fishes even have two or more pairs of ribs per vertebra. The earliest amphibians also had well-developed ribs in both trunk and tail (Fig. 2–19). In contrast, the ribs of modern amphibians are reduced or absent. They are best developed in caecilians, relatively short in salamanders, and (except for the families Ascaphidae, Discoglossidae, and Pipidae) absent in anurans. Free ribs are found only in adult ascaphids and discoglossids; those of pipid larvae later fuse to the diapophyses of the vertebrae and are not distinguishable from them. Large cartilaginous abdominal ribs are present in the myosepta of the rectus abdominis muscle of *Leiopelma* (Family Ascaphidae); Noble (1931) has noted that traces of similar paired cartilages have been found in the ventral musculature of *Bombina* (family Discoglossidae) and *Necturus* (family Proteidae), the mudpuppy.

A true sternum, characteristic of the higher tetrapods, first appears in amphibians. As might be expected, it is poorly developed and differs from that of reptiles, birds, and mammals in that the ribs never attach to it. A sternum is absent in caecilians and some of the more primitive salamanders. It is present in the majority of salamanders as a small cartilaginous plate which articulates with the coracoids of the pectoral girdle. It is well-developed in most anurans, being an ossified plate or rodlike structure in advanced anurans such as those of the family Ranidae. The sternum functions as a site for muscle attachment.

LIMBS AND GIRDLES

The most striking contrast between the earliest amphibians and their crossopterygian ancestors involves the limbs. Early amphibians were not truly terrestrial, being amphibious at best, and spent the majority of their time in water. Their limbs were small and weak and they were unable to raise their bodies off the ground. Only gradually did tetrapod limbs evolve which were capable of lifting the body, enabling rapid running and jumping forms of locomotion to develop. Nevertheless, it is believed that even the most primitive amphibians possessed two pairs of pentadactyl limbs which differed from the fins of crossopterygians primarily in possessing digits and a wrist or ankle (Fig. 2–20).

There is considerable diversity in limbs among modern amphibians as a result of their adaptations to aquatic, burrowing, and arboreal habits. Caecilians, which are primarily forest-dwelling burrowing animals and wormlike in appearance, have neither limbs nor girdles. Most salamanders possess two pairs of relatively weak limbs which do not allow rapid movement on land; usually there are four functional digits on their front feet and five digits on the hind feet. However, salamanders in the family Sirenidae lack hind legs and those in the genus *Pseudobranchus* have

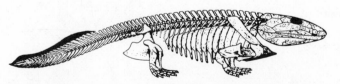

Figure 2–19   Skeleton of oldest known amphibian, *Ichthyostega* of the late Devonian, showing well-developed ribs. (From Romer, A. S.: *Vertebrate Paleontology*, 3rd Ed. Univ. Chicago Press, 1966.)

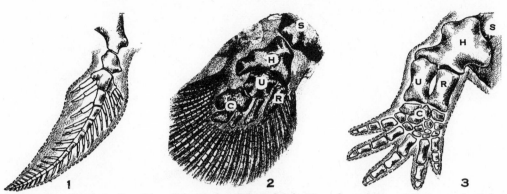

Figure 2–20 Bones in fin of lungfish (1), fringe-finned fish (2), and amphibian (3). *C*, carpal; *H*, humerus; *R*, radius; *S*, scapula; *U*, ulna. (From *Tales Told by Fossils* by Carroll Lane Fenton. Copyright © 1966 by Carroll Lane Fenton. Reprinted by permission of Doubleday & Company, Inc.)

only three digits on their front feet. The aquatic congo eel (*Amphiuma*, family Amphiumidae) of the southeastern United States has extremely small limbs and as few as two or three digits per foot.

The limbs of anurans are specialized for jumping and swimming and are very well-developed in all (Fig. 2–21). Anuran rear legs and feet are enlarged and the tarsus lengthened through elongation of the tibiale (astragalus) and fibulare (calcaneus). Strengthening of both pairs of limbs has occurred through fusion of the radius and ulna, producing the radio-ulna, and of the tibia and fibula, producing the tibio-fibula; the tibiale and fibulare are also occasionally fused. Most anurans, like the majority of salamanders, possess five digits on the rear feet and four functional digits on the front feet. In addition, a prehallux, consisting of one to four bones and commonly supporting a sharp-edged tubercle used for digging by burrowing species, is frequently present on the rear foot. A prepollex is present on the front foot in addition to rudiments of the fifth digit and is frequently enlarged on male anurans and some salamanders, being used to grip the female during amplexus. In anurans, in contrast to salamanders, the digital formula is rarely reduced below the standard four on the front foot and five on the rear.

The upper segment of each limb attaches to a special part of the appendicular skeleton called a girdle. The humerus of each front leg articulates with the pectoral girdle and the femur of the rear legs articulates with the pelvic girdle. Both of these girdles are present in fish, together with a second pectoral girdle that is dermal in origin and really provides the principal support for the fins. The early amphibians also had a dermal shoulder girdle which was similar to that of the crossopterygians except that the two halves of the girdle were joined midventrally by a new dermal element, the interclavicle. The majority of each half consisted of the scapulocoracoid, and on the anterior edge of this were a clavicle and a cleithrum. A supracleithrum and post-temporal were also present in the most primitive Embolomeri (Fig. 2–22). As has happened so consistently in other skeletal structures, there has been a reduction in the dermal elements of the pectoral girdle of advanced amphibians, and theirs are very different from those of their ancestors.

Salamanders have no dermal girdle. The three elements present on the pectoral girdle (scapula, coracoid, and precoracoid) are cartilaginous for the most part; they are only ossified around the glenoid cavity into which the head of the humerus fits. The coracoids overlap medially but are never fused together; they are much broader than in more primitive amphibians. Interestingly, in the family Sirenidae (*Siren* and *Pseudobranchus*) there is an ossification in the posterior half of the coracoid plate paralleling that in anurans. The loss of the clavicle, interclavicle, and cleithrum in salamanders may have been correlated with the greater development of movement in the forelimbs (Noble, 1931).

Compared to the pectoral girdle of sal-

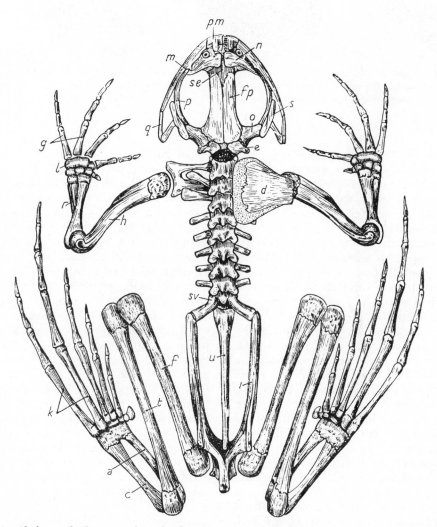

Figure 2–21   Skeleton of a frog, seen from the dorsal surface with the left suprascapular and scapular removed. *a*, astragalus; *c*, calcaneum; *d*, suprascapular; *e*, exoccipital; *f*, femur; *fp*, frontoparietal; *g*, metacarpals; *h*, humerus; *i*, ilium; *k*, metatarsals; *l*, carpus; *m*, maxilla; *n*, nasal; *o*, prootic; *p*, pterygoid; *pm*, premaxilla; *q*, quadratojugal; *r*, radio-ulna; *s*, squamosal; *se*, sphenethmoid; *sv*, sacral vertebra; *t*, tibio-fibula; *u*, urostyle (coccyx). (From Young, J. Z.: *The Life of Vertebrates.* 2nd Ed. Oxford Univ. Press, 1962.)

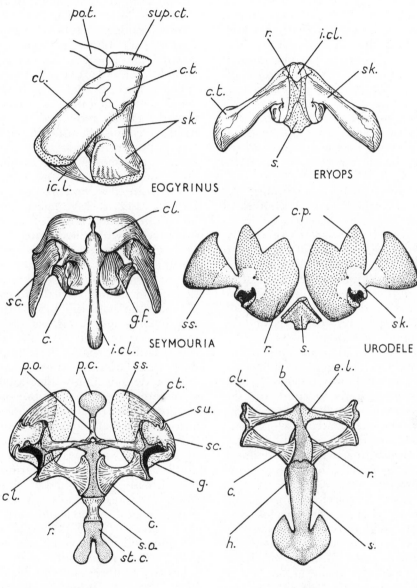

Figure 2–22   Representative amphibian pectoral girdles from embolomere through modern anurans. *b*, precoracoid bridge; *c*, coracoid; *cl*, clavicle; *c.p.*, coracoid process; *c.t.*, cleithrum; *d.b.*, dorsal blade; *e.l.*, precoracoid cartilage; *e.m.*, epicoracoid muscle; *g*, glenoid; *g.f.*, glenoid foss; *h*, epicoracoid horn, *i.cl.*, interclavicle; *l*, left epicoracoid cartilage; *p.c.*, prezonal cartilage; *p.o.*, prezonal bone; *po.t.*, posterior temporal; *r*, epicoracoid cartilage; *s*, sternum; *sc.*, scapula; *sk.*, scapulo-coracoid; *s.o.*, sternal bone; *s.r.*, ventral blade; *ss.*, suprascapular cartilage; *st.c.*, sternal cartilage; *su.*, coraco-cleithral suture; *sup.ct.*, supra-cleithrum. (From Young, J. Z.: *The Life of Vertebrates.* 2nd Ed. Oxford Univ. Press, 1962.)

amanders, that of anurans tends to be more complex and exhibits a greater degree of ossification. Dorsally, the scapula is relatively small but bony. Dorsal to the scapula is a bony cleithrum, which is variable in size, and dorsal to the cleithrum is a relatively large suprascapula which is generally only partially ossified. Ventrally, both the clavicle and coracoid are bony; these connect to a longitudinal median cartilaginous element, the epicoracoid. In all anurans, the epicoracoids are united anteriorly to some degree by a discrete precoracoid bridge, but the precoracoid is variable in size and shape; it is nearly absent in some species but extends from the sternum to the scapula in others.

The pectoral girdle of anurans has played an important role in their taxonomy, and the terms arcifery and firmisterny need to be understood. Cope (1864, 1865) defined two major groups of toothed phaneroglossal (tongued) frogs, the Arcifera and the Raniformia, characterized by the form of their pectoral girdles. The Raniformia were distinguished by nearly abutting clavicles and coracoids and by fused epicoracoids; the Arcifera were characterized by free, overlapping epicoracoid cartilages and by divergent clavicles and coracoids. Subsequently, Cope (1867) used the same criteria to subdivide the toothless Bufoniformia. Boulenger (1882, 1888, 1897, 1910) recognized two primary divisions of phaneroglossids, the Arcifera and the Firmisternia, which he defined, respectively, by exactly the same pectoral girdle characters as Cope had used to separate his Arcifera and Raniformia, but regardless of whether the species concerned were toothed or not. Boulenger's scheme was accepted unquestionably by all authorities until Noble (1922) described the pectoral girdle of the Cuban *Sminthillus limbatus* (family Leptodactylidae) as having epicoracoid cartilages which are fused anteriorly for half their length but are free and overlapping from the level of the coracoids posteriorly. Feeling that this was an intermediate condition between arciferous and firmisternous conditions and noting Parker's (1868) conclusion that all anuran pectoral girdles pass through an arciferous larval condition, Noble concluded that the firmisternous condition could have arisen

independently in different phylogenetic lines and that, therefore, the pectoral girdle was not a valid means to diagnosis of major taxonomic groups. Griffiths (1956, 1963) has noted, however, that basic morphogenetic differences distinguish a range of arciferal families (Bufonidae, Discoglossidae, Hylidae, Pelobatidae, Pipidae, and Leptodactylidae) from firmisternal families (Ranidae, Rhacophoridae, and Microhylidae), even though some species have girdles which do not fit Boulenger's definitions. Consequently, Griffiths (1963) redefined the taxonomic Arcifera as those families in which definitive posteriorly-directed epicoracoid horns are formed and the Firmisternia as those anurans in which no definitive posteriorly-directed epicoracoid horns are formed (Fig. 2–23). These epicoracoid horns are short, blunt elements which intersleeve with the sternal laminae in Discoglossidae and Ascaphidae, whereas in other families they are almost as long as the sternum and their anterior part is housed in lateral sternal grooves; when present, the posterior tips of the epicoracoid horns are insertions for a pair of muscles originating from the abdominal recti (Griffiths, 1963). It is important to note that Griffiths' Arcifera and Firmisternia do not imply that arciferous or firmisternous girdles are present, for he observed (1963) that the epicoracoid horns and their associated musculature persist even when firmisterny is achieved and, as in *Brachycephalus* (family Brachycephalidae), even when the sternum is also lost. Finally, it should be noted that in some anurans (family Microhylidae), clavicles and precoracoids are lost and there are no ossified sternal elements.

Compared to the variations in pectoral girdles, the pelvic girdles of tetrapods are relatively uniform and simple, with three basic elements forming each half. These elements are the ischium, ilium, and pubis. The pelvic girdles of tetrapods differ from those of fish in that they are attached or fused to the vertebral column. In primitive amphibians this attachment was indirect, by way of sacral ribs, but it soon evolved into a direct fusion of the vertebral centra or transverse processes to the girdle, forming the sacrum. The pelvic girdle of salamanders has the shape of a flattened plate. Its anterior region, repre-

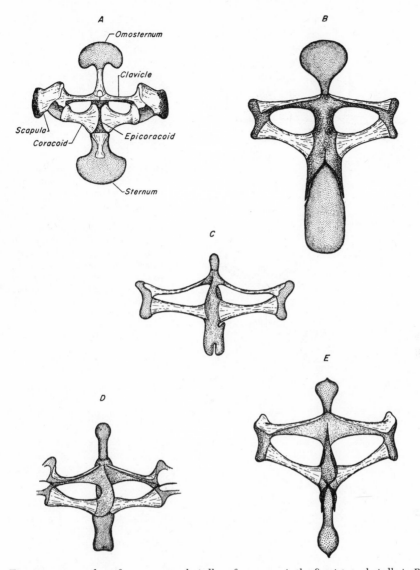

Figure 2–23   Firmisternous and arciferous pectoral girdles of anurans. A, the firmisternal girdle in *Rana tigrina*; and variations in the arciferal type as seen in: B, *Rhinoderma darwini*; C, *Sooglossus seychellensis*; D, *Eleutherodactylus bransfordi*; and E, *Sminthillus limbatus*. (From *Introduction to Herpetology*, Second Edition, by Coleman J. Goin and Olive B. Goin. W. H. Freeman and Company. Copyright © 1971.)

senting the pubic elements, is entirely cartilaginous. In the posterior part of the plate is a pair of rounded ossifications which represent the two ischia. The ilia extend dorsally on each side and are firmly attached to the sacral vertebra. The acetabulum, the socket into which the proximal end of the femur is inserted, is located at the point of junction of the pubic cartilage, ischium, and ilium; it is almost immediately ventral to the sacrum. Many aquatic salamanders have a Y-shaped prepubic (ypsiloid) cartilage attached to

the anterior end of the pubis (Fig. 2–24) which (as shown by Whipple, 1906) functions to control the shape of the inflated lungs; the latter function as hydrostatic organs and facilitate swimming under water. Contraction of the muscle associated with the prepubic cartilage pulls the cartilages dorsally, forcing air anteriorly into the lungs and making the anterior part of the animal more buoyant.

In anurans, the pubis may be either cartilaginous or bony, depending upon the species (Fig. 2–24). The acetabulum

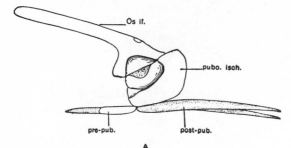

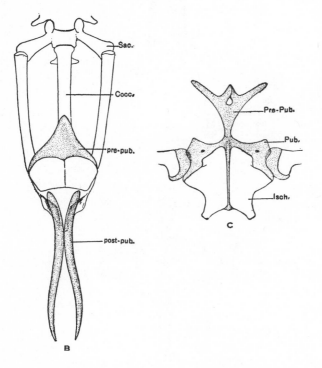

Figure 2–24 Pelvic girdle of an anuran compared with that of a salamander. Pelvis of tailed frog (*Ascaphus truei*) viewed laterally (*A*) and ventrally (*B*); pelvis of salamander (*Tylotriton verrucosus*) viewed ventrally (*C*). In the salamander the pubis is cartilaginous, while in the frog it is fused with the ischium; the ilia are elongated in the anuran. *Cocc.*, coccyx; *Isch.*, ischium; *Os.il.*, ilium; *post-pub.*, postpubis; *pre-pub.*, prepubis; *Pub.*, pubis; *pubo. isch.*, puboischium; *Sac.*, sacrum. (From Noble: *The Biology of the Amphibia.*)

is shifted slightly forward from its position in salamanders so that the pubis forms a greater portion of the socket. The ilia are very much elongated and extend anteriorly parallel to the coccyx (urostyle) and are attached ligamentously to the transverse processes of the sacral vertebra; the sacrum is, thus, far anterior to the acetabulum. Whiting (1961) has shown, contrary to popular belief, that considerable movement occurs at the ilio-sacral articulation and that the movement is of functional importance. In true frogs (*Rana*, family Ranidae), the ilia may rotate through an angle of over 90 degrees on the sacral ribs, in the vertical plane; this vertical bending and unbending is used during jumping and landing. In painted toads (*Discoglossus*, family Discoglossi-dae), the sacrum can be pivoted laterally on the pelvis through an angle of 20 degrees to either side; this movement is used in turning, both to take food and in locomotion. In the aquatic clawed frogs (*Xenopus*, family Pipidae), the sacrum can slide backward and forward on the pelvis, producing a considerable shortening and lengthening of the body; this movement is probably used in driving into underwater mud. Thus, while all three elements of the anuran pelvic girdle are firmly fused to one another and the two halves are similarly fused together, the girdle still functions as a shock absorber and provides flexibility because of its ligamentous attachment with the sacrum.

Caecilians, in addition to lacking limbs, lack pectoral and pelvic girdles.

MUSCULAR SYSTEM

As in other vertebrates, amphibian muscle is of three types: non-striated, involuntary muscle is found in the walls of the digestive tract, blood vessels, and so forth; striated, non-voluntary muscle occurs in the heart; and striated, voluntary muscle is found attaching skeletal elements and functioning in movement of the limbs and body. A detailed description of the muscles of various amphibians is beyond the scope of this book but some general statements can be made.

CARDIAC MUSCLE

Because the heart is an independently functioning organ whose activity is merely influenced by sympathetic and parasympathetic nerve impulses, cardiac muscle will continue to contract when the heart is isolated in a physiologically isotonic salt solution. The rhythmic pattern of beating is due to the refractory period following each contraction during which the muscle cells are incapable of being excited. Because the duration of the refractory period varies interspecifically, the isolated hearts of different species of amphibians will beat at different rates.

BODY MUSCLES

The body musculature of amphibians varies widely; that of aquatic salamanders is similar to the pattern in fishes, whereas that of terrestrial species, especially anurans, is markedly different. In fishes, the body muscles are divided into a dorsal epaxial mass (the definitive back muscles) and a ventral hypaxial mass (the body wall, lower portion of the tail musculature, and ventral throat muscles). The epaxial muscles are innervated by dorsal branches of the spinal nerves, while the hypaxial muscles are innervated by ventral branches. The epaxial muscles of fishes are well-developed and function, together with hypaxial muscles, in bending the body from side to side in swimming. The epaxial muscles of salamanders are basically the same as in fishes, being divided by myosepta into the same number of myotomes as there are vertebrae (Fig. 2–25). Much of the mass of each myotome consists of the *dorsalis trunci*, whose fibers run from myoseptum to myoseptum; each myoseptum, in turn, is attached to a single vertebra. A deeper portion of each myotome is formed by intersegmental muscle fibers running from one vertebra to the next. As in fish, these muscles function primarily in lateral bending of the body. In advanced anurans, the deeper epaxial muscles have become differentiated into *intertransversarii* and *interneurales* muscles which run, respectively, between the transverse processes and the neural arches of adjacent vertebrae; an *ileolumbaris* muscle occurs laterally in several anuran families. Superficial fibers of the *dorsalis trunci* are more or less continuous with the intervertebral fibers in the family Ascaphidae but are distinctly separate in Discoglossidae and, in advanced anurans, form the *longissimus dorsi* muscle which extends from the head to the coccyx. The latter muscle, rather than functioning to bend the body laterally, has assumed the function of holding up the head, especially in jumping, and in some species, such as the European fire-bellied toad (*Bombina bombina*, family Discoglossidae), of bending the body sharply upward in the "unken reflex." Lateral movements of the head are produced by anterior divisions of the *dorsalis trunci* which attach to the skull.

The hypaxial musculature of salamander larvae is fishlike. As with the epaxial muscles, myosepta divide the hypaxial muscles into myotomes, and each of these is composed of two layers of oblique fibers: the *obliquus externus* is the outer layer and runs ventroposteriorly; the inner layer is the *obliquus internus* and runs dorsoanteriorly. Midventrally is the *rectus abdominis*, whose fibers run longitudinally. As the larvae develop, there typically is considerable modification of the hypaxial musculature, and at metamorphosis radical changes may occur. During development, an additional muscle layer, the *obliquus externus superficialis*, differentiates outside of the original outer oblique layer; the latter is then termed the *obliquus externus profundus*. Similarly, another muscle layer, the *transverse*, differentiates inside the *obliquus internus*, but the latter retains its name. The *rectus abdominis* may also become differentiated into an outer *rectus superficialis* and an

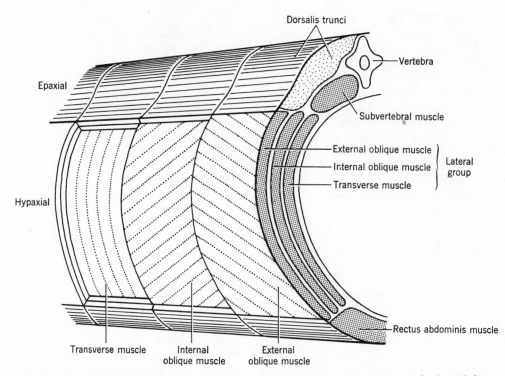

Figure 2–25 Epaxial and hypaxial muscles of salamanders as seen in cross section and a lateral dissection. Epaxial musculature, fine stipple; hypaxial musculature, coarse stipple. (From Torrey, T. W.: *Morphogenesis of the Vertebrates.* 2nd Ed. John Wiley and Sons, 1967.)

inner *rectus profundus.* The *rectus abdominis* is continued in the throat region by the *sternohyoideus,* or *abdominohyoideus,* and the *geniohyoideus;* the latter extends from the hyoid to the anterior margin of the lower jaw and may form a distinct lateral portion, the *omohyoideus.* At metamorphosis, the *rectus lateralis* may disappear and the secondary hypaxial muscles increase in thickness. Variations in this pattern occur among different genera of salamanders. There are also a loss of myosepta and a reduction in metamerism in the body of some groups; the original segmentation is preserved in the hypaxial muscles of the tail. In anuran tadpoles, the *obliquus externus* and *internus* form a single sheet, but just before metamorphosis this sheet is replaced by the *obliquus externus superficialis* and the *transverse.* The *rectus abdominis* develops in the tadpole but remains, even in the metamorphosed adult, as a single undivided muscle; this is also the only muscle to retain myosepta and the primitive pattern of segmentation.

MUSCLES OF THE FORELIMB AND PECTORAL GIRDLE

Variation in the musculature of the pectoral region of amphibians tends to reflect differences in the construction of the pectoral girdle. Salamanders have no clavicle and they lack the *episternocleidohumeralis longus* which runs from the ventral surface of the head of the humerus to the mesial end of the clavicle in anurans. Otherwise, the muscles covering the ventral surface of the pectoral girdle of anurans and salamanders are the same (Fig. 2–26). A broad oblique sheet, the *pectoralis,* extends from the midventral line to the humerus. Anterior fibers of this, the *supracoracoideus,* overlie the rounded coracoid part of the pectoral girdle and insert on the ventral process of the humerus; these cover the *coracoradialis proprius,* a deep muscle which continues as a tendon through the upper arm to the radius. Posterior to the *supracoracoideus,* the *pectoralis* covers the two *coracobrachialis* muscles which extend from the coracoid to the humerus. Both of the latter two muscles flex and adduct the forelimb. Variations in the above pattern occur among anurans. For example, in the family Pelobatidae, the *episternocleidohumeralis longus* fuses with the *supracoracoideus;* in Bufonidae, the posterior portion of the *supracoracoideus* splits off as the *supracoracoideus profundus.* The *pectoralis* is usually split into a sternal and an abdominal portion in anurans, and the abdominal portion may extend to the thighs.

The dorsal musculature of the pectoral girdle is essentially the same in anurans and salamanders and comparable to that of reptiles and mammals. A *dorsalis scapulae,* actually part of the ventrolateral musculature, originates on the suprascapula and has a narrow insertion on the head of the humerus. A deeper muscle, the *subcoracoscapularis,* runs from the posterior edge of the scapula and coracoid to the medial process of the humerus. A large muscle, the *anconeus* or *triceps,* covers the dorsal side of the upper arm; originating from four heads on the scapula, coracoid, and humerus, it merges into a single muscle and inserts on the ulna. The *anconeus* is the primary muscle used in extending the lower arm. The more distal muscles on the forearm and manus are comparable to those of other tetrapods.

MUSCLES OF THE HIND LIMB AND PELVIC GIRDLE

Differences in the pelvic girdles of salamanders and anurans have had a marked effect upon the proximal musculature of the hind limbs. Some of the muscles which are dorsally arranged in salamanders cover the anterior portion of the thigh in anurans. In addition, the loss of the tail in anurans has involved modifications of musculature; a pair of "tail-wagging" muscles of salamanders, the *caudalipuboischiotibialis,* is found only in the primitive *Ascaphus* and *Leiopelma* (family Ascaphidae), neither of which has a tail to wag. Nevertheless, comparison of the musculature of the primitive Olympic salamander (*Rhyacotriton olympicus,* family Ambystomatidae) with that of the primitive tailed frog (*Ascaphus truei,* family Ascaphidae) reveals basic similarities (Fig. 2–27).

On the anterior ventral surface of the thigh of salamanders is the *pubotibialis* muscle, which covers the deeper *puboischiofemoralis internus.* The innervations, origins, and insertions of these two muscles indicate that they are, respectively, homologous with the ventral head of the *adductor magnus* and the deeplying *pectineus* of anurans (Noble, 1931). Posterior to the *pubotibialis,* two large muscles cover the ventral surface of the

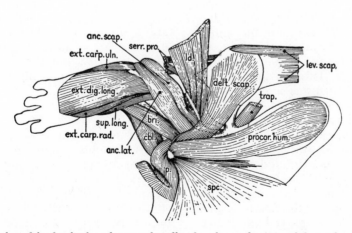

Figure 2–26 Muscles of the forelimb and pectoral girdle of a salamander (*Megalobatrachus*). *anc. lat.,* anconeus lateralis; *anc.scap.,* anconeus scapularis; *bri.,* brachialis inferior; *cb.l.,* coracobrachialis longus; *delt. scap.,* deltoides scapularis-dorsalis scapulae; *ext.carp.rad.,* extensor carpi radialis; *ext.carp.uln.,* extensor carpi ulnaris; *ext.dig.long.,* extensor digitorum longus; *lev.scap.,* levator scapulae; *ld.,* latissimus dorsi; *p.,* pectoralis; *procor.hum.,* procoracohumeralis; *serr.pro.,* serratus profundus; *spc.,* supracoracoideus; *sup.long.,* supinator longus; *trap.,* trapezius. (From Noble: *The Biology of the Amphibia.*)

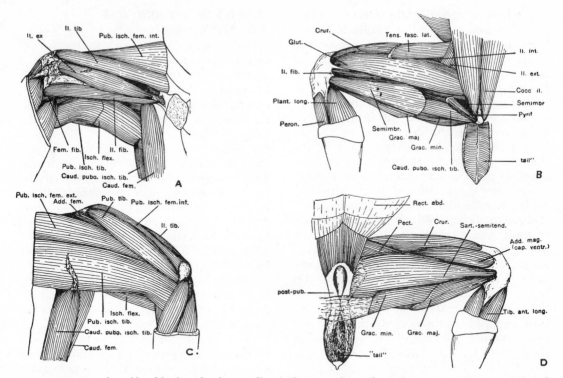

Figure 2–27 Muscles of hind limb and pelvic girdle of Olympic salamander (*Rhyacotriton olympicus*) (*A* and *C*) and of the tailed frog (*Ascaphus truei*) (*B* and *D*). Superficial muscles viewed from dorsal (*A* and *B*) and ventral aspect (*C* and *D*). *Add.fem.*, adductor femoris; *Add.mag.* (*cap. ventr.*) adductor magnus, ventral head; *Caud.fem.*, caudalifemoralis; *Caud.pubo.isch.tib.*, caudalipuboischiotibialis; *Cocc.il*, coccygeoiliacus; *Crur.*, cruralis; *Fem.fib.*, femorofibularis; *Glut.*, glutaeus; *Grac.maj.*, gracilis major; *Grac.min.*, gracilis minor; *Il. ext.*, iliacus externus; *Il.ex.*, ilioextensorius; *Il.fib.*, iliofibularis; *Il.tib.*, iliotibialis; *Il. int.*, iliacus internus; *Isch.flex.*, ischioflexorius; *Pect.*, pectineus; *Peron.*, peroneus; *Plant.long.*, plantaris longus; *Post-pub.*, post-pubis; *Pub.isch.fem.ext.*, puboischiofemoralis externus; *Pub.isch.fem.int.*, puboischiofemoralis internus; *Pub.isch.tib.*, puboischiotibialis; *Pub.tib.*, pubotibialis; *Pyrif.*, pyriformis; *Rect.abd.*, rectus abdominis; *Sart. semitend.*, sartorio-semitendinosus; *Semimbr.*, semimembranosus; *Tens.fasc.lat.*, tensor fasciae latae; *Tib.ant. long.*, tibialis anticus longus. (From Noble: *The Biology of the Amphibia.*)

thigh of salamanders. These are the *pubo-ischiofemoralis externus* and the *pubo-ischiotibialis;* the former is homologous to the dorsal head of the *adductor magnus* of anurans and the latter is homologous to the combined *sartorius* and *semitend-inosus* of advanced anurans (Noble, 1931).

On the dorsal surface of the thigh of anurans, the *puboischiofemoralis internus* (found on the anterior surface of the thigh in the salamander) is extended anteriorly with the ilium as the *iliacus externus*. The *iliotibialis* of salamanders is shortened in anurans and is called the *tensor fasciae latae*. The *ilioextensorius* is homologous with the much more powerful *cruralis* and *gluteus* of anurans (Noble, 1931). Both salamanders and anurans have the *ilio-fibularis* muscle and the *iliofemoralis* muscle. The *ischioflexorius* of the salamander

is comparable to the combination of the large *semimembranosus* on the dorsal side of the femur of anurans plus the *gracilis major* and *gracilis minor* on the posterior margin of the anuran thigh.

VISCERAL MUSCLES

The visceral muscles are striated voluntary muscles which are associated with the gill arches, hyoid, and jaws and are derived from mesenchyme in the wall of the pharynx. During the larval stages of amphibians, the gill arches are controlled by a superficial constrictor muscle and a series of deeper levators and marginales. In addition to these muscles, a *trapezius* arises from the skull or dorsal fascia and inserts on the scapula. All except the *trapezius* muscle generally disappear at metamorphosis, but in anurans the *levatores arcuum* series is retained as the *petrohyoidei* muscle group, which extends from the skull to the hyoid plate and functions to raise the hyoid apparatus and draw it forward (Noble, 1931).

The constrictors of the hyoid arch form a pair of *depressor mandibulae* muscles in amphibians. These run from the dorsolateral surface of the skull to the midventral line of the lower jaw and function to open the mouth; in anurans, the posterior portion of the *depressor mandibulae* may arise from the suprascapula.

In salamanders, the *temporalis* muscle is a jaw constrictor inserting on the coronoid process of the lower jaw and extending back to the first cervical vertebra. In anurans, it originates primarily on the roof of the skull, two slips of it making separate attachments to the squamosal or quadratojugal. These two slips are sometimes termed the *masseter major* and *masseter minor*. Another jaw constrictor of amphibians is the *pterygoideus*, which originates on the pterygoid bone or lateral portion of the braincase and inserts on the lower jaw. Ventrally, the jaw constrictors unite to form the submaxillary or superficial throat muscles. In anurans, an anterior portion differentiates as the *submentalis*, which functions to raise the mentomeckelian bones at the anterior end of the jaw.

DIGESTIVE SYSTEM

MOUTH AND BUCCAL CAVITY

The mouths and buccal cavities of amphibians, reptiles, and mammals form an evolutionary sequence of increased development of lips and tongue and specialization of teeth. Lips are poorly developed in all amphibians compared to their development in many reptiles and mammals. A muscular and more or less protrusible tongue is never present in fishes but is characteristic of most tetrapods, where it plays an important role in ingesting and, sometimes, passing food directly to the pharynx. The definitive tetrapod tongue is formed from three components: the hypoglossal apparatus, which represents the primary tongue of fishes; a thick glandular fold, which develops from the floor of the buccal cavity; and muscle fibers,

which grow under the glandular fold from the *geniohyoideus* muscle (part of the branchial musculature of fishes).

Living amphibians exhibit different degrees of development of the tongue. As might be expected, it tends to be most poorly developed in aquatic species, where it is functionally less important. Only the primary tongue is present in some aquatic salamanders, including the mudpuppy, *Necturus* (family Proteidae). Anurans in the family Pipidae all lack a definitive tongue, and even their primary tongue tends to be degenerate. Aside from these, however, a well-developed definitive tongue is found in the majority of anurans, caecilians, and salamanders. The tongue is generally attached anteriorly in anurans and, at rest, lies on the floor of the mouth with the tip pointed toward the pharynx; some anurans, such as toads (*Bufo*, family Bufonidae), can suddenly flip their tongue out by a forward bend and use it to capture prey. Tongue protrusion, however, is best developed in some plethodontid salamanders, such as *Hydromantes;* their tongue is boletoid and can be rapidly extended several times the length of the head. In general, the tongues of salamanders vary in accordance with whether the animal feeds in the water or in a terrestrial habitat. Ozeti and Wake (1969) have shown, for example, that the tongues of salamanders in the family Salamandridae fall into two distinct classes. Aquatic members of this family use a "gape and suck" method of feeding and have tongues in which there has been an elaboration of the posterior parts of the hyobranchial apparatus and musculature together with a reduction in structure and function of the anterior parts; the hyobranchial apparatus of these tends to be rigid and relatively well ossified. Terrestrial members tend to have an elaboration of the anterior parts of the tongue, especially the tongue pad, and a reduction in the posterior parts; their hyobranchial apparatus is flexible and permits extensive flipping of the tongue pad from the mouth. The tongues of caecilians are generally capable of only limited movement. All amphibians which have protrusible tongues also have special mucous glands in the mouth which secrete a sticky substance that coats the tongue and facilitates capturing prey.

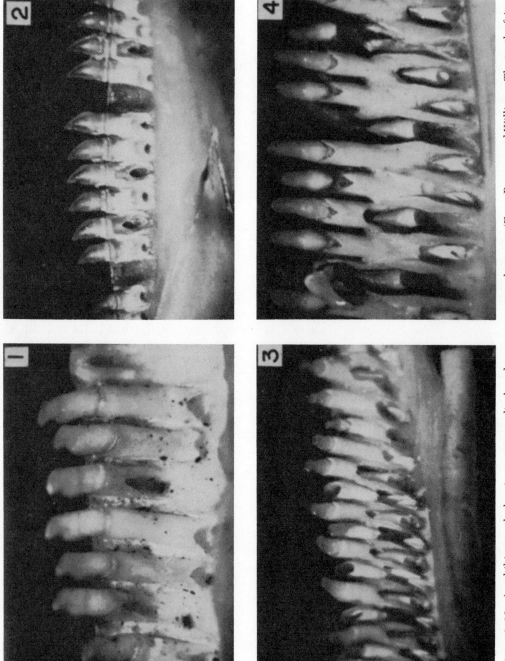

Figure 2–28 Amphibian teeth showing crown, pedicel, and transverse suture between. (From Parsons and Williams: The teeth of Amphibia and their relation to amphibian phylogeny.)

Amphibian teeth tend to be small and, especially in anurans, degenerate. They may be found fused to the premaxillaries and maxillaries of the upper jaw, to the dentaries of the lower jaw, and to the pre-vomers, palatines, and parasphenoid in the roof of the mouth. Those on the jaws are typically pleurodont (attached without sockets to the inner edge of the jaws), but palatal teeth may be essentially acrodont (attached to the ventral surface of the bone without sockets). Parsons and Williams (1962) have shown that in all three orders of living amphibians the teeth are divided into more or less separate crowns and pedicels separated by a transverse line resembling a suture between two bones (Fig. 2–28). The pedicels are tubelike structures firmly fused to the underlying bones; the crowns are capped with well-calcified enamel or durodentine. Because the boundaries between pedicels and crowns are zones of weakness, crowns are lost more commonly than are entire teeth.

Both the number of teeth and their location vary among living amphibians. The Mexican burrowing toad *Rhinophrynus dorsalis* (family Rhinophrynidae) and toads in the genus *Bufo* (family Bufonidae), for example, lack teeth entirely; true teeth are absent in the lower jaw of nearly all frogs; sexual dimorphism in dentition occurs in some salamanders. Apparently amphibian teeth may be replaced an indefinite number of times during an individual's life (polyphyodont dentition). They are homodont (all of the same general kind) and, because amphibians do not masticate their food, are used merely to grasp struggling prey until it can be swallowed.

Fully aquatic amphibians, as with fishes, have no glands in the buccal cavity other than simple mucous cells. However, terrestrial amphibians, like terrestrial vertebrates in general, have buccal glands that produce secretions which are sticky and function in capturing prey; other secretions facilitate swallowing of food by moistening it. In terrestrial salamanders and anurans, a mucous intermaxillary (internasal) gland develops in the nasal septum. Its secretion functions primarily to coat the tongue with adhesive mucus to facilitate capturing prey; this gland is lacking in caecilians. As mentioned, anurans and salamanders with protrusible tongues have numerous mucus-secreting lingual glands on the surface of the tongue and their secretion also functions to make the tongue sticky and aids in the capture of prey. Finally, anurans possess a pharyngeal gland near the internal nares; its secretion is released into these nares and may function to moisten the nasal passages. Although amphibians do not masticate their food, the digestive enzyme ptyalin, which digests starches to maltose, is present in the oral secretions of some anurans.

ALIMENTARY CANAL

The buccal cavity connects by way of the pharynx to the esophagus, which, in all amphibians, is very short and functions to pass food quickly to the stomach (Fig. 2–29). A food mass, represented by an individual prey animal or formed in the pharynx as an aggregation of small prey, is forced into the esophagus against the resistance of the esophageal sphincter muscle by pressure exerted by the tongue, the circumpharyngeal musculature, and the eyeballs which protrude into the buccal cavity. Thus, an amphibian shuts its eyes and constricts its throat and lingual area in swallowing. In most amphibians, the buccal cavity and esophagus are lined with ciliary cells which might help sweep tiny food particles toward the stomach.

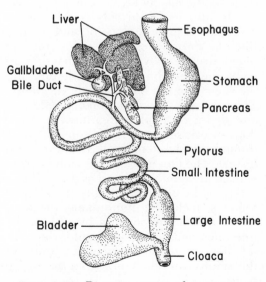

Figure 2–29 Digestive system of an anuran amphibian.

There is no digestion within the buccal cavity, pharynx, or esophagus. However, food may be lubricated by secretions from buccal and esophageal glands. In some anurans, there are esophageal cells which secrete the proteolytic enzyme pepsin but this does not begin to function until it reaches the stomach and is acidified; the buccal and esophageal mucus is alkaline in reaction and prohibits proteolytic digestion.

The esophagus leads into the stomach; this, in amphibians as in tetrapods in general, is where the digestion of protein occurs in an acid medium. The amphibian stomach is generally unspecialized, but in anurans is differentiated into an anterior cardiac portion and a posterior pyloric region; it is generally U-shaped but in some salamanders remains a simple straight tube. The stomach serves as a storage organ and permits amphibians to feed during brief periods of activity when food is available and environmental conditions optimal. The presence of acid in the stomach serves the important functions of inhibiting bacterial decomposition of proteins and hastening the death of prey swallowed alive. The opening from the stomach into the small intestine is constricted, as in all vertebrates, by a sphincter muscle which allows food to pass only when it relaxes; this is important, since the stomach functions both as a storage organ and in initiating digestion.

The small intestine also has two functions. It chemically breaks down food materials into molecules of hexose, amino acid, fatty acid, and glycerol, and it functions to absorb these molecules into the blood stream or lymph. The first is typically the function of the duodenum and the second of the longer, posterior ileum. There is little differentiation of duodenum and ileum of amphibians, however, and both processes occur to some degree throughout the length of the small intestine. The ileum does have an increased surface area for absorption as a result of a series of longitudinal ridges. In all vertebrates, including amphibians, the liver opens into the duodenum by way of the bile duct, and, as is true in all jawed vertebrates, glandular cells in the walls of the amphibian duodenum give rise to the pancreas.

The liver, derived embryonically from the gut endoderm, is important to digestion both because of the bile which it secretes and because of its function in the metabolism of food already digested. Bile acts as an emulsifying agent, breaking fat into minute globules in the lumen of the intestine so that it is more readily acted upon by pancreatic lipase. The liver also functions as an energy reservoir by transforming sugars and proteins into glycogen to be stored in its cells until required by the body. The amphibian liver has a reticular (netlike) structure with many anastomoses among layers (laminae) that are usually two cells thick. The bile capillaries consist of median and lateral capillaries which frequently extend deeply between cells so that less than the width of a cell may divide a bile capillary from a blood capillary. The circulatory system of the amphibian liver is comprised of portal and hepatic veins, hepatic arteries, and many large lymphatic channels. The capillary networks of the hepatic arteries empty into sinusoids formed by the portal vein. Pigment-containing cells sometimes fill the lymphatic channels, and groups of pigmented hepatic cells may also be present. Andrew (1959) has summarized interesting seasonal changes in the liver of Bufo arenarum. These include an enormous increase (100 per cent) in organic matter in the toad liver in the summer and a simultaneous decrease in fat, glycogen, and water. In addition, the liver cell volume of this toad has been observed to be dependent upon fat in December and January and upon glycogen in March and June and has been found to be at a minimum during October and November.

The need for water conservation by terrestrial vertebrates has led to the differentiation of the large intestine. This functions primarily in absorbing water from indigestible matter exiting from the anterior intestinal tract. In most fishes, there is a short rectum which connects the intestine to the anus or cloaca and serves merely as a passage for the elimination of wastes. This is also the situation in caecilians, where there is no real differentiation of small and large intestines. Salamanders possess a large intestine, but it is not well differentiated and consists essentially of an enlarged, short, straight tube which opens into the cloaca.

In anurans, the large intestine is distinctly differentiated from the ileum; as in salamanders, it opens into the cloaca. In all amphibians, there is a bladder which opens from the ventral side of the cloaca and serves the function of receiving and storing urine before it is voided and, thus, also helps to maintain the animal's water balance.

Feeding habits and, consequently, digestive systems change drastically with metamorphosis. In anurans, the larval foregut differentiates into the ciliated esophagus and inflated stomach. Peristaltic movements are not known to occur in the anterior larval gut, and the high development of ciliation in the foregut is the principal means by which food is carried to the larval stomach. Many anuran larvae are herbivorous and, at least in some, the larval stomach (manicotto) lacks proteolytic action and is used strictly for storage. Also associated with the herbivorous diet of anuran larvae is the great length of the larval mid- and hindguts. These together usually have a total length several times greater than the length of the animal. The long intestinal tube is coiled within the abdominal cavity, forcing the stomach and duodenum to the right side of the body. A maximum gut length is reached at the larval stage in which the hindlegs are well-developed. Among different anuran species, those with carnivorous larval diets tend to have shorter guts than those whose larval diets are herbivorous. Interestingly, there is great variation in the ratio of intestinal length and body length among species which appear to have the same or similar diets; the adaptive significance of a particular ratio to each species is not clear. It is also interesting that the length of the larval intestine may be controlled, within limits, by the diet. Larvae fed a vegetable diet have intestines which are relatively longer than individuals of the same species which are fed meat. This hyperextension seems to be due to a chemical influence of the food rather than to the extra bulk and slow digestion of plant material.

At metamorphosis, the anuran gut, posterior to the manicotto, becomes much shorter. This reduction in length apparently takes place at both ends of the intestine through gradual tonic contraction of circular and longitudinal muscles. The shortening and reorganization of the gut requires about 24 hours in toads (*Bufo*, family Bufonidae) and occurs within about 10 days after the front legs break through (Bowers, 1909).

Larval salamanders tend to be carnivorous and feed on relatively large prey compared to anuran tadpoles. As particulate feeders, they lack the epithelial teeth and mouth structures of anuran larvae and have a buccal cavity which is structurally and functionally similar to that of the adult. Food accumulated in the pharynx is swallowed and carried by peristalsis to the stomach, where peptic proteolysis is initiated. At metamorphosis, as in anurans, the salamander foregut undergoes radical change; all layers of the gut experience autolysis, phagocytosis, and reorganization. There is a gross reduction in length of gut and, by contraction of the muscle, a thickening of the intestinal walls. Proteolytic digestion is retarded during a transitional period, but, as gastric glands arise in the pyloric region of the stomach and spread to the cardiac region, proteolysis becomes more effective in the adult than in the larva (Reeder, 1964).

RESPIRATORY SYSTEM

As discussed under integument, modern amphibians utilize cutaneous and buccopharyngeal respiration in addition to respiration through gills or lungs. Typically, larval amphibians are aquatic and utilize gill respiration whereas adult amphibians are terrestrial, or partially so, and utilize lung respiration. However, there are obviously many exceptions to this pattern. Some adult salamanders retain gills and are permanently aquatic; some terrestrial adults lack lungs and utilize cutaneous and buccopharyngeal respiration alone. With the possible exception of cutaneous respiration, the respiratory system of amphibians contains only modifications of basic structures present in some of the more primitive fish and no innovations. External larval gills and lungs were characteristic of most of the fishes during the Devonian and are found in the crossopterygian ancestors of amphibians and lungfish in the class Dipnoi.

GILLS

The gills of larval amphibians closely resemble those of lungfish, but, whereas lungfish possess four pairs of gills, amphibians possess a maximum of three pairs. Four pairs of pharyngeal pouches are present in embryonic anurans and in the majority of salamanders, but a fifth pouch develops in caecilians and primitive salamanders in the suborder Cryptobranchoidea. The pouch corresponding to the spiracle of fishes remains open in caecilians for a short time during larval development but never opens to the exterior in other modern amphibians. In most anurans it develops into the Eustachian tube, which connects the pharynx to the cavity of the middle ear; in other amphibians it simply disappears during early development of the embryo. Depending upon the amphibian group, two or three of the remaining pouches break through to the outside and form the gill slits. Cartilaginous visceral arches develop between the slits and provide support for the external gills which are formed from filaments of vascularized ciliated epithelium. Subsequent development of anuran gills differs from that of salamanders and caecilians.

Shortly after the development of external gills in anuran larvae, an operculum-like layer of tissue grows posteriorly from the hyoid region and fuses to the integument behind and below the gills so that they are enclosed in a space called the atrial chamber (Fig. 2–30). This chamber is connected to the surrounding aquatic medium by an opening called the spiracle. The position of the spiracle is characteristic of each genus and may be on the right, left, or ventral side of the animal; in a few species there are paired spiracles. Shortly after the atrial chamber is formed, new gills develop within the gill slits and the original gills degenerate. The former set of gills remains functional until the animal undergoes metamorphosis. During metamorphosis (which apparently is nearly always complete in anurans, since very few neotenic individuals have ever been found), the gills are absorbed, the gill slits close, and the lungs become functional respiratory organs. Marsupial frogs in the genera *Cryptobatrachus*, *Hemiphractus*, *Gastrotheca*, and *Amphignathodon* (family Hylidae) have distinctive bell-shaped gills which function as vascular wrappings, completely surrounding the larva while it is being carried on the back of the female (Fig. 2–31); those of *Hemiphractus* function to attach the larva to the female.

No atrial chambers develop in caecilians or salamanders. In both groups, what appear to be rudimentary opercular folds form anterior to the gills, but these develop into little more than wrinkles of skin. Three pairs of external gills develop in sirens (*Siren*, family Sirenidae); these, together with gill slits, are retained throughout life; adults possess both functional gills and lungs. Although all adult caecilians lack both gills and gill slits, the larval development and retention of gills is variable among different forms. Those in the family Ichthyophiidae lay their eggs on land, and their embryos possess three pairs of branching lateral gills while still in the egg; the gills are usually absorbed before hatching, but the aquatic larval stage retains one or two pairs of gill slits which are closed at metamorphosis. Caecilians in the family Typhlonectidae are all ovoviviparous. Their embryos possess broad tissue-like gills on the head or neck while in the uterus, but these are absorbed and the gill slits closed before birth. African caecilians in the genus *Geotrypetes* (family Caeciliidae) have long plumose triaxial gills as embryos, but these, too, are resorbed before birth. It appears that in ovoviviparous caecilians the embryonic gills function in the respiratory gas exchange between the embryo and the highly vascularized epithelial lining of the lumen of the oviduct. After loss of the gills, the larvae utilize cutaneous respiration. Other caecilians in the family Caeciliidae are variable in both their mode of reproduction and development of gills.

Salamander larvae tend to have well-developed external gills which vary in form as a result of adaptation to the larval habitat. Certain terrestrial species, including the red-backed salamander *Plethodon cinereus* (family Plethodontidae), lay their eggs on land and are characterized by an abbreviated larval period which is passed within the egg. In these, metamorphosis (such as it is), occurs within the egg, and the young hatch as miniature adults. The very rapid larval development

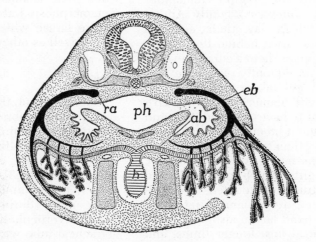

Figure 2–30 Diagram of the relations of external and internal gills in an anuran tadpole. *ab* and *eb*, afferent and efferent branchial arteries; *h*, heart; *o*, ear cavity; *ph*, pharynx; *ra*, radix aortae. (From Orr, R. T.: *Vertebrate Biology*. 3rd Ed. W. B. Saunders Co., 1971.)

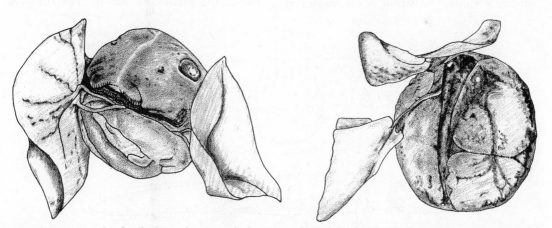

Figure 2–31 Two kinds of anuran larvae with distinctive bell-shaped gills which function while the young are carried on the adult's back. On the left is a larva of the marsupial frog (*Gastrotheca marsupiata*); on the right is a larva of another South American frog, *Cryptobatrachus evansi*. (After Noble: *The Biology of the Amphibia*.)

requires an efficient respiratory system, and, accordingly, the embryos possess large staghorn-shaped gills which are pressed up against the egg membranes (Fig. 2–32A). The gills are absorbed and the gill slits closed before hatching. Salamanders such as the Olympic salamander (*Rhyacotriton olympicus*, family Ambystomatidae), which lay their eggs in streams, have a larval habitat that is rich in oxygen but gives them a current to contend with. Consequently, in these one finds reduced gill rakers and gills with short, broad gill filaments (Fig. 2–32B); such larvae also have reduced fins and short limbs. Pond-breeding salamanders, including the tiger salamander (*Ambystoma tigrinum*, family Ambystomatidae), have larval gills which are elongated feather-like structures adapted to conditions of reduced oxygen and quiet water (Fig. 2–32C); pond larvae also have well-developed dorsal fins, longer limbs, and balancers. Salamanders in the suborder Proteida are permanently aquatic and do not undergo complete metamorphosis. These "perennibranchs" retain gills and two pairs of gill slits as adults but also develop lungs. Congo eels (family Amphiumidae) lose their gills at metamorphosis but retain one pair of gill slits as adults; a similar situation is encountered

in *Cryptobranchus* (family Cryptobranchidae). In both cases lungs are present in adults. Other salamanders lose their gills and gill slits at metamorphosis. However, because of either genetic or environmental conditions, there are occasions when some salamanders do not complete metamorphosis but develop into sexually mature larvae which retain gills and gill slits as well as other larval characteristics.

LUNGS

Where present, the lungs of amphibians are simple membranous bags which lack internal partitioning but may have networks of alveoli-like pockets formed by blood capillaries, connective tissue, and muscle (Fig. 2–33). Amphibian lungs are paired and resemble those of lungfish. They arise during metamorphosis as paired diverticula from an unpaired laryngotracheal tube which opens to the pharyngeal cavity through a laryngeal slit in the ventral wall of the pharynx.

The relative development and form of the lungs vary among living amphibians. Paralleling the situation in snakes and associated with an elongated body plan, the left lung of caecilians is much shorter than the elongated right one; only the right lung possesses alveoli. The lungs of

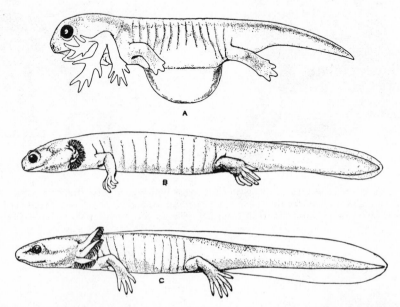

Figure 2–32 Three generalized types of salamander larvae. *A*, terrestrial type (*Plethodon vandykei*); *B*, mountain stream type (*Dicamptodon ensatus*); *C*, pond type (*Ambystoma gracile*). (From Noble: *The Biology of the Amphibia.*)

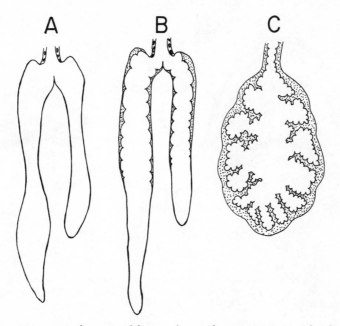

Figure 2-33 Diagrammatic sectional views of lungs of a mudpuppy (*Necturus,* family Proteidae) lacking alveoli (*A*) compared with those of a terrestrial salamander with sacculations in the proximal portion (*B*) and of an anuran with well-developed convolutions and alveoli (*C*). (*A* and *B,* after Kent, G. C., Jr.: *Comparative Anatomy of the Vertebrates.* C. V. Mosby Co., 1969; *C,* after Torrey, T. W.: *Morphogenesis of the Vertebrates.* 2nd Ed. John Wiley and Sons, 1967.)

salamanders in the family Sirenidae are nearly the same length, both extending the length of the body cavity; they are poorly vascularized and lack alveoli. Other salamander lungs are also elongated, but the right one is generally somewhat smaller than the left; depending upon the salamander, alveoli may or may not be present. Both the size of the lungs and the development of alveoli tend to be greatest in pond-dwelling or terrestrial salamanders. Mountain stream-inhabiting salamanders such as the Olympic salamander (*Rhyacotriton olympicus,* family Ambystomatidae) tend to have vestigial lungs in adaptation to both their cold oxygen-rich environment and their need to seek shelter under water. The latter would be made difficult by large lungs, which would act as hydrostatic organs and tend to cause the animals to float. All members of the family Plethodontidae, which includes both aquatic and terrestrial species, lack lungs; and lungs are variously reduced in certain members of other families. Salamanders are the only tetrapods in which the evolutionary loss of lungs has occurred.

Although they are better developed in some species than in others, lungs are always present in adult anurans. Compared to those of other amphibians and correlated with a shortened body plan, anuran lungs are relatively compact and reniform. However, they provide a relatively large surface area for gas exchange as a result of convolutions of the inner walls and the presence of large numbers of alveoli.

TRACHEA AND LARYNX

Air enters the external nares and passes through the nasal passages and internal nares to the buccal cavity, from where it travels to the pharynx. From the pharynx it passes through the laryngeal slit, or glottis, into the trachea, which divides at its posterior end into two bronchial tubes. One of these tubes goes to each lung (Fig. 2–34). The trachea and sometimes the anterior portions of the bronchi are strengthened with cartilaginous rings so that they will not collapse under either negative or positive pressure. The lengths of these tubes are variable among living amphibians. The shortened body of anurans has resulted in lungs extremely close

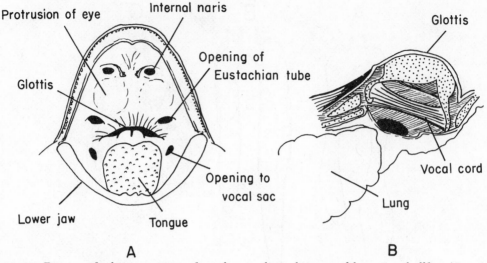

Figure 2–34 Diagram of relative positions of esophagus, glottis, larynx, and lungs in a bullfrog (*Rana cates-beiana*). *A*, view of inside of mouth; *B*, view of interior of left half of larynx. (*B*, after Weichert, C. K.: *Elements of Chordate Anatomy.* 2nd Ed. McGraw-Hill Book Co., 1959.)

to the pharynx so that the trachea and bronchi are almost nonexistent. However, the highly specialized aquatic toads in the family Pipidae have posteriorly positioned lungs which are used as hydrostatic organs, and, accordingly, distinct bronchi and trachea occur. South American members of this family, in the genus *Pipa*, are interesting in that they possess cartilaginous support of the lungs in addition to that of the trachea and bronchial tubes. All salamanders except the lungless forms possess a definite trachea. However, the respiratory passages are relatively short in all but the aquatic congo eels (*Amphiuma*, family Amphiumidae), whose lungs undoubtedly also act as hydrostatic organs and are relatively posterior in position. Salamanders in the family Sirenidae and caecilians possess an elongated trachea which is well strengthened by cartilaginous rings; in caecilians, the lungs are also infiltrated by cartilage. The tracheal and pulmonary cartilages of tetrapods do not appear to be homologous with any structure of fishes, but rather seem to represent innovations in the visceral skeleton.

Sound is produced by air passing through the trachea; this structure has been variously modified in tetrapods to form a sound-producing organ. In amphibians, reptiles, and mammals, the anterior end of the trachea has evolved into a voice box or larynx. In these animals, the last (ninth) visceral arch forms a pair of carti-

laginous arcs which support the larynx at the level of the glottis. In caecilians, they fuse together and form a continuous ring of cartilage anterior to the larynx and around the glottis. In anurans and many salamanders, the paired cartilaginous arcs subdivide to form an anterior pair of spoon-shaped arytenoid cartilages and a posterior pair of cricoid cartilages which form a ring and are often fused together (Fig. 2–14*A* and *B*). As noted earlier, these cartilages are closely associated with the hyobranchial apparatus supporting the tongue.

The sound-producing structures of anurans are contained within the laryngeal chamber formed by the arytenoid cartilages, which, in turn, nestle in the ring formed by the cricoids (Fig. 2–35). The tissue lining the inner surface of each arytenoid is sometimes extended into an arytenoid valve, a fleshy flap on the inner surface of the anterior edge of the arytenoid cartilage which acts to modulate the amplitude of the sound produced. Another extension of the lining is the anterior membrane, or vocal cord, which is a recurved membrane in contact with the arytenoid valve anteriorly and inserted on the cricoid cartilage posteriorly. The anterior membranes are at least partially responsible for producing the dominant frequency of the sound. Each may have a midrib fibrous mass associated with it. The size, shape, and mass of this fibrous structure vary specifically and tend to correlate with the dominant frequency of the sound produced. A pair of posterior membranes, also extensions of the lining of the arytenoids, is normally present parallel to the anterior membranes. The function of the posterior membranes is not clear: they may merely serve to funnel air directly to the vocal cords as it passes from the lungs through the larynx and into the buccal cavity.

The males of most anuran species have vocal sacs. These secondary sex characteristics are resonating chambers formed as diverticula of the lining of the buccal cavity and extending ventrally and laterally to a position directly above the superficial ventral mandibular muscles. They communicate with the buccal cavity by paired or single, round or slitlike apertures in the floor of the cavity at the corner(s) of the mouth. In some species, there is a single vocal sac into which both apertures open. In other species, there is a pair of vocal sacs, each with its own aperture. The partition between the buccal cavity and the vocal sac is partially supported by the anterior cornu of the hyoid and this generally forms the medial margin

of the aperture (Tyler, 1971). The size, shape, and position of the vocal sacs are species-specific. They may be relatively inconspicuous, as in most Ranidae, or nearly as large as the animal's body when inflated, as in chorus frogs (*Pseudacris*, family Hylidae, Fig. 2–36). In general, the larger the vocal sac, the louder the sound produced.

Various salamanders have been reported to make noises, but the sound produced and the mechanism by which it is made vary (Maslin, 1950). Clicking, clucking, or kissing sounds have been noted for a variety of salamanders, including *Salamandra salamandra, Triturus alpestris* (both in family Salamandridae), and *Ambystoma tigrinum* (family Ambystomatidae). The mechanism of these sounds seems to be buccal inspiration rather than expiration: as the animal opens its mouth, the entering air breaks the moist seal of the adpressed lips. Two-toed amphiumas (*Amphiuma means*, family Amphiumidae) make whistling sounds by forcing air through their gill slits while expiring air. Lungless salamanders (family Plethodontidae) produce squeaks by vigorously

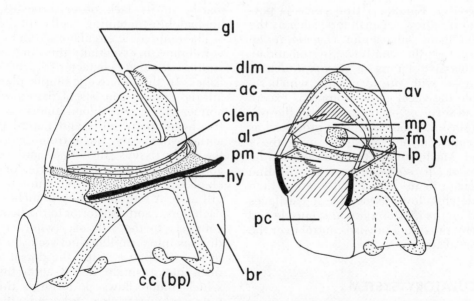

Figure 2–35 Generalized diagram of vocal structures and hyoid apparatus of a toad (*Bufo*). The drawing on the left is of the intact larynx; as drawn, the top of the larynx projects into the buccal cavity. The drawing on the right is viewed from the same angle, but the larynx has been sectioned so as to reveal the sound-producing structures; the hyoid apparatus and *constrictor laryngis externus* muscle have been removed. *ac*, arytenoid cartilage; *al*, arytenoid lumen; *av*, arytenoid valve; *bp*, bronchial process; *br*, bronchus; *cc*, cricoid cartilage; *clem*, *constrictor laryngis externus* muscle; *dlm*, dilator laryngeal muscle; *fm*, fibrous mass; *gl*, glottis; *hy*, hyoid; *lp*, lateral portion; *mp*, medial portion; *pc*, posterior chamber; *pm*, posterior membrane; *vc*, vocal cord. (After Martin: Mechanics of sound production in toads of the genus *Bufo*: passive elements.)

Figure 2-36   Southern chorus frogs (*Pseudacris nigrita*). (From Cochran: *Living Amphibians of the World.*)

contracting the floor of the mouth, forcing air out through the nares or lips (lacking lungs, their air reservoir must be the buccal cavity). Finally, a true voice is possessed by a few salamanders, such as the Pacific giant salamander (*Dicamptodon ensatus*) of the family Ambystomatidae. The massive larynx of *Dicamptodon* is equipped with *plicae vocales* which resemble the vocal cords of anurans except that no recesses occur anterior to the plicae (Maslin, 1950); sound is produced as air from the lungs passes over these plicae. Maslin found the vestibule of the larynx of *Dicamptodon* possesses hyaline cartilages in its lateral walls rather than simple chondroid tissue. These cartilages insert on the *dilatores laryngis* muscle and thereby provide a precise control over the laryngeal apparatus.

CIRCULATORY SYSTEM

All vertebrates have a closed circulatory system in which blood is carried in arteries from the heart to the various tissues, is passed through the tissues in networks of capillaries, and is returned to the heart in veins. Amphibians and higher verte-brates also have lymphatic systems which function to return cellular fluids (lymph) to the heart; lymph differs from blood primarily in its lack of erythrocytes, the hemoglobin-containing cells of blood.

The embryonic circulatory system of all vertebrates is essentially the same and is similar to that characteristic of adult fishes. In this relatively simple plan, the arterial system consists of paired ventral aortae which carry blood anteriorly from the heart to the gills and paired dorsal aortae which carry blood from the gills to the body tissues; the dorsal and ventral aortae are connected by a series of branchial arteries, one passing through each gill arch. A vitelline artery branches from each dorsal aorta posterior to the heart and connects to the yolk sac, where it subdivides into a capillary network. Posterior to the vitelline arteries, the dorsal aortae fuse into a single dorsal aorta through which blood flows posteriorly; this divides into arterioles and gradually into capillaries throughout the body. The embryonic venous system forms as paired longitudinal cardinal veins which extend the length of the body at about the level of the dorsal aorta. A pair of lateral abdominal veins through which blood flows anteriorly is present on the ventral surface of the body posterior to the heart; similarly, a pair of vitelline veins carries blood from the yolk sac through the liver and toward the heart. A branch from each cardinal vein extends ventrally through the transverse septum and these join together to form a sinus, called the duct of Cuvier, into which the lateral abdominal and vitelline veins empty; this common duct returns blood to the heart. The result is a network of tubes in which the general flow of blood is anterior along the ventral surface from the heart to the gills, dorsal through the gills, posterior along the dorsal surface of the body, ventral through the various tissues, and anterior along the ventral surface back to the heart. All blood returning to the heart is low in oxygen and contains much carbon dioxide.

The heart consists of four chambers. From posterior to anterior, these are the sinus venosus, the auricle (or atrium), the ventricle, and the conus arteriosus (or bulbus arteriosus). The sinus venosus is little more than a thin-walled sac which functions primarily as a receptacle for venous blood. The somewhat muscular auricle contracts to force blood into the ventricle so that the latter will be filled to its capacity before contracting. The ventricle is very muscular and, through its contractions, functions as the primary pump forcing blood anteriorly. The conus arteriosus is also muscular and further pressurizes the blood and constricts its flow into the relatively small ventral aortas. Valves preventing the backward flow of blood are normally present between the sinus venosus and the auricle, between the auricle and the ventricle, and within the conus arteriosus. The differentiation of the heart into four chambers is accompanied by an elongation that, because of the limited space available between the transverse septum and the branchial apparatus, causes it to bend into an S-shape; owing to this, the auricle is situated above and a little in front of the ventricle.

From the general circulatory pattern outlined above have evolved the systems which characterize the various classes of vertebrates. The evolutionary trend is for circulatory systems of most fishes to be similar to that just described, whereas tetrapods exhibit an increase in modifica-

tion as one goes from amphibians to reptiles to birds and mammals. Nothing is known of the circulatory systems of extinct crossopterygians or primitive amphibians, so the sequence of evolutionary changes from fishes to primitive amphibians to modern amphibians is not clear: the following discussion of amphibian circulatory systems will, of necessity, be primarily concerned with those of living forms.

HEART

The evolution of lungs as respiratory organs complicated the functioning of the circulatory system, for it meant that both oxygenated and deoxygenated blood returned to the heart, where, in the primitive cardiac plan described above, the two would mix. The mixing of oxygenated and deoxygenated blood streams is obviously disadvantageous, and in all vertebrates that utilize lungs there has evolved at least a partial partitioning of the heart into right and left sides so that mixing of the two blood streams is restricted or prevented entirely (Fig. 2–37). The basic design of the heart in living lungfishes is probably similar to that of the most primitive amphibians, while that of living amphibians appears to be degenerate and not really intermediate between those of fish and reptiles.

Compared to those of more primitive fishes such as the elasmobranch dogfish, the lungfish heart has a more obvious S-bend and the atrium is almost entirely anterior to the ventricle. In addition, each of the anterior three chambers of the heart is incompletely divided by a septum so that at least some separation of oxygenated and deoxygenated blood is achieved. As in the primitive condition, deoxygenated venous blood drains into the sinus venosus, which then empties into the right side of the atrium. Pulmonary veins, which evolved in association with lungs, return oxygenated blood from the lungs directly to the left side of the atrium, by-passing the sinus venosus. The partial partitioning of the atrium and ventricle is on a vertical plane, so in both chambers the oxygenated blood goes through the left side and the deoxygenated blood through the right side. The partition in the conus arteriosus, however, is twisted in a spiral

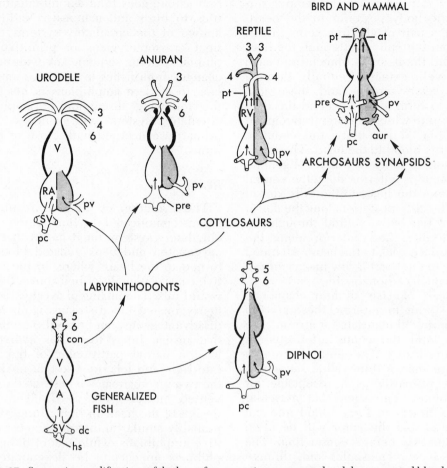

Figure 2-37 Successive modifications of the heart for separating oxygenated and deoxygenated blood. *A,* atrium; *RA,* right atrium; *V,* ventricle; *RV,* right ventricle; *SV,* sinus venosus; *con,* conus arteriosus; *aur,* auricle of mammalian heart; *at,* aortic trunk; 3-6, Third through Sixth aortic arches; *dc,* common cardinal vein; *hs,* hepatic sinus; *pc,* postcava; *pre,* precava; *pv,* pulmonary veins; *pt,* pulmonary trunk. Gray chambers (left side of heart) contain chiefly or only oxygenated blood. (From *Comparative Anatomy of the Vertebrates,* by George C. Kent, Jr. C. V. Mosby Co., 1969.)

manner until the division becomes a horizontal one anteriorly and the oxygenated blood coming from the left side of the ventricle is deflected around to the ventral side of the conus and out through the aortae (systemic and carotid arteries); deoxygenated blood passes from the right side of the ventricle around to the dorsal side of the conus and out the pulmonary artery to the lung, the latter artery having evolved from the sixth branchial artery. The degree to which the heart is partitioned varies among the three groups of living lungfishes and is most complete in South American forms (*Lepidosiren*) which are highly adapted to drought conditions and probably depend almost entirely on their lungs for respiration.

Again living amphibians have hearts which compare generally with those of lungfishes but which tend to be partitioned to a lesser degree, a degeneration reflecting the relative unimportance of lungs as respiratory organs with increased reliance on gill and cutaneous respiration. In anurans, the atrium is completely partitioned into right and left compartments; as in lungfishes and all higher vertebrates, the sinus venosus empties into the right side and the pulmonary veins into the left (Fig. 2-38). Among salamanders, the septum in the atrium is variable: it is complete but very thin in some terrestrial species, is imperfect in European newts in the genus *Salamandra* (family Salamandridae), and, although present in

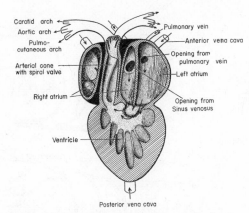

Carotid arch
Aortic arch
Pulmo-
cutaneous arch
Arterial cone
with spiral valve
Right atrium
Ventricle

Pulmonary vein
Anterior vena cava
Opening from
pulmonary vein
Left atrium
Opening from
Sinus venosus

Posterior vena cava

Figure 2–38 Diagrammatic section through the heart of a frog. (From Romer, A. S.: *The Vertebrate Body.* 4th Ed. W. B. Saunders Co., 1970.)

embryos, is completely degenerated in adults of the aquatic perennibranchs in the suborder Proteida. Caecilians also exhibit some degeneration of the atrial septum, for, although apparently possessed by all species, it contains many perforations. No living amphibian is known to have a complete ventricular septum, but some caecilians have trabeculae in the ventricle which may function as a septum (Marcus, 1935; Foxon, 1964); Sharma (1957) described a ridgelike structure in the posterior part of the ventricle of the Indian bullfrog (*Rana tigrina*, family Ranidae). Although its functional significance is questionable, all anurans and some salamanders have a spiral valve in the conus arteriosus which, as in lungfish, diverts one flow of blood ventrally to the carotid and systemic arteries and deflects the remainder dorsally to the pulmonary artery. Other salamanders lack a spiral valve, and it is also missing in caecilians.

Although it is generally agreed that the lack of complete partitions in the amphibian heart must result in some mixing of blood from the right and left sides, differences of opinion exist as to the extent of such mixing. Yapp (1965) states emphatically, but does not document, that studies of the common European frog (*Rana temporaria*, family Ranidae) indicate an almost complete mixing of the blood from the right and left sides and no differences in blood leaving the heart through the carotid, systemic, and pulmonary arteries. In contrast, Foxon (1951, 1964), Simmons (1957, 1959), and de Graaf (1957) have concluded that there is a selective distribution of blood by the anuran heart, that this is brought about by the relative anatomical arrangements of the carotid, systemic, and pulmonary arteries, and that the destination of a particular corpuscle depends on its position while in the heart. Foxon (1951) used radiography to demonstrate that when blood passed from the atria into the ventricle little mixing took place. Simmons used injections of an intense dye (Evans blue) to demonstrate that in *Rana temporaria* blood from the left of the ventricle passes up one side of the spiral valve of the conus into the carotid arteries and the right systemic arch, while blood from the right of the ventricle passes largely up the other side of the spiral valve into the pulmocutaneous arteries and left systemic arch. In this manner, only a small amount of mixing occurs. Simmons also found that blood flows up the two sides of the spiral valve simultaneously; therefore, asymmetrical contraction of the heart does not function in keeping two blood streams separate, as has often been suggested. De Graaf used ultraviolet light to follow flourescine he had injected into the circulation of clawed frogs (*Xenopus laevis*, family Pipidae). He found that blood returning from the lungs to the left atrium passed to all parts of the ventricle and, upon contraction of the ventricle, was distributed to all three arteries (carotid, systemic, and pulmonary). However, blood returning to the right atrium was excluded from the carotid and systemic circulations and flowed through the pulmocutaneous arch; because the cutaneous arteries in this species are very small, most of this blood reached the lungs.

All amphibians utilize cutaneous respiration to some degree, so it is obvious that blood returning from the skin will be oxygenated just as is that from the lungs or gills. Because blood returning from the skin through the cutaneous vein passes into the sinus venosus and right atrial chamber while that returning from the lungs through the pulmonary veins empties into the left atrial chamber, there will be oxygenated blood in both sides of the heart. Depending upon the extent to which cutaneous respiration is being utilized, there may or may not be any advantage in keeping the two blood streams separate.

ARTERIES

The evolutionary development of vertebrate arterial systems includes the replacement of paired ventral and dorsal aortae by single vessels, a shortening of the single ventral aorta until it disappears entirely, and reduction and alteration of the branchial arteries in association with the evolution of lungs and changes in the function of visceral arches. These developments are initiated in modern fishes but become more pronounced in amphibians and higher tetrapods. In larval amphibians a single, short, ventral aorta is joined by four (third, fourth, fifth, and sixth) pairs of branchial arteries to paired dorsal lateral arteries that unite posteriorly to form the dorsal aorta (Fig. 2–39). This system differs from that of primitive fishes primarily in lacking the first and second pairs of branchial arteries. In addition, whereas the flow of blood from the ventral aorta to the dorsal aorta of fishes is interrupted by networks of capillaries in the gills, the branchial arches of amphibians are continuous from the ventral aorta to the dorsal lateral arteries. The gill capillary beds occur on side loops which branch off the third, fourth, and fifth branchial arteries; and the sixth pair of branchial arteries supplies the lungs when they are formed. The loop pattern of gill circulation facilitates the change from gill to lung and cutaneous respiration at metamorphosis, for during the transition the vascular loops merely degenerate while the branchial arteries expand and a continuous flow of blood is maintained from the heart to the dorsal aorta. Neotenic salamanders (which fail to metamorphose, becoming sexually mature in the larval form) and aquatic salamanders with permanent gills retain what is basically a larval arterial system. The exceptions are where the fifth pair of branchial arteries is absent, the dorsal aorta degenerates so that the connections between the third and fourth and between the fourth and fifth branchial arteries are lost; and the anterior extensions of the ventral aortae pass into the head as the external carotid arteries. The result of these changes is that the head receives blood only from the external carotid arteries and the third pair of branchial arteries, the body gets arterial blood directly from the fourth pair of branchial arteries,

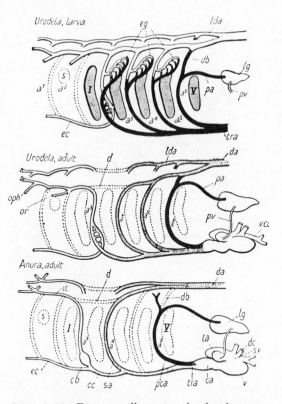

Figure 2–39 Diagrams illustrating the development and fate of the branchial arches in Amphibia, left-side view completed. Vessels carrying mostly oxygenated blood are white, mostly deoxygenated blood black, and mixed blood stippled. a^{1-6}, primary arterial arches; *ca*, conus arteriosus; *cb*, carotid gland; *cc*, common carotid; *da*, median dorsal aorta; *db*, ductus Botalli; *dc*, left ductus Cuvieri; *ec*, external carotid; *eg*, blood-supply to external gill; *ic*, internal carotid; *la*, left auricle; *lda*, lateral dorsal aorta (*d*, obliterated part, ductus caroticus); *lg*, lung; *oph*, ophthalmic; *or*, orbital; *pa*, pulmonary artery; *pca*, pulmo-cutaneous arch; *pv*, pulmonary vein; *s*, closed spiracular slit; *sa*, systemic arch; *sv*, sinus venosus; *tra*, truncus arteriosus (ventral aorta); *v*, ventricle; *vci*, vena cava inferior. (From Goodrich, E. S. 1930. *Studies on the Structure and Development of Vertebrates.* London [Re-issue Dover Books].)

and the lungs and skin receive it from the sixth pair of branchial arteries.

The reduction in branchial arteries is greater in other amphibians and most advanced in frogs (*Rana*, family Ranidae) (Figs. 2–39 and 2–40). When frogs metamorphose, circulation to the gills is stopped; the capillary loops atrophy until they have all disappeared, except those from the third pair of arches (first pair of gills), which form the carotid body. The fifth pair of branchial arches disappears completely and the sixth pair is free of the dorsal aorta as an independent pulmo-

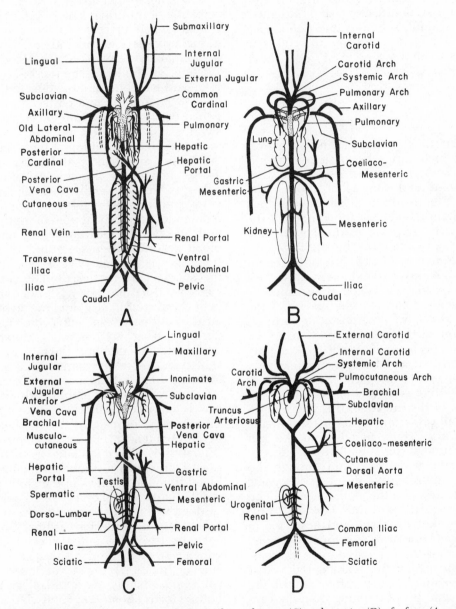

Figure 2–40 Veins (A) and arteries (B) of a salamander and veins (C) and arteries (D) of a frog. (A and C, after Weichert, C. K.: *Elements of Chordate Anatomy.* 2nd Ed. McGraw-Hill Book Co., 1959.)

cutaneous artery. The third pair of branchial arteries, also independent of the dorsal aorta, forms the carotid arch which supplies the head region, while the fourth pair of branchial arteries remains connected to the dorsal aorta as the systemic arch. The ventral aorta is shortened until it is almost nonexistent, and the three arches (carotid, systemic, and pulmonary) branch almost directly from the conus arteriosus. The remainder of the arterial circulation is indicated in Figure 2–40*B* and *D*.

VEINS

The vitellines, cardinals, and abdominal veins which form the basic embryonic venous system also give rise to the principal veins and sinuses of adult vertebrates (Fig. 2–40). The anterior cardinal veins drain blood from the surface of the head to the common cardinal veins and into the sinus venosus. The vitelline veins are infiltrated by liver tissue until a capillary and sinus network separates their anterior segments, which drain into the sinus venosus and become known as the hepatic veins, from the posterior portions. Branching of the posterior portions of the vitellines occurs and, as the yolk supply is diminished and becomes increasingly less important, the principal branch of each vitelline is associated with the ventral side of the intestine; this pair of subintestinal veins join to form the hepatic portal vein. Thus, blood from the gut passes through the subintestinal veins to the hepatic portal vein and through the hepatic sinuses into the hepatic veins which empty into the sinus venosus. The posterior cardinal veins become infiltrated by kidney tissue and here, too, a network of capillaries and sinuses results. Posterior to these renal sinuses the cardinals are called the renal portal veins; these eventually join together to form the caudal vein. Blood draining from the caudal region is, thus, collected in the caudal vein, from which it passes into the renal portal veins. Through the capillaries of the renal portal veins, it flows into the posterior cardinal veins, which join the anterior cardinals in emptying into the common cardinals. Cutaneous veins join the posterior cardinals at their anterior end and return blood from the lateral surfaces of the body. Each of the lateral abdominal (deep lateral) veins sends branches into the anterior (subclavian) and posterior (iliac vein) appendages; then they join together posteriorly to form a loop.

The general plan just described is characteristic of primitive fishes. Evolutionary changes encountered in venous systems of higher vertebrates include the addition of pulmonary veins to return blood from the lungs to the heart, the loss of the common cardinal veins and the duct of Cuvier so that the anterior and posterior cardinal veins separately open into the sinus venosus, the loss of the left posterior cardinal, and a general diminution of the renal portal system and lateral abdominal veins. These changes are initiated in the higher fishes, which generally lack abdominal veins and have reduced renal portal systems so that the caudal vein is continuous with the posterior cardinal vein and the blood generally by-passes the kidney. In lungfishes, pulmonary veins are added, the subclavian empties directly into the sinus venosus, the left posterior cardinal vein is reduced in size, and the right posterior cardinal vein joins the hepatic (forming what is sometimes called the postcaval) and empties directly into the sinus venosus rather than joining the anterior cardinal to form a common cardinal.

Compared to other amphibians and tetrapods in general, the venous system of salamanders, especially those which are perennibranchs, is relatively primitive (Fig. 2–40A). There is a large median postcaval vein dorsal to the gut which carries blood from the kidneys to the sinus venosus, but a pair of relatively small posterior cardinal veins is also present. These, as in fishes, join the anterior cardinals to form common cardinals and a duct of Cuvier which empties into the sinus venosus; a similar situation persists in tailed frogs (*Ascaphus*, family Ascaphidae) and fire-bellied toads (*Bombina*, family Discoglossidae). In other anurans and higher tetrapods, the left posterior cardinal vein, common cardinals, and duct of Cuvier disappear in the adult; the postcaval, formed from portions of the right posterior cardinal and the hepatic vein, becomes medially located and drains both kidneys into the sinus venosus (Fig. 2–40C). All amphibians have a renal portal system like that of fishes: it receives blood from the hind legs and tail or, in the case of anurans, from the hind legs alone. The two lateral abdominal veins of larval amphibians become fused into a single medial ventral abdominal vein in the adult, and this drains blood from the posterior body and appendages into the hepatic portal vein. Thus, all blood returning from the hind legs of amphibians must pass through a capillary network, in either the kidneys or the liver.

The amphibian liver performs important functions related to intermediary metabolism. It removes nutrient material absorbed by the blood from the gut, stores chemical energy in the form of glycogen, and releases glycogen in forms which can be metabolized by cells. The liver also filters out nitrogenous waste products of cellular metabolism from the blood and returns these to the circulating fluid as urea, which is then removed as the blood passes through the renal portal system.

Finally, enlarged cutaneous veins and the pulmonary veins play the important role of returning oxygenated blood to the heart for circulation throughout the body.

LYMPHATIC SYSTEM

Because the vertebrate circulatory system is closed, blood comes into contact with only those cells which line blood vessels. Lymph is the intercellular fluid that carries oxygen and nutrient materials from the blood to cells not in contact with the circulatory system and carries back the waste products of these cells. This fluid

must eventually be returned to the circulatory system: in higher vertebrates, this is accomplished by a network of tubes which comprise the lymphatic system. Primitive fishes such as sharks have a lymphatic system, but it is really part of the venous system. Teleost fishes have an independent lymphatic system that empties into the anterior cardinal veins or the duct of Cuvier, but the movement of lymph through the system is dependent entirely upon compression from the contraction of body muscles. (A few species have a single lymph heart at the base of the tail, but this pumps lymph directly into the caudal vein.)

The lymphatic system of amphibians is well-developed and always includes at least two hearts that are essentially endothelial sacs surrounded by a mesh of striated muscle covered by a layer of connective tissue. In caecilians there may be over 200 of these lymph hearts, and they represent nearly the entire lymphatic system. There is one pair of hearts lying under the skin of each intersegmental fold, and others are scattered throughout the deeper regions of the body; all pump lymph directly into veins. The lymphatic system of salamanders consists of two networks of vessels: one includes subcutaneous sinuses and tubules which collect lymph from superficial regions of the body and return it to the renal portals, posterior cardinals, and cutaneous veins (Fig. 2–41); the second network is deeper, extending parallel to the dorsal aorta, and drains lymph into the subclavian veins. The lymphatic fluid is moved through

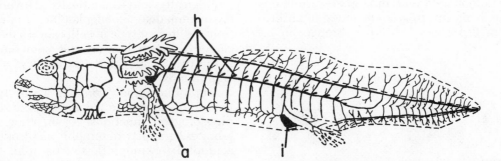

Figure 2–41 The superficial lymphatic system of a salamander, consisting of dorsal, lateral, and ventral longitudinal vessels and their segmental branches. A series of lymph hearts (h) is present along the lateral vessel. Lymph enters the venous circulation from the lateral vessel through an axillary sac (a) whereas lymph from the ventral vessel enters the venous system through an inguinal sac (i). (After Romer, A. S.: *The Vertebrate Body*. 4th Ed. W. B. Saunders Co., 1970.)

these networks by a series of lymphatic hearts. The anuran lymphatic system includes very large sinuses under the skin (these probably function to keep the skin moist) and between layers of muscle. The number of lymphatic hearts varies among anuran species, but it is generally much smaller than in salamanders and caecilians: there may be as few as two pairs. However, there is a rapid circulation of lymph in these amphibians, and it has been estimated that the entire volume of blood plasma passes through the lymphatic system 50 times a day. This reflects its importance in moisture regulation, metabolic exchanges, and cleansing activities.

BLOOD CELLS

The cellular elements normally found in vertebrate blood are red blood corpuscles or erythrocytes, spindle cells or thrombocytes, and white blood corpuscles or leucocytes (Fig. 2–42).

The erythrocytes are the hemoglobin-containing cells which carry nearly all of the oxygen circulated to tissues and are also important in carrying carbon dioxide away. The normal erythrocyte of amphibians is an elliptical nucleated disc, varying in diameter from about 70 microns in congo eels (*Amphiuma*, family Amphiumidae) to less than 10 microns in mudpuppies (*Necturus*, family Proteidae). Those of *Amphiuma* are the largest of any vertebrate. The number of erythrocytes varies in amphibians, both specifically and individually, from about 40,000 to 700,000 per cubic millimeter and is in general much lower than in mammals. The evolutionary trend in higher vertebrates has been for greater efficiency in absorption of oxygen by blood through an increase in the number and a decrease in the size of erythrocytes. This has been accompanied by a reduction and, finally, in mammals, complete elimination of the nucleus, which has nothing to do with gas absorption. A comparable condition is produced in many living amphibians through fragmentation or enucleation of erythrocytes. This occurs in various species, particularly in those salamanders and anurans that are terrestrial, and results in large numbers of enucleated portions of cells which contain hemoglobin; these are called plasmocytes. The proportion of erythrocytes so altered is variable even among closely related species. In the family Plethodontidae, it varies from a little over two per cent in red-backed salamanders (*Plethodon cinereus*) to over 90 per cent in slender salamanders (*Batrachoseps*). Amphibian erythrocytes are relatively long-lived: in leopard frogs (*Rana pipiens*, family Ranidae) the life span of an erythrocyte may be nearly 100 days, whereas an average life span for erythrocytes in man is about 120–128 days.

Thrombocytes are nucleated spindle cells which play a role in blood clotting in non-mammalian vertebrates. In mammals, the same function is performed by platelets. Thrombocytes are very fragile and break down in contact with foreign substances, releasing thrombin. In the presence of blood calcium, thrombin combines with the fibrinogen of plasma to form a net of fibrin. Clotting of blood occurs through the entanglement of blood cells in the fibrin. Fragmentation of thrombocytes occurs in many lungless salamanders (family Plethodontidae), and the resulting portions of cells resemble the blood platelets of mammals both morphologically and functionally. However, the mammalian blood platelets originate from a different type of cell, called a megacaryocyte, and they are not homologous with the fragmented thrombocytes of amphibians.

Amphibian blood contains a variety of leucocytes, but little is known about their physiological functions. As in other vertebrates, granular heterophil (neutrophil), acidophil (eosinophil), and basophil leucocytes are present together with nongranular lymphocytes and monocytes. The white blood cell count of amphibians is generally lower than that of mammals.

Figure 2–42 Blood cells of a frog (*Rana*). *1*, erythrocyte; *2*, lymphocyte; *3*, monocyte; *4*, neutrophil granulocyte; *5*, acidophil granulocyte; *6*, thrombocyte. (From Romer, A. S.: *The Vertebrate Body.* 4th Ed. W. B. Saunders Co., 1970.)

The heterophils are most abundant of the granulocytes and are actively amoeboid and phagocytic. Eosinophils are second in abundance and are actively amoeboid but apparently have little if any phagocytic ability (Andrew, 1959). Basophils are least abundant. Lymphocytes of various types, characterized by different sizes, occur in both blood and lymph. Erythrocytes are derived from medium-sized to large lymphocytes, granulocytes from medium-sized lymphocytes, and thrombocytes from small lymphocytes.

Amphibian red blood cell and thrombocyte formation occur primarily in the spleen, but red cell formation also occurs in the peripheral circulation. In most salamanders, granulocyte formation takes place primarily in the subscapular region of the liver but also occurs to some extent in the submucosa of the intestine. In lungless salamanders (*Plethodon*, family Plethodontidae) white blood cell production also occurs in the bone marrow of the appendages, pelvic girdle, and quadrate bone, and of the thoracic, sacral, and caudal vertebrae (Andrew, 1959).

UROGENITAL SYSTEM

The urogenital system of vertebrates consists of an excretory system and a reproductive system. The excretory system consists of a pair of kidneys plus their ducts. It functions to eliminate nitrogenous metabolic wastes and maintain a proper water balance in the animal. The reproductive system consists of a pair of gonads and their ducts, and functions to produce, store, and release gametes. Functionally, the excretory and reproductive systems obviously have nothing in common. However, in vertebrates the excretory ducts are utilized by the reproductive system for the discharge of gametes to the exterior, and it is impossible to describe one system independently of the other.

KIDNEYS

The kidneys of all vertebrates are paired and consist of masses of urinary tubules. It is generally believed that the most primitive vertebrates had a pair of kidneys extending the length of the body cavity and that these merely drained waste products and excess fluid from the coelomic cavities. This type of kidney, called an archinephros or holonephros, consists of segmentally arranged tubules that open by way of ciliated funnels (nephrostomes) into the body cavities and end in a longitudinal (holonephric) duct emptying into the cloaca. Among living vertebrates, the holonephric type of kidney occurs only in larval hagfishes. In other vertebrates, the tubules tend to be grouped: the anterior group is the pronephros, the middle is the mesonephros, and the posterior is the metanephros. The evolution of the vertebrate kidney has been a succession of changes from pronephric to mesonephric to metanephric kidneys through degeneration of the anterior tubules. The increasingly more compact and efficient kidneys of higher vertebrates lose their connections with the coelomic cavities and become highly vascularized as the circulatory system assumes the function of transporting wastes and excess fluids rather than these merely draining into the coelom.

A pair of pronephric kidneys appears embryonically in all vertebrates, but these are functional only in free-living larvae of fish and amphibians. Each of the pronephric kidneys of anuran and salamander larvae consists of two to four tubules with nephrostomes that drain the coelomic cavities into a common duct. This duct is similar to that of the holonephros but is termed the pronephric duct (Fig. 2–43A). There are two sources of blood flowing to these kidneys. Blood is brought to the kidney tubules by the renal portal vein. In addition, a renal artery branches from the dorsal aorta and forms a ball of capillaries in the peritoneal lining of the body cavity near the opening of each nephrostome. Each of these capillary networks, drained by the posterior cardinal vein, bulges into the coelom and is called a glomus or external glomerulus. This double supply of blood to the kidneys corresponds with both the functioning of the glomi (and glomeruli) as filters for water and crystalloids and the functioning of the tubules as reabsorbers of some of these same items. Thus, the basic rela-

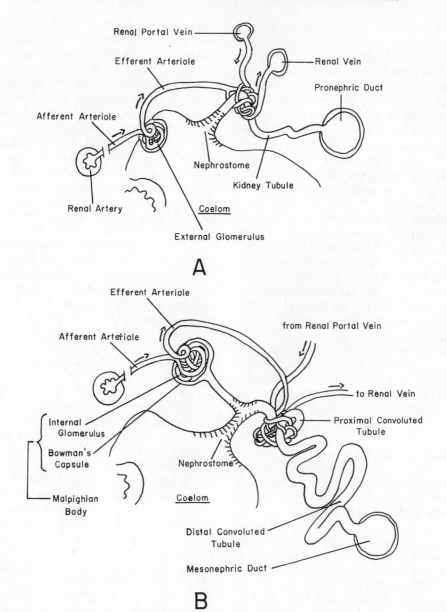

Figure 2–43 Basic structure of an amphibian kidney tubule and Bowman's capsule. *A*, pronephros type in larval amphibians; *B*, mesonephros type in adult amphibians. (Mostly after Kent, G. C., Jr.: *Comparative Anatomy of the Vertebrates.* C. V. Mosby Co., 1969.)

tionship between the circulatory and excretory systems is established in the pronephros, but it is not as intimate as that occurring in mesonephric and metanephric kidneys. The pronephros of larval caecilians is more elongated than that of salamanders and anurans and consists of 10 to 13 tubules which, together with renal corpuscles, and nephrostomes, are arranged in a distinct metameric pattern. The larval pronephros degenerates at or

before metamorphosis and is functionally replaced by the mesonephros (opisthonephros) in adult amphibians.

The initial formation of the mesonephros is the same as that of the pronephros except that the tubules arise farther posteriorly. The mesonephros tubules form segmentally, open by nephrostomes into the body cavity at their proximal end, and connect distally to a common (mesonephric) duct (Fig. 2–43*B*). As in the pronephros, blood vessels extend between the tubules, but in this case a capillary and sinus network forms in the wall of each tubule instead of in the body wall. The hollowed out wall of the tubule forms Bowman's capsule, and the plexus of capillaries and sinuses it surrounds is a glomerulus (internal glomerulus); the combination of a glomerulus and Bowman's capsule is called a Malpighian body or renal corpuscle.

The tubules of caecilian and congo eel (*Amphiuma*, family Amphiumidae) kidneys retain a strict segmental arrangement, but in most other amphibians this metamerism is obscured by the presence of secondary tubules. In salamanders and most anurans, the secondary tubules arise as branches from the primary tubules, but in more advanced anurans (including frogs in the genus *Rana*, family Ranidae) the secondary tubules develop directly from the undifferentiated cells of the blastema. Secondary tubules also have associated glomeruli, and *Rana* may have over 5000 Malpighian bodies. The trend toward independence of the kidneys from the coelom is evidenced in both salamanders and anurans, wherein many of the tubules lack a nephrostome and function entirely by extracting fluid and wastes from blood. In many anurans, the nephrostomes are independent of the kidney tubules and, instead of connecting to tubules, connect to nephric veins; these function to return lymphatic fluid from the coelom to the circulatory system and, thus, really have become part of the circulatory system rather than the excretory system.

Among living amphibians, the gross morphology of the mesonephros differs in a manner reflecting the general trend toward greater compaction seen in higher vertebrates (Fig. 2–44). Those of caecilians are elongated, extend the length of the coelom, and even in the adult show some evidence of segmentation (Wake, 1970). Salamander kidneys are also long, but the mass is shifted posteriorly so that there is a narrow anterior portion and a broad posterior portion. Anuran kidneys are less than half the length of the coelom and are located posteriorly.

GONADS

As in vertebrates in general, the gonads of amphibians are sexually indifferent initially. Their development into either testes or ovaries is affected by hormones; it can also be affected by changes in temperature. If the growth of the developing

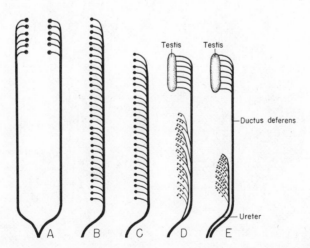

Figure 2–44 Diagrams of different types of kidneys. *A*, pronephros (embryonic); *B*, theoretical holonephros (each trunk segment with a single tubule), as in Apoda; *C*, primitive opisthonephros: pronephros reduced or specialized, tubules segmentally arranged, as in hagfish; *D*, typical opisthonephros, multiplication of tubules in posterior segments, testis usually taking over anterior part of system, trend for development of additional kidney ducts (most anamniotes); *E*, metanephros of amniotes: an opisthonephros with a single additional duct, the ureter, draining all tubules. In *A*, both sides of the body are included; in *B* to *E*, one side only. (From Romer, A. S.: *The Vertebrate Body.* 4th Ed. W. B. Saunders Co., 1970.)

gonad is such that the cortex increases more than the medulla, the germ cells remain in the cortex, become oocytes, and mature into eggs. If the greater growth occurs in the medulla, the germ cells migrate into it, become spermatocytes, and mature into sperm. The relative growth of gonadal cortex and medulla is under hormonal control and, as both male and female determining substances are present, it is the hormonal balance that determines which sex will be expressed.

Amphibians exhibit what is probably the most delicately balanced mechanism for sex determination. Young male leopard frogs (*Rana pipiens*, family Ranidae) exhibit feminine characteristics. Genetic females of the same species may eventually produce sperm in old age; however, when mated with normal females they produce only female offspring, because their genetic constitution for sex has not been changed although its expression has been modified. Temperature changes upset the hormonal balance and, at least in wood frogs (*Rana sylvatica*, family Ranidae), high temperature destroys or inhibits female determining substances and changes genetic females into males. Operative procedures have produced sex reversals in salamanders. Male toads in the family Bufonidae possess a rudimentary ovary (Bidder's organ) on each side in addition to a functional testis (Fig. 2–46); with removal of the testes, Bidder's organ develops into a functional ovary

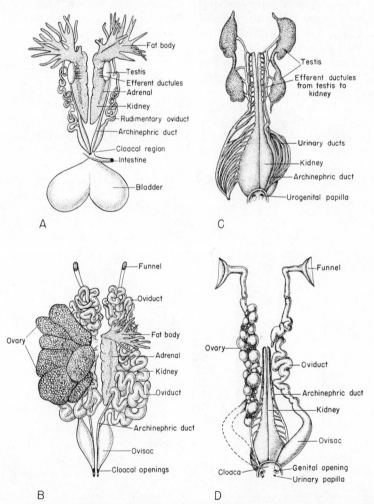

Figure 2–45 Urogenital system of amphibians. *A, B:* male and female organs of frog (*Rana*); in *B*, the ovary (shown only on the right side of the body) is in a condition close to breeding maturity. The bladder and intestine are not shown in *B. C, D:* male and female organs of the salamander *Salamandra;* in *C*, the urinary ducts of the right side are detached and spread out to show their connections with the kidney; in *D*, the ovary is shown only on the right side; the oviduct of the same side is partly removed to show the more posterior urinary ducts. Ventral views. (From Romer, A. S.: *The Vertebrate Body*. 4th Ed. W. B. Saunders Co., 1970.)

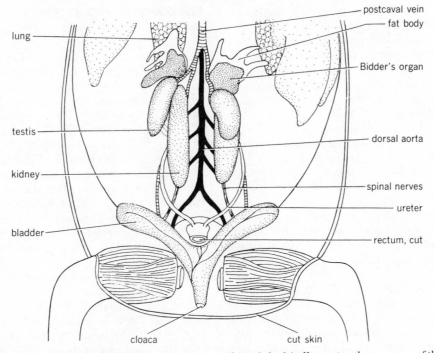

Figure 2–46   The urogenital system of a male common toad (*Bufo bufo*), illustrating the presence of the vestigial ovary called Bidder's organ. (From *Vertebrates: Their Structure and Life*, by W. B. Yapp. © 1965 by Oxford University Press, Inc. Reprinted by permission.)

Amphibian ovaries are saclike and, lacking a solid connective tissue framework, are fundamentally composed of peritoneal folds. Their shape tends to parallel that of the kidneys: they are very elongated and narrow in caecilians, moderately elongated in salamanders, and short and compact in anurans (Fig. 2–45B and D). Within the wall (cortex) of each ovary eggs are found in various stages of development; each is surrounded by a layer of follicle cells and a vascular network. The follicle cells presumably produce the ovarian hormones. During the breeding season eggs ripen under the influence of hormones from the anterior lobe of the pituitary gland. They cause the ovaries to swell to fill the majority of the peritoneal cavity. A small hole appears in the peritoneal lining of the ovary at each point where a ripe egg is attached, and, at ovulation, the egg is squeezed through this hole and released into the peritoneal cavity from which it finds its way into an oviduct. The number of eggs released during ovulation varies from one per female per breeding season in the tiny Cuban frog *Sminthillus limbatus* (family Leptodactylidae) to many thousands per female in most toads (*Bufo*, family Bufonidae) and some frogs in the genus *Rana*, family Ranidae.

The testes of amphibians are connected with the anterior end of the mesonephros and are composed of vast numbers of seminiferous tubules. Two types of cells line the tubules. First, there are the relatively large Sertoli cells; their function is unknown, but the presence of large nucleoli suggests that they are active in protein synthesis. Second are the sex cells, called spermatogonia, which by a complicated series of cell divisions and a metamorphic change (spermiogenesis) develop into spermatozoa or sperm. During a certain stage of spermatogenesis, the developing spermatozoa become attached to the cells of Sertoli. The various seminiferous tubules join together, pass into the efferent vessels of the testis, and connect by means of ducts to the outside.

Aside from producing sperm, the testes are endocrine organs which produce the male hormone testosterone, but the actual tissue which secretes testosterone has not been postively determined in amphibians. As with ovaries, testes have a length which tends to correlate with that of the kidneys (Fig. 2–45A and C). Each caecilian testis consists of an elongated series of segments connected by a longitudinal collecting duct, giving the structure a rosary-like appearance. Each of the segmental swellings represents what could be compared to an individual testis, being a mass of seminiferous tubules; such gonads are sometimes described as multiple testes. Salamander testes are shorter than those of caecilians, and those of lungless forms (family Plethodontidae) and newts (family Salamandridae) are irregularly lobed. Humphrey (1922) found that there is a distinct relationship between the number of lobes on a multiple testis and the age of mature males of dusky salamanders (*Desmognathus fuscus*, family Plethodontidae): a male with two full mature lobes per testis is in its third year of sexual activity, a male with three mature lobes per testis is in its fifth year of sexual activity, and a male with four mature lobes per testis is in its seventh year of sexual activity. Organ (1961) and others have used this structure to differentiate age classes. Each lobe of such multiple-lobed testes represents a center of spermatogenesis and is swollen because of this activity. The areas between lobes are temporarily inactive and so become reduced in size. Anuran testes are compact oval or reniform structures. The entire testis of anurans exhibits a pronounced increase in size during the breeding season, when spermatogenetic activity is at a peak.

Fat bodies are associated with both the ovaries and testes of amphibians and appear to function as a supply of energy for gonadal activity. Their size fluctuates seasonally, being largest in the fall just prior to hibernation and the initiation of gametogenetic cycles and smallest at the end of the breeding season. The fat bodies of caecilians are elongated and lobed, while those of salamanders are elongated bands; both extend alongside and parallel to the gonads. Those of anurans are finger-like and are attached to the anterior end of the gonad.

REPRODUCTIVE AND EXCRETORY DUCTS

During the embryonic development of female salamanders (*Ambystoma*, family Ambystomatidae), each pronephric duct divides longitudinally to form a Müllerian duct and a Wolffian (archinephric) duct. The former is the oviduct for the passage of eggs from the coelom to the cloaca, while the latter retains the function of draining the kidney. The oviducts of anuran females form after the pronephric ducts and, as in higher vertebrates, arise as folds in the epithelium of the dorsal wall of the coelom and later fuse and grow backward to the cloaca ventral to the nephrostromes.

The Müllerian ducts of amphibians are all of the same general pattern. The anterior end, situated well forward in the body cavity, is expanded into a funnel-like opening termed the ostium into which ripened eggs are driven by cilia distributed over the peritoneal lining of the coelmic walls and organs. Posteriorly, each Müllerian duct is slightly enlarged to form a short uterus, which in most amphibians opens directly into the cloaca. In true toads (*Bufo*, family Bufonidae) and midwife toads (*Alytes*, family Discoglossidae), however, the paired uteri sometimes unite just anterior to the cloaca and discharge through a common opening. The fusion of ducts occurs more anteriorly in African live-bearing toads (*Nectophrynoides*, family Bufonidae), which breed on land and have internal fertilization (Angel and Lamotte, 1944). In these, a bicornuate uterus is produced in which larval development takes place. In most other amphibians, the uteri function simply as temporary storage space for eggs prior to oviposition. Prior to the breeding season the Müllerian ducts become greatly enlarged and coiled, and their glandular linings secrete a clear gelatinous material. After entering the ostium, eggs are forced down the ducts by peristaltic contractions and become coated with several layers of the jellylike material. This coating swells when the eggs subsequently enter the water. In instances wherein the eggs are laid on land or carried by an adult, the same material may be whipped into a frothy mass by the female after she has deposited the eggs, as in some species of

Leptodactylus, or may merely function as an adhesive material attaching the eggs to the substrate or adult's skin (see Chapter Eleven for a fuller discussion of differences in reproductive behavior). Internal fertilization occurs in many salamanders as the eggs pass down the oviducts, and in these forms a dorsal diverticulum of the cloaca called the spermatheca is present; this functions as a storage receptacle for spermatozoa prior to ovulation.

The relationship of the reproductive and excretory ducts in male amphibians is closer than in most vertebrates and much more intimate than in females. The Wolffian duct carries both excretory wastes from the kidney and spermatozoa from the testis to the cloaca. A series of up to nine modified kidney tubules, known as efferent ductules or vasa efferentia, grow out from the anterior tissue of the kidney and form a net which joins the longitudinal collecting canal inside the testis or along its medial border. Posteriorly, the vasa efferentia connect to the Wolffian duct; occasionally they empty independently into the cloaca or connect to the longitudinal canal of the opposite side. In salamanders, the vasa efferentia join the narrow longitudinal canal, known as Bidder's canal, extending along the medial edge of the kidney just under the peritoneum. Bidder's canal connects by a series of short ducts to anterior kidney tubules and these connect by other kidney tubules emerging from the lateral side of the kidney to the Wolffian duct. Thus, the anterior part of the Wolffian duct, known as the ductus deferens, functions primarily to transport spermatozoa, while the posterior part serves also for the elimination of urinary wastes. The paired Wolffian ducts enter the cloaca independently. The pattern in caecilians is essentially the same as that in salamanders. The longitudinal collecting canal of the testis connects to Bidder's canal by transverse canals which arise between the lobes of the testis. A second series of transverse canals connect Bidder's canal to the kidney tubules, and these empty into the Wolffian duct. In anurans, Bidder's canal lies within the kidney but close to its medial edge. The vasa efferentia enter the anterior end of the kidney and, depending upon the anuran, may connect directly to the Wolffian duct or join Bidder's canal. From Bidder's canal, spermatozoa are conveyed through kidney tubules to the Wolffian duct. In some anurans, there is a dilation of the Wolffian duct near the cloaca, forming a seminal vesicle in which spermatozoa may be stored temporarily.

As is true of the gonads, the Müllerian duct and ductus deferens vary in size seasonally; they are largest just prior to the breeding season and then may regress to a mere thread. In anurans, the seminal vesicles also are at the height of their development during the breeding season and then undergo regression.

ENDOCRINE SYSTEM

The vertebrate body is controlled and coordinated by the nervous and endocrine systems. The nervous system exerts its control through the conduction of electrical impulses along fibers distributed throughout the body. The endocrine system, on the other hand, operates through the secretion of hormones, chemical substances which exercise a wide variety of regulatory roles over physiological processes. The hormones are released into the blood and lymph, which then passively carry them to all parts of the body. The endocrine organs are special ductless glands which discharge their secretions by diffusion; each gland has its particular and specific function. In addition, some organs, such as the pancreas and the gonads, have a dual role of endocrine function plus other responsibilities.

THYROID GLAND

The thyroid gland, one of several glands derived from the pharynx wall, is histologically and functionally similar in all vertebrates. The thyroid of salamanders appears embryonically as an unpaired structure. This soon divides into two glands which in the adult lie on either side of the throat near the external jugular veins (Fig. 2–47). In anurans, the thyroid develops and remains as a paired gland located on each side of the hyoid apparatus, near the trachea and more medial and deeper than in salamanders. Each gland consists of a number of thyroid follicles, and each follicle consists of a single layer of cuboidal epithelial

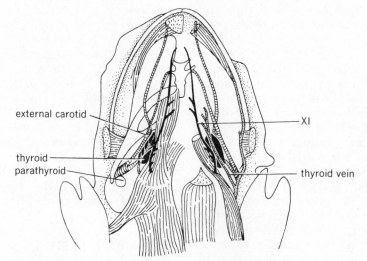

external carotid

XI

thyroid
parathyroid

thyroid vein

Figure 2–47 Dissection of the European salamander (*Salamandra salamandra*), showing the position of the thyroid and parathyroid glands. (From *Vertebrates: Their Structure and Life*, by W. B. Yapp. © 1965 by Oxford University Press, Inc. Reprinted by permission.)

cells surrounding a space filled with a viscous fluid known as colloid. Accessory follicles scattered among the muscles of the lower jaw and neck have been found by Stone and Steinitz (1953) in newts (*Triturus*, family Salamandridae).

Although very few studies have been made regarding the sequence of events in the synthesis of amphibian thyroid hormones, available evidence indicates that these hormones are the same as those of other vertebrates. They are thyroxin, which is produced in large quantities, and triiodothyronine, produced in smaller quantities. These hormones are chemically very similar, consisting of two benzene rings with substituted iodine, and a chain with an amino and carboxyl group. In mammals, and presumably in amphibians as well, the glandular epithelium of the follicles takes up iodide from the blood through active transport. The iodide is oxidized and then combines with tyrosine residues in the protein of the colloid. Thyroxine and triiodothyronine are formed by condensation of pairs of iodinated tyrosines and, as they are still attached to the protein, are stored in the form of large molecules called thyroglobulin. Thyroxine and triiodothyronine must be separated from the other amino acids of the thyroglobulin molecule by enzymatic action before they can leave the follicle; this is done by intracolloidal hydrolysis. The two

hormones then pass into the blood, apparently by diffusion.

The action of thyroid hormones upon amphibian metamorphisis is well known (Chapter Eleven). However, their role in adult amphibians is not clearly understood. Investigations have been restricted largely to the action of thyroid hormone upon oxidative and other types of metaboblism, upon integumentary structure, and upon nervous function (Gorbman, 1964). The following statements are thus merely a summarization of current knowledge.

In mammals, the thyroid hormones control the rate of metabolism of the entire body, but attempts to stimulate oxygen consumption in adult amphibians by treatment with thyroid hormones have been generally unsuccessful. As in all other vertebrates, the structure of the skin of amphibians is partly under the control of the thyroid. Ecdysis, the periodic shedding of the corneal layer of the skin, is inhibited or prevented entirely in most adult amphibians by thyroidectomy; apparently the periodicity of molting is governed by the level of circulating thyroid hormone. The embryonic differentiation and development of the central nervous system in tadpoles is stimulated by thyroid hormones, and in adult frogs thyroid treatments lower the sensitivity of peripheral nerves to electrical stimuli.

Thyroidectomy of juvenile salamanders inhibits limb regeneration; this capability may then be restored by small doses of thyroid hormones. Triiodothyronine and thyroxine both have a stimulatory effect upon the heart rate of frogs. Finally, thyroid treatment of frogs accelerates spermatogenesis, although thyroidectomy of amphibians generally seems to have little effect upon the attainment of sexual maturity (Gorbman, 1964).

PARATHYROID GLANDS

The parathyroid glands (epithelial bodies), as their names implies, lie either alongside or dorsal to the thyroid gland (Fig. 2–47). Correlated with their embryological origin from the epithelium of the pharyngeal pouches, they are first recognizable as discrete structures in amphibians and occur in all higher vertebrates. In anurans, the parathyroids come from the ventral portions of the second pair of pouches, but in other amphibians they are usually derived from the third and fourth pairs of pouches. Little is known of the function of the parathyroids in amphibians but, as in other vertebrates, they seem to be involved in mineral metabolism. It is known that parathyroidectomy of leopard frogs (*Rana pipiens*) results in a reduction of blood calcium and an increase in urinary excretion of this mineral (Cortelyou *et al.*, 1960). In addition, this operation initially causes an increase in plasma phosphorus, but this level declines after several weeks.

ADRENAL GLANDS

Adrenal tissue is found in all classes of vertebrates in the vicinity of the kidneys, and it generally consists of two kinds of cells. One kind of cell, corresponding to the medulla of mammalian adrenal glands, is embryonically derived from the ectodermal neural crest and stains a brownish-yellow when treated with chromic acid and its salts (the chromaffin reaction). Cells of this type compose what is often termed chromaffin tissue. Chromaffin tissue produces two hormones, adrenaline (epinephrine) and noradrenaline. The second type of cell, homologous with the cortex of mammalian adrenals, is derived embryonically from mesodermal mesen-

chyme and constitutes what are called the interrenal bodies in lower vertebrates. Interrenal tissue produces several steroid substances termed adrenocorticoids.

A third type of cell, the "summer cell of Stilling," which has no known equivalent in other classes of vertebrates, has been described in adrenal tissue of the European edible frog (*Rana esculenta*) and other anurans (Stilling, 1898); these cells are not known to occur in salamanders or caecilians. Stilling cells are mesodermal in origin and contain eosinophilic granules which are PAS-positive and therefore probably glycoprotein (Maillet, 1960). The function of Stilling cells is completely unknown. Chromaffin cells and interrenal cells appear to function independently of one another, yet they are closely associated in amphibians, reptiles, birds, and mammals; they are separated in lower vertebrates. The adrenals of salamanders are islets of tissue extending between the kidneys, lying on the ventral surfaces of the kidneys, or actually embedded in kidney tissue. As is true of other structures, the adrenals of anurans are more compact and exist as orange or yellow bands on the ventral surface of each kidney.

An extensive volume of literature exists on the action of adrenaline in amphibians but little is known of the action of noradrenaline. In summary, injections of adrenaline cause a rise in blood pressure, a stronger heart beat, an elevation of blood sugar, and a dilation of respiratory passages; there is a blanching of the skin through melanin concentration, except in clawed frogs (*Xenopus laevis*, family Pipidae), which unexplainably give the opposite reaction (Burgers *et al.*, 1953); and, finally, there are a dilation of the pupils of frog eyes, hyperglycemia in anurans and salamanders, and a reduction in fatiguability of frog muscle.

Although they have not all been identified, it appears the same adrenocorticoids may be present in amphibians as in other vertebrates. Corticosterone and hydrocortisone have been found in venous blood from the adrenals of the congo eel (*Amphiuma*, family Amphiumidae), and hydrocortisone in the blood of clawed frogs (*Xenopus*), by Chester Jones *et al.* (1959). Cultures of bullfrog (*Rana catesbeiana*) adrenal tissue form the mineralocorticoid aldosterone abundantly and form corticosterone less abundantly (Carstenson *et al.*, 1959). Little is known of the actions of individual adrenocorticoids in amphibians. Most available information is based upon studies of adrenalectomized animals or those subjected to hormone injections. In summary, such studies have shown that adrenocorticoids increase the work capacity of frog muscle, reduce the tendency of amphibians to become edematous when immersed in water, promote the inward flow of sodium through the skin, reduce the permeability of blood vessels, help regulate carbohydrate metabolism, inhibit metamorphosis of young frog tadpoles but stimulate metamorphosis of older tadpoles, and synergize the destructive phases of metamorphosis (see Chapter Eleven). Evidence is accumulating that temperature and the season of the year affect amphibian responses to these hormones.

ISLETS OF LANGERHANS

The pancreas is both a ducted and ductless gland and is composed of two distinct, functionally independent kinds of tissues. The exocrine portion consists of acinar cells that secrete several types of digestive enzymes through ducts into the duodenum. The endocrine portion consists of irregular masses of cells called islets of Langerhans. These islets are highly vascularized and in amphibians, reptiles, birds, and mammals are scattered throughout the pancreas. Amphibian islet tissue develops in the typical vertebrate manner from primitive pancreatic ducts and becomes functional approximately at the time of metamorphosis.

In birds and mammals, the islets are composed of two kinds of cells, called alpha and beta cells: the alpha cells are believed to secrete glucagon, a blood sugar-raising hormone; and beta cells are thought to produce insulin, which has the effect of lowering blood sugar. Of amphibians, several species of anurans have been described as having these two types of cells, but only beta cells are known to occur in the newt (*Taricha torosa*, family Salamandridae).

PITUITARY GLAND

The pituitary gland, or hypophysis cerebri, is basically the same in all vertebrates other than the most primitive fish. It is formed by the union of the infundibulum, a ventral diverticulum from the diencephalon region of the brain, with the adenohypophysis, a dorsal evagination from the roof of the buccal cavity. The adenohypophysis of amphibians (at least in anurans and salamanders) arises as a solid fingerlike projection of ectoderm that is homologous with the hollow structure called Rathke's pouch in other vertebrates. The amphibian pituitary is composed of three or four lobes that, although closely associated, are functionally independent. The neurohypophysis originates from the floor of the infundibulum and gives rise to the posterior lobe of the pituitary. The anterior (pars distalis) and intermediate (pars intermedia) lobes are derived from the adenohypophysis. The fourth component, missing in many salamanders, is the pars tuberalis, a bilobed structure derived from the adenohypophysis and located primarily on the anterior surface of the pituitary stalk.

All of the hormones known to occur in the pars distalis of other vertebrates have been found in amphibians as well. Among these are the gonadotropins FSH (follicle-stimulating hormone), controlling the development and endocrine function of ovarian follicles in females and sperm maturation in males, and LH (luteinizing hormone), controlling the development of the corpora lutea in ovaries and activity of interstitial cells of testes; TSH (thyroid-stimulating hormone or thyrotropin), which stimulates the thyroid to secrete thyroxin; ACTH (adrenocorticotropic hormone or adrenocorticotropin), which stimulates the adrenal cortex to secrete adrenocorticoids; the lactogenic hormone (or prolactin), affecting reproductive behavior in amphibians (e.g., the "water drive" of *Triturus*); and the growth hormone (somatotropin), which is necessary for growth to occur and which may stimulate the islets of Langerhans to secrete glucagon. Intermedin, the hormone of the pars intermedia, typically causes a dispersion of melanin granules in the melanophores and a concentration of guanophores; the result is a darkening of the skin. Despite the occurrence of a pars tuberalis in the majority of vertebrates, it is not known to have any hormonal function. Three neurohypophyseal hormones have been positively identified from the neurohypophysis of the European edible frog (*Rana esculenta*); they are oxytocin, arginine vasotocin, and arginine vaso-pressin (Acher *et al.*, 1960). Tests on a variety of amphibians have shown that the neurohypophyseal hormones have an antidiuretic action, causing an increase in the passive permeability of the skin, mesonephros, and urinary bladder to water, salts, and urea. They also stimulate the active transport of sodium through the skin and urinary bladder. It is significant that terrestrial amphibians are more responsive to neurohypophyseal hormones than aquatic amphibians (Gorbman, 1964).

GONADS

Testes and ovaries, in addition to producing spermatozoa and ova, are endocrine organs that cause the development and maintenance of the secondary sex characters and seasonal breeding activity.

Testosterone is produced by the stromal cells surrounding the lobules of the testis. This hormone at least partially controls such sexually dimorphic characters as the swelling of the thumb pads, enlargement of the mental gland, changes in color pattern, development of a dorsal crest in some salamanders, and enlargement of cloacal glands. Similarly, ovarian hormones induce the development of the oviducts, stimulate mating behavior in females, and probably affect cutaneous secretions.

NERVOUS SYSTEM

The general organization of the central nervous system (brain and spinal cord) of amphibians does not differ greatly from that of fishes (Fig. 2–48). The tectum (roof of the midbrain) is dominant in controlling the body. It receives sensory impulses from olfactory, auditory, optic, and other receptors and, upon correlating these,

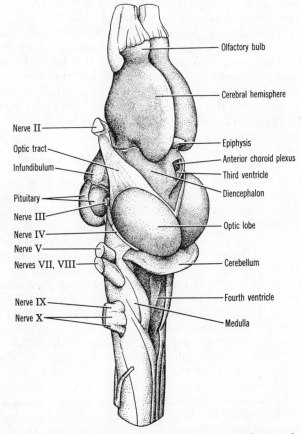

Olfactory bulb

Cerebral hemisphere

Nerve II

Optic tract

Infundibulum

Epiphysis

Anterior choroid plexus

Third ventricle

Pituitary

Diencephalon

Nerve III

Nerve IV

Optic lobe

Nerve V

Nerves VII, VIII

Cerebellum

Nerve IX

Fourth ventricle

Nerve X

Medulla

Figure 2–48   The brain of a frog. (From Torrey, T. W.: *Morphogenesis of the Vertebrates.* 2nd Ed. John Wiley and Sons, 1962.)

transmits motor impulses controlling the muscles. In higher vertebrates, the cerebral hemispheres are progressively more important until they are dominant, forming the greatest bulk of the brain in mammals.

The cerebral hemispheres of amphibians are, as in lower vertebrates, mainly olfactory centers, but their differentiation has begun. Their ventral basal nuclei have become a correlation center, receiving impulses from the thalamus, the ventral part of the midbrain, and correlating them with olfactory sensory impulses. The middle part of the hemispheres remains purely olfactory and unaltered from the primitive condition. The dorsal part is partly olfactory but is also a correlation center that receives impulses from the tectum and transmits impulses back to it. The bodies of nerve cells, the gray matter, are, as in fishes, primarily deep in the hemisphere walls near the central cavity, but some are scattered near the surface of the dorsal parts and are precursors of the cerebral cortex of higher vertebrates. Since both lungfishes and coelacanths (*Latimeria*) exhibit the same sort of differentiation of the hemispheres, it appears that amphibians inherited it from their rhipidistian ancestors. Impulses from the lateral line system are correlated in the cerebellum; among those amphibians in whom this system is reduced the cerebellum is also very poorly developed.

As is true of fishes, amphibians have only 10 pairs of cranial nerves in addition to the terminal nerve (cranial nerve 0), rather than 12 pairs as in higher vertebrates; the spinal accessory nerve (XI) and hypoglossal nerve (XII) are not developed as separate entities in amphibians or lower vertebrates.

The peripheral nervous system of amphibians differs from that of fishes primarily in that the nerve supply goes to the paired limbs. Like fins, both pairs of limbs are derived from several body segments and are supplied by several spinal nerves. However, the movements of limbs are more complicated than those of fins, and the several spinal nerves innervating the limbs are enlarged and form plexuses between the limbs and the spinal cord. The spinal cord itself is enlarged in the cervical and lumbar regions; no such enlargements of the spinal cord occur in fishes.

The autonomic nervous system (sympathetic and parasympathetic nerve fibers innervating the smooth muscles of the digestive, respiratory, urogenital, and circulatory systems) is well-developed in amphibians. It is simpler than that of mammals but of the same general pattern, with antagonistic actions of sympathetic and parasympathetic systems. The autonomic system consists of a double series of connected ganglia extending most of the length of the body on each side of the dorsal aorta. Mudpuppies (*Necturus*, family Proteidae) are unusual in having an accessory trunk of sympathetic nerves lying lateral and parallel to the primary sympathetic trunk.

RECEPTOR ORGANS

Receptor organs are either cells or groups of cells which respond to environmental changes. Upon being stimulated, they transmit impulses by nerve fibers to the central nervous system, where they are interpreted as sensations. With few exceptions, receptor organs are formed from epidermal cells and are constructed in such a way as to be particularly sensitive to one kind of stimulus, more so than any other cells of the body. In addition to receptor organs for the familiar senses of sight, hearing, smell, taste, and touch, there are receptors for pain, temperature, equilibrium, hunger, thirst, fatigue, and muscle positions.

Lateral Line System

The lateral line system represents a sensory mechanism connected to the skin. This system is not found in reptiles, birds, or mammals but occurs in all fishes and in most aquatic amphibians and aquatic larvae of terrestrial amphibians. The organs that compose the lateral line system are clusters of free neuromast cells (Fig. 2–49). These usually occur in definite rows of little pits on the head and body; in water, they respond to low-frequency vibrations of the water. Presumably, these organs function in the maintenance of equilibrium and posture. There is also some evidence that they may be used in "distant touch" of objects, in which the animal produces waves that reflect off objects and are then perceived by the lateral line system.

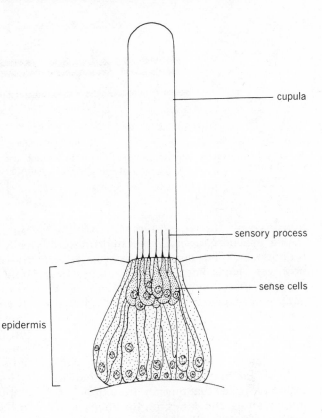

Figure 2–49 Diagram of a neuromast organ. (From *Vertebrates: Their Structure and Life*, by W. B. Yapp. © 1965 by Oxford University Press, Inc. Reprinted by permission.)

The lateral line organs of amphibians are less specialized than those of fishes and lie entirely within the epidermis in species-specific patterns (Figs. 2–50 and 2–51). When metamorphosis from an aquatic larva to a terrestrial adult occurs, the lateral line system is generally lost. An exception to this involves the American newts in the genus *Notophthalmus* (family Salamandridae); these metamorphose into a terrestrial stage in which they are called efts, spend up to three years on land, and then return to water to develop a tail fin and become sexually mature aquatic newts. In these newts, the lateral line system is fully developed in the aquatic larval stage, becomes partially atrophied and covered with corneum during the terrestrial stage of the eft, and then reappears as a functional system in the aquatic adult stage.

Completely terrestrial amphibians lack lateral line systems, and those aquatic species which lack them, the caecilian *Typhlonectes* (family Typhlonectidae), for example, are presumed to have evolved from terrestrial ancestors in which the system was lost because it did not provide any selective advantage out of water.

The larval development of the European salamander *Salamandra atra* (family Salamandridae) is completed within the

Figure 2–50 Dorsal (left) and ventral (right) outlines of an adult female clawed frog (*Xenopus laevis*), showing the lateral line canals. (From *Vertebrates: Their Structure and Life*, by W. B. Yapp. © 1965 by Oxford University Press, Inc. Reprinted by permission.)

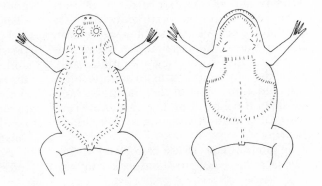

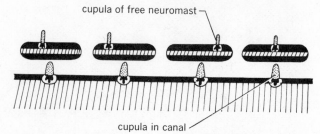

Figure 2–51   Diagrammatic longitudinal section through a lateral line canal. Black represents epidermis. (From *Vertebrates: Their Structure and Life*, by W. B. Yapp. © 1965 by Oxford University Press, Inc. Reprinted by permission.)

female reproductive system so that the young are born as fully metamorphosed individuals. The larvae of this species, however, have lateral line systems, indicating that they are descendants of a *Salamandra* which had an aquatic larval stage.

Ear

The lateral line system forms embryonically from thickenings of the ectoderm called placodes that migrate from the head region along the sides of the animal. A similar placode forms at the side of the head behind the eye, sinks into the head, and develops into the inner ear (properly termed the membranous labyrinth) and the ganglion of the auditory (VIII) cranial nerve. The labyrinth is composed basically of the utriculus, sacculus, and semicircular canals and functions as an equilibrium organ in much the same manner as the lateral line system. The actual receptors used in this equilibratory function are patches of sensory cells located in the ampullae of the semicircular canals (cristae ampullares), the floor of the utriculus (macula utriculi), and the wall of the sacculus (macula sacculi). Movement of the fluid (endolymph) contained in the system stimulates hair cells in these sensory patches and causes transmission of nervous impulses to the brain along the vestibular branch of the auditory nerve. In higher fishes, the saccular macula is sensitive to low frequency vibrations and, thus, the inner ear begins to assume auditory function. Most of the changes which occur in the ears of higher vertebrates are associated with this secondary function of hearing.

In amphibians (Figs. 2–52 and 2–54), the sacculus has a small outpocketing, the

lagena, with its own patch of sensory cells called the macula lagenae. In some forms, the macula lagena gives rise to an additional patch of sensory cells called the papilla basilaris which generally occurs in another outpocketing from the wall of the sacculus. Both the macula lagena and the papilla basilaris are characteristically found in all higher vertebrates, and the papilla basilaris becomes the dominant factor in the development of hearing. The amphibian sacculus is unique, however, in that it often has still another outpocketing with its individual patch of sensory cells, the papilla amphibiorum. The significance of this peculiar characteristic is uncertain, but it seems to be concerned with hearing. The endolymphatic ducts of anurans are very much enlarged and extend around the brain, meeting both dorsally and ventrally, and along the spinal cord within the vertebral canal. These ducts contain considerable quantities of calcium carbonate that appears to be used during metamorphosis for the formation of the bony skeleton.

All tetrapods possess one or more bones that conduct sound to the labyrinth from the surface of the animal. A single bone, the columella auris, is present in amphibians, reptiles, and birds; this is homologous with the hyomandibular bone or cartilage of fishes and the stapes of mammals. In primitive fishes, the hyomandibula provides both auxiliary support to the jaw and, on its posterior margin, support for the gills. During the evolution of tetrapods, the transition from gill respiration to pulmonary and pharyngeal respiration resulted in a separation of feeding mechanisms from respiratory mechanisms. As a consequence, the hyomandibula was left functionally unimportant in both so

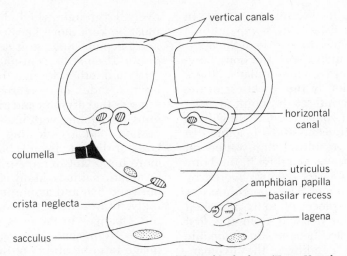

Figure 2–52 Diagram of the internal ear (membranous labyrinth) of a frog. (From *Vertebrates: Their Structure and Life*, by W. B. Yapp. © 1965 by Oxford University Press, Inc. Reprinted by permission.)

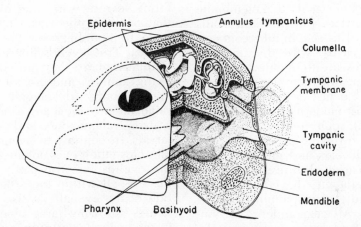

Figure 2–53 The auditory apparatus of an anuran. The stereogram exposes a narrow vertical band of the animal's sagittal plane and the left half of a transverse section which has been swung several degrees backward on its sagittal axis to bring it into a favorable angle for the observer. Both sections cut through the pharynx. The transverse section bisects the tympanic membrane and shows the left auditory nerve (VIII) passing from the medulla oblongata to the internal ear (membranous labyrinth). The sagittal view shows the forked anterior tip of the retracted tongue pointing backward into the pharynx. (From Rand, H. W.: *The Chordates*. McGraw-Hill Book Co., 1950.)

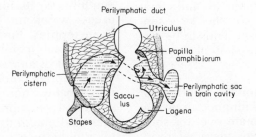

Figure 2–54 Schematic section of the internal ear in a salamander. The basilar papilla is absent here, but is found in addition to the papilla amphibiorum in anurans. (From Romer, A. S.: *The Vertebrate Body*. 4th Ed. W. B. Saunders Co., 1970.)

that it was free to assume a new role. In crossopterygian fishes, the hyomandibula is connected to the inner side of the operculum and articulates with the bony labyrinth. In primitive amphibians, there began a reduction in the thickness of the wall of the labyrinth at the point of articulation of the hyomandibula, allowing a more perfect transmission of sound. In later forms the resulting thin-walled pit evolved into an oval window in the bony labyrinth, the fenestra ovalis, covered by a resilient membrane. The operculum was replaced by a piece of skin stretched across the space derived from the first gill slit; this space became the tympanic cavity. In this condition, the hyomandibula, which evolved into the columella auris, transmits sound directly from one membrane with a large surface (the former operculum, now the tympanic membrane) across the tympanic cavity to a membrane with a small surface (the fenestra ovalis membrane). An Eustachian tube connects the tympanic cavity with the pharynx, allowing air pressure to equalize on both sides of the tympanic membrane.

The middle ear just described is characteristic of most modern anuran amphibians (Fig. 2–53), some reptiles, and some birds. Some burrowing toads have degenerate middle ears and probably have very limited hearing abilities; the middle ear and columella auris have been lost in many frogs, and the tympanum is absent in many others. All salamanders lack tympanic cavities and membranes, and the columella is often degenerate. In some salamanders, however, the distal end of the columella comes into contact with the squamosal bone, indicating that it may be used to transmit ground-borne vibrations to the labyrinth from the lower jaw or forelimbs. Caecilians lack middle ears and, as is true of salamanders, appear to hear only very low frequencies and vibrations. In general, the presence of a sensitive hearing mechanism tends to be correlated with the possession of a voice.

Eye

The transition from aquatic to terrestrial life had a significant influence upon the evolution of the vertebrate eye, for solutions to several problems were required. On land, the external surface of the cornea, the transparent anterior layer of the eyeball through which light rays enter, must be kept moist because it is damaged and becomes opaque if desiccated; in addition, it must be continually cleansed of dust and other foreign bodies. The refractive index of the cornea is almost the same as that of water and presents no problems for aquatic vertebrates but is an important factor affecting aerial vision. Greater distant vision is both possible and more important in terrestrial habitats than in aquatic habitats. Many of the differences between fish and amphibian eyes, accordingly, represent adaptations to terrestrial life. Changes in the eyes of higher vertebrates represent improvements in the image-forming ability of this organ.

In most terrestrial vertebrates, the external surface of the cornea is kept moist and clean by eyelids that periodically sweep over it; glands in the lids of higher forms secrete water and oily substances that facilitate the cleansing and also reduce friction between the lids and the cornea. True eyelids first evolved in amphibians where they are characteristic of living anurans and are present but generally poorly developed in terrestrial salamanders; they are absent in purely aquatic salamanders and all amphibian larvae. In contrast to mammals, amphibians close their eyes not merely by moving the eyelids but also partly by retracting the eyeball within the orbit. This retraction of the eyeballs is also used to facilitate swallowing large objects: it results in the eyeballs protruding into the buccal cavity, forcing food into the esophagus.

Harderian glands, one of three kinds of eye glands present in mammals and secretors of an oily substance, are present in amphibians for the first time. The lacrimal duct, a canal connecting the space inside each lower lid with the nasal passage, is another structure that first appears in amphibians; its function is to drain away excess water and secretions. In anurans, as in reptiles, birds, and some mammals, a palpebral membrane is present; however, that of anurans is actually the transparent upper border of the lower lid and is probably not homologous with the palpebral membranes of higher vertebrates, although its function is the same — to provide the cornea with protection from foreign bodies. All of these structures (lids, glands, and lacrimal ducts) are formed in amphibians at or just prior to metamorphosis.

Accommodation, the adjustment of the eye to vision at different distances, is accomplished in modern amphibians in a different manner than in other terrestrial vertebrates (Fig. 2–55). The amphibian lens is similar to that of fishes, being hard and almost spherical; it is incapable of changing shape. When relaxed, the amphibian eye is focused on distant objects and accommodation for near vision is accomplished by moving the lens forward (as in elasmobranch fishes but not as in teleosts, where the lens is moved backward for distant vision). The relaxed eye is focused on distant objects in other terrestrial vertebrates and is focused on closer objects by deformation of the lens.

Amphibians have four types of photoreceptor cells in their retinas: red rods, green rods, single cones, and double cones. Red rods contain visual purple (rhodopsin) and green rods contain visual green; the two may function in the same manner to facilitate vision in dim light. It is not known whether anurans have color vision, but the presence of two kinds of cones implies at least some discrimination of wavelength must be possible;

salamanders and newts appear to have good color vision.

Among living amphibians, eyes are best developed in anurans. Lettvin and Maturana (1959) have described three classes of ganglion cells in the frog retina and have provided definitions of their operations. In the superficial layer of the frog's optic tectum are cells whose response is a function of the sharpness and extent of a boundary between "lighter" and "darker" in their receptive fields, and their response is sustained. Some of these cells measure target size and they respond to movement in proportion to the velocity and size of the moving object. In the second layer are cells that respond to movement with on-off bursts rather than a sustained response; these can precisely signal movement of a target either lateral to, toward, or away from the anuran. The deepest stratum comprises cells which respond to changes in brightness in the receptive field. Ingle (1968) has found that there is only a partial correspondence between the stimulus optima for the retinal cells and the "preference" (as indicated by orientation and snapping behavior) of the whole frog; he has interpreted this as indicating that the retina

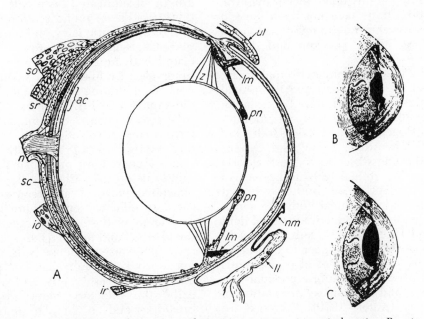

Figure 2–55 The amphibian eye and its accommodation. A, anuran eye in vertical section; B, anterior segment of *Bufo* in relaxation; C, same as B only in accommodation; note forward movement of lens. *ac*, area centralis; *io*, inferior oblique; *ir*, inferior rectus; *ll*, lower lid; *lm*, lens muscles (protractors); *n*, optic nerve; *nm*, nictitating membrane; *pn*, pupillary nodules; *sc*, scleral cartilage; *so*, superior oblique; *sr*, superior rectus; *ul*, upper lid; *z*, zonula Zinnii. (From Walls, G. L.: *The Vertebrate Eye.* The Cranbrook Press, 1942. After Franz and Beers.)

is only the first-stage filter of sensory information and the response of retinal cells does not commit the frog to any particular perception nor does it alone lead the frog to a particular decision.

Salamander eyes are poorly studied but, when present, are generally smaller and less complex appearing than those of anurans. Eyes are poorly developed in *Typhlomolge, Thphlotriton,* and *Haideotriton* (family Plethodontidae), which occur in caves and underground waters.

Caecilians have very small eyes, less than a millimeter in diameter, that lie beneath and are fused to a transparent area of skin; sometimes they are covered by bone. The eye of young African caecilians in the genus *Scolecomorphus* (family Scolecomorphidae) is covered solidly by bone and is imbedded in the internal structure of the tentacle, but as the tentacle grows forward, the eye is pulled forward into the tentacular groove and often completely out from under the bone (Taylor, 1967).

Extraoptic Photoreceptors

Wavelength-sensitive photoreceptors are known to occur in the skin of leopard frogs (*Rana pipiens*) and salamanders (*Ambystoma mexicanum*). Although these receptors apparently do not occur over the entire body, they have not been located anatomically and no function has been ascribed to them (Becker and Cone, 1966).

Both the pineal end organ (frontal organ, stirnorgan, or parapineal) and the pineal body (epiphysis cerebri) have been shown experimentally to function as photoreceptors (Bogenschütz, 1965; Dodt and Morita, 1967; Adler, 1971); and, where known, the fine structure of these closely resembles the retinal rods and cones of the eyes (Adler, 1970). Both the pineal end organ and the pineal body are sensitive to wavelength and to intensity of light (Dodt and Heerd, 1962; Dodt and Jacobson, 1963), and they appear to be important in the establishment of endogenous rhythms, thermoregulation, gonadal development, and in compass orientation of amphibians (Adler, 1970). Anurans are the only amphibians to possess the pineal end organ and, in them, it and the pineal body (found in all amphibians) develop from a common dorsal outpocketing of the diencephalon; each is hollow and the

photoreceptive segments protrude into the lumen (Adler, 1970). The pineal end organ is located in the dermis on the dorsal surface of the head between the eyes and is homologous with the parietal eye of lizards and the tuatara. In salamanders, the pineal body forms as a hollow outpocketing of the roof of the brain, as in anurans, but the lumen is obliterated during development.

Chemical Sense Organs

The chemical senses include taste, smell, and sensitivity to chemical irritants. In fishes, receptor organs for chemicals are widely distributed in the moist skin, whereas terrestrial vertebrates with dry and impervious skins have chemoreceptors concentrated in the nose, mouth, and other mucous membranes that are more or less exposed to the environment. The situation in living amphibians is generally like that of fishes, with a wide distribution of chemical-sensitive nerve endings over the body surface. Thus, an amphibian skin responds to chemical stimuli that would affect only the mouth or nose of man. However, organs of taste are found only in epithelium of the amphibian mouth and not on the outer surface of the body as in some fishes. Taste buds, isolated groups of elongated taste cells, and supporting cells are found scattered over the roof of the mouth, jaws, and tongue of amphibians, where they are usually associated with fungiform papillae. They are innervated by fibers from several cranial nerves, including the facial (VII) and glossopharyngeal (IX) nerves.

Olfactory organs, the organs of smell, have a close association with the respiratory system in all air-breathing vertebrates. Whereas in most fish they are in blind pits on the surface of the snout, the olfactoreceptors of air-breathers are located in olfactory epithelium lining sensory areas in the nasal passages between the external and internal nares. Individual receptors consist of sense cells, each bearing up to 12 olfactory hairs on its tip, together with basal cells and supporting cells; the olfactory hairs are exposed brushlike ends of bipolar neurons from the olfactory (I) cranial nerve. The hairs are covered with mucus produced by Bowman's glands; this serves as a moist medium in which air-borne chemical substances can be dissolved in order to stimulate the olfactoreceptors. Olfactory hairs of amphibians have been shown to remain functional when the animal is in or out of water. The distribution of olfactoreceptors varies among amphibians. In perennibranchs, the nasal passages are lined with folds and the receptors lie in the depressions between folds. Larval salamanders also have a divided sensory area, but anuran tadpoles and terrestrial adult amphibians have the receptor cells concentrated into a single area.

An accessory olfactory organ, the vomeronasal or Jacobson's organ, first appears in amphibians and is present in many of the higher vertebrates (Fig. 2–56). This organ is a blind sac, on the medial side of the nasal passage in metamorphosed anurans and on the lateral side of each passage in terrestrial salamanders, with an epithelial lining that includes sensory cells similar to those of ordinary nasal epithelium and is innervated with fibers from the olfactory nerve. In amphibians, it apparently is used to test substances held in the mouth. This organ is associated with terrestrialism and is absent in such aquatic salamanders as the mudpuppy (*Necturus*) and the olm (*Proteus*), both in the family Proteidae; it is particularly well-developed in anurans and caecilians. In the latter, it is associated with a short retractile tentacle on either side of the head; movements of the tentacle bring odors into contact with the sensory cells of Jacobson's organ.

Cutaneous Receptors

As mentioned previously, the entire amphibian skin is sensitive to irritating chemicals. Free nerve endings from spinal and cranial nerves are widely distributed in the epidermis and are the receptors for these stimuli. While these do not appear to be as sensitive as the olfactory organs, they undoubtedly play an important role in amphibian behavior, particularly by affecting movements and selection of resting sites.

Similar but different nerve endings, called tangoreceptors, are sensitive to mechanical stimulations; the sense of touch is generally well-developed in amphibians. Tangoreceptors are abundantly distributed over the amphibian skin and are located between epidermal cells and are associated with connective tissue capsules or with specialized groups of cells. Such groups of cells occur locally in the corneum layer of the integument, where they produce raised sensory papillae. Sensory papillae occur on the head, back, and feet of many anurans and along the lips of some salamanders. The tentacle of caecilians is a tactile organ as are the tentacles of larval clawed frogs (*Xenopus laevis*, family Pipidae).

Finally, amphibians are sensitive to temperature changes, and there is evidence that cold receptors are present in the epidermis and heat receptors lie in deeper layers of the skin.

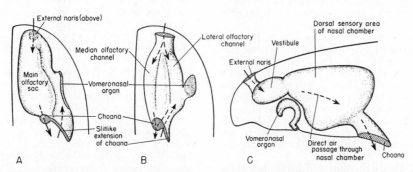

Figure 2–56 Palatal regions of a salamander, anuran, and lizard. *A*, ventral view of anterior part of the left side of the palatal region of the salamander *Triton*, with the nasal channels shown as solid objects, remainder as transparent structures. *B*, similar view of the toad *Pipa*. *C*, longitudinal section of the nasal region of a lizard, cut somewhat to the right of the midline, to show the cavities of the nasal apparatus. In the embryo lizard, the vomeronasal organ was a lateral pocket of the main nasal channels, as in an amphibian; in the adult (as in many mammals), this organ has separated to open independently into the roof of the mouth by a nasopalatine duct. Arrows show the main air flow inward in all figures and the outward flow toward the vomeronasal organ in the amphibians. (From Romer, A. S.: *The Vertebrate Body*. 4th Ed. W. B. Saunders Co., 1970.)

References

Acher, R., J. Chauvet, M. T. Lenci, F. Morel, and J. Maetz. 1960. Présence d'une vasotocine dans la neurohypophyse de la grenouille (*Rana esculenta* L.). Biochim. Biophys. Acta *42*:379–380.

Adler, K. 1970. The role of extraoptic photoreceptors in amphibian rhythms and orientation: a review. J. Herpetol. *4*(3-4):99–112.

——— 1971. Pineal end organ: role in extraoptic entrainment of circadian locomotor rhythm in frogs. *In:* M. Menaker (ed.), Biochronometry. Washington, D.C., United States National Academy of Sciences. Pp. 342–350.

Andrew, W. 1959. Textbook of Comparative Histology. New York, Oxford Univ. Press, 652 pp.

Angel, F., and M. Lamotte. 1944. Un crapaud vivipare d'Afrique occidentale: *Nectophrynoides occidentalis* Angel. Ann. Sci. Nat. Zool. 6:63–89.

Becker, H. E., and R. A. Cone. 1966. Light-stimulated electrical responses from skin. Science *154*:1051–1053.

Bogenschütz, H. 1965. Untersuchungen über den lichbedingten Farbwechsel der Kaulquappen. Zeitschr. f. Vergl. Physiol. *50*:598–614.

Boulenger, G. A. 1882. Catalogue of the Batrachia Salientia s. Ecaudata in the Collection of the British Museum. London.

——— 1888. Descriptions of new Brazilian batrachians. Ann. Mag. Nat. Hist. *1*:187–189.

——— 1897. The tailless batrachians of Europe. London.

——— 1910. Les batraciens et principalement ceux d'Europe. Paris.

Bowers, M. A. 1909. Histogenesis and histolysis of the intestinal epithelium of *Bufo lentiginosus*. Am. J. Anat. 9:263–280.

Burgers, A. C. J., T. A. Boschman, and J. C. van de Kamer. 1953. Excitement darkening and the effect of adrenaline on the melanophores of *Xenopus laevis*. Acta Endocrinol. *14*:72–82.

Carstenson, H., A. C. J. Burgers, and C. H. Li. 1959. Isolation of aldosterone from incubates of adrenals of the American bullfrog and stimulation of its production by mammalian adrenocorticotrophin. J. Am. Chem. Soc. *81*:4109–4110.

Carter, G. S. 1967. Structure and Habit In Vertebrate Evolution. Seattle, Univ. Washington Press. 520 pp.

Chester Jones, I., J. G. Philips, and W. N. Holmes. 1959. Comparative physiology of the adrenal cortex. Comp. Endocrinol., Proc. Columbia Univ. Symp. Cold Spring Harbor, N. Y. *1958*:582–612.

Cochran, D. M. 1961. Living Amphibians of the World. Garden City, N. Y., Doubleday. 199 pp.

Cope E. D. 1864. On the limits of the Raniformes. Proc. Acad. Nat. Sci. Philadelphia *16*:181–183.

——— 1865. Sketch of the primary groups of Batrachia Salientia. Nat. Hist. Rev., London January: 97–120.

——— 1867. On the structure and distribution of the genera of arciferous Anura. J. Acad. Nat. Sci. Philadelphia 6:67–97.

Cortelyou, J. R., A. Hibner-Owerko, and J. Mulroy. 1960. Blood and urine calcium changes in totally parathyroidectomized *Rana pipiens*. Endocrinology 66:441–450.

de Graaf, A. R. 1957. Investigations into the distribution of blood in the heart and aortic arches of *Xenopus laevis* (Daud.). J. Exp. Biol. *34*:143–172.

Dodt, E., and E. Heerd. 1962. Mode of action of pineal nerve fibers in frogs. J. Neurophysiol. *25*: 405–429.

———, and M. Jacobson. 1963. Photosensitivity of a localized region of the frog diencephalon. J. Neurophysiol. *26*:752–758.

———, and Y. Morita. 1967. Conduction of nerve impulses within the pineal system of frog. Pflugers Arch. Ges. Physiol. *293*:184–192.

Eaton, T. H., Jr. 1959. The ancestry of modern Amphibia: a review of the evidence. Univ. Kansas Publ. Mus. Nat. Hist. *12*:155–180.

Foxon, G. E. H. 1951. A radiographic study of the passage of blood through the heart in the frog and the toad. Proc. Zool. Soc. London *121*:529–538.

——— 1964. Blood and respiration. *In:* J. A. Moore (ed.), Physiology of the Amphibia. New York, Academic Press. Pp. 151–209.

Goin, C. J., and O. B. Goin. 1961. Introduction To Herpetology. San Francisco, W. H. Freeman. 341 pp.

Gorbman, A. 1964. Endocrinology of the Amphibia. *In:* J. A. Moore (ed.), Physiology of the Amphibia. New York, Academic Press. Pp. 371–425.

Griffiths, I. 1956. The structure and development of the breast-shoulder apparatus in Amphibia, Salientia. (Ph.D. thesis, Univ. London.)

——— 1963. Phylogeny of the Salientia. Biol. Rev. Cambridge Philos. Soc. *38*:241–292.

Hecht, M. K. 1963. A reevaluation of the early history of the frogs. Part II. Systematic Zool. *12*:20–35.

Humphrey, R. R. 1922. The multiple testis in urodeles. Biol. Bull. *43*:45–67.

Ingle, D. 1968. Visual releasers of prey-catching behavior in frogs and toads. Brain, Behav. Evol. *1*: 500–518.

Jones, D. A. 1967. Green pigmentation in neotropical frogs. Diss. Abstr. *29*:1213-B.

Lettvin, J. Y., and H. R. Maturana. 1959. Frog vision. M.I.T. Quart. Progr. Report *53*:191–196.

Maillet, M. 1960. Action de l'hormone adrénocorticotrope hypophysaire sur les cellules positives à la réaction de MacManus de la surrénale de *Rana esculenta*. Compt. Rend. Soc. Biol. *154*:582–583.

Marcus, H. 1935. Zur Stammesgeschichte des Herzens. Morphol. Jahr. 76:92–103.

Martin, W. F. 1971. Mechanics of sound production in toads of the genus *Bufo:* passive elements. J. Exp. Zool. *176*:273–294.

Maslin, T. P. 1950. The production of sound in caudate amphibia. Univ. Colorado Studies, Series in Biology *1*:29–45.

Noble, G. K. 1922. The phylogeny of the Salientia. I. The osteology and thigh musculature; their bearing on classification and phylogeny. Bull. Amer. Mus. Nat. Hist. *46*:1–87.

——— 1931. Biology of the Amphibia. New York, McGraw-Hill. 577 pp.

Organ, J. A. 1961. Studies of the local distribution, life history, and population dynamics of the salamander genus *Desmognathus* in Virginia. Ecol. Monographs *31(2)*:189–220.

Ozeti, N., and D. B. Wake. 1969. The morphology and evolution of the tongue and associated structures in salamanders and newts (family Salamandridae). Copeia *1969*:91–123.

Parker, W. K. 1868. A monograph on the structure and development of the shoulder girdle and sternum in the vertebrates. London.

Parsons, T., and E. Williams. 1962. The teeth of Amphibia and their relation to amphibian phylogeny. J. Morphol. *110*:375–389.

Reeder, W. G. 1964. The digestive system. *In:* J. A. Moore (ed.), Physiology of the Amphibia. New York, Academic Press. Pp. 99–149.

Ridgewood, W. G. 1898. On the structure of the hyobranchial skeleton and larynx in *Xenopus* and *Pipa;* with remarks on the affinities of the Aglossa. J. Linn. Soc. (Zool.) *26*:53–128.

Sharma, H. L. 1957. The anatomy and mode of action of the heart of the frog, *Rana tigrina* Daud. J. Morphol. *100*:313–343.

Simmons, J. R. 1957. The blood pressure and the pressure pulses in the arterial arches of the frog (*Rana temporaria*) and the toad (*Bufo bufo*). J. Physiol. (London) *137*:12–21.

——— 1959. The distribution of the blood from the heart in some amphibia. Proc. Zool. Soc. London *132*:51–64.

Stilling, H. 1898. Zur Anatomie der Nebennieren. Arch. Mikroskop. Anat. Entw. *52*:176–195.

Stone, L. S., and H. Steinitz. 1953. Effects of hypophysectomy and thyroidectomy on lens regeneration in the adult newt, *Triturus viridescens*. J. Exp. Zool. *124*:469–504.

Taylor, E. H. 1967. The Caecilians Of The World. Lawrence, Univ. Kansas Press. 848 pp.

Trewavas, E. 1933. The hyoid and larynx of the Anura. Philos. Trans. Roy. Soc. (B) *222*:401–527.

Trueb, Linda. 1966. Morphology and development of the skull in the frog *Hyla septentrionalis*. Copeia *1966*:562–573.

——— 1970. Evolutionary relationships of casque-headed tree frogs with co-ossified skulls (family Hylidae). Univ. Kansas Publ. Mus. Nat. Hist. *18*: 547–716.

Tyler, M. J. 1971. The phylogenetic significance of vocal sac structure in hylid frogs. Univ. Kansas Pub. Mus. Nat. Hist. *19*:319–360.

Weichert, C. K. 1959. Elements of Chordate Anatomy, 2nd ed. New York, McGraw-Hill. 503 pp.

Whipple, I. L. 1906. The ypsiloid apparatus of urodeles. Biol. Bull. (Mar. Biol. Lab.) *10*:255–297.

Whiting, H. P. 1961. Pelvic girdle in amphibian locomotion. *In:* Zoological Society of London, Vertebrate Locomotion. Symposia *5*:43–57.

Williams, E. E. 1959. Gadow's arcualia and the development of tetrapod vertebrae. Quart. Rev. Biol. *34*:1–32.

Yapp, W. B. 1964. Vertebrates: Their Structure and Life. New York, Oxford Univ. Press. 525 pp.

CHAPTER 3

THE ORIGIN AND PHYLOGENETIC RELATIONSHIPS OF AMPHIBIA

It is certain that amphibians evolved from lobe-finned bony fishes (subclass Crossopterygii); this conclusion is based on the morphological similarity of the two groups. The primitive crossopterygians (Fig. 3–1) possessed lungs, and some even had internal nares. Their skeleton was well-ossified, a prerequisite for life on land, and their fins were supported by bones corresponding to those of tetrapod limbs (see Figure 2–20). Despite the certainty of a crossopterygian-amphibian relationship, the evolutionary pattern of the origin and radiation of amphibians is clouded by scanty data linking them to particular crossopterygians, a lack of evidence connecting major extinct groups, and great gaps in the fossil record between extinct groups and modern amphibians. Thus, at the present time this subject can only be discussed in general terms. It must be realized that differences of opinion exist on many points because the evidence supporting any one theory is insufficient to cause rejection of others. Undoubtedly, the picture will be clarified by future paleontological findings, and it is to be expected that systems of clas-

Figure 3–1   Devonian fringe-finned (crossopterygian) fish. (From *Tales Told by Fossils,* by Carroll Lane Fenton. Copyright, © 1966 by Carroll Lane Fenton. Reprinted by permission of Doubleday & Company, Inc.)

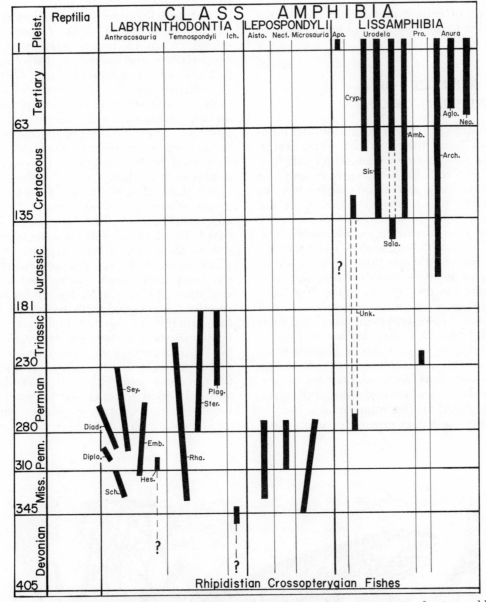

Figure 3–2 Temporal and generally accepted phylogenetic positions of the major groups of extinct and living amphibians relative to one another and to crossopterygian fishes and reptiles. Abbreviations: *Aglo.*, Aglossa; *Aisto.*, Aistopoda; *Amb.*, Ambystomatoidea; *Apo.*, Apoda; *Arch.*, Archaeobatrachia; *Cryp.*, Cryptobranchoidea; *Diad.*, Diadectidae; *Diplo.*, Diplomeri; *Emb.*, Embolomeri; *Hes.*, Hesperoherpeton; *Ich.*, Ichthyostegalia; *Nect.*, Nectridea; *Neo.*, Neobatrachia; *Plag.*, Plagiosauria; *Pro.*, Proanura; *Rha.*, Rhachitomi; *Sala.*, Salamandroidea; *Sch.*, Schizomeri; *Sey.*, Seymouriamorpha; *Sir.*, Sirenoidea; *Ster.*, Stereospondyli; *Unk.*, unknown suborder. (Partially after Olson: Evolution and relationships of the Amphibia.)

sification will change accordingly, but the fossil record will never be so complete as to eliminate conjecture and speculation. Figure 3–2 is a schematic diagram summarizing the temporal and generally accepted phylogenetic positions of the major groups of extinct and living amphibians relative to one another and to crossopterygian fishes and reptiles.

THE EVIDENCE FOR A RHIPIDISTIAN ORIGIN

All lines of evidence indicate that lobe-finned bony fishes in the order Rhipidistia are the group from which amphibians evolved. Early remains of amphibians are always associated with fresh-water deposits, which suggests that their ancestors were inhabitants of fresh water rather than of marine environments. The earliest amphibians known were the carnivorous Ichthyostegidae (*Ichthyostega, Ichthyostegopsis,* and *Acanthostega*) that appeared in the late Devonian (Fig. 3–3). These were in many ways transitional between crossopterygian fishes and later amphibians; it is thus apparent that the evolution of amphibians, a gradual process involving morphological, physiological, and behavioral adaptations to terrestrial life, had its beginning during the early to middle Devonian time.

The rhipidistian crossopterygians were strictly fresh-water carnivorous fishes that abounded on the bottom of relatively shallow bodies of water during the Devonian and Carboniferous periods, the last of them dying out early in the Permian. Rhipidistians, therefore, satisfy the ecological and temporal prerequisites of an amphibian ancestral stock. Importantly, numerous studies by paleontologists and comparative anatomists on osteological features of rhipidistians and ichthyostegids have produced an overwhelming body of evidence implicating rhipidistian crossopterygians as the ancestors of amphibians. The caudal fin of *Ichthyostega*, the best known of the ichthyostegids, was supported by both dermal fin rays and a double series of bones, just as in the crossopterygians; dermal fin rays do not occur in any other group of amphibians. The tail and ventral surfaces of *Ichthyostega* were covered with small scales that appear similar to those possessed by crossopterygians; as mentioned in the previous chapter, bony scales were quickly lost in the evolution of amphibians. Except for modifications associated with terrestrial life, the structure of the vertebral column of *Ichthyostega* was very similar to that of the rhipidistians. In both groups, the notochord provided the principal support for the trunk and tail, since it was large and only partially enclosed by rudimentary vertebral centra; in higher forms, greater ossification of the vertebrae occurs,

Figure 3–3 *Ichthyostega* from the late Devonian. (From *Tales Told by Fossils,* by Carroll Lane Fenton. Copyright, © 1966 by Carroll Lane Fenton. Reprinted by permission of Doubleday & Company, Inc.)

and they provide all the support. The three features of the vertebral columns of ichthyostegids that distinguished them from those of rhipidistians are characteristics of tetrapods but not of fish. First, the anterior neural spines, processes extending dorsally from the vertebrae, were vertical and were enlarged for the attachment of muscles supporting the head. Second, successive vertebrae were connected to one another by anterior and posterior zygapophyses, processes extending from the neural arch of one vertebra to articulate with that of the next, in such a way that the intervertebral joints were strengthened but the column was still allowed to bend. Third, each of the long broad trunk ribs had a double articulation with its vertebra, on the neural arch and on the hypocentrum; the space between was the vertebrarterial canal. In rhipidistians, the neural spines were smaller and sloped posteriorly, there were no zygapophyses, and there was a single articulation of rib to vertebra with no vertebrarterial canal.

The cranial anatomy of *Ichthyostega* closely resembled that of *Eusthenopteron*, a relatively large rhipidistian (compare Figures 3–4 and 3–5). The dermal roof components of the two were essentially the same, except that the parietal foramen of *Ichthyostega* was farther back, as it is in more advanced vertebrates. The palates of *Ichthyostega* and *Eusthenopteron* were composed of the same bones, including a short and narrow parasphenoid that is characteristic of crossopterygians. Rhipidistian crossopterygian skulls had a transverse hinge that allowed limited movement of the anterior and posterior parts on one another. It is signficant that such a division does not occur in any other vertebrate group, but traces of it, including a suture on the ventral side, are evident in the skull that was possessed by *Ichthyostega*. The dermal bones of the lower jaw (laterally the dentary, splenial, postsplenial, angular, and supra-angular; medially the prearticular and three coronoids) are entirely comparable in the two. Vestigial preopercular and subopercular bones were present at the posterior lateral angles of the skull of *Ichthyostega*, indicating the presence of a movable gill cover that enclosed the internal gills and

regulated the passage of water over them; these opercular elements do not occur in any other amphibian group. As in the rhipidistians, the lateral line system of the ichthyostegids passed within the dermal bones of the skull and connected to the surface by pores, whereas in later amphibians the system came to be housed in open grooves on the surface, not completely enclosed. Another rhipidistian character of *Ichthyostega* not found in later amphibians was the presence of rostrolaterals, a pair of small bones that crossed over the external nares and through which the infraorbital canal passed. Two cranial features of *Ichthyostega* represented initial steps in evolutionary trends of terrestrial vertebrates. One of these trends was a reduction in the number of bones making up the skull: in *Ichthyostega* this occurred in the broad roof over the snout (ethmoid shield), it being composed of a smaller number of larger bones than in crossopterygians. The other trend was for the posterior roof of the skull to be reduced in length compared to the anterior portions; compared to crossopterygians, this too was apparent in *Ichthyostega*. As is true of other early amphibians, the teeth of both ichthyostegids and rhipidistians were numerous and large, with greatly convoluted folds of dentine (labyrinthodont dentition).

As discussed in the preceding chapter, the stucture of the limbs and girdles of ichthyostegids is a major difference distinguishing them from crossopterygians. In contrast to the fins of crossopterygians, the appendages of even these very primitive amphibians were pentadactyl limbs. The greater development of the anterior limbs indicates they functioned primarily in crawling on shore, whereas the relatively small flipperlike posterior limbs functioned as fins for swimming; both kinds of locomotion were accomplished mostly by lateral undulations of the body and tail. The pectoral girdle lacked the connection to the skull occurring in fishes and also differed in including an interclavicle that is not found in fishes. The pelvic girdle consisted of a single bone on each side, as in fishes, but differed in that it surrounded the body and attached to a single sacral rib.

Although it is certain that rhipidistian

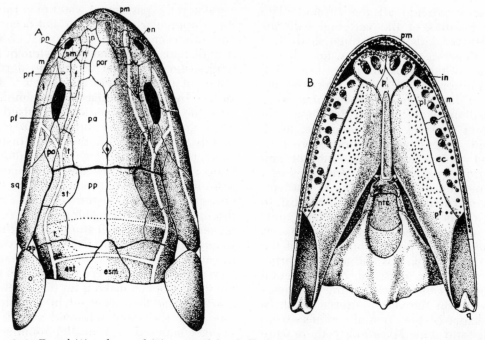

Figure 3–4 Dorsal (A) and ventral (B) views of the skull of *Eusthenopteron*, an Upper Devonian rhipidistian crossopterygian from North America. *ec*, ectopterygoid; *en*, external naris; *esl*, lateral extrascapular; *esm*, medial extrascapular; *f*, frontal (nasal); *in*, internal naris; *it*, intertemporal (postorbital or dermosphenotic); *j*, jugal; *m*, maxilla; *n*, nasal; *ntc*, area occupied by notochord; *o*, opercular; *p*, parasphenoid; *pa*, parietal (frontal); *pf*, postfrontal (supraorbital); *pl*, palatine; *pm*, premaxilla; *pn*, postnarial; *po*, postorbital; *por*, postrostral or internasal; *prf*, prefrontal; *pt*, pterygoid; *q*, quadrate; *r*, rostral; *sm*, supramaxilla; *sp*, spiracular cleft; *sq*, squamosal; *st*, supratemporal (dermal sphenotic); *v*, vomer. (From Romer: *Vertebrate Paleontology*.)

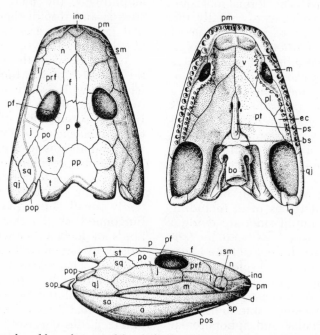

Figure 3–5 Dorsal, ventral, and lateral views of the skull of *Ichthyostega;* length of original about eight inches. *a*, angular; *bo*, basioccipital; *bs*, basisphenoid; *d*, dentary; *ec*, ectopterygoid; *f*, frontal; *ina*, internasal; *j*, jugal; *l*, lacrimal; *m*, maxilla; *n*, nasal; *p*, parietal; *pf*, postfrontal; *pl*, palatine; *pm*, premaxilla; *po*, postorbital; *pop*, postsplenial; *pp*, postparietal; *prf*, prefrontal; *ps*, parasphenoid; *pt*, pterygoid; *q*, quadrate; *qj*, quadratojugal; *sa*, surangular; *sm*, septomaxillary; *sop*, subopercular; *sp*, splenial; *sq*, squamosal; *st*, supratemporal; *t*, tabular; *v*, vomer. (From Romer: *Vertebrate Paleontology*.)

crossopterygians gave rise to amphibians, it is not known from which group, or groups, of these fishes they originated. Four closely related families of rhipidistians are currently recognized: they are Porolepidae, Holoptychidae, Osteolepidae, and Rhizodontidae. It appears that parallel, divergent, and possibly convergent evolution occurred in these families, but there seem to be two lines of descent, in which holoptychids evolved from porolepids and rhizodontids evolved from osteolepids. Accordingly, the rhipidistians have been treated as two suborders, of which Holoptychioidei (=Porolepiformes) includes the former two families and Osteolepioidei (=Osteolepiformes) the latter two. Although all taxonomic groups are classically treated as being monophyletic, some workers have suggested that amphibians are polyphyletic in origin and that the two suborders of rhipidistians represent two different ancestral groups. Jarvik (1942), emphasizing differences in the structure of the snout between Osteolepioidei and Holoptychioidei and similar differences between Urodela and Anura, concluded there is a connection between Anura and Osteolepioidei and between Urodela and Holoptychioidei. He recently (1960) has proposed that there may have been 10 independent origins of tetrapods, including separate origins of Apoda and Urodela from Holoptychioidei and separate origins of Anura and seven other tetrapod groups from the Osteolepioidei. In contrast, Schmalhausen (1968) has argued for a monophyletic origin for amphibians, from the Osteolepidae. For the most part, Jarvik's radical conclusions have been disputed, but the question of a monophyletic versus polyphyletic origin of amphibians is not settled by any means.

The two major groups of extinct amphibians, the labyrinthodonts and the lepospondyls, were fully distinct at the times of their first appearances in the fossil record. Furthermore, within the labyrinthodont assemblage, the ichthyostegids and *Hesperoherpeton* (Plesiopoda) seem to represent different modifications of crossopterygians, and neither is on a main line of evolution; the major temnospondyl and anthracosaur groupings differ in many ways from one another. The lepospondyls, which are most easily separated from labyrinthodonts by the structure of their vertebrae but which also had different ways of life, represent a diversified grouping of three orders for which there is no real evidence of a common ancestor. The diversification which characterizes the early amphibians could possibly reflect different origins, but it is also possible that it reflects the rapid, divergent evolution one expects when organisms are adapting to a new environment. Much of the issue of a monophyletic versus a polyphyletic origin revolves around the definition of an amphibian and how one treats transitional forms. Finally, if one accepts G. G. Simpson's definition of monophyly as the derivation of a taxon through one or more lineages from one immediately ancestral taxon of the same or lower rank, the class Amphibia must be monophyletic since there is no doubt as to its origin from the class Teleostome, the subclass Crossopterygii, and the order Rhipidistia. The question of monophyly versus polyphyly, then, is pertinent only when applied to particular groups of amphibians, living or extinct, rather than to amphibians as a class.

The rhipidistian ancestors of amphibians were better adapted to aquatic conditions than were even the most primitive amphibians, yet all evidence indicates that early tetrapods were primarily aquatic. This raises a question: Why did tetrapods become tetrapods with terrestrial adaptations when they spent most of their time in the same aquatic habitats as their rhipidistian ancestors? A number of ingenious theories have been proposed to explain the circumstances under which natural selection would favor adaptations to a terrestrial environment in what were basically aquatic animals. One school of thought is that there was a vast terrestrial niche available for vertebrates and that competition in the aquatic habitats was so severe that it placed a premium on the ability to become at least partially terrestrial. Another school of thought is that terrestrial adaptations evolved because they increased the probability for survival of their possessors as aquatic animals and not as terrestrial animals. This seemingly paradoxical conclusion becomes logical when one considers the environmental conditions that must have prevailed in fresh water habitats during the time that amphibians were evolving from rhipidistians. The Devonian period was one of great climatic fluctuation in which exceedingly wet periods were followed by severe droughts. Thus, there were alternations of abundance and scarcity of water such as occur in many areas of the tropics today. During droughts, obviously, many bodies of water must have become dry, stagnant, or overcrowded with competitors and predators so that the survival of an aquatic animal would be greatly enhanced if it could travel overland to reach larger and fresher bodies of water. The time required to produce the gradual change by natural selection from rhipidistian to amphibian indicates that conditions promoting terrestrial movements must have prevailed for many millions of years, during which amphibians progressively became more capable of spending longer and longer periods of time out of water. It should be noted, however, that amphibians have never become completely independent of the water. This was accomplished only with the evolution of reptiles.

CLASSIFICATION OF AMPHIBIANS

Living amphibians constitute a mere remnant of what was once a dominant and abundant group. This situation, combined with the enormous gaps in the fossil record, make the classification of Amphibia a particularly difficult problem. For years, great emphasis has been placed on the fact that two contrasting types of vertebral development occurred among extinct amphibians, and, accordingly, two subclasses have been recognized. The subclass Lepospondyli included a diversified group with "husk vertebrae" characterized by centra lacking a cartilaginous stage of development but rather being formed through direct deposition of cylinders of bone around the embryonic notochord. This type of vertebral development also occurs in living salamanders and caecilians. In contrast, the subclass Apsidospondyli contained the labyrinthodonts, all of which had "arch vertebrae" formed

embryonically as blocks of cartilage; in these, two sets of ossified arch-shaped structures formed in the central region of each vertebra, the hypocentra (intercentra) anteriorly and the pleurocentra posteriorly. Various modifications of the apsidospondyl type of development occurred among labyrinthodonts, and these differences have generally been used to define superorders within the subclass. Although there is disagreement among accounts of different workers, the vertebral development in anurans appears to be a modification of the apsidospondylous pattern, in which ossifications of the notochordal sheath and/or extensions of the bases of the neural arches down and around the notochord form the centra and replace the hypocentra and pleurocentra. Because of the two apparent patterns of vertebral development, in older systems of classification the class Amphibia comprised two subclasses: the subclass Lepospondyli included both living and extinct amphibians with "husk vertebrae;" and the subclass Apsidospondyli included anurans and labyrinthodonts with "arch vertebrae."

In recent years, greater attention has been paid to those features which are common to the three orders of living amphibians and less to differences in their vertebral development. There are numerous cranial characteristics which particularly suggest a common ancestry for modern amphibians. For example, there has been a great and comparable reduction in the number of bones in the skulls of all, the pineal opening is always lost, and the teeth, where present, are different from those of any other tetrapod in having the weak uncalcified segment between the base and crown. These and other similarities, plus the broad temporal and evolutionary gap between modern amphibians and the two major extinct groups have caused modern systematists to consider the three orders of existing amphibians plus the extinct order Proanura as constituting a natural group, the subclass Lissamphibia, and the class Amphibia to be composed of three subclasses. The subclass Labyrinthodontia includes the extinct amphibians with apsidospondylous vertebrae; the subclass Lepospondyli includes the fossil orders with lepospondylous vertebrae; the subclass Lissamphibia includes all modern orders.

The following is a compendium of the class Amphibia, based on Romer's (1966) classification. This is one of several taxonomic arrangements to be found in current literature, others may be found in the references listed at the end of this chapter.

CLASS AMPHIBIA

Subclass Labyrinthodontia

Order Ichthyostegalia—Caudal fin partially supported by dermal rays; snout short and rounded; posterior roof of skull relatively long; intertemporal absent. These oldest-known amphibians were primarily aquatic and in many ways intermediate between crossopterygians and later tetrapods. Their vertebrae are like those of the Temnospondyli, but specializations, such as the lack of an intertemporal which is present in many of the later amphibians, indicate that ichthyostegals were not on the main line of tetrapod evolution. Five genera in three families are recognized; they are Elpistostegidae (*Elpistostege* from the Upper Devonian of North America), Ichthyostegidae (*Ichthyostega* and *Ichthyostegopsis* from the Upper Devonian of Greenland, Fig. 3–3), and Otocratiidae (*Acanthostega* from the Upper Devonian of Greenland and *Otocratia* from the early Lower Carboniferous of Europe).

Order Temnospondyli—Pleurocentra reduced or absent, never forming complete discs around the notochord; intercentra larger, sometimes forming complete rings; tabular not contiguous with parietal; internal nares widely separated by broad vomers; laterosphenoid portion of braincase completely walled; four digits on front feet. The most extensive radiation of amphibians occurred in this group, which was widespread and abundant from the Devonian until the end of the Triassic, when it died out. The most primitive were fresh-water aquatic forms; subsequent evolution produced terrestrial and secondary aquatic kinds, at least some of which were probably marine in the Triassic (see superfamily Tremeatosauroidea).

Suborder Rhachitomi—Moderate pleurocentra present on either side between neural arches and intercentra; intercentra crescentic median ventral elements not surrounding notochord but forming majority of centrum; intertemporal present in early forms but absent in later members; one, two, or three occipital condyles present. These are the most primitive of the Temnospondyli.

Superfamily Loxommatoidea—Skull high and narrow; orbits enlarged and keyhole-shaped; palate movably attached to braincase and nearly lacking vacuities; intertemporal present; a single occipital condyle. This superfamily is represented by a single family (Loxommatidae) containing four genera known only from skull remains from Carboniferous deposits of Europe and North America. They are the most primitive of the Rhachitomi but appear to be an early side branch from the main line of rhachitomous evolution.

Superfamily Edopoidea—Skull depressed; palate with moderate interpterygoid openings; intertemporal present in some but absent in others. This grouping includes the following families: Edopidae (one or two genera from the lower Permian of North America and Europe); Dendrerpetontidae (three genera from the early Pennsylvania of North America and from the early Upper Carboniferous of Europe); Cochleosauridae (one genus from lower Permian of North America and eastern Europe and three genera from middle Pennsylvanian deposits of North America and Upper Carboniferous deposits of Europe); and Colosteidae (two genera from the middle Pennsylvania of North America). These were mainly aquatic forms that resembled the ichthyostegals.

Superfamily Trimerorhachoidea—Skull much flattened; face short, postorbital portion of skull elongated; palatal openings relatively large; intertemporal present; basal articulation of palate movable. This taxon includes the family Trimerorhachidae, with three genera from Pennsylvanian and Permian deposits of North America and Europe, and the family Dvinosauridae, with *Dvinosaurus* from the Permian of eastern Europe. These appear to have been purely aquatic forms with well-developed lateral line systems. External gills were retained in adults of *Dvinosaurus*, indicating that they were neotenic (sexually mature larval forms). In older accounts, these and other gilled forms were called "branchiosaurs" and placed in a separate order, the Phyllospondyli. The order Phyllospondyli is no longer recognized, since it obviously contained larval forms of a variety of amphibians and was not a natural group.

Superfamily Eryopoidea—Skull moderately flat; palate fused rigidly to braincase; relatively large openings in palate; double occipital condyle; limbs well-developed. Families included here are Eryopidae (eight genera from Pennsylvanian and Permian deposits of North America and Europe); Dissorophidae (14 genera from Upper Carboniferous, Permian, and Triassic deposits of North America, Europe, and South Africa; see Figure 3–8); Trematopsidae (three genera from the lower Permian of North America); Parioxycidae (*Parioxys* from lower Permian deposits of North America); Zatracheidae (four genera from the Upper Carboniferous and Permian of Europe and North America); Archeogosauridae (three genera from the Permian of Europe, Asia, and South America); Melanosauridae (three genera from the middle Permian

of Europe); and Intasuchidae (two genera from the middle Permian of eastern Europe). This was an advanced group of Rhachitomi which, in general, consisted of very predaceous forms with long snouts and powerful tusklike teeth. *Eryops*, probably the most common and best known member, was a large crocodile-like animal that attained lengths of six to eight feet and was truly amphibious (Fig. 3–6). It was a swamp-dweller but had stout limbs and girdles that certainly enabled it to move about on land. On the other hand, the archegosaurids were large fish-eating predators and secondarily aquatic Rhachitomi.

Superfamily Trematosauroidea—Skull high and narrow, triangular in shape; snout elongated and pointed; intertemporal absent; body long and slender; limbs reduced; vertebrae transitional between rhachitomous and stereospondylous types. This superfamily, known only from lower Triassic deposits, includes the family Trematosauridae, with 12 genera from North America, Europe, Asia, Africa, Madagascar, and Australia, and the family Rhytidosteidae, with three genera from Australia, Europe, and South Africa.

Figure 3–6 *Eryops*, one of the more advanced and abundant Rhachitomi from the Permian red beds of Texas, and *Diplocaulus*, a pond-dwelling nectridean. (From *Tales Told by Fossils*, by Carroll Lane Fenton. Copyright, © 1966 by Carroll Lane Fenton. Reprinted by permission of Doubleday & Company, Inc.)

Worldwide in distribution, the trematosaurs were specialized, secondarily aquatic fish-eaters. Trematosaur remains are generally associated with marine deposits, and it is thought for this reason that they were possibly marine animals. If this is true, they were certainly unique: no other amphibian group has been able to tolerate salt water, although a variety of living amphibians can tolerate brackish water and the crab-eating frog (*Rana cancrivora*) survives in full-strength sea water. Rather than being marine animals, it is also possible that the trematosaurs lived in large fresh-water rivers and that their bodies were carried by currents into salt water, where they were preserved.

Suborder Stereospondyli—Pleurocentra reduced or absent, seldom ossified; intercentra wall developed, forming complete rings in advanced forms; skull flat; intertemporal absent; palate attached to braincase by broad bar on either side of parasphenoid; braincase largely cartilaginous; occipital condyle double; limbs reduced (where known). The stereospondyls were a degenerate group of aquatic bottom-dwellers and form a continuum with the main line of advanced temnospondyls from which they obviously evolved.

Superfamily Rhinesuchoidea—Pleurocentra present but reduced; intercentra well developed but not forming complete rings; skull flattened with many unossified elements in braincase. There are three families recognized. They are Rhinesuchidae (five or six genera from the Permian of Africa, southern Asia, and eastern Europe); Lydekkerinidae (three or four genera from the Lower Triassic of South Africa); and Uranocentrodontidae (two or three genera from the Lower Triassic of Europe and South Africa). This group, frequently referred to as the "neorhachitomes," is not sharply distinct from advanced Rhachitomi but rather represents a transitional stage between the Rhachitomi and other Stereospondyli.

Superfamily Capitosauroidea—Pleurocentra absent; intercentra form complete rings in advanced members; snout long. The two families included here are Benthosuchidae (seven genera from the Lower Triassic of eastern Europe, southern Asia, and Madagascar) and Capitosauridae (16 genera from throughout the Triassic in Europe, southern Asia, Africa, Mada-gascar, Australia, North America, Greenland, and South America). This widespread group was one of the most common of the Triassic labyrinthodonts. It consisted of benthic forms with huge broad heads, orbits that were close to each other on the roof of the skull, and wide short bodies. In all their characteristics, these were adapted for lying quietly on the bottom where they could ambush prey rather than catch it by pursuit. The family Capitosauridae includes the largest known amphibian, *Mastodonsaurus*. This giant European labyrinthodont had a skull that was up to one meter in length.

Superfamily Brachyopoidea—Skull extremely broad and flat; snout short, postorbital part of skull elongated; vertebrae primitive stereospondyl type, with intercentra that do not form complete rings. This superfamily is a divergent group of about 14 genera, all in the family Brachyopidae, from late Permian and Lower Triassic deposits. Brachyopids were abundant throughout the Southern Hemisphere during the Triassic and have also been found in eastern Europe and North America. They apparently represent a side branch that split off from the main line of primitive Stereospondyli and died out near the end of the Triassic.

Superfamily Metoposauroidea—Skull broad and flat; snout short, postorbital part of skull elongated; vertebrae advanced stereospondyl type with intercentra that form rings. This group includes only the family Metoposauridae and the genus *Metoposaurus* from the Upper Triassic of Europe, southern Asia, and North America. The metoposaurs were an advanced group of "short-faced" stereospondyls that superficially resembled the trimerorhachid rhachitomes and appear to have been adapted to the same sort of an aquatic existence.

Suborder Plagiosauria—Skull very broad and short; body (where known) broad and flat and at least partially armored; vertebrae highly specialized, centrum a single cylindrical imperforate structure. This aberrant group includes the family Peltobatrachidae (*Peltobatrachus* from the later Permian of East Africa) and the family Plagiosauridae (four genera from the Upper Triassic of Europe). The plagiosaurs, relatively common in the late Triassic, were similar to the brachyopids and have been classified with them in the past. However, their highly specialized vertebrae indicate that they evolved for a long time as an independent group of Temnospondyli. *Peltobatrachus* is probably representative of the ancestral group that gave rise to the plagiosaurids, but their origin from other Temnospondyli is unknown. The fully armored *Gerrothorax*, in the family Plagiosauridae, reached lengths of three feet and had persistent external gills in the adult stage, indicating that it was neotenic.

Order Anthracosauria — Pleurocentra not reduced, generally forming majority of centrum — in advanced forms, complete discs enclosing the notochord; intercentra generally present but variable in size; tabular contiguous with parietal; internal nares close together, separated by narrow vomers; laterosphenoid portions of braincase not completely walled; where known, five digits on front feet. The anthracosaurs form a relatively small and poorly-known group compared to the temnospondyls. However, they are of particular interest since they are the group from which reptiles evolved and, accordingly, were ancestral to all higher vertebrates. The order Anthracosauria was itself short-lived. It first appeared during the Mississippian and became extinct late in the Permian. The relationship of anthracosaurians to temnospondyls is unknown.

Suborder Schizomeri — Pleurocentra of adult in form of paired half-rings on either side of notochord; intercentra large and functionally important; iliac element among rib muscles, free of pelvic girdle. This poorly known suborder is comprised of the oldest known anthracosaurians, *Papposaurus* and *Pholidogaster* in the family Pholidogasteridae, from the late Lower Carboniferous of Europe. It is presumed to be ancestral to the other suborders. The skeleton is known only for *Pholidogaster*; this had the proportions of a modern salamander, with a long trunk and tail and reduced limbs; it reached lengths of over a meter.

Suborder Diplomeri — Centrum of adult a complete ring formed by fused pleurocentra; intercentra large, forming incomplete rings around the notochord. The only member of this suborder is *Diplovertebron* of the family Diplovertebrontidae, known from the middle Upper Carboniferous of

Europe and North America. This was evidently a terrestrial amphibian with well-developed limbs and lizard-like proportions. It has sometimes been classified as a seymourian but its vertebrae clearly indicate it belonged to a more primitive group with a schizomer ancestry.

Suborder Embolomeri — Pleurocentra and intercentra both forming complete rings in adult; skull relatively high and narrow, frequently elongated; otic notch slitlike; neural arches normal, not broad and swollen. The two families included here are Anthracosauridae (*Anthracosaurus* from the Upper Carboniferous of Europe and North America and *Crassigyrinus* from the late Lower Carboniferous of Europe) and Cricotidae (about 12 genera from the Upper Carboniferous and Permian of North America and Europe). The embolomeres were quite abundant during the Carboniferous. They were aquatic fish-eaters with elongated bodies, strong tails, and reduced limbs and represented a divergent group of anthracosaurs that branched off at about the diplomer level.

Suborder Seymouriamorpha — Centrum of adult a complete ring formed by fused pleurocentra; intercentra especially large, wedge-shaped; skull relatively short; otic notch large and extended anteriorly; neural arches broad and swollen. This diversified suborder includes nine families. They are Seymouriidae (three genera from the lower and middle Permian of western Asia, eastern Europe, and North America); Kotlassiidae (three to five genera from the middle and upper Permian of eastern Europe); Discosauriscidae (three to four genera from the late Upper Carboniferous and lower to middle Permian of Europe); Chronisuchidae (*Chronisuchus* from the upper Permian of eastern Europe): Nycteroleteridae (*Nycteroleter* from the middle Permian of eastern Europe); Waggoneriidae (*Helodestes* and *Waggoneria* from the lower Permian of North America); Lanthanosuchidae (*Lanthanosuchus* from the middle Permian of eastern Europe); Tseajaiidae (*Tseajaia* from the lower Permian of North America); and Diadectidae (four genera from the upper Pennsylvanian and lower Permian of North America and the lower Permian of Europe). Typically, the seymouriamorphs were rather heavy-bodied with stout limbs. Although they clearly evolved from the embolomeres, they exhibit a combination of amphibian and reptilian characteristics; accordingly, their classification as amphibians has often been disputed. For example, *Seymouria* (family Seymouriidae, Fig. 3–7) had a skull that was identical with that of anthracosaurs and included an intertemporal; this bone is not present in reptiles. It had a single occipital condyle, as occurs in both primitive amphibians and in reptiles. The pectoral girdle included a coracoid; this bone is present in reptiles but not in typical amphibians. The broad and swollen neural arches are characteristic of primitive reptiles, not amphibians. *Seymouria* truly provides the evidence that there is no sharp morphological gap separating amphibians from reptiles.

The Diadectidae, highly specialized seymouriamorphs that were probably the first terrestrial vertebrate herbivores (Fig. 3–8), also possessed a combination of amphibian and reptilian characteristics and were borderline between the two classes. As stated earlier, the real differences between amphibians and reptiles lie in the kinds of eggs produced and their patterns of development: reptiles have amniotic eggs that are resistant to desiccation and may be laid on land; amphibians have anamniotic eggs that cannot resist desiccation and remain viable only in a moist environment. Thus, no matter what the adult form, amphibians typically have an aquatic larval stage, whereas reptiles lack

a distinct larval stage and develop directly into the adult form. The kind of egg and pattern of development possessed by *Seymouria* are not known, but the remains of gilled larvae of *Discosauriscus* (family Discosauriscidae) have been found. These indicate that the seymouriamorphs were definitely amphibians even though they were very close to the reptilian phylogenetic line.

Before leaving the subclass Labyrinthodontia, mention must be made of *Hesperoherpeton garnettense* Peabody. On the basis of a scapulocoracoid, neural arch, and rib fragment discovered in Pennsylvanian deposits of Kansas, this fishlike amphibian was first described as an anthracosaurian in the suborder Embolomeri, family Cricotidae. A second and more complete specimen, found later at the type locality, shows *Hesperoherpeton* had a combination of rhipidistian and embolomerous characteristics. It had forelimbs that ended in digits but retained posterior flanges on the axial bones as in Rhipidistia. The braincase, also as in rhipidistians, was composed of two sections. As was true of rhipidistians, *Hesperoherpeton* lacked an occipital condyle. Its amphibian characters include nares separated from the edge of the jaw, an anthracosaurian type of pectoral girdle, vertebrae like those of the ichthyostegids except for larger pleurocentra, and a stapes with an external process that probably articulated with a tympanic membrane and transmitted

Figure 3–7 *Seymouria*, an anthracosaurian which possessed a combination of reptilian and amphibian characteristics. (From *Tales Told by Fossils*, by Carroll Lane Fenton. Copyright, © 1966 by Carroll Lane Fenton. Reprinted by permission of Doubleday & Company, Inc.)

Figure 3–8 *Diadectes*, a highly specialized seymouriamorph (2) with *Varanosaurus*, a pelycosaurian reptile (1) and *Cacops*, a member of the Rhachitomi amphibian family Dissorophidae (3). (From *Tales Told by Fossils*, by Carroll Lane Fenton. Copyright, © 1966 by Carroll Lane Fenton. Reprinted by permission of Doubleday & Company, Inc.)

sound. As a result of this new information, *Hesperoherpeton* was placed in a new order, Plesiopoda, and a new family, Hesperoherpetonidae, by Eaton and Stewart (1960); this classification, however, has not been widely accepted, and the taxonomic status of this enigmatic amphibian is unsettled.

Subclass Lepospondyli

Order Nectridea — Tail adapted for swimming with high fin; caudal vertebrae with expanded fan-shaped neural and hemal spines; accessory articulations present between vertebrae; limbs and girdles reduced or entirely absent; when present, four digits on front feet and five digits on hind feet. This somewhat varied order includes three families: Urocordylidae (five genera from the lower Permian and middle Pennsylvanian of North America and from the early and middle Upper Carboniferous of Europe); Lepterpetontidae (*Lepterpeton* from the early Upper Carboniferous of Europe); and Keraterpetontidae (seven genera from throughout the Upper Carboniferous of Europe and North America and from the lower Permian of North America). The nectrideans were pond-dwelling amphibians that were

particularly abundant during the Pennsylvanian. Two morphologically distinct types were present: those in the family Urocordylidae, exemplified by *Urocordylus* and *Sauropleura*, were slender eel-shaped animals with long pointed heads and very reduced or no limbs; and those in the family Keraterpetontidae, exemplified by *Diplocaulus* (see Figure 3–6) and *Diploceraspis*, were larger animals (up to two feet in length) adapted for living on the bottom, with broad flat bodies and grotesque skulls in which the tabulars were extended posteriorly and laterally to form huge hornlike processes, giving the head the shape of a broad arrowhead. Although, as a group, these were the least specialized of the Lepospondyli, they provide no indication of a relationship to the Labyrinthodontia.

Order Aistopoda — Body elongate, snake-like, with up to 200 vertebrae and 100 ribs; ribs single-headed but forked; limbs and girdles completely missing, not even a rudiment present. This order includes two families and three described genera. The family Phlegethontiidae includes *Dolichosoma* (from the early Upper Carboniferous of Europe) and *Phlegethontia* (from middle Pennsylvanian to lower Permian deposits of North America). The family

Ophiderpetontidae includes only *Ophiderpeton* (from the early and middle Upper Carboniferous of Europe and the middle Pennsylvanian of North America). In addition, an undescribed aistopod has been found in Lower Carboniferous deposits of Scotland. The aistopods are puzzling in that their lack of limbs and girdles indicates a high degree of specialization, yet they are among the oldest known amphibians. The presence of the limbless aistopod in the Lower Carboniferous means that the ancestral form, presumably with four legs, must have been living in the Devonian, a period from which very few amphibians of any kind have been found and from which not one Lepospondyli is known. Aistopods were all aquatic and, in view of their age, represent a branch of amphibian evolution which departed from the transition toward terrestrial adaptation extremely early in the history of the class Amphibia. The relationships of this order to others are obscure. Of the Lepospondyli, only the nectrideans show any affinities, and they seem to be only distantly related.

Order Microsauria—Body with moderately elongated trunk but otherwise normal proportions; limbs generally feeble; front feet not known to have more than three digits each; skull roof primitive, with numerous dermal bones, but tabular is absent and very large supratemporals widely separate parietal and squamosal; postorbital portion of skull greatly elongated; otic notches absent; two occipital condyles present. There is considerable variation among microsaurs and this is reflected in the classification of the relatively small number of genera into six families. The family Adelogyrinidae includes two genera from the Lower Mississippian of Europe. Molgophidae includes three genera from the early Lower Carboniferous of Europe and the middle Pennsylvanian and lower Permian of North America. Lysorophidae includes two genera of wormlike microsaurs from middle Pennsylvanian to lower Permian deposits of North America; these have sometimes been classified as a separate order. Microbrachidae includes three genera from the lower Permian of Europe. Gymnarthridae, the largest family, is comprised of 10 genera from the lower Pennsylvanian and lower Permian of North America and the middle and late Upper Carboniferous

of Europe. Finally, Tuditanidae includes two to four genera from the early to middle Upper Carboniferous of Europe. The name Microsauria, meaning "little reptiles," alludes to the fact that these small generally salamander-like amphibians paralleled early reptiles and have sometimes been considered to be related to them, a belief that is no longer accepted. The confusion between microsaurs and primitive captorhinomorph reptiles (see Chapter Five) arose from superficial similarities in the skulls and the vertebrae of these two groups. Among the skull similarities are the general proportions, the shapes of many individual bones making up the skull roof, and a movable basal articulation of the palate. However, among other characters, captorhinomorphs differed from microsaurs in having small supratemporals and a single occipital condyle as well as fundamental differences in the structure of the otic region and the atlas. As discussed in the symposium "Evolution and Relationships of the Amphibia" (Olson, 1965), any similitude of microsaurs and captorhinomorph reptiles is now believed to reflect evolutionary convergence rather than developmental relationship. As a group, microsaurs were among the most abundant of swamp-dwelling amphibians during Carboniferous times. As in aistopods and some nectrideans, some of the microsaurs had reduced limbs. The extreme in this sort of specialization occurred in the lysorophids, which were about the size of and had the proportions of large angleworms; they also resembled living caecilians. It appears possible, although it cannot be demonstrated with existing evidence, that the microsaurs were ancestral to both modern salamanders and caecilians.

Subclass Lissamphibia

SUPERORDER SALIENTIA—Vertebral centra reduced or even absent, functionally replaced by downgrowths of the neural arches; skull with reduced number of bones; trunk shortened; iliac portion of pelvic girdle elongated.

Order Proanura—More than nine presacral vertebrae present; postsacral vertebrae not fused into coccyx (urostyle); radius and ulna not fused, tibia and fibula not fused; ilia elongated, but sacral diapophyses neither enlarged nor fused to sacrum;

elongated tarsals present which appear to represent distal, rather than proximal, elements (Hecht, 1962); skull essentially anuran-like with enlarged orbits and a large "frontoparietal" bone more complex and extensive than that of living anurans (Hecht, 1962). This order contains only *Triadobatrachus* (*Protobatrachus*), in the family Triadobatrachidae (Protobatrachidae), from the Lower Triassic of Madagascar (Fig. 3–9).

Triadobatrachus is an enigma since it includes a combination of primitive and advanced anuran characteristics as well as the features of a long tail and salientian-type hind limbs. It may be an adult aquatic form with saltatorial preadaptations, it may be an adult terrestrial form with saltatorial adaptations, or it may be a tadpole or metamorphosing tadpole (Hecht, 1962). As proposed by Piveteau (1937), it has been classically described as an evolutionary intermediate stage between the ancestral amphibians and modern anurans; however, this is now being questioned. Piveteau considered *Triadobatrachus* an adult terrestrial form which represented a group ancestral to modern anurans. Griffiths (1956) has noted the incompatibility of the tail and hind limbs and, further, the inconsistency between hind limbs adapted for transmitting thrust against the vertebral column and the absence of a functional sacrum; he has concluded that the specimen is a metamorphosing tadpole and that this is why the tail is present and the pelvic girdle is not attached to the sacral vertebra. Hecht (1962) considers the tadpole interpretation of *Triadobatrachus* unlikely because of the well-ossified skeleton and suggested that, if it was ancestral to modern anurans, it probably was an aquatic form which swam by a sculling type of movement of the hind limbs. Because of the questionable homologies of the enlarged tarsals of *Triadobatrachus* with those of modern anurans (and other differences), Hecht (1962, 1963) questions whether or not *Triadobatrachus* is even on the ancestral line of anurans. Romer (1966, 1968), emphasizing similarities of the skull to that of anurans, believes that *Triadobatrachus* was related to anuran ancestry whether or not it was a direct progenitor. It does seem to represent a distinct group, but it gives no indication of a relationship to any older group of amphibians.

Order Anura—Tail absent in adult; postsacral vertebrae fused to form coccyx; five to nine presacral vertebrae present; radius fused to ulna, tibia fused to fibula; no distinct neck region; hind legs adapted for swimming and/or jumping, considerably larger than forelimbs; tympanum usually well-developed; voice generally well-developed as result of presence of larynx; eyes well-developed and possess movable eyelids, lower lid usually more movable than upper; fertilization generally external; reproduction oviparous in all members except *Nectophrynoides* (family Bufonidae), which is viviparous. The anurans have been the most successful of the lissamphibians, both in terms of numbers and in terms of adaptive radiation (Fig. 3–10). The 17 families presently recognized are tentatively arranged into three suborders, primarily on the

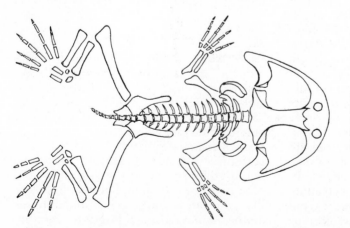

Figure 3–9 A sketch restoration of *Triadobatrachus* (*Protobatrachus*), a Triassic froglike form; data from Piveteau; feet, front of skull, etc., restored. (From Romer: *Vertebrate Paleontology.*)

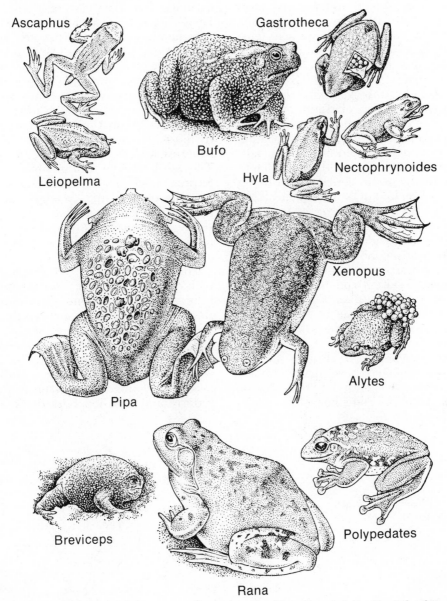

Figure 3–10 Various anuran amphibians, not all to one scale. (From Young, J. Z.: *The Life of Vertebrates.*)

basis of tadpole morphology and osteological characters.

Suborder Archaeobatrachia—Free ribs (not fused to vertebrae) present; eight or nine presacral vertebrae present; tadpole (not known for fossil forms) with elaborate mouth parts, including wide lips, horny beaks, and labial teeth in transverse rows which generally include more than one line of teeth; single median spiracle. This suborder is believed to be the most primitive and includes three families, one of which is extinct. The family Notobatrachidae includes only *Notobatrachus* from the Middle Jurassic of South America. Ascaphidae, distinguished from all other Recent families by the presence of nine presacral vertebrae and a pair of tail-wagging muscles (caudalipubioschiotibialis) in the adult, contains two living genera, *Leiopelma* of New Zealand highlands and *Ascaphus* of North American mountains in the Pacific Northwest. The family Discoglossidae is identified by a triradiate cartilaginous sternum and a complex, interleaving zonosternal mechanism (Fig. 3–11); it contains four living Old World genera plus *Eodiscoglossus* from the Upper Jurassic of Europe and four other genera from the Miocene and later Oligocene of Europe.

Suborder Aglossa—Definitive tongue lacking or attached posteriorly and not fully protrusible; sternum reduced or absent; vertebrae opisthocoelous (cen-trum convex anteriorly and concave posteriorly); eight presacral vertebrae present; ribs not free in adult, tending toward reduction; tadpole with a pair of spiracles, right and left gill chambers separate; tadpole mouth a simple slit with narrow labial folds and long barbels; Meckel's cartilage long, supporting entire lower jaw. This suborder includes two primitive families that have become specialized in opposite ways. The family Pipidae, distinguished by the lack of a tongue and the possession of a single median opening of the eustachian tubes, includes four living genera of aquatic "toads" which sometimes retain the lateral line system in the adult; one genus occurs in South America and three genera are in Africa. The most familiar of the pipids are the Surinam toad, *Pipa pipa*, of South America and the clawed frog, *Xenopus laevis*, of Africa; the latter has been used extensively in pregnancy tests and is becoming more and more popular as a laboratory animal.

The second family, Rhinophrynidae, includes the extinct *Eorhinophrynus* from the middle Eocene of Wyoming and *Rhinophrynus canadensis* from the lower Oligocene of Canada plus one living species, the burrowing toad, *Rhinophrynus dorsalis*. Rhinophrynids are the only anurans to have an arciferal girdle and lack a sternum; their tongue is free anteriorly and attached posteriorly. *Rhino-*

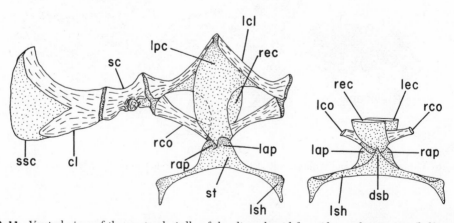

Figure 3–11 Ventral view of the pectoral girdle of the discoglossid frog *Alytes obstetricans* (left) and dorsal view of reconstruction of the zonosternal junction of the same girdle (right). *cl*, cleithrum; *dsb*, dorsal sternal blade; *lap*, left alary process; *lcl*, left clavicle; *lco*, left coracoid; *lec*, left epicoracoid cartilage; *lpc*, left precoracoid cartilage; *lsh*, left sternal horn; *rap*, right alary process; *rco*, right coracoid; *rec*, right epicoracoid cartilage; *sc*, scapula; *sf*, sternal flange; *ssc*, suprascapular cartilage; *st*, sternum. (After Griffiths: The phylogeny of the Salientia.)

phrynus dorsalis exists today in dry regions between the Rio Grande Valley of Texas and Costa Rica; it is a peculiar burrowing form that apparently feeds almost exclusively on termites and returns to water only to breed.

Suborder Neobatrachia — Ribs never present at any stage in life; vertebral column procoelous (centra all concave anteriorly and convex posteriorly), amphicoelous (centra all concave both anteriorly and posteriorly), or diplasiocoelous (the first presacral vertebra is amphicoelous and the remaining trunk vertebrae are procoelous); tadpole with a single spiracle draining both gill chambers and opening midventrally or on the left side; tadpole mouth parts variable, relatively simple or complex. Twelve families are included in the suborder Neobatrachia, and it contains the majority of living amphibians.

The family Pelobatidae includes about 16 genera, of which eight are living. Living members are widely distributed in the Northern Hemisphere and also extend into Southeast Asia and the East Indies. Fossil pelobatids are known from the Miocene, Oligocene, and Eocene of Europe and from the Pliocene, Miocene, Oligocene, and Eocene of North America. Members of this family possess a combination of primitive and advanced characteristics and are often suggested to be phylogenetically intermediate between the two evolutionary lines represented by the "primitive frogs" in the first two suborders and the "advanced frogs" that compose the remainder of this suborder. Estes (1970) has suggested that the Pelobatidae probably differentiated from a discoglossid-like ancestor in the Holarctic middle-latitude tropics. Pelobatids have true teeth in the upper jaw, an arciferal pectoral girdle, expanded sacral diapophyses, a coccyx which is either fused to the sacral vertebra or articulates with it by a single condyle, and separate astragalus and calcaneum tarsal bones.

The family Pelodytidae seems to be more closely related to the pelobatids than to any of the other living families; it is considered by Griffiths (1963) to be a subfamily of Pelobatidae. This small taxon includes only three genera: the living European *Pelodytes; Miopelodytes* from the Miocene of North America; and *Pro-*

pelodytes from the Eocene of Europe. Pelodytidae is distinguished from Pelobatidae by a calcaneum and an astragalus that are fused into a single bone and a coccyx that articulates with the sacral vertebra by a double condyle.

There is considerable difference of opinion about the phylogeny of the remainder of the anurans and there is little point in further speculation about their relationships until more evidence is available.

The family Leptodactylidae includes about 57 genera of southern frogs, of which eight are known as fossils dating back to the Eocene of India and South America. Members of this family typically have procoelous vertebral columns and a cartilaginous prezonal element to the pectoral girdle; the pectoral girdle is generally arciferal with posteriorly projecting epicoracoid horns; maxillary teeth may or may not be present; and Bidder's organ is lacking. The leptodactylids are a very diversified group and share anuran dominance with the Hylidae in both the Australian region and South America; five species are in southern United States and *Helophryne* is restricted to South Africa. The adaptive radiation of leptodactylids in Australia and South America has resulted in their occupying nearly every ecological niche available to anurans. Some of the South American workers would split the Leptodactylidae into two or more families.

In contrast to the Leptodactylidae, the family Bufonidae has not diversified greatly, so its members are generally easy to recognize. Bufonids are arciferous with well-developed epicoracoid processes and have procoelous vertebrae, as in Leptodactylidae, but Bidder's organ is present. The family Bufonidae includes 13 living genera and three to eight fossil genera of true toads. Fossil bufonids are known from Europe, Asia, Africa, North America, and South America; the oldest are from the Eocene of southern Asia, Europe, and South America. Living bufonids are native all over the world except in New Guinea, the Polynesian islands, Australia, and Antarctica, but only the genus *Bufo* approaches a cosmopolitan distribution and is the only one in North America; *Bufo marinus*, the marine toad, is now abundant where introduced into

Australia, New Guinea, and Pacific islands. *Atelopus, Dendrophryniscus, Melanophryniscus,* and *Oreophrynella* occur in South America; the remaining genera are in Africa, southern Asia, and the East Indies.

The family Pseudidae includes two living genera and five species of aquatic South American frogs; no fossils have been assigned to this family. These frogs have an extra phalanx on each digit, an apparent adaptation for swimming, and opposable thumbs; their vertebral columns are procoelous; the sacral diapophyses are cylindrical; maxillary teeth are present. The paradoxical frog, *Pseudis paradoxa,* is the best known of the pseudids; it has attracted considerable attention because of the huge size of its tadpoles. The paradox alluded to by the specific name is that the tadpoles are known to reach lengths of more than 180 millimeters but metamorphose into frogs that, as adults, are only about 75 millimeters in length. It should be noted that some authorities consider the Pseudidae merely a group within the Hylidae.

The family Hylidae contains about 32 genera and several hundred species of tree frogs; 3 genera are known as fossils, the oldest dating back to the Miocene of North America. Although living tree frogs occur in Europe, Asia, North Africa, New Guinea (Papua), Australia, North America, and South America, the family is primarily a New World grouping. Thirty of the genera are found in the New World, and the family is especially abundant and diversified in South America; only *Hyla,* which occurs in Eurasia, and *Litoria,* known only from Papua and Australia, are found outside the New World. As their vernacular name implies, most hylids are arboreal and have enlarged discs on the tips of all digits that act as suction cups and enable them to climb almost any surface. Others, such as the cricket frogs (*Acris*) and chorus frogs (*Pseudacris*) of North America and some Australopapuan *Litoria* have reverted to a terrestrial life, living in grasslands and marshlands, and have reduced digital discs; Hylids are procoelous and share many characteristics with leptodactylids; they differ in possessing intercalary cartilages. They are one of the most interesting groups of amphibians; they are great acrobats, are frequently brilliantly colored, and many have very specialized reproductive patterns.

Another small group of arboreal frogs constitutes the family Centrolenidae, considered to be a subfamily within Hylidae by some workers. This family includes two living genera of small, bright green frogs that are distributed from southern Mexico through Central America to Brazil and Bolivia. No fossils have been assigned to this family. Centrolenids are similar in appearance to tree frogs and were originally included in the family Hylidae. However, they have some unique osteological characters, including fused tarsal bones, which form the basis for their placement in a separate family.

The family Brachycephalidae (McDiarmid, 1969) includes only the tiny gold frog, *Brachycephalus ephippium,* that is less than 20 millimeters long and is known only from eastern Brazil. As mentioned in Chapter Two, this species is characterized by a broad dorsal bony shield which is confluent with the processes of the second to seventh vertebrae. It lacks the omosternum, has a cartilaginous sternum, and has weakly dilated sacral diapophyses.

The family Ranidae, the true frogs, is a very large grouping that occurs on all major land masses except Antarctica, but only the genus *Rana* approaches a cosmopolitan distribution; it is the only genus in North America. Four or five fossil genera of ranids are known, and the oldest of these, also *Rana,* dates back to the Miocene of Europe and North America. Thirty-six genera of living ranids are recognized, and these include many hundreds of species; there are about 400 species of *Rana* alone. Ranids are diplasiocoelous, have a firmisternous pectoral girdle (epicoracoid processes are never present) with an ossified prezonal element, and lack intercalary cartilages. The following eight subfamilies have been recognized by various authors: Arthroleptinae (about six genera of African ranids); Sooglossinae (*Sooglossus* and *Nesomantis* of the Seychelles Islands); Dendrobatinae (*Dendrobates, Phyllobates,* and *Prostherapis* of tropical America); Astylosterninae (*Astylosternus, Gampsosteonyx, Nyctibates,* and *Scotobleps,* all of West Africa); Phrynopsinae (*Leptodactylodon* and *Phrynopsis* of Cameroun, Africa); Raninae (*Rana* and about six closely related genera); Petropedetinae (*Arthroleptides* and *Petropedetes*, both of Africa); and Cornuferinae (about 10 genera distributed from Africa across the Indo-Australian Archipelago to the Solomon and Fiji Islands and from China to northern Australia). The Dendrobatinae are sometimes treated as a separate family, Dendrobatidae. Griffiths (1963) recognized the Sooglossinae as a separate family, Sooglossidae.

The Family Rhacophoridae is an Old World group inhabiting Southeast Asia, Japan, the East Indies, Africa, and Madagascar. There are 14 genera and 89 species of living rhacophorids (Liem, 1970). The only fossils known are from the Pleistocene of Asia; they have been assigned to the living genus *Rhacophorus*. Rhacophorids are mostly arboreal and are similar to tree frogs in their adaptations. They are distinguished by the following characteristics: a firmisternous pectoral girdle, a metasternum consisting of a bony style, and intercalary cartilages between the two distal phalanges of both fingers and toes; vocal sacs are generally present, the base of the omosternum is generally forked, and three labial tooth rows are generally present on the lower lip of tadpoles.

The family Hyperoliidae is distributed in Africa south of the Sahara, Madagascar, and the Seychelles Islands. There are 14 genera and 52 species in this grouping and they are recognized by a firmisternous pectoral girdle, a broad cartilaginous platelike metasternum, an omosternum which is broadly forked at the base, and intercalary cartilages between the two distal phalanges of all digits; vocal sacs are generally present, and there are generally one uninterrupted row of labial teeth on the upper lip (rarely more than two rows) and up to three rows on the lower lip of tadpoles.

The family Microhylidae includes 56 genera of terrestrial and arboreal frogs which are widely distributed throughout both the Old and New World tropics. Microhylids are firmisternal, have a reduced pectoral girdle with neither prezonal nor sternal ossifications, have a paired ethmoid in the adult, and frequently lack teeth. Intercalary cartilages are present in the subfamily Phrynomerinae but are lacking in others. Sacral diapophyses are more or less dilated in all members. Fossils assigned to this family date back to the Miocene of Europe, Africa, and North America.

SUPERORDER CAUDATA—Lissamphibians with functional lepospondylous vertebral centra formed by direct deposition of bone around the notochord and not preceded by cartilaginous elements; frontal and parietal not fused together; orbits not enlarged; trunk not shortened; when present, iliac portion of pelvic girdle not elongated; when present, hind legs about the same length as forelegs, no greatly elongated tarsal elements present.

Order Urodela (Caudata)—Limbs present; no tentacular organ; no dermal scales. This order includes eight families of living urodeles and four extinct named families; three genera of fossil forms are presently assigned to two unknown families. Fossil representatives are known for all of the living families.

Compared to anurans and apodans, the urodeles have remained unspecialized; their gross anatomy bears the closest resemblance to that of primitive amphibians (Fig. 3–12). They have much in common with the family Dissorophidae of the Rhachitomi superfamily Edopoidea but, as with the rest of the Lissamphibia, there is insufficient evidence at this time to positively identify any urodele ancestor from among the Paleozoic groups. Although some species have become arboreal and others fossorial, modern salamanders are basically terrestrial, aquatic, or get along well in either kind of habitat. It, therefore, seems logical to assume that they have descended from a metamorphosing ancestral stock that was primarily aquatic and that terrestrial adaptations have evolved independently in a number of different groups.

Noble (1931) considered the urodele families he know to constitute five natural groups, which he designated suborders. His scheme, which has formed the foundation for subsequent classifications, is as follows:

Suborder Cryptobranchoidea—Angular and prearticular bones of lower jaw not fused; premaxillary spine short, not separating nasals; the second epibranchial is retained at metamorphosis; fertilization external (not known for extinct forms). The three families placed here are Crypto-

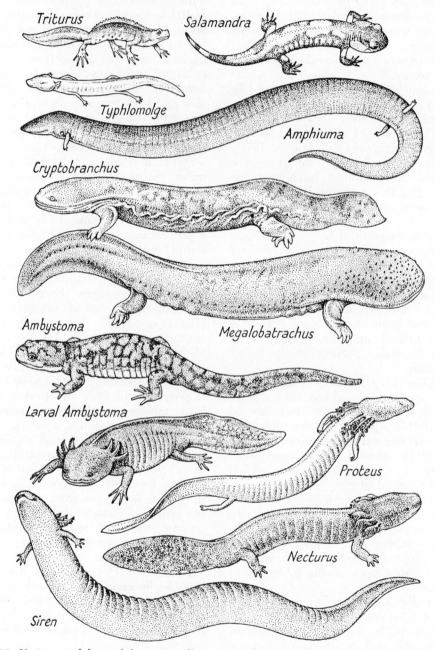

Figure 3-12   Various urodele amphibians, not all to same scale. (From Young, J. Z.: *The Life of Vertebrates*.)

branchidae, Scapherpetontidae (extinct), and Hynobiidae.

Suborder Ambystomatoidea — Angular fused to prearticular; premaxillary spines elongated; second epibranchial lost at metamorphosis; prevomers short and lacking posterior processes; vertebrae amphicoelous; fertilization internal (not known for extinct forms). Following Noble (1931), this suborder includes only the family Ambystomatidae.

Suborder Salamandroidea — Metamorphosed adults with teeth on roof of palate behind the internal nares, these being on processes of the prevomers lying on either side of the parasphenoid or else in one or two groups lying directly on the parasphenoid; vertebrae opisthocoelous; fertilization internal (not known for extinct forms). This is the most diversified and widely distributed of Noble's suborders, and more recent workers have questioned whether or not it represents a natural grouping. Noble included four families here: Salamandridae, Amphiumidae, Plethodontidae, and the extinct Batrachosauroididae.

Suborder Proteida — Permanently larval aquatic salamanders that develop lungs but retain three pairs of elongated gills and two pairs of gill openings as adults; ypsiloid apparatus, maxillary bones, and eyelids absent; premaxillary spines long; angular and prearticular fused; fertilization internal. In Noble's scheme, this includes only the family Proteidae.

Suborder Meantes — Permanently larval aquatic salamanders that develop lungs but retain three pairs of external gills and one or three pairs of gill openings as adults; maxillary bones, eyelids, and cloacal glands lacking; hind legs entirely lacking, front legs greatly reduced; caudal vertebrae with expanded neural arches; upper and lower jaw toothless, covered with horn; eyes small; fertilization probably internal. This grouping includes only the family Sirenidae.

Classifications of urodeles which have been proposed since Noble's time reflect increased knowledge of relationships, such as has come from the osteological studies of ambystomatids (Tihen, 1958) and plethodontids (Wake, 1966), and the discovery of additional fossil forms. The most modern and complete classification of Recent and fossil urodeles is that of

Brame (1967), who groups 12 living and extinct families into five suborders. The following annotated classification of the order Urodela is based upon the scheme proposed by Brame.

Suborder Unknown — Brame has added this unknown suborder to include the family Vaughniellidae (*Vaughniella urodeloides*) from the early Permian of New Mexico and the family Hylaeobatrachidae (*Hylaeobatrachus croyi*) from the Lower Cretaceous of Belgium. Romer (1966) has placed the former in the suborder Rhachitomi, superfamily Eryopoidea, family Dissorophidae and the latter in the urodele family Proteidae, both questionably.

Suborder Cryptobranchoidea — Down to family, this suborder is as constituted by Noble (1931). The family Hynobiidae includes five genera of living Asiatic salamanders plus *Wolterstorffiella* from the Paleocene of Germany; the latter has been tentatively placed in the family Ambystomatidae by Romer (1966). *Hynobius* has the widest distribution of the genera and is found from Japan to northern Siberia, Turkestan, and the Ural Mountains of Russia. The other living genera occur within the range of *Hynobius* and apparently all evolved from it. Hynobiids undergo a more complete metamorphosis than cryptobranchids. Adults develop eyelids and mature teeth but, at least in *Hynobius shihi*, rudiments of gills may be retained.

The family Cryptobranchidae contains two living species of giant salamanders (*Andrias*), one in China and the other in Japan, and *Cryptobranchus alleganiensis*, the hellbender of eastern United States. Although the family is now only represented by three species, it was once more extensive. Fossils assigned to the genus *Andrias* have been commonly found from the Oligocene, Miocene, and Pliocene of Europe, the Miocene of North America, and the Pleistocene of eastern Asia. *Zaisanurus* is known from the lower Miocene and possibly the Oligocene of Kazakstan, U.S.S.R. The oldest known fossil representative of the family is *Piceoerpeton willwoodensis* from the lower Eocene of Wyoming. Cryptobranchids are aquatic animals that never complete metamorphosis: although gills are lost in the adult, larval teeth are retained and eyelids never develop.

The family Scapherpetontidae, considered by some to be part of Cryptobranchidae, includes only two extinct genera, *Scapherpeton* and *Lisserpeton*, both of which are abundant in Upper Cretaceous and Paleocene deposits of North America. These had poorly ossified skulls and also may have only partially metamorphosed.

Suborder Sirenoidea—This name has been applied to Noble's suborder Meantes and the grouping includes only the family Sirenidae with two living genera: *Siren*, with two species, and *Pseudobranchus*, with one species, the dwarf siren. Sirens are distributed from eastern United States into Mexico; the dwarf siren (*Pseudobranchus striatus*) is known only from Florida, Georgia, and South Carolina. Fossils assigned to this family (*Habrosaurus, Prosiren, Pseudobranchus*, and *Siren*) are known from North America and date back to the Cretaceous. Goin and Goin (1962) have suggested that the sirenids be placed in a separate order, Trachystomata, because of peculiarities of the skull and vertebrae, but the validity of this assignment is widely questioned.

Suborder Salamandroidea—This grouping has been revised considerably. It now includes five families, three of which are living; the family Plethodontidae has been removed from it and placed in the suborder Ambystomatoidea; the family Proteidae, rather than being in a separate suborder, has been assigned to Salamandroidea. Proteidae includes two living genera, and three other genera are known only as fossils. The living genera are *Proteus*, the European olm, and *Necturus*, the mudpuppies and waterdogs of eastern United States. There is only one species of olm (*Proteus anguineus*), and this is a blind cave salamander lacking skin pigment. Five species of *Necturus* are currently recognized; these inhabit Atlantic and Gulf drainages and are pigmented. Fossil proteids are known from the Eocene and Miocene of Europe.

Comonecturoides marshi, from the upper Jurassic of Wyoming, has been placed in a separate but unknown family by Brame; Romer (1966) questionably has placed this fossil species in the family Proteidae.

The family Salamandridae, the newts, includes 15 living genera plus 18 genera known only as fossils. The living genera, represented by about 42 species, occur in North America, Europe, Asia, and North Africa; fossils are known from Upper Cretaceous, Paleocene, Eocene, Oligocene, and Miocene to Recent deposits of Europe, from the upper Miocene of China, and from Oligocene and Miocene deposits of North America. Newts are all generalized salamanders with well-developed limbs and tail; the large number of genera reflects their wide distribution more than a high degree of diversification. Among other characters, newts differ from other members of the suborder in possessing both an ypsiloid cartilage and lungs.

The family Batrachosauroididae includes only the extinct *Batrachosauroides* from the upper Miocene of Texas. Limited evidence indicates that this genus was permanently larval.

The family Amphiumidae includes only one genus, *Amphiuma*, the congo eels of southeastern United States. There are three living species currently recognized, and Brame lists the fossil *Amphiuma norica* from the Pleistocene of Germany. Congo eels are semilarval aquatic animals with long slender bodies and very tiny limbs. At metamorphosis, adults lose their gills but one pair of gill slits remains open and eyelids fail to develop. Amphiumids have lungs but no ypsiloid cartilage.

Suborder Ambystomatoidea—In addition to the family Ambystomatidae, Brame includes an unknown family of fossils and the family Plethodontidae in this suborder. The family Ambystomatidae, the mole salamanders, includes four living genera and two extinct genera. *Ambystomichnus montanensis* is known from the Paleocene of Montana, fossil *Ambystoma* are known from Pliocene deposits of North America, and *Bargmannia wettsteini* is known from the Miocene of Europe. Living members of the family occur only in North America, and the tiger salamander (*Ambystoma tigrinum*) is the only widely distributed species; it is found from southern Canada into Mexico and over most of the United States. The other species and genera are concentrated in the Pacific Northwest (*Dicamptodon, Rhyacotriton, Ambystoma*), eastern United States (*Ambystoma*), and the central plateau of Mexico (*Ambystoma, Rhyacosiredon*). Populations of mole salamanders are often neotenic, and this has confused their taxonomy. For years, the Mexican axolotls were considered to be a distinct genus, *Siredon*, of permanently larval salamanders, but they have been found to be typical ambystomas when induced to metamorphose. *Dicamptodon* and *Ambystoma* are commonly neotenic in local populations, and *Rhyacosiredon* may have neotenic populations. Neotenic individuals are of course aquatic, but generally metamorphosed adults of all species are terrestrial. Tihen (1958) has suggested that the Ambystomatidae evolved from Asiatic Hynobiidae in the late Mesozoic or very early Tertiary. Ambystomatid salamanders are characterized by having vomerine teeth in transverse series; parasphenoid teeth are absent; the ypsiloid cartilage is present; the carpus and tarsus are ossified.

Dehmiella schindewolfi, from the Miocene of Germany, and *Geyeriella mertensi*, from the Paleocene of Germany, are placed in a separate, unknown, extinct family by Brame. Romer (1966) places the former questionably in the family Plethodontidae and the latter questionably in the family Ambystomatidae.

The family Plethodontidae, the lungless salamanders, has the largest number of genera (25, two of which are extinct) and species (183 living) of all salamander families and is so diversified that the two subfamilies, Plethodontinae and Desmognathinae, have sometimes been considered to be different families. Except for the relict *Hydromantes genei*, which occurs on the Italian Island of Sardinia, and *Hydromantes italicus*, which survives in southeastern France and Italy, all living plethodontids occur in the New World; three other species of *Hydromantes* occur at moderate to high elevations in California. Fossils assigned to the extinct *Prodesmodon* (Desmognathinae) are known from the Miocene of Europe and the Lower to Upper Cretaceous of North America (Romer, 1966); all other fossil plethodontids (*Opisthotriton* and *Plethodon*) are from North America and date back to the Upper Cretaceous. Many plethodontids are terrestrial and many are aquatic, adapted especially to swift streams, a few species live in subterranean water and caves, and some are arboreal. As their vernacular name implies, plethodontids lack lungs in the adult form; they

also lack the ypsiloid cartilage. Wake (1966) proposes that the plethodontids arose from ambystomatid ancestors in eastern North America and that the subfamily Desmognathinae is a specialized early derivative of the ancestral stock.

Order Apoda (Caecilia, Gymnophiona)— Elongated burrowing amphibians in which all traces of limbs and girdles have been lost; dermal scales embedded in skin of some genera; tail very reduced, vent near posterior end of body; eyes reduced and generally buried under pigmented skin, sometimes under bones of skull; sensory tentacular organ arising from side of brain and protruding through orbit or an aperture separate from eye; length ranging from 65 millimeters in smallest species to 1500 millimeters in largest; fertilization internal, cloaca of male modified into protrusible copulatory organ; reproduction oviparous, ovoviviparous, or viviparous. Only two fossil apodans have been described (*Ichthyophis muelleri* and *Prohypogeophis tunariensis*), and their relationship to living or extinct forms is uncertain. Following Taylor (1969), living members of the order comprise four families.

The family Typhlonectidae includes four genera (*Typhlonectes, Potomotyphlus, Nectocaecilia, Chthonerpeton*) and about 12 species of aquatic forms inhabiting both streams and ponds; no tail is present; no scales are present; reproduction is either ovoviviparous or viviparous; young develop broad baglike external gills while in the uterus (all larvae of other families have three pairs of branched gills); gill slits close before birth. This family is known only from tropical South America.

The family Caeciliidae comprises 19 genera of apodans and may not represent a natural grouping. Members of this family lack septomaxillae, prefrontals, and oculars; their premaxillae and nasals are fused to form two nasopremaxillae; the stapes is present; scales may or may not be present; reproduction is oviparous or ovoviviparous, and there may or may not be a free-swimming aquatic larval stage. Members of this family are found in the tropics and subtropics of both the Old World and the New World, throughout the range of the order.

The family Ichthyophiidae includes

Figure 3–13 Above, Panamanian caecilian (*Caecilia ochrocephala*) and, below, a South American caecilian (*Siphonops annulatus*). (From Cochran, D. M.: *Living Amphibians of the World;* photographs by Lorus and Margery Milne [above] and by Carl Gans [below].)

| Order | Families | | Genera | | Species |
	Extinct	Living	Extinct	Living	Living
Apoda	0	4	1	20	150
Urodela	5	8	35	54	316
Anura	1	16	38	218	2600
Proanura	1	0	1	0	0
Microsauria	6	0	21	0	0
Aistopoda	2	0	3	0	0
Nectridea	3	0	13	0	0
Anthracosauria	12	0	39	0	0
Temnospondyli	26	0	128	0	0
Ichthyostegalia	3	0	5	0	0

four genera (*Caudacaecilia, Rhinatrema, Ichthyophis,* and *Epicrinops*) in which the septomaxillae, premaxillae, and nasals are present as separate bones; prefrontals, oculars, and stapes are present; a tail is present; a relatively well-developed eye is present in the socket; scales are present. Adults are terrestrial and have numerous splenial teeth. Reproduction is oviparous, eggs are deposited on land, and larvae have a free-swimming aquatic stage. The family includes about 43 species distributed in the New World tropics, Southeast Asia, the southern Philippines, and the western Indo-Australian Archipelago.

The family Scolecomophidae includes only the genus *Scolecomorphus* with six species which are found only in Africa (Kenya, Tanzania, Malawi, Zambia, and Cameroun). These are characterized by septomaxillae, premaxillae, and nasals that are present as separate bones; they lack prefrontals, oculars, and stapes, no tail is present, eyes develop under bone and there is no socket, and splenial teeth are absent.

Some groups of amphibians have been studied more extensively than others, and this has had an effect upon their classification. Comparisons of extinct and living groupings are difficult because the extinct groups are classified primarily on the bases of osteological and temporal data and the osteology of living groupings is poorly known. Consequently, the precise numbers of families, genera, and species are not too meaningful, but, nevertheless, their distribution through time is significant. Above is a tally of the extinct and living families and genera plus the approximate number of living species for

each order. The numbers for Apoda are based on Taylor (1968, 1969); those for Urodela are from Brame (1967) and Wake (1970); those for Anura have been compiled from a variety of references; those on the remaining orders are from Romer (1966).

Obviously, the class Amphibia is represented by a relatively small number of living members, and it is a group that was dominant in many areas of the world in the past. The large number of genera and species of living anurans reflects the fact that they are the most successful of the Lissamphibia and have undergone a great deal of adaptive radiation. The order Temnospondyli was the dominant group of extinct amphibians.

References

Brame, A. H., Jr. 1967. A list of the world's recent and fossil salamanders. Herpeton 2:1–26.

Eaton, T. H., Jr., and P. L. Stewart. 1960. A new order of fishlike amphibia from the Pennsylvanian of Kansas. Univ. Kansas Publ. Mus. Nat. Hist. 12:217–240.

Estes, R. 1970. New fossil pelobatid frogs and a review of the genus *Eopelobates*. Bull. Mus. Comp. Zool. 139(6):293–339.

Goin, C. J., and O. B. Goin. 1962. Introduction to Herpetology. San Francisco, W. H. Freeman. 341 pp.

Griffiths, I. 1956. Status of *Protobatrachus massinoti*. Nature 177:342–343.

_____ 1963. The phylogeny of the Salientia. Biol. Rev. Cambridge Philos. Soc. 38:241–292.

Hecht, M. K. 1962. A reevaluation of the early history of the frogs. Part I. Systematic Zool. 11(1):39–44.

_____ 1963. A reevaluation of the early history of the frogs. Part II. Systematic Zool. 12(1):20–35.

Jarvik, E. 1942. On the structure of the snout of crossopterygians and lower gnathostomes in general. Zool. Bidrag Uppsala 21:235–675.

_____ 1960. Théories de l'Évolution des Vertèbres, Reconsiderées à la Lumière des Recentes Découvertes sur les Vertèbres Inférieures. Paris, Masson. 104 pp.

Liem, S. S. 1970. The morphology, systematics, and evolution of the Old World treefrogs (Rhacophoridae and Hyperoliidae). Fieldiana Zool. 57:1–145.

McDiarmid, R. W. 1969. Comparative morphology and evolution of the Neotropical frog genera *Atelopus*, *Dendrophryniscus*, *Melanophryniscus*, *Oreophrynella*, and *Brachycephalus*. Diss. Abstr. 29:4895-B.

Noble, G. K. 1931. Biology of the Amphibia. New York, McGraw-Hill. 577 pp.

Olson, E. C. (ed.). 1965. Evolution and relationships of the Amphibia; Symposium of the American Society of Zoologists. Am. Zool. 5:263–334.

Piveteau, J. 1937. Un amphibien du Trias inférieur; essai sur l'origine et l'évolution des amphibiens anoures. Ann. Paléont. 26:135–176.

Romer, A. S. 1966. Vertebrate Paleontology, 3rd ed. Chicago, Univ. Chicago Press, 468 pp.

_____ 1968. Notes and Comments on Vertebrate Paleontology. Chicago, Univ. Chicago Press. 304 pp.

Schmalhausen, I. I. 1968. The Origin of Terrestrial Vertebrates. New York, Academic Press. 314 pp.

Szarski, H. 1962. The origin of the Amphibia. Quart. Rev. Biol. 37:189–241.

Taylor, E. H. 1968. The Caecilians of the World. Lawrence, Univ. Kansas Press. 848 pp.

_____ 1969. A new family of African Gymnophiona. Univ. Kansas Sci. Bull. 48(10):297–305.

Tihen, J. A. 1958. Comments on the osteology and phylogeny of ambystomatid salamanders. Bull. Florida State Mus., Biol. Ser. 3(1):1–50.

Wake, D. B. 1966. Comparative osteology and evolution of the lungless salamanders, Family Plethodontidae. Mem. Southern California Acad. Sci. 4:1–111.

_____ 1970. The abundance and diversity of tropical salamanders. Am. Naturalist 104:211–213.

CHAPTER 4

STRUCTURAL AND FUNCTIONAL CHARACTERISTICS OF LIVING REPTILES

The class Reptilia evolved from an-thracosaurian amphibians, became the dominant class of vertebrates during the Mesozoic, gave rise to mammals during the Triassic and to birds during the Jurassic, and has been declining ever since. Only four of the 17 orders of reptiles are represented by living members. As is true of the class Amphibia, the class Reptilia as a whole is difficult to characterize. There was a gradual transition from amphibians to reptiles and another from reptiles to mammals; birds, in reality, have so many reptilian characteristics that they may be thought of as specialized reptiles.

The four living orders of reptiles, represented by turtles, crocodiles, the tuatara, and lizards and snakes, are easily differentiated from other contemporary vertebrates, for they are the only amniotes (having an amniotic egg) that lack both hair and feathers. The most obvious differences between living reptiles and amphibians involve features that allow reptiles to be relatively independent of free water and to be truly terrestrial animals. Other differences involve improvements of organs and structures so that they are more efficient than they are in amphibians.

INTEGUMENT

A major requirement of terrestrial organisms is that they be able to maintain a proper internal water balance in a generally dehydrating environment. In amphibians, the respiratory function of the integument requires that it be kept moist; this results in continuous evaporation from the body surface when the animal is out of water, a major reason why amphibians require habitats that provide a continual supply of free water. In reptiles, on the other hand, the integument does not have to be kept moist since it has no respiratory function. Instead, it plays an important role in conserving body fluids by forming a protective barrier between them and the dehydrating environment. In general, the characteristics that distinguish the integument of a reptile from that of an amphibian represent adaptations facilitating the conservation of water.

EPIDERMIS

The horny layer (stratum corneum) is a much thicker portion of the epidermis in reptiles than in amphibians (Fig. 4–1).

115

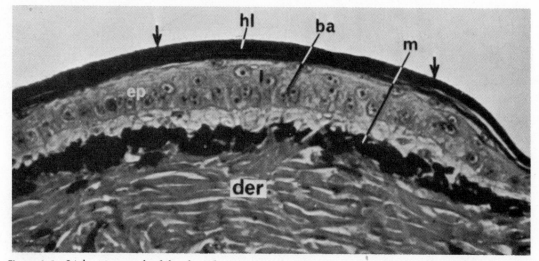

Figure 4–1 Light micrograph of the skin of a viviparous lizard (*Lacerta vivipara*) showing epidermis, dermis, and horny epidermal layer; the latter can be seen on the right to consist of inner and outer keratin layers as is apparent in Figure 4–2; × 500. *ba*, basal layer cells; *der*, dermis; *ep*, epidermis; *hl*, horny epidermal layer; *l*, lipophore; *m*, dermal melanophores. (From Bellairs: *The Life of Reptiles*.)

Being composed of compacted layers of thin, flat, horny, dead cells filled with keratin, it forms a covering over the reptilian body surface which is the primary barrier to the passage of water. Cornified epithelium is being formed continually by the underlying stratum germinativum and in reptiles, as in amphibians, the outer layers are replaced through time by those formed beneath (Figs. 4–2 and 4–3). As the underlying layers develop, nerve and circulatory connections with the outer layer are lost and there is a general loosening of the outer layer; the loosening is accomplished through the diffusion of lymph between old and new layers and through the pressure exerted by minute rugosities in the new layer. In snakes, separation of the layers of transparent plates over the eye is what causes clouding of the eyes prior to shedding. The shedding of the outer corneum is generally initiated at the head region. A "swell mechanism" causes an enlargement of the head through an abrupt increase in the blood pressure in the veins of the head. This is primarily the result of contraction of the *jugularis* muscle, a small sphincter surrounding the internal jugular vein, which is the major vessel draining the head. Contraction of the *jugularis* closes the internal jugular and traps blood in the head as it continues entering through arteries, resulting in an increase in blood pressure in head sinuses and veins. Swelling of the head both loosens and ruptures the layers being shed so that they may be rubbed or scratched off. In most snakes and a few lizards, the sloughed layers come off intact during a definite period of ecdysis. In the majority of lizards, there is a definite period of ecdysis, but the outer layer is shed in pieces ranging in size from the covering of an entire region (head, body, tail) to individual scales. The frequency of molting varies with the species and may be as great as twice a month; rapidly growing juveniles gen-

Figure 4–2 Electron micrograph of epidermis of tail skin of the viviparous lizard (*Lacerta vivipara*) one day after sloughing; × 5800. A similar stage is shown diagrammatically in Figure 4–3A. *A*, *B*, A- and B-keratin layers of horny layer; *ba*, basal layer cell; *der*, dermis; *iz*, intermediate zone; *ob*, serrations of *Oberhautchen*. (From Bellairs: *The Life of Reptiles;* after Bryant *et al.*: J. Zool. London 152:209–219.)

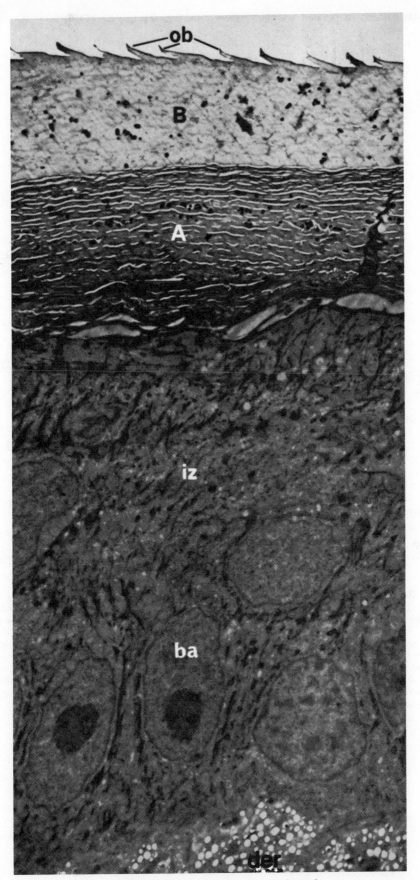

Figure 4–2 *See opposite page for legend.*

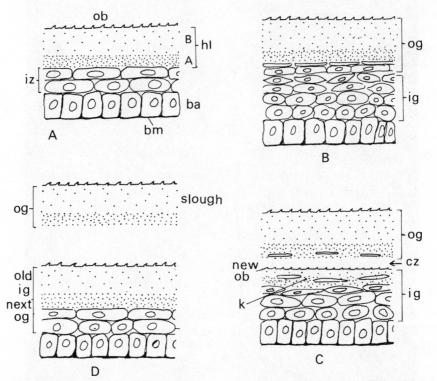

Figure 4–3 Simplified diagrams showing changes in epidermis during the sloughing cycle of a snake. *A*, resting stage. *B*, before sloughing but after basal cells have divided to form a new inner epidermal generation (*ig*); the snake's color is dulled. *C*, shortly before sloughing; a cleavage zone appears between the two generations and the superficial part of the inner generation is becoming keratinized (*k*); a new serrated *Oberhautchen* is being formed; this stage probably coincides with the clearing of the skin. *D*, sloughing; the original outer generation is shed and the old inner generation becomes the next outer generation. *A, B,* A- and B-keratin layers which together comprise the horny layer (stratum corneum); *ba,* basal cell layer (stratum germinativum); *bm,* basement (basal) membrane between basal cells of epidermis and dermis; *cz,* cleavage zone; *hl,* horny layer; *iz,* intermediate zone cells; *ob, Oberhautchen; ig, og,* outer and inner epidermal generations. (From Bellairs: *The Life of Reptiles.*)

erally shed more often than mature adults. In some turtles such as painted turtles (*Chrysemys*, family Testudinidae), the outer shell scutes are shed annually, whereas in other turtles, including box turtles (*Terrapene*, family Testudinidae) and crocodilians, the outer layers are not shed regularly but gradually wear off and are replaced by those accumulating underneath.

A conspicuous characteristic of the integument of all living reptiles other than a few turtles is a covering of well-developed epidermal scales. Those of lizards and snakes are formed through folding of the epidermis and outermost dermal layers (Fig. 4–4). They vary from small granular scales to large rectangular or spiny structures and occur in definite patterns; these provide a convenient means of identification. On the body, they generally occur in longitudinal, transverse, or diagonal rows; the size, shape, and pattern of cephalic scales of Squamata are usually distinctively different from those of body scales (Fig. 4–5). The scales of the tuatara are relatively uniform and granular in shape. Those of adult turtles and crocodilians are less specialized than in the other living reptiles and form large flat horny shields, each consisting of accumulated layers of cornified scales. The epidermal scales of turtles form individually and generally do not overlap. They comprise the lam-

inae, relatively thin layers covering but not conforming in pattern to the underlying bony dermal plates of the carapace and plastron (Fig. 4–6). The leatherback turtles (family Dermochelyidae) and a few other turtles lack epidermal laminae and have smooth skins. Crocodilians always have a complete covering of horny epidermal shields. Scales on their sides, belly, and tail each have a small pitlike depression in which a small sensory capsule is located.

Many scales have become secondarily specialized for functions other than waterproofing the surface, and their size and shape vary accordingly. Tips of the characteristic "horns" of horned lizards (*Phrynosoma*, family Iguanidae) are epidermal scales functioning as protective devices and modified into elongated spines. The rattles of rattlesnakes (*Crotalus*, *Sistrurus* in family Viperidae) represent an effective warning device formed by modified scales that adhere to the base of the tail. They are not shed at ecdysis but are often broken off accidentally in older snakes (Fig. 4–7). The belly scales of most snakes and some lizards are broad rectangular plates with a locomotive function; these are called gastrosteges. The egg caruncle of turtle, crocodile, and tuatara embryos is a transitory epithelial structure that develops on the snout at the time of hatching and functions to rupture the embryonic membranes and shell; the egg

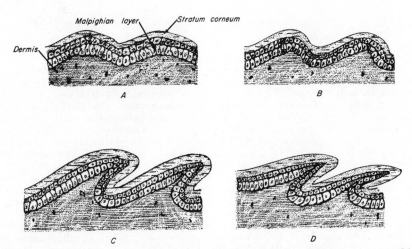

Figure 4–4 Diagrammatic stages in the development of epidermal scales of the type possessed by lizards and snakes. (From *Introduction to Herpetology*, second edition, by Coleman J. Goin and Olive B. Goin. W. H. Freeman and Company. Copyright © 1971.)

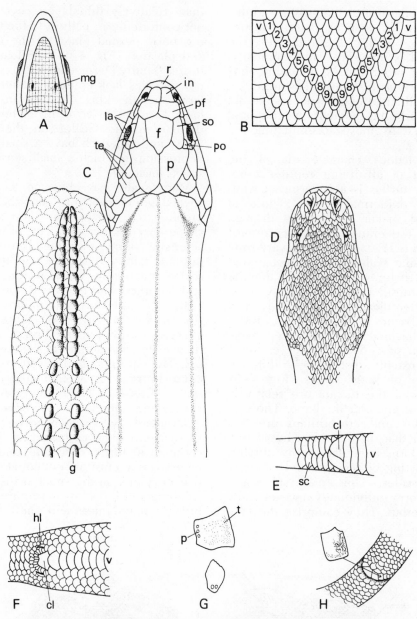

Figure 4–5 *A*, underside of head of young alligator showing openings of musk glands. *B*, body scales of smooth snake (*Coronella austriaca*, family Colubridae) showing method of counting dorsal scales in oblique transverse series. *C*, anterior body and head of a natricine snake (*Natrix nuchalis*) with the dorsal skin of the neck folded back to show the attached nucho-dorsal glands. The nomenclature of some of the head shields is also given. *D*, head of Russell's viper (*Vipera russelli*, family Viperidae), covered by small scales instead of large shields. *E*, ventral surface of cloacal region of the colubrid snake *Natrix stolata*, illustrating broad ventrals, divided cloacal or anal shield, and paired subcaudals. *F*, ventral surface of cloacal region of the Indian python (*Python molurus*, family Boidae), illustrating comparatively small ventral shields and rudimentary hind limb claws. *G, above:* upper labial shield of the colubrid snake *Alsophis leucomelas sanctorum* showing scale tubercles (tiny dots) and pits; these are probably sense organs; *below:* dorsal neck scale of same snake showing pair of pits. *H*, cloacal region of slowworm (*Anguis fragilis*, family Anguinidae), illustrating small ventral scales. One of the shields bordering the cloaca is drawn separately to show the patchy distribution of melanin pigment; in life, the dots and enclosed areas appear pale against the general dark background. *cl*, cloacal or anal scale or shield; *f*, frontal; *g*, nuchodorsal gland; *hl*, hind limb claw; *in*, internasal; *la*, upper labial; *mg*, musk gland opening; *p*, pit, or parietal; *pf*, prefrontal; *po*, postocular; *r*, rostral; *sc*, subcaudals; *so*, supraocular; *t*, tubercle; *te*, temporal; *v*, ventrals. (From Bellairs: *The Life of Reptiles*; *A*, after Reese: J. Morph. 35:609; *B*, after Smith: *The British Amphibians and Reptiles*; *C*, after Smith: Proc. Zool. Soc. London 107:578; *D–F*, after Wall: *Ophidia taprobanica or the Snakes of Ceylon*; *G*, after Underwood: *A Contribution to the Classification of Snakes*.)

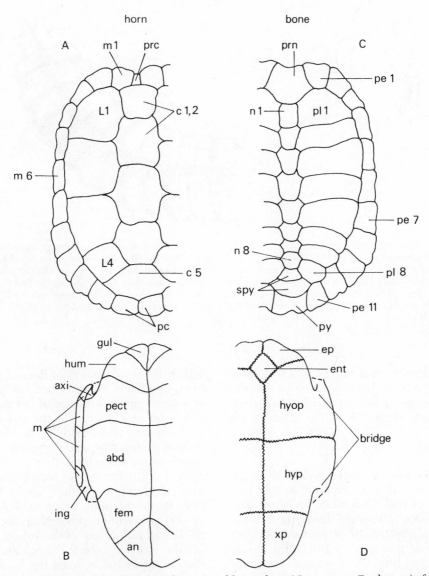

Figure 4–6 Horny laminae (*A*, carapace; *B*, plastron) and bony plates (*C*, carapace; *D*, plastron) of the yellow–bellied turtle (*Pseudemys s. scripta*). Horny laminae: *abd*, abdominal; *an*, anal; *axi*, axillary; *c*, central; *fem*, femoral; *gul*, gular; *hum*, humeral; *ing*, inguinal; *l*, lateral; *m*, marginal; *pc*, postcentral; *pect*, pectoral; *prc*, precentral. Bony plates: *ent*, entoplastron; *ep*, epiplastron; *hyop*, hyoplastron; *hyp*, hypoplastron; *n*, neural; *pe*, peripheral; *pl*, pleural; *prn*, proneural (nuchal); *py*, pygal; *spy*, suprapygal; *xp*, xiphiplastron. (From Bellairs: *The Life of Reptiles*; after Carr: *Handbook of Turtles*.)

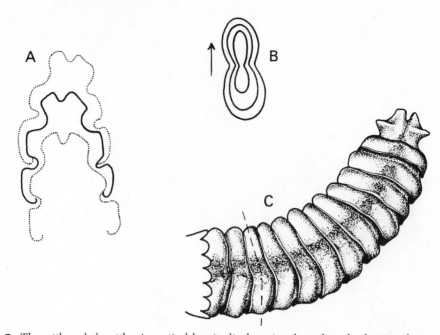

Figure 4–7 The rattlesnake's rattle. *A*, vertical longitudinal section through rattle showing three interlocking segments. *B*, transverse section showing three interlocking lobes; the fit between the lobes is tighter dorsally than ventrally, an arrangement which allows the rattle to bend dorsally farther (shown by arrow) than ventrally. *C*, side view of rattle showing upward tilting caused by arrangement shown in *B* and by the fact that the part of each lobe dorsal to the furrow (stippled) slants slightly forward toward rattle base, as shown by broken line. (From Bellairs: *The Life of Reptiles; A* and *B* after Klauber: *Rattlesnakes: Their Habits, Life Histories, and Influence on Mankind.*)

tooth of lizards and snakes is a true dentine-covered tooth that develops on the anterior edge of the premaxilla and is not homologous with that of the other living reptiles.

DERMIS

The dermis is well-developed in all reptiles, but that of crocodilians is particularly valued as leather for shoes and pocketbooks. Although the dermal armor that was characteristic of the fish ancestors of tetrapods was quickly lost in the evolution of amphibians, a strong tendency toward the redevelopment of dermal armor is found early in the evolution of the Reptilia. The seymouriamorph amphibian *Kotlassia*, one of several forms borderline between amphibian and reptile, was almost entirely covered dorsally by a series of rectangular bony plates. Rounded or oval bony ossicles were present on the neck, trunk, and base of the tail of pareiasaurs, an advanced group of stem reptiles (order Cotylosauria). Turtles (order Chelonia) are an ancient line of reptiles in

which the exploitation of dermal armor is obvious. Their ventral shell, or plastron, is formed through fusion of dermal plates with one another, while the dorsal shell, or carapace, is formed by the fusion of dermal plates, ribs, and vertebrae; the two portions are joined together on each side by a bridge formed by upgrowth of the plastral elements (Romer, 1956). In addition to their shell, some turtles have a series of ossicles protecting the limbs, neck, and tail. Many of the ruling reptiles in the extinct order Thecodontia had dermal armor on at least the back of the neck, trunk, and tail; some were almost entirely enclosed by bony plates. The order Crocodilia as a whole is characterized by highly developed dermal ossifications which, although present in living species, are relatively reduced in modern compared to primitive forms. Most crocodilians, living and extinct, have a dermal ossification in each upper eyelid, the palpebral bone, that protects the protruding eye. The same bone is also found in some of the extinct orders. Osteoderms, bony

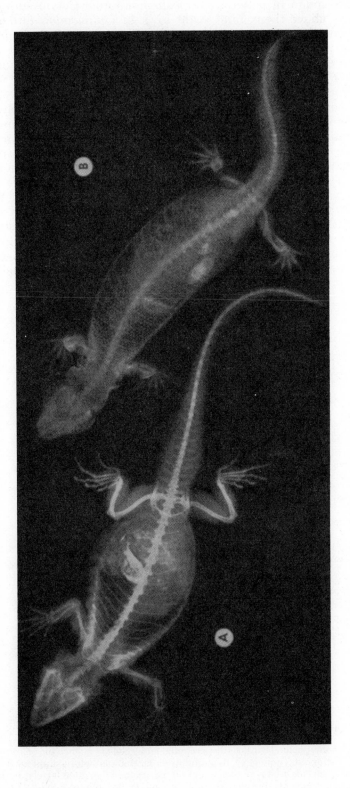

Figure 4–8 Radiograph of northern chuckwalla (*Sauromalus obesus*), a species lacking osteoderms (A), compared to a radiograph of Gila monster (*Heloderma suspectum*), a species possessing osteoderms (B). (Reprinted from Hobart M. Smith: *Handbook of Lizards: Lizards of the United States and Canada.* Copyright 1946 by Comstock Publishing Company, Inc. Used by permission of Cornell University Press.)

ossicles buried in the dermis, are frequently present in lizards; often these are the same size and correspond in pattern to the epidermal scales above them, but sometimes there is no correlation of either sizes or numbers. The presence and pattern of osteoderms are easily determined by means of X rays (Fig. 4–8). Dermal armor is not present in the tuatara nor is it known for any of the extinct rhynchocephalians; it is also unknown for a number of other extinct reptilian orders.

Numerous primitive amphibians, including the seymouriamorph *Discosauriscus*, possessed a series of ventral dermal scales called gastralia that protected the belly and strengthened the abdominal wall. In reptiles, these evolved into slender rodlike structures supporting the ventral abdominal wall posterior to the true ribs and anterior to the pelvic girdle. Gastralia were present in a variety of extinct reptiles and occurred in numbers ranging from only a few rows in *Procolophon* (order Cotylosauria) to about 200 rows in *Ophiacodon* (order Pelycosauria). Those of *Ophiacodon* were closely-packed chevron-shaped structures that formed a complete abdominal covering. Gastralia occur in at least two of the four living orders of reptiles. Seven or eight rows of riblike gastralia are present in living crocodilians; each row consisting of four elements, a median pair that forms a V and a lateral pair of splints that extend the arms of the V. There are 25 rows of gastralia in the tuatara (*Sphenodon*, order Rhynchocephalia), each row consisting of a fused median V and a pair of lateral splints (Fig. 4–9). Gastralia have probably been incorporated into the plastron of turtles (order Chelonia). They have not been found in lizards or snakes (order Squamata).

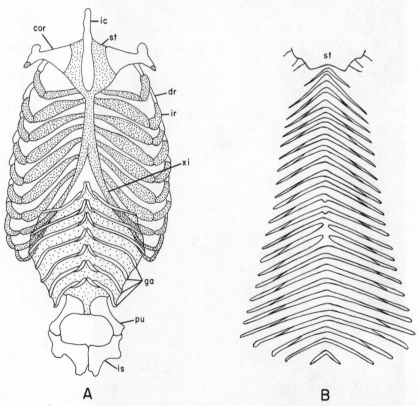

A   **B**

Figure 4–9 *A*, ventral view of the pectoral girdle, ribs, gastralia, and pelvis of a crocodile (*Crocodylus*); cartilaginous regions are darkly stippled, connective tissue between gastralia is lightly stippled. *B*, gastralia of the tuatara (*Sphenodon*). *cor*, coracoid; *dr*, dorsal segment of rib; *ga*, gastralia; *ic*, interclavicle; *ir*, intermediate segment of rib; *is*, ischium; *pu*, pubis; *st*, sternum; *xi*, xiphisternum. (*A* and *B* after Romer: *Osteology of the Reptiles*.)

GLANDS

In contrast to amphibians, which have an abundance of integumentary glands, reptiles almost totally lack them. This difference is, again, associated with the fact that the amphibian skin must be kept moist in order to function as a respiratory surface, whereas that of reptiles lacks this function and the surface may be dry. The general elimination of integumentary glands in reptiles removes a need for body fluids at the surface of the integument and prevents unnecessary evaporation of water from the body. The few integumentary glands which do occur in reptiles are generally scent glands that seemingly function to ward off predators or that function in sex discrimination and species recognition during the breeding season.

Crocodilians of both sexes have two pairs of integumentary musk glands; one pair is located on the throat (Fig. 4–5) and the other within the cloacal opening. These seem to be comparable in function to the hedonic glands of amphibians; they play a role in courtship and mating activities. Crocodilians also have two rows of glands of unknown function running along the middle of the back between the first and second rows of scales.

Many groups of lizards have preanal or femoral pores that are openings of follicular glands in the dermis (Fig. 4–10). Femoral pores typically occur in a single series on the undersurfaces of the thighs, whereas the preanal pores generally form a V-shaped series or are in clusters in the preanal area. The number of pores present is usually the same in both sexes, but those of males are generally larger than those of females; in some species, pores are only present in males. Cole (1966a) has described the anatomy of femoral glands and (1966b) summarized the knowledge of them. The femoral glands are composed of branching tubes and tubules in both sexes, but those of males increase in size and complexity as the individual matures; little ontogenetic change occurs in the glands of females. The relative length of male glands appears to vary seasonally, suggesting a seasonal variation in their activity, and the greatest sizes occur during the breeding season. During the breeding season, the preanal and femoral pores become filled with a horny yellow material that may project as small cones from the surface. Internal to part of the secretion plug, a stratum corneum continuous with that of the integument forms in the duct of the gland. This stratum corneum appears to form in the autumn, prior to hibernation, and when it is shed at the next ecdysis part of the secretion plug is removed; meanwhile, the production of a new secretion plug has been initiated. Little is known about the chemistry of the secretions of preanal and

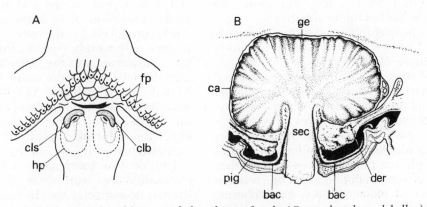

Figure 4–10   A, femoral pores, cloacal bones, and cloacal sacs of gecko (*Gymnodactylus pulchellus*). B, microscopic section through middle of the femoral pore and gland of a male collared lizard (*Crotaphytus collaris*). *bac*, basal cell layer of surrounding epidermis; *ca*, capsule of gland; *clb*, cloacal bone; *cls*, opening of cloacal sac, outline of sac dotted; *der*, dermis; *fp*, femoral pore; *hp*, hemipenis (outline); *ge*, germinative epithelium of gland; *pig*, pigment in dermis; *sec*, plug of secretion in duct of pore. (From Bellairs: *The Life of Reptiles*. A, after Smith: *The Fauna of British India, Ceylon and Burma, Including the Whole of the Indo-Chinese Sub-region: Reptilia and Amphibia*. B, After Cole: Femoral glands of the lizard, *Crotaphytus collaris*.)

femoral glands, and, although at least five or six hypotheses have been proposed, their biological significance is unknown.

Some snakes, including the common garter snakes (*Thamnophis*, family Colubridae), have cloacal glands that seem to be protective in function: they secrete a nauseating fluid. Nuchodorsal glands (on the dorsal skin of the neck) occur in some snakes and may play a role in species and sex recognition (see Figure 4–5).

A number of turtles have musk glands along the edge of the lower jaw and the sides of the body between the plastron and carapace (axial and inguinal regions) which secrete odoriferous substances. These may be merely protective in function, or they may be involved in species recognition.

PIGMENTATION

Among reptiles are found varying degrees of complexity in mechanisms producing color and color changes. The true chameleons (family Chamaeleontidae) are famous for their ability to change both the shade and pattern of their skin coloration; they can go from nearly white to blotchy to a new, almost patternless coloration. A similar but more limited ability is possessed by some of the New World lizards in the genus *Anolis* (family Iguanidae), often sold in pet stores as chameleons, which may change from dark brown through shades of yellow to bright green. Many other reptiles have the ability to darken or lighten their coloration.

The chromatophore systems of lizards are best known; relatively little is known about those of other reptiles. The integument of some lizards contains at least five combinations of four chromatophores: melanophores only, melanophores plus guanophores, melanophores plus guanophores and lipophores, melanophores plus guanophores and allophores, and melanophores plus all three of the other kinds of chromatophores. The relative positions of melanophores, guanophores, and lipophores are as in amphibians (Fig. 4–11). The cell bodies of the melanophores are normally in the dermis, beneath all the other chromatophores. These tend to be large cells with elongated pseudopods that extend outward between the

other chromatophores to the stratum germinativum of the epidermis. Reptilian guanophores (leucophores) are small cells that generally form a relatively thick layer immediately peripheral to the melanophore cell bodies. If present, lipophores (xanthophores) form the most superficial layer, between the guanophores and the stratum germinativum, but they may be covered externally by pseudopods from the melanophores. Allophores contain red, yellow, or violet pigment that differs from the pigment in lipophores in that it is insoluble in alcohol. Allophores are not commonly found in reptiles; when present, they are usually scattered among the guanophores.

Alligators, lizards, and snakes are known to have physiologically active melanophores capable of darkening or lightening the skin. When the melanin is fully dispersed, it fills the distal ends of the melanophore pseudopods and may almost completely conceal the other chromatophores; this will render the skin very dark or, in extreme cases, completely black. When the melanin is withdrawn into the bodies of the melanophores, their pseudopods are transparent, and the color of the skin will largely depend upon the presence of other chromatophores. If the only other chromatophores present are guanophores, the skin will appear iridescent blue, blue-white, or white. If lipophores are also present, the skin will appear yellow, yellow-green, or green. If allophores are present and not concealed, the skin will appear orange, red, or violet. If slight dispersion of melanin occurs, the colors produced by the other chromatophores will merely darken, but as more melanin passes into the peripheral pseudopods the color of the animal will turn to dark brown or black. As with amphibians, it is likely that interference phenomena in the outer layers of the skin may affect the color of a reptile, particularly in regard to the blue wavelengths.

The control mechanism for reptilian chromatophores varies, at least among lizards, in a species-specific manner (see reviews by Parker, 1948, and Waring, 1963). The variability ranges from what is probably the most primitive mechanism, in which there is a hormonal control of chromatophores, through those in which there is a coordinated nervous and hor-

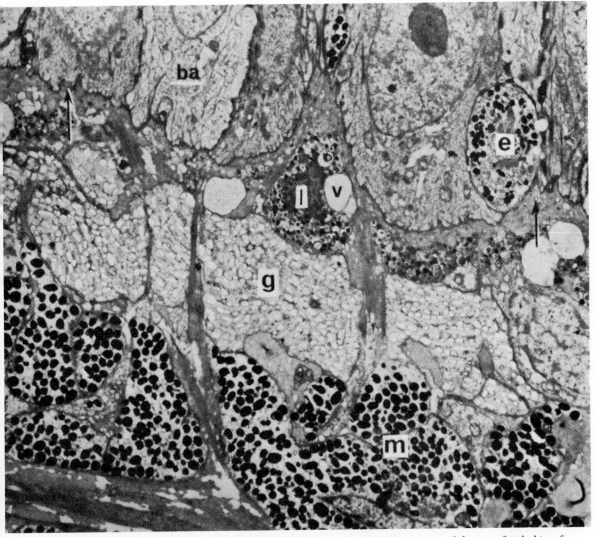

Figure 4–11 Electron micrograph of basal layer of epidermis and underlying part of dermis of tail skin of viviparous lizard (*Lacerta vivipara*), showing pigment cells. The lipophores (*l*) contain numerous cytoplasmic granules and large vesicles (*v*). The guanophores (*g*) show a honeycomb appearance due to the loss of guanine crystals from the cytoplasm during processing. The dermal melanophores (*m*) are packed with melanin granules; basal cells (*ba*) and an epidermal melanocyte (*e*) are also visible; arrows point to the basal lamina ('basement membrane') of epidermis (× 5500). (From Breathnach, A. S., and S. V. Poyntz: Electron microscopy of pigment cells in tail skin of *Lacerta vivipara*. J. Anatomy *100*:549–569, 1966.)

monal control to those in which there is control primarily by direct nervous innervations. In some species, melanophores may react to direct light or temperature stimulation or both. For example, the melanophore response of American "chameleons" (*Anolis*, family Iguanidae) has been shown conclusively to be independent of any nervous control. In these lizards, melanin is dispersed under the influence of intermedine, a hormone produced by the intermediate lobe of the pituitary; concentration of the melanin in melanophore bodies is caused by adrenaline. Available evidence indicates that nerves and hormones both play a role in the control of melanophores in horned lizards (*Phrynosoma*, family Iguanidae). In these, concentration of the melanin can be induced by either nervous stimulation or injection of adrenaline. Dispersion of melanin in *Phrynosoma* appears to be entirely under hormonal (intermedine) control for studies have revealed no pigment-dispersing nerve fibers. The melanophores of *Phrynosoma* will also respond directly to the influence of temperature and light intensity. Dispersion of melanin occurs in response to low temperatures and to very strong illumination; concentration of melanin occurs at high temperatures and in darkness. Finally, the melanophores of true chameleons (*Chamaeleo*, family Chamaeleontidae) appear to be exclusively controlled by the autonomic nervous system. Electrical stimulation of these nerves will cause concentration of the pigment; dispersion of the pigment occurs either passively because of the absence of tonic impulses to the melanophores or actively by means of a second set of nerve fibers which have yet to be discovered. The melanophores of *Chamaeleo* will respond to light intensities only in the intact animal; no response can be obtained in denervated regions of the skin.

The biological significance of reptilian coloration and color changes is discussed in Chapter Nine.

SKELETAL SYSTEM

SKULL

The most obvious features of reptilian skulls that distinguish them from those of amphibians are a single occipital condyle, a greater degree of ossification, a higher and narrower shape, a reduction or loss of certain bones, and closure of the characteristic amphibian otic notch. The skulls of the earliest reptiles are very similar to those of labyrinthodont amphibians except that they have the single occipital condyle, reduced postparietals, tabulars, and supratemporals, and lack the amphibian otic notch and intertemporal bone; there are only two or three bones in the lower jaw, a single splenial plus one or two coronoids. From this generalized primitive reptilian skull pattern, many variations evolved. As classified by Romer (1956), these evolutionary divergences involve seven categories of cranial characteristics: (1) a reduction, fusion, modification, or increase in individual dermal elements; (2) changes in skull proportions; (3) changes in skull openings; (4) changes in the temporal (cheek) region associated with the action of the temporal muscles; (5) modifications of the palate; (6) variations in the occipital region; and (7) variations in the ossification of the braincase.

Of the above variations, differences in the temporal region have played a particularly important role in the classification of major groups. The most primitive reptiles (order Cotylosauria) and most turtles (order Chelonia) lack temporal openings (fenestrae) in the skull behind the eye (Figs. 4–12A and 4–13). In most later reptiles, one or two openings appear on each side between various bones in this region of the skull, apparently to accommodate bulging of the temporal muscles during their contraction. In some extinct forms (order Araeoscelidia and order Sauropterygia), there is a single opening on each side high up on the skull that is bordered below by the postorbital and squamosal bones (Figure 4–12B). In others (order Pelycosauria, order Mesosauria, and order Therapsida, all extinct), there is a single opening low on each side bordered above by the same two bones, (Fig. 4–12C). Members of the order Ichthyosauria, extinct fishlike reptiles, had a single temporal opening very high on each side bordered below by the postfrontal and supratemporal bones (Fig. 4–12D). Finally, in a large variety of both living and extinct reptiles (order Eosuchia,

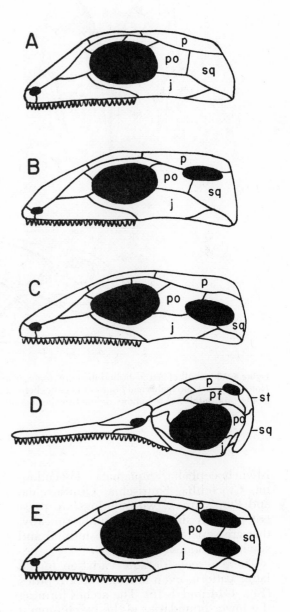

Figure 4–12 Diagrammatic lateral views of reptilian skulls showing five types of temporal openings. *A*, anapsid type, lacking temporal openings and characteristic of order Cotylosauria and of some chelonians; *B*, euryapsid type, characteristic of orders Araeoscelidia, Sauropterygia, and Placodontia; *C*, synapsid type, characteristic of orders Pelycosauria and Therapsida; *D*, ichthyopterygian type, characteristic of order Ichthyosauria; *E*, diapsid type, characteristic of the orders Eosuchia, Rhynchocephalia, Squamata, Thecodontia, Crocodilia, Saurischia, Ornithischia, and Pterosauria. *j*, jugal; *p*, parietal; *pf*, postfrontal; *po*, postorbital; *sq*, squamosal; *st*, supratemporal (*A, B, C, E* after Romer, A. S.: *The Vertebrate Body*. 4th Ed. W. B. Saunders Co., 1970; *D*, after Romer: *Osteology of the Reptiles.*)

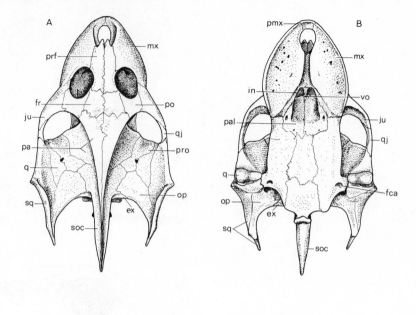

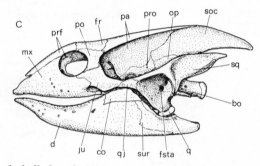

Figure 4–13   Skull of a soft-shelled turtle (*Trionyx triunguis*). *A*, dorsal view; *B*, ventral view; *C*, lateral view. Note: the nasal and lachrymal bones of Recent chelonians may be absent or incorporated in what is interpreted as the prefrontal. Abbreviations defined in the legend for Figure 4–16. (From Bellairs: *The Life of Reptiles.*)

Rhynchocephalia, Squamata, Thecodontia, Crocodilia, Saurischia, Ornithischia, and Pterosauria), two openings are present on each side, at least embryonically, that are separated by the postorbital and squamosal bones (Fig. 4–12*E*).

There is considerable variation in the latter pattern, even among living reptiles (Fig. 4–14 and 4–15). The arches forming the lower boundaries of the two temporal openings of tuataras (order Rhynchocephalia) are both retained during ontogeny, so the basic diapsid pattern of the temporal region is present in adults and the quadrate is fixed in position (Fig. 4–14*B*). The temporal region of Squamata is variable. The roofing bones are reduced so as to typically form one (upper) arch in lizards (Fig. 4–16) and no arch in snakes (Fig. 4–17). However, in some lizards the remaining temporal opening is overgrown by neighboring elements, and in others, particularly burrowing forms, the remaining arch is lost through reduction or disappearance of the postorbital and squamosal bones (Romer, 1956). Associated with the loss of the lower arch, there is a widely opened area on the side of the cheek and a movable quadrate in Squamata; the latter is particularly important for snakes that swallow very large prey, the quadrate being the bone with which the lower jaw articulates. Among living crocodilians, both temporal open-

(*Text continued on page 135.*)

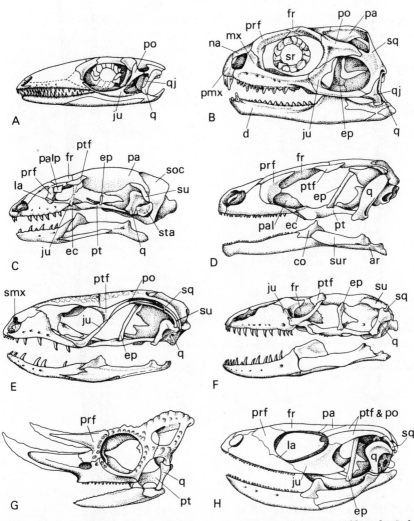

Figure 4–14 Lateral views of the skulls of the tuatara (*Sphenodon*) and a variety of lizards. Only some of the bones are labeled. *A*, the small eosuchian *Prolacerta* from the Lower Triassic of South Africa; *B, Sphenodon*; *C*, the burrowing anguoid *Anniella pulchra*; *D*, the banded gecko *Coleonyx variegatus*; *E*, the Slowworm *Anguis fragilis*; *F*, the earless monitor *Lanthanotus borneensis*; *G*, a horned chamaeleon, probably *Chamaeleo jacksoni*, showing bony horn cores; *H*, the green lizard *Lacerta viridis*, in which the superior temporal opening is covered over by the postorbital and neighboring bones. Abbreviations defined in the legend for Figure 4–16. (From Bellairs: *The Life of Reptiles; A*, after Parrington, Camp, and Romer: *Osteology of the Reptiles; B*, after Romer: *Osteology of the Reptiles; C, D, F*, after McDowell and Bogert: Bull. Amer. Mus. Nat. Hist. 105:1–142; *E*, after Smith: *The British Amphibians and Reptiles; G*, after Williston: *Osteology of the Reptiles; H*, after Weidersheim: *Vergleichende Anatomie der Wirbeltiere*.)

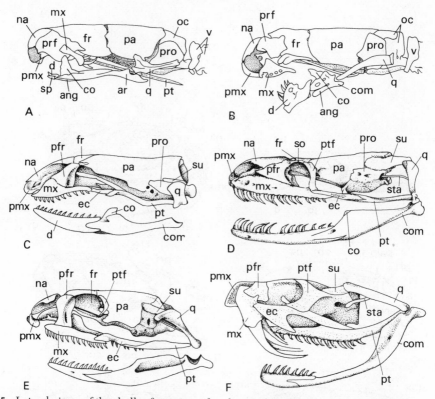

Figure 4–15 Lateral views of the skulls of a variety of snakes. Only some of the bones are labeled. *A*, the blind snake *Typhlops braminus; B*, the blind snake *Leptotyphlops dulcis dissectus; C*, the pipe snake *Cylindrophis rufus; D*, the reticulated python *Python reticulatus; E*, the smooth snake *Coronella austriaca; F*, the puff adder *Bitis lachesis*. Abbreviations defined in the legend for Figure 4–16. (From Bellairs: *The Life of Reptiles; A* and *B*, after List: *Comparative Osteology of the Snake Families Typhlopidae and Leptotyphlopidae; C* and *E*, after Smith: *The British Amphibians and Reptiles; F*, after Fitzsimons: *Snakes of Southern Africa*.)

Figure 4–16 Skull of a black iguana lizard (*Ctenosaura pectinata*). From top to bottom: lateral view of skull; outer side of lower jaw; inner side of lower jaw; posterior view of skull. The interorbital septum and other cartilaginous regions of the braincase are not shown. Abbreviations for Figures 4–13 through 4–18: *ang*, angular; *apr*, angular process; *ar*, articular; *bo*, basioccipital; *bp*, basipterygoid process; *bs*, basisphenoid; *co*, coronoid; *com*, composite bone consisting of prearticular, articular, etc., fused; *con*, occipital condyle; *d*, dentary; *ec*, ectopterygoid (transpalatine); *ep*, epipterygoid (columella cranii); *Eut*, opening for median Eustachian tube; *ex*, exoccipital; *f*, fossa for articulation with condyle of quadrate at jaw-joint; *fa*, fang; *fca*; foramen for internal carotid artery; *fjo*, foramen for duct of Jacobson's organ; *fm*, foramen magnum; *fr*, frontal; *fsta*, foramen for stapes or columella auris; *in*, internal nostril; *itf*, inferior temporal fossa; *ju*, jugal; *la*, lacrimal; *mc*, groove for Meckel's cartilage; *mx*, maxilla; *na*, nasal; *n*, notch for fourth lower tooth; *oc*, exoccipital and supraoccipital; *op*, opisthotic; *pa*, parietal; *pal*, palatine; *palp*, palpebral; *palt*, palatine teeth; *pif*, pineal foramen; *pmx*, premaxilla; *po*, postorbital;° *pop*, paraoccipital process; *prf*, prefrontal; *pro*, prootic; *ps*, parasphenoid; *pt*, pterygoid; *ptf*, postfrontal; *ptt*, pterygoid teeth; *q*, quadrate; *qc*, quadrate condyle; *qj*, quadratojugal; *rp*, retroarticular process of articular; *smx*, septomaxilla; *so*, supraorbital; *soc*, supraoccipital; *sp*, splenial; *sph*, compound sphenoid bone; *sq*, squamosal;° *sta*, stapes (columella auris); *stf*, superior temporal fossa; *su*, supratemporal;° *sur*, surangular; *t*, median tooth; *tyc*, tympanic crest of quadrate; *v*, first or second vertebra (atlas or axis); *vo*, vomer (prevomer); *I*, region of kinetic joint between basipterygoid process of basisphenoid and pterygoid; *2*, metakinetic joint between paroccipital process, parietal and adjacent bones (the pointer in Figure 4–16, bottom, ends on an intercalary cartilage); *3*, median metakinetic joint between supraoccipital

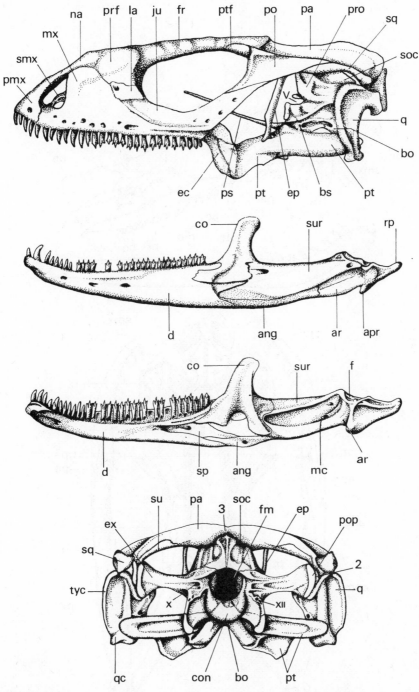

Figure 4–16 *Continued.*
and parietal; *II*, fenestra for optic nerves, etc.; *V*, trigeminal notch lying around the trigeminal ganglia and transmitting branches of the fifth cranial nerve; *X, XII*, foramina for vagus and hypoglossal nerves.

°In some lizards and in many snakes, there is a single bone only behind each orbit which may be the post-orbital, the postfrontal, or the two fused together. It is labeled postfrontal here, but some workers call it the postorbital. In crocodilians and chelonians, the single bone behind the orbit is here called the postorbital.

The bone labeled supratemporal here in lizards is called by some workers the tabular; in snakes the supra-temporal here may be called either tabular or squamosal. All three bones were present in the skulls of some primitive reptiles, and it is difficult to be certain of the identities of the one or two which remain in modern forms. (From Bellairs: *The Life of Reptiles;* after Oelrich: Misc. Publ. Mus. Zool. Univ. Michigan, No. 94: 1–122.)

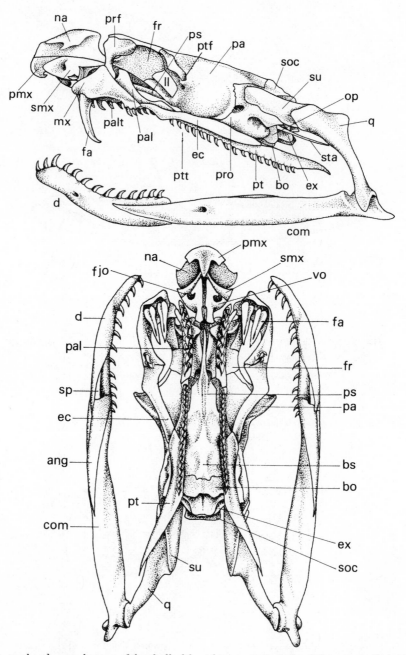

Figure 4–17 Lateral and ventral views of the skull of the African spitting cobra (*Naja nigricollis*). Abbreviations defined in legend for Figure 4–16. (From Bellairs: *The Life of Reptiles;* after Bogert: Bull. Amer. Mus. Nat. Hist. *81*:285–360.)

ings are generally present, but the upper opening is typically small and sometimes closed (Fig. 4–18).

Modifications of individual elements comprising the skull are generally associated with changes in its proportions. In contrast to the flattened shape characteristic of amphibian skulls, the trend in reptilian skulls is for the shape to become progressively higher and narrower. This reflects a generality in vertebrate evolution: As vertebrates become more and more terrestrial, the face becomes rela-

tively longer. For example, the relative position of the orbits is progressively farther posterior, the jaws increase in length, and the parietals and postparietals are simultaneously reduced and shifted posteriorly. It is probable that most extinct, like modern, amphibians forced air into the lungs by contracting the floor of the mouth. Accordingly, the possession by amphibians of broadly flattened skulls is advantageous, in that this shape enlarges the capacity of the mouth and facilitates respiration. On the other hand, high

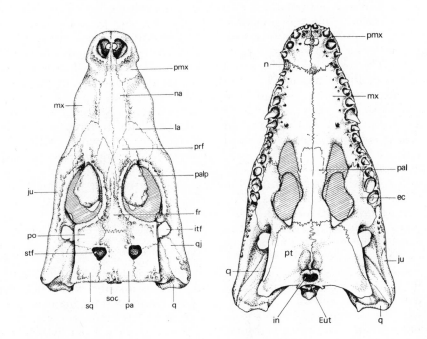

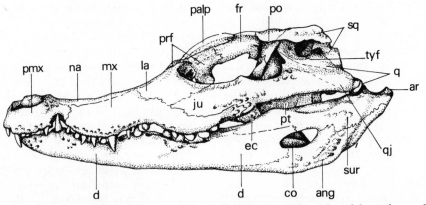

Figure 4–18 Dorsal, ventral, and lateral views of the skull of the West African broad-fronted crocodile (*Osteolaemus tetraspis*), not all drawn to the same scale. Abbreviations defined in legend for Figure 4–16. (From Bellairs: *The Life of Reptiles.*)

arched skulls provide greater strength. The fact that even the most primitive reptiles generally had relatively high skulls may be taken as an indication that they had already acquired thoracic breathing, in which expansion and contraction of the rib cage fills and empties the lungs, and that selection favored greater strengthening of the skull.

The maxillary arch has been greatly modified in snakes. The premaxillae typically form a single small median bone and the maxillae form moderately elongate and toothed elements that are loosely connected to the rest of the cranium. Each maxilla has a movable articulation with the prefrontal as its primary connection, and connects, in various ways in different species, with the ectopterygoid at its posterior end. The maxillae are reduced and sometimes toothless in some burrowing snakes and very short in solenoglyph venomous snakes; the venom-delivering mechanism of the latter is discussed in detail in a later section on the mouth and buccal cavity.

The evolution of a secondary palate, separating the nasal passages from the mouth cavity, is seen in reptiles. This shelf of bones increases the length of the nasal passages and, in doing so, moves the internal nares posteriorly to the back of the mouth. A secondary palate is developed to various degrees in turtles, well-developed in crocodilians, and lacking in tuataras, lizards, and snakes.

The lower jaw of reptiles is generally more complex than that of amphibians. The posterior portion of Meckel's cartilage almost always ossifies as the endochondral articular bone articulating with the quadrate. In addition, reptiles typically have six dermal elements in their mandible. The major tooth-bearing element is the dentary. Posterior and primarily external to the dentary are, in anteroposterior order, a splenial, angular, and surangular, while internally are a prearticular and coronoid. The symphysis of the anterior ends of the two dentaries is a rigid suture in some reptiles, whereas in others, particularly snakes and many lizards, it is merely a ligamentous connection. The articular and prearticular are fused in most reptile groups, and frequently the surangular fuses with them; the angular may also fuse to the articular-prearticular combina-

tion. Anteriorly, the splenial commonly fuses to the dentary. In snakes and some lizards, there is a flexible joint between the anterior (dentary and splenial) and posterior elements of the jaw.

Cranial kinesis, independent movement of the upper jaw on the braincase, and general flexibility of the skull vary among reptiles: due to the solidly fused structures in Chelonia, Crocodilia, and *Sphenodon*, there is very little if any kineticism, whereas other structural situations in lizards permit the presence of different types of kinesis (Fig. 4–19); and in advanced snakes (Fig. 4–20) an extremely flexible situation, in which the palatal apparatus and jaws are loosely connected to the skull, allows each half of the upper and lower jaws to be moved independently of the other (this provides the great distensibility of the mouth necessary for swallowing large prey). In contrast to advanced snakes, burrowing lizards and snakes tend to have solidly bound anterior skull elements, with an expanded nasal region and braincase, and reduced eyes and orbits; in these, the snout region is the burrowing tool and the elongate braincase is the "handle" of the tool (Romer, 1956; Fig. 4–21).

HYOBRANCHIAL SKELETON

The hyoid apparatus of reptiles is concerned primarily with the support and movement of the tongue and support of the larynx. It typically consists of a median ventral copula from which the lingual process extends anteriorly into the tongue and paired horns extend posteriorly and dorsally around the throat region to end freely or attach to the cranium. However, there is much variation in the structure and ossification of the hyoid apparatus in different reptilian groups (Fig. 4–22). Three pairs of well-developed cornua are present in the hyoid apparatus of tuataras and only the first ceratobranchial pair is ossified. That of primitive lizards is similar to the hyoid apparatus of tuataras, but in advanced members the third pair of cornua may be reduced or lost; the third epibranchial portion may be lost while the ceratobranchial persists, the epibranchial may be present while the ceratobranchial is reduced or lost, or both epibranchial and ceratobranchial may be

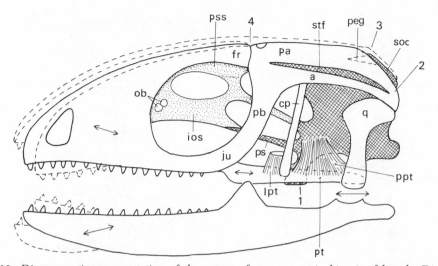

Figure 4–19 Diagrammatic representation of the extent of movements in kinesis of lizards. Directions of movement are shown by arrows and positions of jaws by broken lines. The 'fixed' occipital segment is shown in cross-hatching and the cartilaginous anterior parts of the inner shell are stippled. The numbers refer to the positions of the principal kinetic joints. *a*, superior temporal arch; *cp*, epipterygoid; *fr*, frontal; *ios*, interorbital septum; *ju*, jugal; *lpt*, levator pterygoidei muscle (only lower part shown); *ob*, origins of superior and inferior oblique muscles of eye; *pa*, parietal; *pb*, postorbital arch or bar; *peg*, cartilaginous peg projecting from front of supraoccipital into back of parietal in midline; *ppt*, protractor pterygoidei muscle; *ps*, parasphenoid rostrum; *pss*, planum supraseptale surrounding olfactory stalks of brain; *pt*, pterygoid; *q*, quadrate; *soc*, supraoccipital; *stf*, superior temporal fossa; *1*, joint between basi-pterygoid process (outline in broken lines) and pterygoid; *2*, region of joint between paroccipital process (not visible) and parietal and supratemporal; *3*, joint between supraoccipital peg and parietal. These three joints are functional in metakinetic forms. *4*, joint between frontal and parietal, found in mesokinetic forms, where the postorbital bar and superior temporal arch tend to disappear. Mesokinesis + metakinesis = amphikinesis, as in *Varanus*. (From Bellairs: *The Life of Reptiles*.)

lost. Lizards in the families Iguanidae and Agamidae having neck frills have well-developed second ceratobranchial cornua which extend posteriorly as a support for the frill. The lingual process is always slender and elongate in lizards but is particularly long in true chameleons (family Chamaeleontidae), which have extremely extensible tongues. The hyoid apparatus of snakes and a few lizards is reduced to a tuning-fork design and is cartilaginous. That of adult crocodilians consists of a broad plate and a single pair (first cerato-branchial) of ossified cornua; the lingual process and hyoid cornu develop embryonically but are resorbed ontogenetically. A pair of ossifications, which have been interpreted as second ceratobranchials (Romer, 1956), develop in the posterior plate of old crocodilians. The hyoid apparatus of turtles is massive, and most of it is ossified. An extra cartilaginous element, the entoglossum, lies beneath the tongue in the floor of the mouth and is

unique to turtles; it appears to be a neomorph (Romer, 1956).

The reptilian larynx is supported by paired arytenoid cartilages and a ring-shaped cricoid cartilage, much as in anuran amphibians.

VERTEBRAL COLUMN, STERNUM, AND RIBS

As a group, reptiles vary widely in habitats and habits; consequently, they exhibit a wide range of body forms and have many different types and arrangements of vertebrae. It is possible, however, to make a few general remarks concerning their axial skeleton. The regional differentiation of the vertebral column, initiated in amphibians, is much more apparent in the reptiles. All reptiles have a distinct neck characterized by cervical vertebrae with short ribs that do not reach the sternum or, in some cases, seem to be lacking. As in amphibians, the vertebra that articulates with the

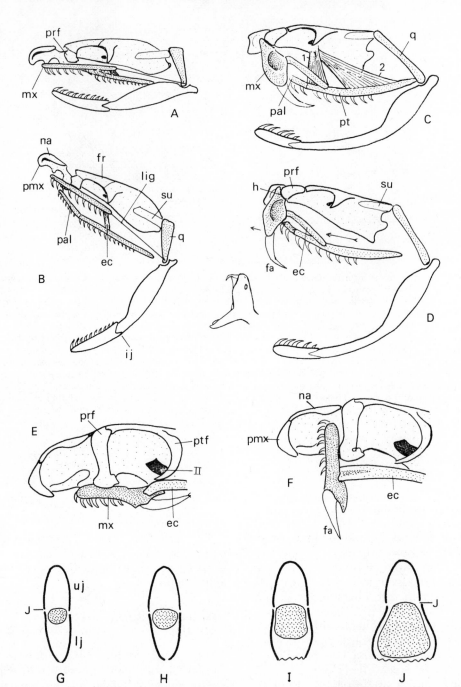

Figure 4–20 Kinesis in snakes. The movable bones of the upper jaw are stippled. *A*, *B*, jaw movements in a generalized colubrid during closing and opening of mouth; *C*, *D*, method of fang erection in viperid snake showing quadrate and pterygoid moving independently. The muscles shown in *C* are the levator (1) and the protractor (2) pterygoidei. Arrows in *D* show direction of movement of bones. The small inset shows a viper about to deliver its stabbing strike with the head drawn back, the fangs fully erected, and the jaws almost at full gape. *E*, *F*, front of skull of the aglyphous colubrid *Xenodon* sp. showing the non-poisonous fang at the rear of the maxilla folded back in *E* and erected in *F*. *G–J*, diagrams showing stages in evolution of mechanism for eating larger prey. The mouth is seen from in front, wide open. The thick black lines show the margins of the upper and lower jaw bones, the stippled areas represent the prey. The lines (*J*) indicate the transverse axes of the jaw joints. *G*, tips of lower jaw attached firmly to each other; *H*, slight separation of lower jaw tips and greater mobility of quadrate, which allows some increase of distance between jaw joints on two sides of head; *I*, freeing of lower jaw tips so that skin between them (corrugated lines) is stretched; *J*, final stage (more

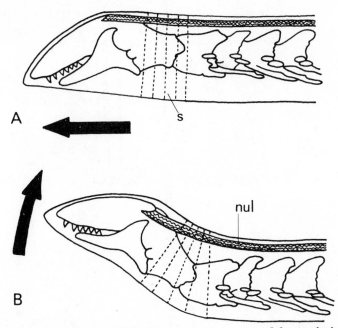

Figure 4–21   Diagrammatic representation of the burrowing movements of the amphisbaenid *Rhineura flori-dana*. *A*, ramming stroke; *B*, digging or elevating stroke, as shown by arrows. The dashed lines indicate the arrangement of the folds between the rings of scale (*s*). *nul*, nuchal ligament. (From Bellairs: *The Life of Reptiles*; after Gans: Bull. Amer. Mus. Nat. Hist. *119*:129--204.)

head is modified as the atlas. In reptiles, as in birds and mammals, the second cervical vertebra is also modified as the axis. The atlas is formed primarily by basidorsals, with only a small contribution from the basiventrals. The basidorsals form a neural arch and the basiventrals join to form a small hypocentrum ventral to the nerve cord. The interdorsals from the first vertebra (atlas) do not attach to it but, rather, join with the centrum of the axis to form an anterior projection, called the odontoid process. The mechanism formed is such that the atlas functions as a split washer between the axis and the skull. The single occipital condyle on the skull articulates with the hollow formed by the base of the neural arch and the hypocentrum.

The mammalian trunk vertebrae are divided into an anterior thoracic set which bears ribs and a posterior lumbar set which does not. In reptiles, the terms thoracic and lumbar are not too meaningful, for generally all trunk vertebrae bear ribs and there is never a well-defined lumbar region. However, the ribs in the posterior region of the trunk of reptiles other than snakes are usually short and capable of little, if any, movement. Often they seem to fuse with the vertebrae and give the impression of being merely transverse processes of the vertebrae, so it is some-

Figure 4–20 *Continued.*
or less realized in some snakes) where the tips of the lower jaws can spread wider than the distance between the jaw joints. The widest part of the prey can pass ventral to the jaw joint axis, and the tissues of the throat are also distensible. *h*, hinge between prefrontal and maxilla; *ij*, intramandibular joint (movements not shown); *lig*, ligament (quadratomaxillary); *lj*, *uj*, lower and upper jaws; other abbreviations are defined in legend for Figure 4–16. (From Bellairs: *The Life of Reptiles*; *A* and *B*, after Albright and Nelson: J. Morpholgy *105*:193–240, 241–292; *E* and *F*, after Anthony: Ann. Sci. Nat. Zool. *17*:7; *G–J* after Gans: American Zoologist, *1*:217–227.)

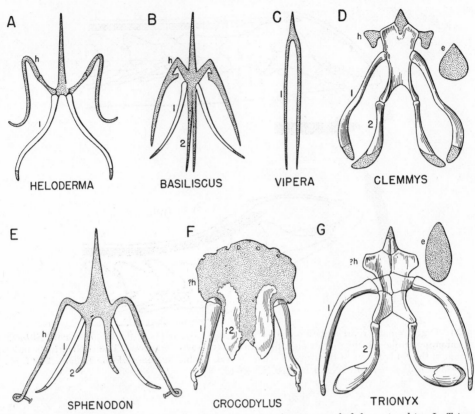

Figure 4–22 The hyoid apparatus of a variety of reptiles. Cartilage is stippled; bone is white. In *Trionyx* (soft-shelled turtles) the body of the hyoid is unusual in its ossification from a number of centers. *h*, hyoid horn; *1, 2*, first and second branchial horns; *e*, entoglossal element of chelonians (not in relative position). (From Romer: *Osteology of the Reptiles.*)

times impossible to determine whether or not true ribs are present in this "lumbar" region. In snakes, the ribs are powerful, semicircular, and connected distally with the enlarged belly scales. They play an important role in rectilinear locomotion but, contrary to the "rib-walking theory," remain stationary relative to the vertebrae (Bogert, 1947). During such locomotion the skin is pulled forward, the belly scales contact the surface and provide stationary points, and then the body is pulled by them (Gans, 1962; Fig. 4–23). In reptiles other than snakes, there may be one or two "lumbar" vertebrae (lizards) or more than two (crocodilians).

Whereas amphibians have a single sacral vertebra, even the most primitive reptiles had two, and in many extinct forms, among dionosaurs especially, there were up to six; living reptiles generally have two sacral vertebrae. The exact number of sacrals is sometimes difficult to determine

in reptiles, for, although the beginning of the sacral series is generally obvious, there may be transitional vertebrae between what are clearly sacrals and caudals; fusion of sacral vertebrae also occurs to various degrees among reptiles.

All reptiles have a distinct tail, and the vertebral column terminates with the caudal region. Small chevron bones on each side of the caudal vertebra seem to represent the basiventrals and, if this is true, are homologous with the hemal arches of fishes and amphibians. Elsewhere in the vertebral column, there is a tendency for the reduction of the basiventrals in all reptiles. In many reptiles, they are present only as small ventral crescents wedged between the centra of successive vertebrae; in others, they disappear completely.

A variety of types of centra are found among reptiles. They are amphicoelous in the most primitive species and typically procoelous in later groups, but there are exceptions. They may be amphiplatyan, in which case the ends of the centra are flat rather than concave or convex; they may be biconvex, in which case both ends of the centrum are convex; or they may be opisthocoelous. Most living reptiles have procoelous vertebrae, but again there are exceptions. For example, some of the vertebrae of turtles are opisthocoelous, those of the tuatara and geckonid lizards are amphicoelous, and the caudal vertebrae of crocodilians are biconvex.

The tuatara and many lizards are capable of caudal autotomy. This is facilitated by a vertical plane of fracture that passes through the centrum and at least part of the neural arch of caudal vertebrae. The plane of fracture is formed by a septum of cartilage or connective tissue that develops after ossification has taken place (Etheridge, 1967); the vertebra is easily broken at this point, so a portion of the tail may be easily snapped off in escape maneuvers. Natural selection would act against autotomy of the anterior portion of the tail, for it accommodates the hemipenes, slips of pelvic musculature, fat deposits, and so forth; consequently, the base of the tail is always supported by a

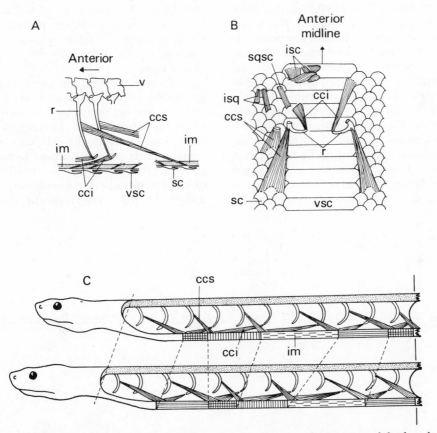

Figure 4–23 *A,* diagrammatic representation of the ventral scale musculature in a colubrid snake (*Natrix*) as viewed from the left side; *B,* dorsal view of muscles associated with ventral scales after removal of rest of body. Only a few of the individual muscles are shown. *C,* diagrammatic illustration of mechanism of rectilinear locomotion in a boa. The superior costocutaneous muscles (*ccs*) hitch the ventral skin with its broad scales forward at each 'step,' while the intrinsic muscles of the ventral scales relax. Then the scales are engaged against the substrate and the inferior costocutaneous muscles contract, pulling the ribs, vertebrae, and body as a whole forward in relation to the ventral skin; the intrinsic ventral muscles contract at the same time. The rhythm of these activities follows a wave pattern passing down the body. Differences in the state of contraction of the muscles in several different sections of the body are shown. *cci,* inferior costocutaneous muscles; *ccs,* superior costocutaneous muscles; *im,* intrinsic muscles of scales, including *isc,* interscutali, *isq,* short and long intersquamali; *sqsc,* squamoscutali; *r,* rib; *sc,* body scales adjacent to ventrals; *v,* vertebra; *vsc,* ventral scales (gastrosteges). (From Bellairs: *The Life of Reptiles; A* and *B,* after Bellairs and Underwood: Biol. Rev. 26:193–237; *C,* after Lissmann: J. Exp. Biol. 26:368–379.)

series of non-autotomic caudal vertebrae. These are distinct from the autotomic vertebrae: they are functionally more closely associated with the sacral and posterior trunk vertebrae than with the autotomic caudal vertebrae and differ from the latter in bearing a single pair of transverse processes which are usually oriented laterally or posterolaterally. The beginning of the autotomic series of vertebrae is indicated by an abrupt modification in the pattern of these transverse processes; they may have a different orientation, shape, and size, they may disappear completely, or a second pair of processes may appear (Etheridge, 1967; Fig. 4–24). When present, fracture planes are usually found in all vertebrae posterior to the non-autotomic series. However, they may be restricted ontogenetically to a few vertebrae, or they may disappear entirely through replacement by bone of the cartilaginous septum as the individual matures; the former occurs in chuckwallas (*Sauromalus obesus*, family Iguanidae), while the latter situation prevails in the common iguana (*Iguana iguana*, family Iguanidae). It should be noted that, rather than having fracture planes, a few lizards (*Agama*, family Agamidae, for example) and some snakes have fragile

tails that will break off intervertebrally. Extensive regeneration, including replacement of the vertebral column by a calcified cartilaginous rod, normally occurs after autotomy at a fracture plane, but regeneration following an intervertebral break is rare (Etheridge, 1967). Commonly, a regenerated tail has a different scutellation and color pattern from the original, and it may be shorter and more blunt. In other instances, however, only X ray will distinguish a regenerated tail from the original. The selective advantage of caudal autotomy lies in the distraction of the predator by a brightly colored tail wriggling about after being broken off or escape from the predator's grasp when caught by the tail. This advantage seems to be very high in lizard populations, for the frequency of adults with lost tails has been generally reported to be greater than 30 per cent.

As in salamanders, anurans, and higher tetrapods, the intervertebral joints of reptiles are strengthened by articulating prezygapophyses and postzygapophyses. Snakes and limbless lizards have two additional points of articulation between adjacent vertebrae. In these reptiles, a pair of structures called zygosphenes protrude anteriorly from the base of the neural

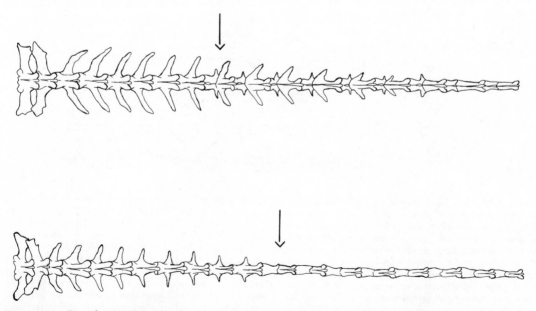

Figure 4–24 Dorsal view of the sacral and anterior caudal vertebrae of *Basiliscus vittatus* (left) and *Dipsosaurus dorsalis* (right). Arrows indicate fracture planes of the most anterior vertebrae possessing them. Remaining vertebrae in each column are similar to the most posterior vertebra illustrated. (After Etheridge: Lizard caudal vertebrae.)

arch and fit into grooves called zygantra on the posterior face of the next vertebra forward. This additional articulation strengthens the vertebral column while allowing it to remain flexible in the horizontal plane, a definite advantage in these animals that depend upon body movements for locomotion.

Processes from the vertebrae have become associated with an armored exoskeleton in more than one group of reptiles. In turtles, the neural spines of the thoracic, sacral, and first caudal vertebrae are expanded and become fused with a row of dermal bones to form the neural plates that comprise the middorsal axis of the shell. The 10 thoracic vertebrae of turtles lack transverse processes. In some members of the extinct order Pelycosauria the neural spines were huge structures, up to three feet in length, that presumably supported a web of skin or a sail-like structure, the function of which is not clear (Fig. 4–25).

Reptiles generally have ribs on all precaudal vertebrae; some also have caudal ribs. The ribs vary greatly in both shape and vertebral attachment among different groups. The evolution of a distinct neck in most reptiles was accompanied by a reduction of the ribs in this region, and modification of the sacral region in very large or bipedal forms has often included an increase in the number of sacral ribs. As in amphibians, most primitive reptiles had bicapitate (two-headed) ribs, with the capitulum (the head proper, representing the proximal end of the rib) articulating with the centrum of the vertebra and the tuberculum (a proximodorsal projection) articulating with the transverse process of the neural arch. In later groups, the attachments of the two heads to the vertebra frequently changed position or there was a reduction to a single head. For example, in the subclass Archosauria, the capitulum frequently attaches to the neural arch, although the tuberculum remains attached to the transverse process. In many Permian reptiles, the anterior ribs were distinctly two-headed, but in posterior ribs the two heads approach each other more and more closely until the most posterior ones are essentially single-headed. Of living reptiles, the tuatara, snakes, and most lizards have single-headed ribs which seem to have evolved through the fusion of the capitulum with the tuberculum to form a single knob that articulates with the centrum. The ribs of turtles are distinctive in several ways. They are double-headed, and those of the trunk are greatly expanded and support the carapace. Short ribs are present in the sacral and caudal regions, those in the tail frequently being fused to the vertebrae. Crocodilians have the archosaurian type of double-headed rib. Both the tuatara and the crocodilians have uncinate processes on their ribs, cartilaginous projections extending posteriorly from one rib to articulate with the next rib posterior. Uncinate processes are also characteristic of birds, where they are explained as an adaptation to strengthen the thorax so that it will not crush under the force exerted by the downbeat of the wings during flight; however, their presence in the two groups of reptiles is unexplainable.

Typically, reptilian ribs attach ventrally to a sternum in the anterior part of the trunk. In lizards and the tuatara, more posterior trunk ribs may attach to a parasternum situated between the sternum and the pelvic girdle. Turtles and snakes lack a sternum, turtle ribs being fused to the carapace and those of snakes being attached to the belly scales. What is undoubtedly the greatest specialization in

Figure 4–25 Skeleton of a herbivorous pelycosaur (*Edaphosaurus*) with long neural spines. (From Romer, A. S.: *Vertebrate Paleontology*. 3rd Ed. Univ. Chicago Press, 1966.)

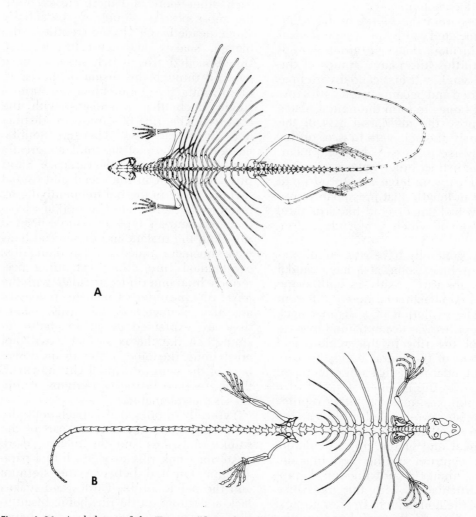

Figure 4–26 A, skeleton of the Triassic "flying lizard" (*Kuehneosaurus*). B, skeleton of the "Flying Dragon" (*Draco volans*). (A, from Robinson, P. L.: Gliding lizards from the Upper Keuper of Great Britain. Proc. Geol. Soc. London, No. 1601: 137–146; B, from Bellairs: *The Life of Reptiles*, after Owen: *Comparative Anatomy and Physiology of Vertebrates*.)

function of ribs in any vertebrate has occurred in two groups of lizards where they have been used to support flaps of skin used as gliding membranes (Fig. 4–26). In the extinct "flying lizards" (family Kuehneosauridae), the trunk ribs were greatly elongated and rigidly extended horizontally to support what seem certain to have been horizontal sails used in gliding. In the living "flying dragons," *Draco volans* (family Agamidae), the flight membranes are supported by five, six, or seven elongated posterior trunk ribs; spreading of the membrane is accomplished by contraction of iliocostal and intercostal muscles, while the arching and stiffening of the ribs are performed by very long, slender muscle slips that run along the length of each rib (Colbert, 1967). The flight membranes of *Draco* allow them to

glide up to 60 meters. When the lizard is resting and the ribs are pressed against the body, the flight membranes are scarcely evident.

LIMBS AND GIRDLES

There is much variation in the limb structure of reptiles as a result of adaptive radiation to burrowing, terrestrial, arboreal, and aquatic habits and to modes of locomotion including body undulation, bipedal and quadrupedal gaits, flight, and swimming. Reptilian limbs are typically pentadactyl, five digits normally being present on both the front and rear feet, and the forelimbs and hind limbs are fundamentally the same (Figs. 4–27 and 4–28). An interesting feature of the rep-

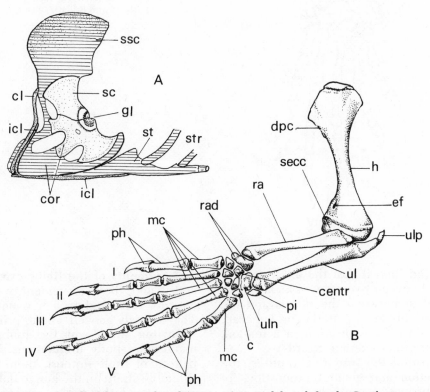

Figure 4–27 *A,* pectoral girdle of monitor lizard (*Varanus*) viewed from left side. Cartilaginous regions are lined, bone stippled, and fenestrae unshaded. *B,* left forelimb of young *Varanus,* lateral view. The carpal bones are shown slightly separated from one another and the head of the radius is slightly displaced. *c,* distal carpal of fifth digit; the other four distal carpels are unlabeled; *centr,* centrale; *cl,* clavicle; *cor,* coracoid; *dpc,* front of deltopectoral crest; *ef,* ectepicondylar foramen; *gl,* glenoid fossa for head of humerus (*h*); *icl,* interclavicle; *mc,* metacarpels; *ph,* phalanges; *pi,* pisiform; *ra,* radius; *rad,* radiale (perhaps fused with intermedium); *sc,* scapula; *secc,* secondary bony center (bony epiphysis); there is another at the head of the humerus; *ssc,* suprascapula; *st,* sternum; *str,* sternal segment of rib; *ul,* ulna; *uln,* ulnare; *ulp,* ulnar patella; *I–V,* digits. (From Bellairs: *The Life of Reptiles; A,* after Parker: *A Monograph on the Structure and Development of the Shoulder-girdle and Sternum in the Vertebrata,* Ray Soc.)

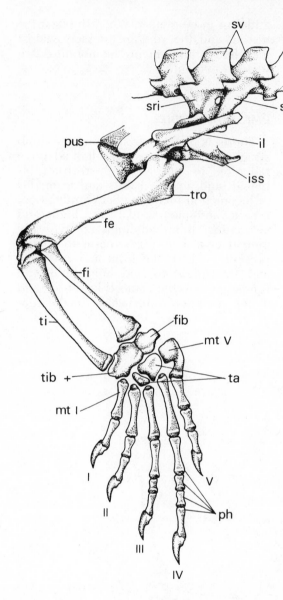

Figure 4–28 Pelvis and left hind limb of monitor lizard (*Varanus*), seen from the side. Tarsal bones are shown slightly separated. *fe*, femur; *fib*, fibulare or calcaneum; *il*, ilium; *iss*, ischial symphysis; *mt*, metatarsals; *ph*, phalanges; *pus*, pubic symphysis; *sri*, sacral ribs; *sv*, spines of vertebrae (sacral); *ta*, tarsals; *ti*, tibia; *tib +*, a compound bone which probably represents the tibiale, intermedium, and one of the centralia, and is often called the astragalus; *tro*, trochanter; *I-V*, digits. (From Bellairs: *The Life of Reptiles*.)

tilian hind limb is that the ankle joint is located between two rows of tarsal bones rather than between the tarsals and the tibia and fibula. A patella, or kneecap, formed through ossification in a tendon, is first present in certain lizards.

In most of the primitive reptiles, the construction and general proportions of the limbs were essentially the same as in many of the extinct amphibians. The rear legs were often moderately longer than the front legs, but both pairs were short and powerful; the body was close to the surface and the length of the stride was short. The feet turned almost directly forward but, as in amphibians, the proxi-

mal segments of the limbs extended laterally and their movement was almost entirely restricted to a horizontal plane. The animals moved about primarily by sprawling thrusts of the limbs supplemented by body undulations. With this pattern of limb support, there was considerable strain on the horizontal proximal segments of the limbs, explaining their short length and stout construction. The primitive sprawled pose was characteristic of many of the cotylosaurs, the turtles, pelycosaurs, and other large quadrupeds. An evolution of more elongated, slender limbs took place in faster and smaller quadrupedal forms. In two specialized groups, the pareiasaurs (order Cotylo-sauria) and the mammal-like reptiles (order Therapsida), the structure of the girdles and limbs changed so that the elbow rotated backward and the knee forward. This resulted in limbs that were brought closer to the body, so that much of the weight of the body was vertically supported by the limbs rather than suspended between horizontal elements. This important locomotor advance led to the very efficient gaits characteristic of mammals and solved much of the problem of weight-bearing by the limbs.

The order Squamata is the only group of reptiles in which there has been an evolutionary loss of limbs and redevelopment of undulatory body movements; lateral undulation is supplemented in some cases by rectilinear, concertina-like, and side-winding modes of locomotion. In various families of lizards different stages in the reduction and loss of limbs may be found, from those in which there are two pairs of well-developed limbs to those of partial reductions (generally of the forelimbs) to those in which there has been complete loss of limbs. All snakes lack front limbs,

but in the more primitive families, such as the boas (family Boidae), vestigial rear legs are present; in more advanced forms, all vestiges of limbs are gone. This evolutionary loss of limbs is generally associated with burrowing habits or densely-vegetated habitats such as grasslands.

Gans (1962) has discussed the evolution, mechanics, and effectiveness of limbless movement: he notes that lateral undulatory motion (Fig. 4–29A) seems to be the first step for species evolving toward limblessness and that all species of limbless tetrapods make use of this form of locomotion; some short-bodied, long-tailed reptiles also utilize lateral undulatory motion. It is used in traveling both on the surface of the ground and (by arboreal forms) over and amid twigs and branches of trees. The effectiveness of lateral undulation is reduced when the substratum lacks fixed surfaces or when these fixed surfaces are too widely spaced. Many snakes living where these conditions prevail utilize concertina-like movement (Fig. 4–29B), in which the stationary portion of the body is bent into a series of S-coils, from which the moving portion

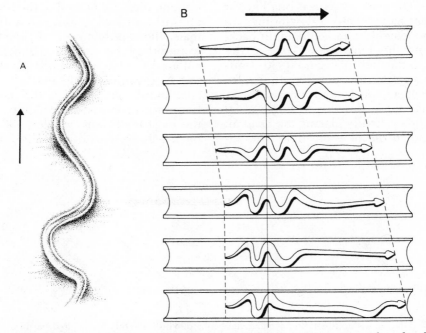

Figure 4–29 *A*, the track pattern left by a snake using lateral undulations to move over soft sand in the direction of the arrow; thrust of the body against each outside curve causes the sand to be piled at that point. *B*, diagrammatic illustration of a snake crawling through a narrow trough by means of concertina movement. (From Bellairs: *The Life of Reptiles; B*, after Gans: Nat. Hist. New York 75(2):10–17; (3):36–41.)

straightens and then bends again; the basic pattern involves bending, stretching, and rebending. The concertina pattern is the primary mode of movement of several groups of heavy-bodied arboreal snakes, including many boids and rat snakes (*Elaphe*, family Colubridae).

Rectilinear locomotion, described earlier, has evolved in both snakes and amphisbaenid lizards through liberation of the skin from its non-elastic connection with the skeleton and through the development of dermal mobility. Amphisbaenids are adapted for moving in underground tunnels: they have symmetrical rings of scales that provide friction points for gripping the tunnel walls and are capable of going forward and backward; snakes can only move forward through rectilinear locomotion. Rectilinear movement in snakes is advantageous in that it permits stout snakes to traverse smooth surfaces, allows passage through narrow crevices and along branches, and gives short snakes the ability to approach a prey in a straight line with movement less obvious than that in other kinds of locomotion.

Sidewinding is unique to snakes and provides the advantage of travel over a smooth substratum at a higher speed than could be attained by concertina-like or rectilinear movement. Sidewinding involves movement at an acute angle backward along the longitudinal axis of the body. As indicated in Figure 4–30, starting from an extended position, the snake raises its head, neck, and anterior body and swings these through an angle of 90 to 120 degrees, finally placing the head on the ground in the new position. The an-

terior body spans from old to new contact position but remains slightly off the substratum. The snake next lifts its body section by section from the old line of rest and starts to roll down the neck end of the suspended loop. The downward movement starts from the resting position of the neck and continues along a track more or less parallel to the old one; the head normally starts swinging to a third track before the tail comes to rest on the second (Gans, 1962).

A variety of reptiles has utilized bipedal locomotion. Sometimes, as in the subclass Archosauria, this has involved modification of the pelvic girdle and hind limbs in such a way that the knee is rotated forward and the limbs brought close to the body for vertical support of it; the extinct Therapsida evidence this modification. There is characteristically associated with bipedal locomotion a long tail used to balance the weight of the body. Among very specialized bipedal species, in which there is complete dependency upon the hind legs for locomotion, there tends to be a reduction in the size of the front legs and a reduction in the number of toes on the front feet. The principal advantage of a bipedal hopping mode of locomotion is greater speed for catching prey or escaping enemies. Compared to the bipedalism of birds, primates, and saltatorial mammals, the reptilian type of bipedalism as exemplified by modern lizards such as collared and leopard lizards (*Crotaphytus*, family Iguanidae) and basilisks (*Basiliscus*, family Iguanidae) is the most primitive and, in terms of economy of action, the most inefficient (Snyder, 1962); it has been achieved with only slight modifica-

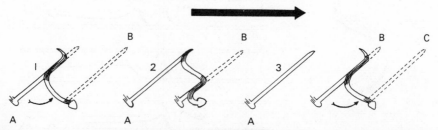

Figure 4–30 Diagrammatic illustration of movement by sidewinding. Only the shaded portions of the snake are in contact with the surface; solid lines indicate impressions already made while broken lines indicate future impressions. *A* and *B* identify the same tracks in three successive phases of movement; these phases are numbered *1*, *2*, and *3*; *C* in phase *3* identifies the next track which will be made. Movement is in the direction of the dark arrow. (From Bellairs: *The Life of Reptiles*; after Gans: Nat. Hist. New York 75(2):10–17; (3):36–41.)

tions of the pelvic or appendicular structure (Snyder, 1954). Compared to quadrupedal lizards, bipedal species have a slightly narrower pelvis, a longer ilium with a better developed preacetabular process, and heavier, more solidly fused transverse processes of the sacral vertebrae (Snyder, 1962). Reduction of the forelimb is relatively greatest in the manus; elongation of the hind leg of bipedal iguanids involves each limb segment and is most marked in the proximal segments, whereas in bipedal agamids there is either no elongation of the hindlimbs or an increase primarily in the foot (Snyder, 1962).

The evolution of bipedal locomotion freed the forelimbs for other functions, leading to the modification of the forelimbs into wings in two independent lines of descent stemming from the order Thecodontia: the order Pterosauria, which includes the extinct flying reptiles, and the class Aves (birds), which, while also derived from the thecodonts, represents a phylogenetic line different from that of the pterosaurs. As in the bat, the pterosaur wing consisted of a membrane of skin, but it was supported entirely by one greatly elongated digit; the other four digits were small hooklike structures that perhaps were used for clutching or for hanging from branches or ledges. The relatively weak structure of the pterosaur wing and its very elongate shape indicate these reptiles were primarily gliders that depended upon soaring rather than flapping of their wings.

Modification of the limb has also occurred in aquatic reptiles. Members of the extinct order Ichthyosauria, porpoiselike reptiles, possessed a well-developed tail that was their primary organ for locomotion; their limbs were reduced to flippers used for steering. In another group of specialized aquatic reptiles, the order Plesiosauria, swimming was accomplished by long oarlike limbs with greatly elongated paddle feet. The Upper Cretaceous marine lizards (mosasaurs, family Mosasauridae) used their tail as the main swimming organ but had paddlelike limbs with shortened proximal bones and elongated digits which were probably webbed. Highly aquatic turtles such as the living marine varieties utilize their limbs for swimming, and their forelimbs are especially enlarged and paddle-shaped. There is generally little or no modification of limbs in more amphibious or terrestrial species.

The girdles of cotylosaurs were very similar to those of the rhachitomous amphibians. The pectoral girdle was a massive structure that included both dermal (clavicles, cleithra, and interclavicle) and endochondral (scapula, coracoid, and procoracoid) elements. It differed from that of most amphibians in having a separate coracoid component and a cleithrum that was reduced to an elongated splint lying adjacent to the scapular blade. The cleithra are completely lost in more advanced reptiles. The pelvic girdle of cotylosaurs was composed of the typical three elements (pubis, ischium, and ilium) and differed from those of amphibians in having a greater development of the ilium. In modern reptiles, girdles are well-developed, except in the limbless lizards and snakes. No trace of the pectoral girdle may be found in snakes, but those species having vestiges of hind limbs also have vestiges of the pelvic girdle. The girdles of turtles are unique in that they are located internal to the ribs, but nevertheless they are composed of elements comparable to those of other reptiles.

MUSCULAR SYSTEM

The muscular system of reptiles is generally better adapted to terrestrial life than that of amphibians, primarily because of modification of the muscles used in conjunction with the vertebral column to support the weight of the body, of those used in pulmonary respiration, and of those used in various modes of locomotion.

BODY MUSCLES

The epaxial muscles of reptiles are less modified than the hypaxial (Fig. 4–31). The principal epaxial muscles of the trunk are the *longissimus dorsi,* lying dorsal to the transverse process of the vertebrae, and the *transversospinalis* muscles, a series of smaller muscles between the vertebrae. These continue anteriorly as, respectively, the *longissimus cervicocapitis* and the *spinalis capitis.* A third

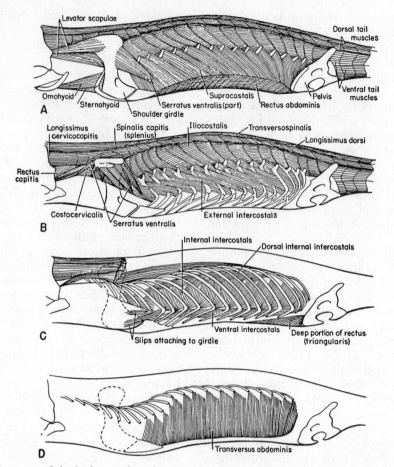

Figure 4–31 Anatomy of the body muscles of *Sphenodon* as they would be revealed by dissections. In *A*, a thin superficial sheet of the external oblique has been removed. In *B*, the supracostals, rectus, throat muscles, and more superficial muscles to the scapula have been removed. In *C*, the epaxial muscles are cut posteriorly, and the internal intercostals and triangularis (not shown in the last figure) are indicated. In *D*, the ribs have been cut and all other muscles removed to show the transversus abdominis. (From Romer, A. S.: *The Vertebrate Body*. 4th Ed. W. B. Saunders Co., 1970.)

epaxial muscle, the *iliocostalis*, extends ventrally as a thin sheet over the dorsal flank of the trunk region. These epaxial muscles, especially the *longissimus dorsi*, control vertical bending of the vertebral column. They are opposed by the *subvertebralis*, located ventral to the vertebral column; this muscle is derived from the hypaxial musculature. The *longissimus cervicocapitis* and *spinalis capitis* function primarily to support and move the head.

The hypaxial muscles are more complicated in reptiles than in amphibians because of the development of ribs on most, if not all, trunk vertebrae. In addition to a subcutaneous sheet of muscles, the external oblique muscle of reptiles is split into a superficial layer and a deep layer. The outer layer of the external oblique lies external to the ribs and forms the supracostal muscles. The deep layer of the external oblique forms the external intercostal layer, while the internal oblique forms the internal intercostal layer. The *transversus abdominis* persists as a continuous thin sheet internal to the ribs. Ventrally, the *rectus abdominis* runs longitudinally from the shoulder region to the pelvis. By controlling rib movement and reinforcing the body wall, the hypaxial musculature functions both in breathing and in supporting the viscera.

The development of the trunk musculature varies considerably from one kind of reptile to another. At one extreme, turtles with well-developed bony shells have very reduced dorsal and ventral trunk muscles. At the other extreme, locomotion in snakes is dependent upon muscles that lie between vertebrae, between vertebrae and ribs, between ribs, and between vertebrae or ribs and integument (Auffenberg, 1962). Consequently, trunk muscles of snakes are well-developed and complex and allow a greater variety of movements than is possible in the trunk of limbed forms. Auffenberg (1962) has recognized three evolutionary tendencies that lead from simpler to more complex musculature in snakes: (1) a tendency for median myological elements to become associated with more lateral ones; (2) a tendency for anterior elements to become morphologically associated with more posterior ones, thereby lengthening considerably certain functional units through fusion of shorter muscles; and (3) a tendency for myological chains to form through linkage of different muscles so that one unit arises from the terminal insertion of another. He also has noted that certain minor modifications in musculature are associated with burrowing, terrestrial, and arboreal habits. Fossorial snakes generally have lower neural spines and arches than surface-dwelling species; their dorsal epaxial muscles (*spinalis* complex and *multifidus*) insert proportionately closer to the longitudinal axis of the centrum and are relatively reduced in dorsal-ventral thickness. These muscles are primarily involved in moving anterior parts of the body dorsally and laterally. In contrast, the *longissimus* muscles are better developed in fossorial snakes than in nonburrowers; these probably are involved largely in lateral movements. Arboreal snakes that depend upon short radius curves for movement, such as constrictors (*Constrictor*, family Boidae), are very dependent upon vertical control of parts of the body; compared to surface dwellers, their neural spines and arches are usually relatively high and their *spinalis* complex and *multifidus* muscles well-developed. Arboreal snakes such as green tree snakes (*Dryophis*, family Colubridae), which depend upon body rigidity for bridging or spanning rather than for short radius curves, tend to have long slim bodies with thin vertebrae and low neural arches and spines; these modifications result in dorsal epaxial muscles that lie in a horizontal rather than vertical plane.

Crocodilians have retained a relatively primitive distribution of dorsal and subvertebral muscles and use them for lateral bending of the tail when swimming or doing the undulatory "belly walk."

MUSCLES OF THE FORELIMB AND PECTORAL GIRDLE

As in all tetrapods, the reptilian body is slung from the scapulae by muscles; these are the *serratus ventralis* ventrally and the *levator scapulae* dorsally. Both of these are derived from the external oblique muscle. In addition, the *trapezius* muscle, derived from the visceral musculature, passes between the scapula and the vertebral column of the neck.

In quadrupedal locomotion, the muscles attached to the reptilian humerus and femur have two primary functions: (1) they must hold the bones steady at the appropriate angle to the horizontal so that the height of the body is maintained; and (2) they must rotate the bones backward and forward (Carter, 1967). As indicated in Figure 4–32, the *latissimus dorsi* is inserted on the posterior side of the humerus and passes behind the girdle to the flank; this supports the humerus in a vertical plane and rotates it posteriorly. The *subcoracoscapularis,* a deeper muscle, also inserts on the humerus and passes to the scapula; it has the same function as the *latissimus dorsi.* The *deltoideus* runs from the dorsal side of the humerus to the scapula and clavicle; thus, it both holds the humerus up and pulls it forward. Lying under the scapular deltoid is the *scapulohumeralis;* passing from the dorsal side of the humerus to the scapula, it also holds the humerus up. Ventral muscles opposing those mentioned are the *pectoralis*, which passes from the ventral side of the head of the humerus to the sternum and ribs, and the *supracoracoideus*, which extends from the ventral side of the humerus to the clavicle and interclavicle. Both of these function to prevent the humerus from rising; the *pectoralis* also pulls it posteriorly, while the *supracoracoideus* pulls it anteriorly.

The dorsal surface of the humerus is covered by the stout *triceps*, which extends from the olecranon process of the

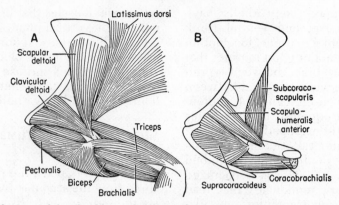

Figure 4-32   Lateral views of the shoulder and upper arm muscles of a lizard. Superficial muscles are shown in A; in B, the latissimus, deltoid, pectoralis, and long muscles (triceps, biceps, brachialis) have been removed. (From Romer, A. S.: *The Vertebrate Body.* 4th Ed. W. B. Saunders Co., 1970.)

ulna to attach by one or more heads to the pectoral girdle; by pulling on the olecranon process, this muscle extends the forearm. The ventral side of the humerus is covered by the *coracobrachialis;* this attaches to the coracoid and functions to extend the upper arm. Two muscles, the *brachialis* and the *biceps,* flex the forearm: the *brachialis* originates on the anterolateral surface of the humerus and inserts on the ulna, while the *biceps* originates on the coracoid, passes over the ventral surface of the humerus, and inserts on the radius.

Flexor muscles of the forearm and digits (Fig. 4-33) originate on the entepicondyle of the humerus and form three groups: two, the radial and ulnar groups, are partially rotary in function and are marginal groups; the third group is central and is concerned with flexion of the digits. The radial group includes the *pronator teres* and *flexor carpi radialis;* the ulnar group includes the *epitrochleoanconeus* and *flexor carpi ulnaris;* the *flexor palmaris superficialis* flexes the digits and is the central grouping. All of these flexors are located on the ventral side of the forearm and manus.

Extensor muscles of the forearm and manus are located on the dorsal surface. These also form three groups—radial, ulnar, and intermediate. The radial and ulnar groups function to rotate or prevent

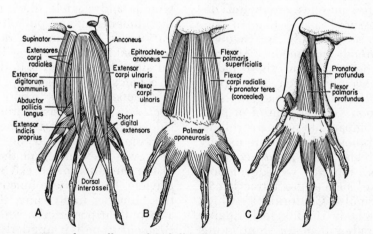

Figure 4-33   Diagrammatic and partially simplified illustration of the muscles of the forearm and manus of a lizard. A, superficial muscles of the extensor surface; B, superficial muscles of the flexor surface; C, deeper flexor muscles. Deep, short muscles of the digits are not shown. (From Romer, A. S.: *The Vertebrate Body.* 4th Ed. W. B. Saunders Co., 1970.)

rotation; the intermediate group functions only in extension of the digits. The radial group consists of a *supinator* and *extensores carpi radiales;* all originate on the humerus and insert on the radius and adjacent part of the carpus. The ulnar group includes a short *anconeus* and an elongate *extensor carpi ulnaris.* The *anconeus* originates on the humerus and inserts on the ulna while the *extensor carpi ulnaris* originates on both the humerus and ulna and inserts on the fifth metacarpal. A series of short forearm extensors lie beneath the long extensors; these are the *extensores digitorum breves,* which originate on the ulna and upper carpus and insert onto the digits. In addition, there is a small dorsal interosseous muscle on each digit running from the metacarpal to the base of the first phalanx.

MUSCLES OF THE HIND LIMB AND PELVIC GIRDLE

The muscles of the hind limb and pelvic girdle (Figs. 4–34 and 4–35) also consist of dorsal extensors and ventral flexors. On the dorsal surface, the *iliofemoralis* and *puboischiofemoralis internus* function to raise the femur or keep it from falling. Four long muscles function to extend the lower leg. These are the *iliotibialis,* the *femorotibialis,* the *ambiens,* and the *iliofibularis.* Distal to these, another set of dorsal muscles function to extend the tarsal region and digits: the *tibialis anterior,* the *extensor digitorum communis,* the peroneal muscles, the *extensores*

digitorum breves, and, as in the manus, dorsal interosseous muscles that extend over the digits.

The ventral muscles of the hind limb and pelvic girdle function to lower the femur and flex the knee joint, which means they raise the body and push it forward during locomotion. Three deep muscles originate on the pelvis and insert on the femur; they are the *puboischiofemoralis externus,* the *ischiotrochantericus,* and the *adductor femoris.* Four superficial muscles cover the ventral surface of the thigh and function as flexors of the lower leg; they are the *puboischiotibialis,* the *flexor tibialis internus,* the *flexor tibialis externus,* and the *pubotibialis.* Two *caudifemorales* (long and short) originate on the ventral side of the proximal caudal vertebrae and insert on the femur; by pulling the femur back they play an important role in forward movement.

The pattern of the ventral muscles in the lower leg and foot differs somewhat from that in the forelimb, as there are no lateral groups of long flexors. One large muscle, the *gastrocnemius,* occurs superficially, and beneath this is the *flexor digitorum longis;* both of these originate on the plantar aponeurosis and insert on the toes through a complex set of tendons. Other deep muscles run from the tarsus to the toes.

The general pattern of girdle and limb musculature described above is typical of the tuatara and of lizards. In crocodilians, most of the muscles are comparable to those of lizards, but some changes have occurred in the pelvic region and ventral muscles of the thigh in association with

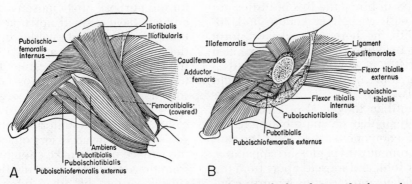

Figure 4–34 Lateral views of the muscles of the pelvis and thigh of a lizard. Superficial muscles are shown in *A;* deeper muscles are indicated in *B.* (From Romer, A. S.: *The Vertebrate Body.* 4th Ed. W. B. Saunders Co., 1970.)

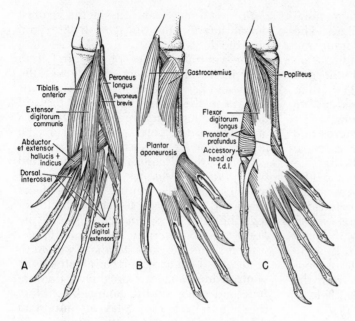

Figure 4–35 Diagrammatic and partially simplified illustration of the muscles of the lower leg and foot of a lizard. *A,* superficial muscles of the extensor surface; *B,* superficial muscles of the flexor surface; *C,* deeper muscles of the flexor surface. Deep short digit muscles are not shown. (From Romer, A. S. *The Vertebrate Body.* 4th Ed. W. B. Saunders Co., 1970.)

ancestral bipedal adaptations. The presence of a shell in turtles has imposed limitations on the movement of limbs and necessitated modifications of the musculature, particularly of the pectoral girdle, but the basic pattern is similar to that described.

DIGESTIVE SYSTEM

MOUTH AND BUCCAL CAVITY

The mouth parts of living reptiles are generally more complex than those of amphibians. Reptilian lips, although nonmuscular and not movable, are often thickened and better developed than amphibian lips. As in birds, the lips of turtles are covered with a shell of keratin and, together with the horn-covered jaws, form a beak. Beaklike structures were also characteristic of several extinct reptiles, including some of the ornithischians and therapsids.

The reptilian tongue also is better developed than that of amphibians. It has the three components that form the amphibian tongue (hypoglossal apparatus, glandular fold, and muscle fibers from the *geniohyoideus* muscle) plus a fourth component—paired lateral lingual folds from above the mandibular arch. The tongue

in turtles and crocodilians is non-protrusible, but in the tuatara, lizards, and snakes it is better developed and, depending upon the species, can be extended and retracted to various degrees. In some lizards, particularly the insectivorous varieties, the tongue is the primary instrument for capturing prey. African chameleons, a highly specialized group of insectivores, have the best developed tongue of all reptiles. By extending and retracting it at high speeds, they feed while perched motionless in ambush (Fig. 4–36); their tongue may be extended to a length exceeding that of their body, and it has a thickened sticky tip to which insects adhere. The long narrow forked tongue of snakes and possibly some lizards is extended and retracted repeatedly for the purpose of chemosensation. These animals have a small notch at the end of their upper jaw that enables them to flick the tongue out and in again without opening their mouth. Each time the forked tongue is retracted the tips are inserted into a paired vomeronasal organ (Jacobson's organ) which opens into the buccal cavity through the anterior palate. Jacobson's organ is an accessory olfactory organ; the surface of the snake's tongue picks up from the air and transfers volatile chemical substances into this organ, where they may be detected by olfactory sensory cells. Thus, the commonly observed flicking in

Figure 4-36 A chameleon capturing a fly. This photograph has caught the chameleon at the precise moment between extension and withdrawal of the tongue. Notice the prehensile tail being used to grip the twig. (Photograph by R. Van Nostrand. Courtesy San Diego Zoo.)

and out of a snake's tongue merely indicates that the animal is attempting to "taste" the air.

Teeth are present in the embryos of soft-shelled turtles (*Trionyx,* family Trionychidae), but otherwise all turtles and tortoises (order Chelonia) are completely toothless; some of the extinct reptiles with beaks were also toothless. In chelonians, teeth are functionally replaced by the sharp horny beak. All other reptiles have well-developed marginal teeth, usually in a single linear series, on the premaxillae and maxillae of the upper jaws and on the dentaries of the lower jaws. Some groups of reptiles have, in addition, teeth on dermal bones in the roof of the mouth and, rarely, on dermal bones on the inner surface of the lower jaw. In crocodilians, there are only marginal teeth that are nearly uniform in conical shape (homodont dentition). These are set, as are the teeth of mammals, in sockets in the jawbones (thecodont attachment); this thecodont type of attachment was also characteristic in several groups of extinct reptiles but does not occur in any other living reptilian group. In addition to marginal teeth, some lizards possess teeth on the palatine and pterygoid bones of the palate; the marginal teeth of most lizards are uniform, but a few species have teeth which are differentiated into incisors, canines, and molars (heterodont dentition), as in mammals. In most lizards, the marginal teeth attach, without sockets, to the biting edges of the jaws (acrodont attachment). Some lizards, however, have a pleurodont type of tooth attachment, in which teeth rest on a ledge on the inner side of the jaw and are attached on both the inner side and the biting edge of the jaw. The marginal teeth of the only known poisonous lizards (the Gila monster and the Mexican beaded lizard of southwestern United States and Mexico, both in the genus *Heloderma,* family Helodermatidae) are large, pleurodont, and deeply embedded in the gums. Each tooth has a groove on its anteromedial surface and, occasionally, also on its posterior surface. Although they themselves have no connection with the labial venom glands, the poison secreted may trickle through these grooves as the lizard works its jaws back and forth in grasping the prey.

Most snakes have backwardly curved or slanted, uniform, pleurodont teeth arranged in rows on the upper and lower jaws. Many species also have longitudinal rows of teeth attached to the palatines and pterygoids. Venomous snakes fall into three categories in regard to their teeth: *opisthoglyphs, proteroglyphs,* and *solenoglyphs.* Opisthoglyphs lack hollow fangs but have grooved teeth on the back of the upper jaw which are much like those of the Gila monster. As with the Gila monster, these "rear-fanged" snakes inject poison by chewing the prey after it has been taken into the mouth. They cannot inject the poison by merely striking the victim and, accordingly, most opisthoglyphs are relatively harmless. However, some, including the African boomslang (*Dispholidus typus,* family Colubridae), have caused human fatalities.

Both proteroglyphs and solenoglyphs have a pair of grooved or hollow elongated fangs attached to the maxillary bone in the front of the upper jaw. The base of each of the fangs is surrounded by a sheath of mucous membrane which, in turn, is connected by the venom duct to the venom gland. The entire mechanism functions much like a hypodermic needle and syringe and allows the venom to be injected upon merely striking the victim. When the snake strikes, a part of the *adductor mandibulae superficialis* muscle surrounding the venom gland contracts, forcing the venom along the length of the fang and into the victim. Proteroglyphs, including coral snakes (family Elapidae) and sea snakes (family Hydrophidae), have fangs which are rigidly attached in an erect position and which fit into a pocket in the outer gum of the lower jaw. Proteroglyph fangs are basically grooved teeth that are elongated, but in the elapids the groove is more or less closed over with calcium, forming a longitudinal canal within the tooth.

Solenoglyphs, the vipers and pit vipers in the family Viperidae, have the most highly specialized mechanism for the injection of venom. In these, the maxillary is very short but deep and is movably attached to the prefrontal and ectopterygoid bones so that it rotates like a hinge on the anterior end of the prefrontal (Figs. 4–37 and 4–38). No teeth other than the

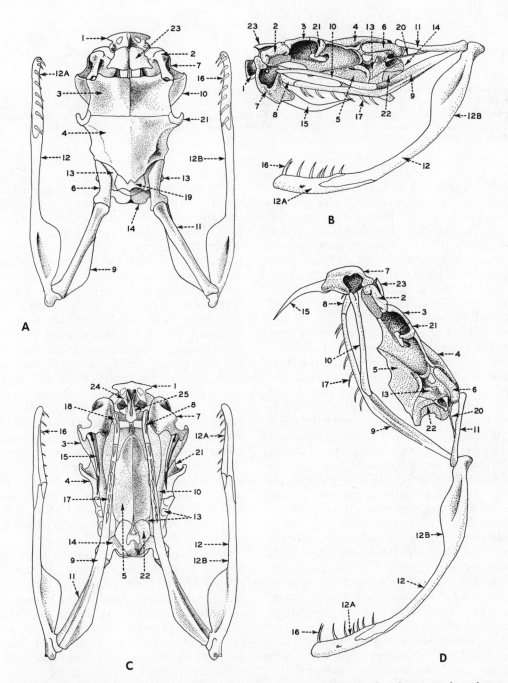

Figure 4–37 Skull of the red diamond rattlesnake (*Crotalus ruber ruber*). *A*, dorsal view; *B*, lateral view; *C*, ventral view; and *D*, lateral view in striking position. *1*, premaxilla; *2*, prefrontal; *3*, frontal; *4*, parietal; *5*, basisphenoid; *6*, squamosal; *7*, maxilla; *8*, palatine; *9*, pterygoid; *10*, ectopterygoid; *11*, quadrate; *12*, mandible; *12A*, dentary; *12B*, articular; *13*, prootic; *14*, exoccipital; *15*, poison fang; *16*, mandibular teeth; *17*, pterygoid teeth; *18*, palatine teeth; *19*, supraoccipital; *20*, stapes; *21*, postfrontal; *22*, basioccipital; *23*, nasal; *24*, turbinal; *25*, vomer. (From Klauber: *Rattlesnakes: Their Habits, Life Histories, and Influence on Mankind.*)

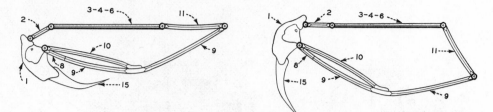

Figure 4–38 Diagrammatic representation of the fang-tilting mechanism of a solenoglyph. In the figure on the left, the fang is folded back against the roof of the mouth and is out of use; on the right, the fang is advanced to the striking position. *1*, maxillary; *2*, prefrontal; *3*, frontal; *4*, parietal; *6*, squamosal; *8*, palatine; *9*, pterygoid; *10*, ectopterygoid; *11*, quadrate; *15*, fang. (From Klauber: *Rattlesnakes: Their Habits, Life Histories, and Influence on Mankind*.)

fangs are attached to the maxillary bone. When the mouth is closed, the greatly elongated fangs are folded back into the mouth so that they lie along the upper jaw. When the snake opens its mouth to strike, the complicated set of bones that forms the palatomaxillary arch articulates in such a way as to rotate the maxillary and cause the fangs to swing down and forward through an arc of about 90 degrees. This linkage and its operation have been described by Klauber (1939). Briefly, from its normal resting position, the quadrate rotates downward, pushing the pterygoid and ectopterygoid forward, and these in turn rotate the maxilla downward. The two maxillae on the opposite sides of the head can be operated independently of one another when, for instance, the snake is using the fangs to aid in engulfing prey (Klauber, 1956). Solenoglyph fangs are hollow, with no external indication of a groove and may, literally, inject venom into the victim. Since the fangs protrude out of the mouth, solenoglyphs may strike an object of any size. This latter feature, together with the more efficient means of delivering the venom, makes solenoglyphs potentially more dangerous than either proteroglyphs or opisthoglyphs.

The tuatara has acrodont dentition, the marginal teeth appearing almost like serrations in the jawbones covered by enamel-like material. It is also characterized by a pair of much-enlarged incisor-like teeth in the front of the upper jaw. Another unusual feature of these primitive reptiles is that they have a second row of teeth attached to the vomerine bones of the palate and running parallel to the jawbone for about half its length.

The fact that old individuals often have teeth that are worn excessively indicates that the tuatara has monophyodont dentition. That is, there is no replacement of teeth during an individual's life. However, in most reptiles there is a constant and steady replacement of teeth throughout life (polyphyodont dentition). This is accomplished through the development of anywhere from one to a whole series of replacement teeth in a special fold of the mouth lining, the *vagina dentis*, beneath the functional teeth. There is periodic resorption of the base of the old teeth, and studies indicate that, in general, tooth replacement occurs in a regular alternating pattern. Thus, adjacent teeth are at different stages in their developmental history and are shed at different times (Edmund, 1969), so old teeth and young teeth alternate along the jaw. Because the replacement is regular the animal always has a functional set of teeth, for, at the very least, half of the teeth will always be present throughout the jaw.

Reptiles possess more differentiated oral glands than do amphibians. A palatine gland which corresponds to the intermaxillary gland of amphibians is present. In addition, lingual, sublingual, and labial glands are present. The venom gland of poisonous snakes is a modified labial gland in the upper jaw and is homologous with the parotid salivary gland of mammals. The two principal constituents secreted by these venom glands, which vary from species to species, are "hemolytic" agents, which break up the blood corpuscles and attack the lining of blood vessels, and "neurotoxic" agents, which attack the nerve centers, causing paralysis and generally affecting the respiratory center. Hemolytic agents may be hemorrhagins (destroying the endothelial cells lining the blood vessels), thrombase (producing intravascular thrombosis), hemolysins (destroying red blood corpuscles), cytolysins (acting on blood corpuscles, leucocytes, and tissue cells), or antifibrins and anticoagulins (retarding the coagulation of blood) (Klauber, 1956). Very few venoms, if any, are purely neurotoxins or hemotoxins. Instead, for a given venom, one type of an effect is generally more serious than the other. In this sense, venom of elapids is primarily a neurotoxin, while that of the vipers is primarily a hemolytic poison. Venoms function in the digestion of prey as well as in the procurement of it, and this purpose is fulfilled by the presence of ferments and kinases which prepare the prey for pancreatic digestion.

The venom gland of *Heloderma* is a modified sublingual gland with four ducts penetrating the bone of the lower jaw and emerging in the mouth cavity in front of the grooved teeth. The venom of the Gila monster is primarily a neurotoxin affecting the respiratory and circulatory centers.

ALIMENTARY CANAL

The esophagus of reptiles is generally relatively longer than that of amphibians and is clearly differentiated from the stomach (Fig. 4–39). Its walls are folded into a series of longitudinal pleats that, upon unfolding, greatly increase the diameter of the esophagus to allow the passage of large objects; this is particularly advantageous to snakes which tend to feed on relatively large prey. Another modification occurs in fish-eating marine turtles. In these, the lining of the esophagus is covered with cornified papillae which point posteriorly and aid in the swallowing of slippery prey. At least three taxa of egg-eating snakes in the family Colubridae (*Dasypeltis, Elachistodon,* and *Elaphe climacophora*) have esophageal "teeth" formed by anterior vertebral hypapophyses that project into the esophagus to form an eggshell-cutting apparatus (Fig. 4–40). The neck and mouth of these snakes are enormously distensible so that the egg may be swallowed whole. When it reaches the esophageal "teeth," compression of the neck muscles and a back-and-forth sawing motion break the shell. The contents of the egg are then carried into the stomach by peristalsis, and the shell is regurgitated.

Reptilian stomachs tend to vary in shape with their body form, those of lizards and snakes being particularly elongated and spindle-shaped. Otherwise, only those of crocodilians show much specialization. The posterior part of the crocodilian stomach has been modified into a muscular gizzardlike compartment, much like that of birds. The reptilian small intestine is relatively longer and generally more coiled than that of amphibians. It therefore has an increased surface for the absorption of nutrients. The diameter of the small intestine is relatively uniform throughout its length. The large intestine, larger in diameter than the small intestine, is generally straight; it empties into a cloaca which opens to the outside through the vent. An ileocolic valve is present at the junction of the small and large intestines to regulate the passage of materials from small to large intestine. Reptiles are the first vertebrates to have a colic cecum. This blind outpocketing, homologous with the vermiform appendix of man, is present in many reptiles (other than crocodiles) at the point of junction between the small and large intestine.

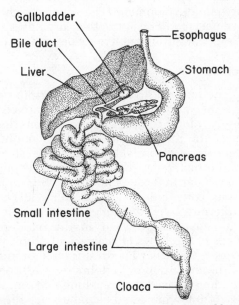

Figure 4–39 The digestive system of a horned lizard (*Phrynosoma*), as seen in ventral view. (After Romer, A. S.: *The Vertebrate Body.* 4th Ed. W. B. Saunders Co., 1970.)

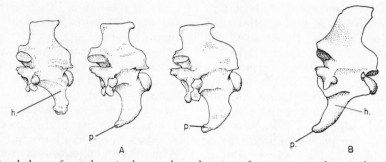

Figure 4–40 Morphology of vertebrae in the esophageal region of egg-eating snakes. A, three anterior vertebrae of *Elaphe climacophora;* B, left lateral view of the posterior vertebra in the modified series of *Dasypeltis. h,* hypapophysis; *p,* very hard process which projects through the esophageal wall. (From Hoffstetter, R., and J. P. Gasc: Vertebrae and ribs of modern reptiles, *Biologia of the Reptilia,* Vol. 1, pp. 201–310, Academic press, 1969.)

RESPIRATORY SYSTEM

Reptiles breathe primarily by lungs, but some aquatic turtles have supplemental cloacal, pharyngeal, or cutaneous respiration. Pharyngeal respiration also occurs in at least one skink and may be important to other reptiles. As in amphibians, there are five pairs of pharyngeal pouches in the reptilian embryo, but gills never develop in association with the pouches and the normal adult never has open gill slits. The first pharyngeal pouch persists in the adult as the Eustachian tube and middle ear, and in reptiles (and birds) this temporarily opens to the outside during embryonic development. The other pouches become modified into other structures or disappear during embryonic development.

LUNGS

The lungs of living reptiles vary in degree of complexity (Fig. 4–41) but are generally more complex and efficient than those of amphibians and less complex than those of birds and mammals. Those of the tuatara are the most primitive and are similar to those of anurans, with numerous convolutions and alveoli scattered uniformly over the inner surfaces. Similar convolutions occur in the basal regions of the lungs of snakes and primitive lizards. In higher lizards, for the first time in the evolution of vertebrates, partitions dividing the cavity of each lung into a number of interconnected chambers are present. The surfaces of the chambers are convoluted and are lined with alveoli, so that the lungs, rather than being saclike, are more like sponges. The lungs of chameleons (family Chamaeleontidae) have several membranous saclike diverticula projecting from their basal portions, and only the anterior portions of the lungs are spongy. The diverticula increase the air capacity and respiratory surface area of the lungs and functionally parallel the development of air sacs in birds. They also may be important in increasing the efficiency of the anterior respiratory surfaces, for in all lizards the tendency is for the anterior portion of each lung to be the most vascularized and effective respiratory region since the airstream is drawn across it at both inspiration and expiration. Many lizards and all snakes, in correlation with their elongate body form, exhibit various degrees of reduction of the left lung. In the most primitive snakes, such as the boas and pythons (family Boidae), the left lung is large in comparison to that of the majority of snakes: it is about 30 to 80 per cent as large as the right lung. Pipe snakes (family Aniliidae) have a small left lung, about 15 per cent of the right in size. Finally, many snakes completely lack a left lung. Among reptiles, the internal partitioning of lungs is best developed in the crocodilians, whose lungs most nearly resemble those of mammals. Turtle lungs are more saclike but have a complex pattern of interior folds and subdivisions.

RESPIRATORY PASSAGES

The reptilian larynx is not much different from that of amphibians. Its skeletal support consists of paired arytenoid cartilages and an incomplete cricoid ring. A third skeletal element, the thyroid cartilage, is present in crocodilians. Mammals are the only other group possessing the thyroid cartilage. Some lizards and turtles have a fold of tissue just in front of the glottis that may be homologous with the epiglottis of mammals. Although most reptiles are voiceless, some turtles, crocodilians, and lizards possess vocal cords and are able to make sounds which vary from the squeaking of geckos to the roaring of crocodilians and the grunting of turtles. All snakes are capable of hissing by forcefully exhaling, and some of the larger species, with a large lung capacity, can do so loudly enough to be heard at some distance. A few species of snakes, including bull snakes (*Pituophis melanoleucus*, family Colubridae), have a vertical laminiform epiglottis that vibrates as the airstream passes over it and magnifies the hiss.

Correlated with the presence of a distinct neck, the trachea of reptiles is better developed than in amphibians. Its length varies directly with that of the neck: it is shortest in lizards and longest in turtles and crocodilians. The trachea of turtles may actually be longer than the neck and, in such cases, is convoluted. The trachea is more completely supported by cartilages in reptiles than in amphibians, and the anterior cartilages may even form rings. However, the posterior cartilages do not completely surround the trachea, the gap being on the dorsal side. The trachea divides within the thorax into two bronchi which are also supported by cartilages. Snakes in which the left lung is absent have only a single (right) bronchus. In contrast, turtles, crocodilians, and the more advanced lizards have secondary and tertiary branching of the bronchi, with smaller and smaller branches that eventually end in the alveoli-lined convolutions of the lung.

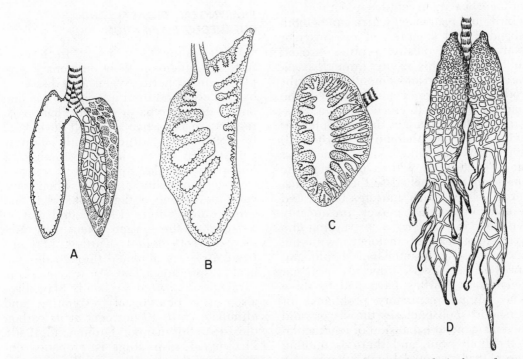

Figure 4–41 Diagrammatic representations of reptilian lungs. *A*, the tuatara (*Sphenodon*); the surface over the right lung is partially removed to reveal details of the inner surface, and the left lung is shown in a sectional view to indicate poorly developed folding of the respiratory surface. *B*, a monitor lizard (*Varanus*); this is a sectional view showing the development of folds increasing the respiratory surface area. *C*, a loggerhead turtle (*Caretta*); this is also a sectional view showing the folding and partial partitioning of the respiratory surface. *D*, a chameleon (*Chamaeleo*); external view showing the development of saclike diverticula. (*A* and *D*, after Kent, G. C., Jr.: *Comparative Anatomy of the Vertebrates*, C. V. Mosby Co., 1965; *B* and *C*, after Bellairs: *The Life of Reptiles*.)

The ventilation mechanism of reptilian respiratory systems is more advanced than that of amphibians. In air-breathing amphibians, the lungs are filled under pressure by acts of swallowing and are emptied by the release of this pressure through the open glottis. Although the swallowing mechanism persists in some living reptiles, it is generally used only in an emergency or, in chameleons, to inflate the lungs and air sacs to an abnormally large volume. The rhythmic swallowing movement frequently observed in reptiles other than snakes does result in the ventilation of the buccopharyngeal cavity, but studies have shown that this causes no significant uptake of oxygen and only a slight elimination of carbon dioxide. The main function of these throat movements seems to be olfaction rather than respiration.

The normal ventilation of lungs in living reptiles is, as in birds and mammals, inspiration by active suction and expiration by pressure. This is possible because of lungs that lie in a closed cavity (pleuroperitoneal cavity in reptiles, pleural cavity in birds and mammals) which can be both expanded and compressed by muscular contractions. Typically, reptiles expand and compress this cavity through backward and forward movements of the ribs produced through contractions of the attached muscles. In crocodilians, a transverse partition separates the lung (pleural) cavity from the abdominal (peritoneal) cavity. This "diaphragm" is non-muscular but is continuous with a diaphragmatic muscle that is attached to the abdominal sternal plates. This muscle, innervated by abdominal spinal nerves, presumably facilitates inspiration by expanding the cavity in a manner analogous with the function of the mammalian diaphragm; however, the two structures are not homologous. Turtles have had to solve some unique respiratory problems, for their rigid shell encloses the viscera and prevents general expansion or contraction of the body wall, and there is nothing present that is comparable to a diaphragm. Expiration in turtles occurs through the contraction of paired muscular membranes enclosing the viscera, the *diaphragmaticus* anteriorly and the *transversus abdominis* posteriorly. These muscles are connected by a tendinous band continuous across the midventral line to their homologues on the other side. Air is forced out as the contraction of these muscles compresses the viscera against the lungs; the effect may be augmented by pulling in of the legs and neck, which further decreases the size of the body cavity. Inspiration is the result of contraction of another pair of muscular membranes enclosing the flanks, the *serratus magnus* anteriorly and the *obliquus abdominis* posteriorly. These muscles act like the mammalian diaphragm by enlarging the body cavity, thereby causing the sucking of air into the lungs. The glottis is normally closed except during inspiration and expiration.

In addition to the modifications of the lung and tracheal system in snakes, their glottis can be protruded so that the air passage is kept clear while large prey is being slowly swallowed. In some species, part of the trachea is lined with respiratory epithelium and is, in reality, a specialized tracheal lung.

PHARYNGEAL, CLOACAL, AND CUTANEOUS RESPIRATION

While submerged, many fresh-water turtles supplement their pulmonary supply of atmospheric oxygen with dissolved oxygen obtained from water drawn in and out of the pharynx or cloaca. Pharyngeal respiration is particularly well-developed in soft-shelled turtles (family Trionychidae). In these, highly vascularized fingerlike papillae line the pharyngeal cavity, functioning like gills when muscular movements of the throat suck in and expel water. Girgis (1961) found that 30 per cent of the oxygen taken in by submerged soft-shelled Turtles (*Trionyx triunguis*) was obtained through pharyngeal respiration. The Australian skink (*Trachysaurus rugosus*, family Scincidae) also uses pharyngeal respiration and eliminates 6 to 10 per cent of its carbon dioxide in this manner (Drummond, 1946). Pharyngeal respiration is probably important to other reptiles as well.

Many fresh-water aquatic and amphibious turtles in the families Chelydridae, Testudinidae, and Pelomedusidae possess a pair of accessory bladders which connect to the dorsal wall of the cloaca and appear to function as respiratory organs when the animals are submerged. As these bladders are filled and emptied by water pumped in and out of the vent, oxygen diffuses in through their thin walls and carbon dioxide passes out. Accessory bladders are absent in marine turtles and land tortoises, and they are lacking in the fresh-water Trionychidae which utilize cutaneous respiration (Smith and James, 1958; Bellairs, 1970).

In contrast to its importance in amphibians, cutaneous respiration is probably of little significance to the majority of reptiles. However, when submerged, the highly aquatic soft-shelled turtles (family Trionychidae) have been found to take up about 70 per cent of their oxygen through the leathery skin covering the carapace and plastron (Girgis, 1961).

While no turtle can depend entirely upon pharyngeal, cloacal, and cutaneous respiration, they do provide oxygen sufficient to materially increase the animal's ability to remain submerged when the metabolic rate is minimal, as during hibernation.

CIRCULATORY SYSTEM

Reptiles are the first truly terrestrial vertebrates, and lungs are the only respiratory organs present in the majority of species. Therefore, many of the differences between reptilian and amphibian circulatory systems are associated with the loss of functional gills and the need for an efficient pulmonary circulation to and from the lungs. Other circulatory changes first appearing in reptiles are related to the evolution of the metanephric kidney and the need for conservation of body fluids in a generally dehydrating terrestrial environment.

HEART

Compared to that of amphibians and fish, the reptilian heart has been shortened at both ends. The conus arteriosus is never present as such, except in the tuatara, where it is very short. The sinus venosus is large in turtles. In other reptiles, although it may retain the same general form as in amphibians, it is reduced and much of it is incorporated into the wall of the right auricle, with some ves-

tiges present as valves where veins empty into this auricle. The right and left auricles are completely separate in all reptiles, and, as in the higher vertebrates, oxygenated blood returning to the heart from the lungs flows through the pulmonary veins and into the left auricle; deoxygenated blood flows into the right auricle. There are nearly all degrees of partitioning of the ventricle in living reptiles, varying from the situation (in some lizards) of having practically no interventricular septum to the situation (in crocodilians) of possessing a complete interventricular septum. Crocodilians differ from other reptiles in the formation and position of the interventricular septum.

In reptiles other than crocodilians, the septum is in an essentially horizontal plane, dividing the ventricle into dorsal and ventral chambers. Embryological studies indicate that there is a twisting of the heart during development and that the dorsal chamber represents the original left side of the ventricle while the ventral chamber represents the right. In these reptiles, the septum always originates from the posterior wall of the ventricle, is incomplete anteriorly, and is positioned in such a way that both auricles open into the dorsal chamber. Thus, all blood enters the same (dorsal) ventricle, and that entering the ventral chamber must flow across the free edge of the interventricular septum.

Again, the conus arteriosus no longer exists as such. Its anterior portion and the ventral aorta basically split to produce a pulmonary trunk leading to the sixth branchial arches, represented by the pulmonary arteries, and a systemic trunk leading to the fourth branchial arches, represented by the right and left aortas. However, there is a subsplitting of the systemic trunk into two vessels, one going to the right aorta and one to the left aorta, so that three trunks branch directly from the ventricle, each with a row of semilunar valves at its base. In non-crocodilian reptiles, the pulmonary trunk always branches from the ventral chamber and the right systemic trunk from the dorsal ventricular chamber. The ventricular opening into the left systemic trunk is always nearly opposite the free edge of the interventricular septum, or slightly to one or the other side of it. In Squamata, it is dorsal to the septum; in Chelonia, it is ventral. From their ventricular origins, the right aortic trunk crosses to the right side and the left aortic trunk crosses to the left side. A small aperture, the foramen of Panizzae, is located at the point where the right and left aortas are in contact, and some mixing of blood may occur at this point. However, the physical relationships of the auricles, ventricular chambers, and arterial trunks are such that the blood most likely to enter the ventral chamber is the deoxygenated blood from the right auricle; this, therefore, is the predominant blood to be carried to the lungs in the pulmonary arteries. Because of its limited size, the ventral chamber cannot hold all of the venous blood. Some remains in the dorsal chamber, from which most of it passes into the left aorta and, mixed with oxygenated blood, is distributed throughout the body. Oxygenated blood from the left auricle mostly enters the dorsal chamber and, therefore, comprises most of the blood going into the right aorta and to the head region and the heart itself.

The separation of the blood streams to the right and left aortae becomes more complete, in certain lizards at least (the European lizard *Lacerta viridis* in the family Lacertidae, for example), by the presence of another incomplete interventricular septum; this secondary septum vertically divides the dorsal chamber into left and right halves, the cavum arteriosum and the cavum venosum, respectively (Fig. 4–42). The right aorta drains the cavum arteriosum and the left aorta drains the cavum venosum. When the auricles contract, the cavum venosum becomes filled with deoxygenated blood from the right auricle and the cavum arteriosum with oxygenated blood from the left auricle. Because the cavum venosum is nearer the incomplete horizontal interventricular septum, most of the deoxygenated blood in it flows over the free edge

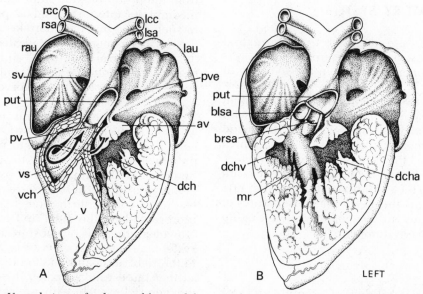

Figure 4–42 Ventral views of a dissected heart of the green lizard (*Lacerta viridis*). *A*, ventral wall removed from both auricles and the majority of the ventricle to reveal the ventral chamber and origin of the pulmonary trunk. *B*, ventral wall, interventricular septum, and base of pulmonary trunk removed to show the entire dorsal chamber and the base of the left and right systemic arches. *av*, left auriculo-ventricular valve; *blsa*, base of left systemic arch (with valve); *brsa*, base of right systemic arch (with valve); *dch*, dorsal chamber; *dcha*, part of dorsal chamber (cavum arteriosum) on left of muscular ridge; *dchv*, part of dorsal chamber (cavum venosum) on right of muscular ridge; *lau*, left auricle; *lcc*, left common carotid artery (carotid arch); *lsa*, left systemic (aortic) arch; *mr*, muscular ridge or secondary septum; *put*, pulmonary trunk; *pv*, pulmonary valve; *pve*, pulmonary vein; *rau*, right auricle; *rcc*, right common carotid artery; *rsa*, right systemic (aortic) arch; *sv*, opening of sinus venosus; *v*, ventricle; *vch*, ventral chamber; *vs*, ventricular septum. (From Bellairs: *The Life of Reptiles*; after Foxon, Griffith, and Price: Proc. Zool. Soc. London *126*:145–157.)

of the septum and into the ventral chamber. When the ventricle contracts, its walls come in contact with the free edge of the horizontal septum, thus temporarily separating the dorsal and ventral chambers of the ventricle. The result is that the deoxygenated blood in the ventral chamber is forced out the pulmonary trunk; in Chelonia, it is pumped out the right systemic trunk as well. The oxygenated blood in the dorsal chamber, mixed with some deoxygenated blood, is forced through both systemic trunks (in Squamata) or the left systemic trunk (in Chelonia). In contrast to the primary horizontal interventricular septum, which forms by an ingrowth of a ridge from the ventricular wall, the secondary septum forms by the fusion of trabeculae; these are small bundles of muscle stretching across the ventricle from one wall to another in most reptilian and amphibian hearts, giving the ventricle a spongy character. It is this vertical interventricular septum which is evolutionarily important, for it, and not the primary septum, becomes the definitive interventricular septum of crocodilians and birds.

In Crocodilia, possibly as a carry-over of an ancestral adaptation to the upright posture associated with bipedal locomotion, the secondary vertical interventricular septum completely partitions the ventricle, while the primary horizontal septum disappears during development. The result is that the crocodilian heart is completely divided into right and left halves. The right auricle leads into the right ventricle and the left auricle into the left ventricle. The ventricular division is such that the pulmonary trunk and the left systemic trunk arise from the right ventricle and the right systemic trunk arises from the left ventricle. Consequently, the left aorta, like the pulmonary arteries, carries only deoxygenated blood from the heart, there being no mixing of oxygenated and deoxygenated blood within the heart. The significance of this separation is reduced by the fact that posterior to the heart the left aorta, carrying only deoxygenated blood, is joined to the right aorta, carrying oxygenated blood. However, the left aorta is much the smaller of the two, so the major volume of blood to be distributed posteriorly is oxygenated. The foramen of Panizzae is too small to allow much mixing of the two blood streams.

Reptilian hearts, particularly those of turtles, have been used extensively in studies of cardiac physiology because of their capacity to beat for many hours after death of the individual. As is also true for amphibians, the heart rate for reptiles is extremely variable and is affected by such factors as temperature, muscular activity, and digestion of food.

ARTERIES

As in advanced amphibians, the primary arterial trunks of reptiles represent the third, fourth, and sixth aortic arches (Fig. 4–43). A carotid duct (ductus caroticus), representing remains of the dorsal aorta between the third and fourth aortic arches, is present in reptiles such as the tuatara and lizards in the genus Lacerta (family Lacertidae). An arterial duct (ductus arteriosus), representing the dorsal portion of the sixth aortic arch, is present in many reptiles, including the tuatara, American alligator, and some turtles. Immediately anterior to the heart, the right systemic trunk of all reptiles gives off a large artery, the brachiocephalic artery, which represents the union of the bases of the third pair of aortic arches; it is the common trunk for the carotid system. The brachiocephalic artery typically gives rise to paired (left and right) common carotid arteries, from each of which branches an external and internal carotid artery. However, the external carotid arteries of reptiles are frequently very small; and in Crocodilia the internal carotids join during development and then the left one disappears. Both the right and left subclavian arteries, which supply blood to the forelimbs, receive blood from the right systemic trunk, but their pattern of branching varies among living reptiles. In Crocodilia, the right subclavian originates from the left common carotid. In Chelonia, both subclavian arteries branch directly from the brachiocephalic. In other reptiles, both subclavians come off the right aorta just anterior to where it is joined to the left aorta.

The carotid and subclavian arterial systems provide for the distribution of oxygenated blood to the head and anterior

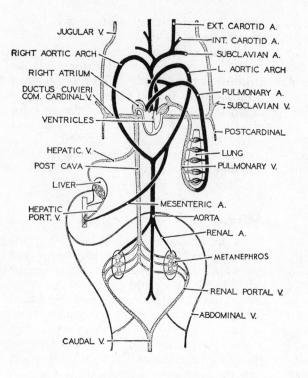

EXT. CAROTID A.
INT. CAROTID A.
JUGULAR V.
SUBCLAVIAN A.
RIGHT AORTIC ARCH
L. AORTIC ARCH
RIGHT ATRIUM
DUCTUS CUVIERI
COM. CARDINAL V.
PULMONARY A.
SUBCLAVIAN V.
VENTRICLES
POSTCARDINAL
HEPATIC. V.
LUNG
POST CAVA
PULMONARY V.
LIVER
HEPATIC
PORT. V.
MESENTERIC A.
AORTA
RENAL A.
METANEPHROS
RENAL PORTAL V.
ABDOMINAL V.
CAUDAL V.

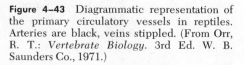

Figure 4–43 Diagrammatic representation of the primary circulatory vessels in reptiles. Arteries are black, veins stippled. (From Orr, R. T.: *Vertebrate Biology.* 3rd Ed. W. B. Saunders Co., 1971.)

parts of the body. Equally significant is the presence of several small coronary arteries that branch from either the base of the brachiocephalic artery or, occasionally, the right aorta, carrying oxygenated blood to the muscular wall of the reptilian heart itself: a special coronary circulation is lacking in many, if not all, modern amphibians. (Although anurans have a small coronary artery arising from the right carotid, it does not reach all of the heart.) The amphibian heart is apparently supplied with oxygen by the blood which is being pumped through it. This is possible only because the skin of amphibians acts as a respiratory organ, so that much of the blood returning to the heart is highly oxygenated; such is not the case in reptiles, and the selective advantage of the coronary arterial system is obvious. The reptilian right aorta has no other branches but continues posteriorly as the dorsal aorta.

The left aorta, much smaller than the right and carrying deoxygenated blood, supplies the abdominal viscera. It runs to the left of the esophagus and dorsal to the stomach, where it gives rise to three major branches: the gastric artery, which goes to the stomach; the coeliac artery, which goes to the posterior stomach, pan-

creas, liver, and small intestine; and the superior mesenteric artery, which goes to the small intestine, large intestine, and cloacal region. Posterior to these three branches, what remains of the left aorta joins the right aorta to form the dorsal aorta. The dorsal aorta sends numerous renal arteries into the kidneys and genital arteries to the gonads. Posterior to these, it forms into the right and left common iliac arteries, supplying the hind limbs, and the caudal artery, which terminates in the tail.

It seems probable that the arrangement of aortic arches in the amphibian ancestors of reptiles was similar to that of modern amphibians, where the aortic arches branch from a single trunk (the ventral aorta) draining the heart. This is also the pattern that develops initially in the mammalian embryo, but the right aorta is lost before birth. In modern reptiles and in bird embryos, however, the two aortae are always separated right down to the points of origin of the systemic trunks; the difference is that in birds the foramen of Panizzae and the left aorta are eliminated from the crocodilian arrangement during development. Thus, birds and mammals, both descended from reptiles but representing two distinct lines, have achieved separation of arterial and venous blood in comparable but different ways. The retention of the left aorta in modern reptiles, a carry-over from the symmetrical arterial pattern of lower vertebrates, is frequently regarded as nothing but a disadvantage and a hindrance to the development of an efficient circulatory system. It is possible, however, that it may be advantageous, since the left aorta provides an auxillary drain for excess venous blood that cannot be accommodated in the pulmonary system. The foramen of Panizzae, which according to some authors is a secondary perforation, could also have been selected, for it is an additional device that serves as a safety valve when the pulmonary circulation is congested. Such a congestion occurs whenever breathing is stopped for a time, as when crocodilians and turtles are submerged.

VEINS

The general plan of the reptilian venous system (Fig. 4–43) does not differ greatly from that of amphibians, but some modifications have occurred in association with changes in the heart and kidneys and with the elimination of cutaneous respiration. In reptiles, the anterior cardinals (precavals, according to many authors) and the postcaval vein, returning deoxygenated blood to the heart from the systemic circulation, have shifted toward the right side and empty into the right auricle; in amphibians, these empty into the medially-positioned sinus venosus. The anterior cardinals receive blood from the head by way of external and internal jugular veins, from the forelimbs by way of the subclavian veins (reduced or missing in snakes and limbless lizards), and from the body by way of vertebral veins (representing reduced remains of the anterior portions of the postcardinal veins).

The posterior veins of reptiles have changed very little from the amphibian scheme, but there are some differences from one group to another. In reptiles other than Squamata, as in amphibians, venous blood from the hind limbs and tail may return to the heart either by entering the renal portal vein and being shunted into the renal portal system and the postcaval vein or by entering the ventral abdominal vein and flowing through the hepatic portal system and hepatic veins directly into the sinus venosus or the postcaval vein as it passes through the liver. In Squamata, however, the ventral abdominal vein receives blood only from the body wall and does not connect to the veins draining the hind limbs and tail. The reptilian postcaval vein has the same origin as that of amphibians, being formed from portions of both the subcardinals and the vitelline veins, but is generally better developed. It carries blood forward from between the kidneys, passing through the liver on its way to the sinus venosus.

Compared to that of amphibians, there has been a considerable reduction in the renal portal system of reptiles. This reduction has been accomplished largely through the development of major vessels that carry blood directly from the renal portal veins to the postcaval vein, allowing it to by-pass the capillary network of the kidneys. The posterior portion of the postcaval vein is split into two vessels in at least some lizards and the tuatara; crocodilians have a single postcaval vein but have paired ventral abdominal veins.

The elimination of cutaneous respiration in reptiles, as previously mentioned, places greater importance on the pulmonary circulation; this is again reflected in larger pulmonary veins and smaller cutaneous veins. Squamata with asymmetrical development of the lungs have comparable asymmetrical development of the pulmonary veins. There is only one pulmonary vein in those snakes completely lacking the left lung.

LYMPHATIC SYSTEM

The lymphatic system of reptiles resembles that of salamanders. In addition to smaller vessels, a large thin-walled subvertebral vessel extends the length of the body, divides anteriorly, and empties into the anterior cardinal veins. Lymphatic vessels and sinuses are particularly well-developed in snakes. A pair of muscular contractile lymph hearts is present in the pelvic region of most reptiles, and these return fluid to the circulatory system by pumping it into the iliac veins. In possessing lymph hearts, reptiles differ from most birds and all mammals. Lymph hearts develop embryonically in birds but generally become non-functional in adults; they never form in mammals. A thymus gland and a spleen, both involved in the production of lymphocytes, occur in all reptiles, and a pair of tonsils is present beneath the pharyngeal mucous membrane of some species.

BLOOD CELLS

Reptiles appear to have the same kinds of blood cells as amphibians, but the information regarding them is scanty. Reptilian erythrocytes are oval and nucleated, and they tend to be smaller and more numerous than those of amphibians. They range in length from about 15 microns to about 23 microns and in width from about 5 microns to about 12 microns; length-width ratios vary from about 1.44 to 2.06. Although relatively few reptiles have been studied, the blood cells of snakes seem to generally be larger than those of other reptiles; those of pit vipers in the genus *Bothrops* (family Viperidae) are the largest known (about 21 to 23 microns in length). The nuclei of snake erythrocytes tend to be longer than those of lizards, but because the role of the nucleus in red blood cells is not clear the significance of this is unknown.

Erythrocyte counts for reptiles range from about 466,500 to 1,500,000 per cubic millimeter of blood, generally higher than for elasmobranch fish and amphibians but lower than for many bony fishes, birds, and mammals. Erythrocytes normally seem to be about 100 times more numerous than leucocytes, but the latter have received very little attention in reptiles. There are some data suggesting that the red blood counts vary geographically within species, and it seems logical to assume that they are higher in high altitude populations than in those at lower elevations. The fragmentation of erythrocytes into plasmocytes, so conspicuous in amphibians, apparently is not a characteristic of reptiles. Studies of the coagulation role of crocodilian thrombocytes have shown that, like mammalian platelets but unlike fish thrombocytes, they require calcium to effect acceleration of clotting.

UROGENITAL SYSTEM

The evolution of reptiles has produced significant differences between their urogenital system (Fig. 4–44) and that of amphibians. These differences, for the most part, reflect adaptations for conserving body water in a generally dehydrating terrestrial environment, the evolutionary trend toward separation of excretory and reproductive tracts, and development of more precise control over the animal's internal environment.

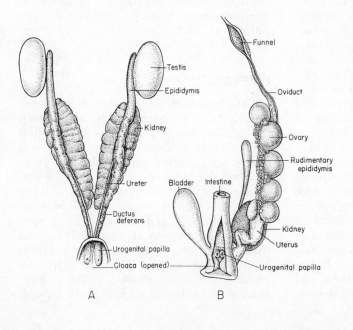

Figure 4–44 Ventral views of the urogenital organs in reptiles. *A*, male organs of a monitor lizard (*Varanus*) with the bladder omitted. *B*, female organs of a tuatara (*Sphenodon*) with the bladder turned to the right and the organs omitted on that side. (From Romer, A. S.: *The Vertebrate Body*. 4th Ed. W. B. Saunders Co., 1970.)

KIDNEYS

During the embryonic development of reptiles, birds, and mammals the pronephric, mesonephric, and metanephric regions of the holonephric primordia develop sequentially, with only the metanephros persisting as the adult kidney. The pronephros of reptiles forms very early in embryonic development and is only functional for a short time. It soon undergoes degeneration and is replaced by the mesonephros which has developed posterior to it. The mesonephric kidney (Wolffian bodies of many authors) becomes well-developed in both reptiles and birds and is comprised of large numbers of tubules; anteriorly, there may be nephrostomes draining the peritoneal cavity. Eventually, a metanephric kidney develops on each side posterior to the mesonephros; during later stages in embryonic development both function simultaneously, until the mesonephros degenerates and its excretory function is completely assumed by the metanephros. The simultaneous excretory functioning of the mesonephros and metanephros, which in reptiles may last until after hatching or birth, is comparable to the opisthonephros of adult amphibians. Portions of the mesonephros become incorporated into the male reproductive system, and other remnants are persistent as vestigial structures lacking any known function.

Although metanephric and mesonephric kidneys consist of the same basic units (renal corpuscles, secretory tubules, and collecting tubules), the metanephros is more compact than the mesonephros. Metanephric tubules never have any trace of segmentation and, always lacking nephrostomes, never have an opening into the peritoneal cavity. Compared to those of amphibians, there is a definite reduction in the size of the glomeruli in reptiles. This is an adaptation which conserves water by reducing the flow of urine passed into the tubule. Although they are frequently very small, glomeruli are present in the majority of reptiles, but they are missing in some lizards and snakes, as in some bony fishes, so that aglomerular tubules are present. Reptilian kidney tubules differ from those of mammals and some birds in lacking the loop of Henle between the proximal and distal convoluted tubules. In both birds and mammals, the loop of Henle functions to concentrate urine by the reabsorption of water. The posterior tubules of some snakes and lizards are sexually dimorphic, for those in males have swollen portions, called sexual segments, which appear to undergo seasonal cellular changes and produce an albuminous secretion which is apparently a component of the seminal fluid in which spermatozoa are suspended. The gross morphology and degree of symmetry of the kidneys vary among living reptiles, but they are generally smaller and more lobulated than those of amphibians. Limbless lizards and snakes have unusually elongated kidneys, and generally the right kidney is anterior to the left, causing an asymmetry of blood vessels and ureters. Occasionally the two kidneys of lizards may fuse posteriorly.

The nitrogenous excretions of reptiles may be in the form of ammonia, urea, or uric acid. The relative amount normally excreted of each of these three products, which are decreasingly toxic in the order given, is a reflection of adaptation to aquatic, semiaquatic, or terrestrial modes of life. Ammonia is a direct product of the cellular metabolism of carbohydrate, fat, and protein and, being extremely toxic, must either be eliminated rapidly from the body or be converted into some less toxic compound. Urea is relatively harmless and may be retained in the body for some time, but in high concentrations it may dangerously upset the osmotic balance of the cells. Both ammonia and urea are highly soluble and may be easily eliminated by the kidneys if there is sufficient water available. If, however, the animal lives in an arid environment where it is critically important to conserve the internal water supply, it becomes advantageous to excrete primarily uric acid because it is both relatively inert and insoluble and, thus, can be concentrated and eliminated with very little waste of water.

In view of these facts it is not surprising that the proportions of these products in the excretions of aquatic and terrestrial reptiles are not the same. Marine and highly aquatic fresh-water turtles and crocodilians excrete primarily ammonia (up to 75 per cent of their nitrogenous wastes). Amphibious pond and swamp turtles, such as the painted turtle (*Chrysemys picta,* family Testudinidae), excrete approximately twice as much urea as ammonia or uric acid. Tortoises, which inhabit arid areas, such as *Testudo graeca* (family Testudinidae), excrete primarily uric acid (up to 90 per cent of their nitrogenous wastes), with only traces of ammonia. All lizards and snakes excrete primarily uric acid (80 to 98 per cent), but the relative amounts of ammonia and urea vary. The excretion of uric acid by reptiles, in addition to promoting the conservation of water in the adult, is also a prerequisite for the long embryonic life in an amniotic egg during which nitrogenous wastes are stored in the allantois. If urea were the principal excretory product, it would be impossible for the allantois to store the volume of urine required to contain all the urea produced during embryonic development.

SALT GLANDS

When marine animals eat food with a high salt content or drink sea water, they face a problem of getting rid of excess salt with which their kidneys are unable to cope. In marine turtles and the marine iguana (*Amblyrhynchus cristatus*, family Iguanidae), as in marine birds, this problem has been solved by the presence of salt-secreting glands located on the head. The salt gland of the turtles appears to be the lachrymal gland, while that of the marine iguana is the nasal gland. These glands, which are only active when the osmotic concentration of the plasma is too high, have ducts that carry their secretions to the corners of the eyes (turtles), from which they are shed like tears, or into the nasal passages (marine iguana), from which they are expelled during exhalation. Marine sea snakes (*Laticauda*, family Hydrophidae) have a natrial gland in the palate which also seems to function as a salt gland (Dunson and Taub, 1967). The effectiveness of these glands is indicated by the fact that Schmidt-Nielsen and Fange (1958) have found the sodium content of the secretions from salt glands of the marine iguana to be over 50 per cent higher than that of sea water and many times higher than that of the animal's blood.

Several herbivorous terrestrial iguanid lizards (*Ctenosaura, Dipsosaurus, Sauromalus, Iguana*) have also been found to excrete salt by means of their nasal glands (Templeton, 1966). Among terrestrial reptiles, extrarenal salt secretion would theoretically be most advantageous to desert-dwelling species, since it would increase their ability to conserve water. Although uric acid readily precipitates the nitrogenous wastes so that they can be eliminated in a semisolid form, salt would remain in solution in the urine; its elimination by the kidneys would require the wastage of water, since the osmotic concentration of the urine cannot exceed that of the plasma. Thus, elimination of the salt (primarily potassium in herbivores) by the nasal gland allows greater reabsorption in the cloaca of water from the urine.

GONADS

Reptilian gonads (testes or ovaries) are, as in amphibians, paired and situated within the abdominal cavity. Typically, they lie opposite one another, but in limbless lizards and snakes with elongated bodies one of the pair is often anterior to the other. The testes resemble those of anuran amphibians in their compact round, oval, or pear-shaped structure. They are comprised of long convoluted seminiferous tubules. Reptiles typically have a definite breeding season, and their testes undergo marked seasonal fluctuations in size, being swollen when spermatogenetic activity is at its peak.

The ovaries of snakes and lizards are sacular with hollow lymph-filled cavities and, thus, are similar to those of amphibians. In contrast, those of turtles, the tuatara, and crocodilians are solid structures, like those of mammals and birds, with an inner medullary region composed of connective tissue, blood and lymphatic vessels, smooth muscle, and nerve fibers. The shape of the ovaries is correlated with the body form of the reptile. Those of the tuatara, turtles, and crocodilians are relatively broad, while those of lizards and snakes are variously elongate. The right ovary of many snakes is larger than the left.

As is characteristic of ovaries is general, ova at various stages of development are contained in Graafian (ovarian) follicles in the walls of the ovaries. In reptiles the number maturing at one time is much smaller than in most amphibians, generally less than 100. The size of a mature ovum varies from about 5 by 5 millimeters in small lizards, such as *Draco* (family Agamidae), to about 60 by 120 millimeters in the large pythons. There is much greater growth in reptiles prior to hatching than occurs in amphibians, and this requires a nutrient supply. Accordingly, the relatively large size of reptilian eggs compared to those of fish and amphibians is due primarily to a large amount of "yolk," a complex of proteins, phospholipids, and fats within the ovum that sustains the embryo during its development. Of the various components of the reptilian egg, only the ovum with its content of "yolk" is formed in the ovaries. Like those of amphibians, mature reptilian Graafian follicles cause the wall of the ovary to rupture, and the ova are released into the coelomic cavity.

The ovaries of at least some viviparous Squamata produce corpora lutea, masses of yellowish tissue which fill the cavities of Graafian follicles after each has ruptured and released its mature ovum. In mammals, the corpora lutea secrete progesterone, which acts to maintain pregnancy through its action on the uterine wall. Presumably those of reptiles also have a hormonal function, but studies have indicated that their hormone is not always necessary for the maintenance of pregnancy (Bellairs, 1970).

REPRODUCTIVE AND EXCRETORY DUCTS

Associated with the evolution of the metanephric kidney in amniotes (reptiles, birds, and mammals) has been a complete separation of the excretory and reproductive tracts above the level of the cloaca. The oviducts of reptiles are basically the same as those of amphibians. In females, the Wolffian (archinephric) ducts degenerate and remain in the adult only as strands of tissue (Gartner's ducts) that end blindly posterior to each ovary. Paired Müllerian ducts, as in amphibians, open into the coelom by long slitlike ostia; they are the functional oviducts. Ovulation occurs by the same process as in amphibians: a mature ovum is released into the coelom through rupture of its Graafian follicle and the ovarian wall, it enters the ostium, and it is forced down the Müllerian duct by a combination of ciliary action and muscular contractions of the duct wall. Fertilization is internal in all reptiles; it must take place in the upper portion of the oviducts soon after the ova pass through the ostia and before deposition of the egg envelopes by glands lining the oviduct. This means that, at the same time that the ova are being driven down the oviducts and away from the ostia, sperm must be able to move up the oviducts in numbers sufficient to fertilize the ova almost as soon as they enter the ostia. In turtles, and presumably in other reptiles as well, there is a narrow band of cilia along the side of each oviduct that beat toward, instead of away from, the ostium; apparently these represent a mechanism that helps the sperm move upward to meet the ova.

Each oviduct is differentiated longitudinally into glandular regions, and as the fertilized eggs move along they are coated with secretions that form the envelopes surrounding them when they are deposited. Glands in the anterior portions of tuatara, turtle, and crocodilian oviducts secrete albumen about the egg. Comparable glands are missing in snakes and lizards, and, accordingly, their eggs lack albumen. Following deposition of the albumen, the reptilian egg is covered by a thin double "shell membrane," also secreted by glands in the oviduct wall. As frequently occurs in bird eggs, the shell membranes of reptilian eggs other than those of the tuatara and turtles separate at one pole of the egg, producing a space which becomes filled with air during early development. The posterior segments of the reptilian oviducts, referred to as the uteri, are specialized shell glands that secrete the eggshell over the other layers. Reptilian eggshells are generally a tough but flexible parchmentlike material without much calcareous matter. However, those of some lizards and crocodilians are hard and calcareous, like those of birds. In all cases, the shell is porous enough, as it must be, to allow respiratory gas exchanges to take place through it. The paired uteri (shell glands) of reptiles empty separately into the cloaca rather than joining together as in many mammals. As might be expected, the size of the oviducts changes with the season and is greatest during the breeding period.

When the mesonephros degenerates in male reptiles, some of the mesonephric tubules persist as efferent ductules which, as in many fish and amphibians, carry spermatozoa from the seminiferous tubules of the testis to the Wolffian duct. The Wolffian duct itself loses its excretory function with degeneration of the mesonephros but, rather than degenerating as it does in females, persists as a reproductive duct to conduct the spermatozoa from the efferent ductules to the outside. The portion of the Wolffian duct closest to the testis becomes highly coiled and convoluted and is thereafter referred to as the epididymis. In some reptiles the tubular mass formed by the epididymis is larger than the testis. That portion of the Wolffian duct extending between the epididymis and the cloaca is called the ductus deferens and may be either straight or convoluted. The entire Wolffian duct (epididymis and ductus deferens), as with the testis, undergoes regular changes in size and is noticeably swollen during the breeding season. These seasonal changes are under endocrine control.

During the embryonic development of the metanephros, a diverticulum called the ureteric bud grows anteriorly on each side from the posterior end of the Wolffian duct. Anteriorly, this diverticulum branches repeatedly, ultimately forming large numbers of small collecting tubules in the kidney. The remainder of the diverticulum forms a hollow tube, the ureter, which drains the collecting tubules and is the excretory duct of the metanephros.

The reptilian cloaca is more complex than that of amphibians and shows the initial steps toward separation of the openings of the digestive, excretory, and reproductive tracts that is achieved in higher mammals. To some degree in all reptiles, but especially in turtles, the cloaca is divided by low ridges into three chambers: the *coprodeum*, the *urodeum*, and the *proctodeum*. The coprodeum is the anterior chamber and is essentially the terminal portion of the large intestine where feces are collected. The urodeum is the middle chamber and is where the excretory and genital ducts open. A urinary bladder is present in the tuatara, turtles, and most lizards but is absent in snakes and crocodilians. This develops as a median and ventral outpocketing from the urodeum and is generally a bilobed structure. In turtles, the ureters empty into the stalk of the bladder near its opening into the dorsal side of the urodeum. Sometimes the ureter and ductus deferens on each side join together just anterior to the cloaca so that they discharge into the urodeum through a common opening. Frequently, this common opening is on a urogenital papilla protruding into the cavity of the urodeum. In other instances, the ureters and ductus deferens open separately into the urodeum. This means that in reptiles other than turtles, as in amphibians, urine must pass into the cloaca (urodeum) and then into the bladder. In some turtles, as mentioned, a pair of highly vascularized thin-walled accessory bladders is also present; these open into the dorsolateral wall of the urodeum. In addition to acting as respiratory organs, the accessory bladders may act at times as water reservoirs. The posterior chamber of the cloaca, the proctodeum, is a short tube closed by a sphincter at the anus. Both the coprodeum and the urodeum are endodermal in origin, whereas the proctodeum is ectodermal.

Internal fertilization, characteristic of all reptiles, has led to the possession of some form of an intromittent organ by all modern male reptiles except the tuatara; in the latter species, insemination must occur as it does in most birds, through pressing together of the male and female cloacas. Among other reptiles, two different kinds of male copulatory organs exist which appear to have evolved independently of one another. One kind, found in turtles and crocodilians, is a single structure and seems to represent a primitive stage in the evolution of a penis similar to that of the mammals. A pair of ridges composed of erectile and connective tissue are on the medial ventral wall of the urodeum posterior to the opening of the bladder. These ridges, the *corpora cavernosa*, can be expanded by internal blood pressure. When expanded, they form a canal through which the semen can be ejaculated. Just posterior to these ridges is another projection of spongy erectile tissue, the *glans penis*. Both the glans penis and the posterior portions of the ridges can be extruded through the cloacal opening when the blood pressure is increased in the corpora cavernosa, thus forming an intromittent organ. This kind of an intromittent organ could not function in those reptiles in which the Wolffian duct openings are on the dorsal side of the urodeum.

In lizards and snakes, the intromittent organ is a pair of saclike structures called the hemipenes. The hemipenes lack erectile tissue and when relaxed lie under the skin at the base of the tail, adjacent to the cloaca. They sometimes produce a thickening of the region which makes it possible to determine the individual's sex externally. The hemipenes are tubular, and each has a groove, the *sulcus spermaticus,* traversing its length for the conduction of the spermatozoa. The lining of the hemipenis is frequently folded into pleats, and in many snakes it has spines and fingerlike projections which help to anchor it in the oviduct during copulation. The hemipenes are erected by being turned inside out and everted, like the finger of a glove. The mechanism for this involves the combined action of propulsor and retractor muscles and the filling of blood sinuses in the hemipenes. Only one hemipenis is inserted during copulation, depending upon which side the male happens to be on at the time. After copulation, the distal end of the hemipenis can be drawn back by long retractor muscles.

The shape of the hemipenes is extremely variable, especially among snakes (Dowling and Savage, 1960) and is of great importance in systematic work. In snakes, it may be cylindrical, conical, bulbous, or corrugated, and it may be divided into two lobes, each with its own cavity, or (as in pit vipers) deeply forked (Fig. 4–45). African burrowing snakes in the genus *Prosymna* (family Colubridae) have wormlike hemipenes which are almost as long as the tail (Angel, 1950). The hemipenes of lizards may be forked or knoblike and their surfaces may be pleated or possess small papillae near the tip.

ENDOCRINE SYSTEM

The endocrine system of reptiles is in great need of study, but although it is probable that less is known about endocrinology for reptiles than for any other class of vertebrates, the system appears to be basically the same as in amphibians.

THYROID GLAND

There is a single thyroid gland in all reptiles other than lizards; this lies ventral to the trachea and just anterior to the heart. In turtles and snakes, the thyroid is roughly spherical, whereas that of the tuatara is a narrow, transversely elongate structure. In crocodilians, the thyroid is bilobed, and the lobes, on each side of the trachea, are connected by a narrow isthmus. The thyroid of lizards is variable in shape, and, although a particular form is always characteristic of a given species and usually for the genus as well, it may be unpaired, bilobed, or completely paired among different members of the same family (Lynn, 1970). Some species of worm lizards (family Amphisbaenidae) have threadlike thyroids which stretch alongside the trachea and, though a millimeter or less in diameter, may be several centimeters in length. The thyroid gland of worm lizards also lies farther anterior than in other reptiles. The ultrastructure of the reptilian thyroid glands appears to be similar to that of mammals. They contain a substance comparable to thyroxine and they respond morphologically in a typical way to thiouracil and to thyroid

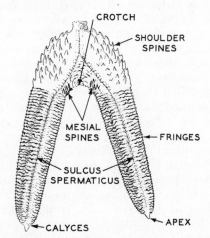

Figure 4–45 A generalized pit viper hemipenis; each male has a pair of these forked structures. When extruded, the sulcus aspects of the pair face one another across the vent. (From Klauber, L. M.: *Rattlesnakes: Their Habits, Life Histories, and Influence on Mankind,* 1956. Originally published by the University of California Press; reprinted by permission of the regents of the University of California.)

stimulating hormone (TSH). They are under pituitary control.

Very few studies have been made of the physiological effects of thyroxine in reptiles, but it is known to play an important role in ecdysis, just as it does in amphibians. Studies of lizards in the genus *Lacerta* (family Lacertidae) by Eggert (1933, 1936) showed that the thyroid is relatively inactive during the formation and cornification of new epidermis, but during the brief period of ecdysis it becomes active. This activity continues for several days after ecdysis but then gradually declines until the next molt period. Thyroidectomy causes cessation of ecdysis, and implantation of thyroid tissue into the thyroidectomized *Lacerta* results in their successfully molting several times. Variations of these results have been obtained by other workers using different species of lizards. Noble and Bradley (1933) reported that either thyroidectomy or hypophysectomy of house geckos (*Hemidactylus*, family Gekkonidae) resulted in a lengthening of the period between molts, an effect which could be eliminated by thyroxin treatments, but that neither caused cessation of molting. Ratzersdorfer *et al.* (1949) found that the administration of thiourea to the American anole (*Anolis carolinensis*, family Iguanidae) inhibited the thyroid but caused no noticeable effect on ecdysis. Chiu *et al.* (1967) have found that thyroidectomy decreases the frequency of molting in the tokay (*Gekko gecko*, family Gekkonidae) and injections of thyroxine into intact animals increase the frequency of molting. They have concluded that thyroid secretions only affect the resting phase of the molting cycle and have no effect upon the keratinization phase, since no hyperkeratosis has occurred after thyroidectomy.

Interestingly, experimentation with a variety of species indicates that thyroidectomy of snakes enhances their shedding of the corneal layer and that the injection of thyroxine inhibits ecdysis (several references are cited by Lynn, 1970). This is just the reverse of what happens in lizards and amphibians, suggesting that other endocrine mechanisms may also be concerned with this process.

Drzewicki (1929) and Giusti (1931) provided limited evidence that thyroidectomy causes growth inhibition in young lizards (*Lacerta agilis*, family Lacertidae) and turtles (*Clemmys caspica*, family Testudinidae). Dimond (1954) found that inhibiting the production of thyroxine in embryonic snapping turtles (*Chelydra serpentina*, family Chelydridae) resulted in decreased growth rate, abnormalities of the carapace, delayed hatching, and inhibition of the retraction of the yolk sac into the body.

Differing results have been obtained in studies of the effect of thyroid hormones on reptilian metabolic processes. Drexler and von Issekutz (1935) found that extended feeding of thyroid to turtles produced no change in oxygen consumption, but Scott (1935), using isolated erythrocytes from a thyroxine-injected alligator, and Haarmann (1936), using isolated pieces of thyroxine-treated snake muscle, reported that it stimulated respiration. In a more recent study, Maher and Levedahl (1959) have found that neither thyroidectomy nor the injection of thyroxine affects oxygen consumption in American anoles (*Anolis carolinensis*, family Iguanidae) at temperatures of 21 to 24 degrees centigrade, but that these treatments cause an effect at a temperature of 30 degrees. This suggests that the tissues may be more responsive to thyroxine at 30 degrees centigrade (near the optimal temperature for this species) than they are at lower temperatures. Studies of other species of lizards have also shown a similar relationship between temperature and response to thyroid hormone (see Lynn, 1970, and Jankowsky, 1964).

Tests using radiotracers indicate the maximum uptake of iodine by the thyroid is relatively low in reptiles, usually less than 20 per cent, and that thyroidal radioactivity decreases very slowly, indicating a slow rate of turnover of iodine in the thyroid. Probably associated with this slow rate of turnover is the low rate of thyroxine production. Usually less than 10 per cent of the radioiodine in the thyroid is in the form of thyroxine after 8 to 10 days. An interesting response to dehydration in turtles is the reabsorption of excreted radioiodine and a progressive accumulation of it in the thyroid until, after 8 days, about 70 to 80 per cent is found there (Shellaburger *et al.*, 1956). Lizards do not show this response to dehydration.

PARATHYROID GLANDS

Parathyroid glands (epithelial bodies) have been found in the few species of lizards, snakes, crocodilians, and turtles that have been examined. They are present in the tuatara and are assumed to be present in all of the reptiles. There are usually two pairs of these glands, which develop embryonically from the third and fourth pairs of pharyngeal pouches. However, the number and location of the glands are variable, even within a species. Most frequently, they are found posterior and lateral to the thyroid gland and closely associated with the carotid and systemic arteries. In some lizards (*Anguis* and *Lacerta*), two or more pairs may develop embryonically, but then only the tissue from the third pharyngeal pouch may persist in the adult. The occasional persistence of parathyroids from the fourth pouch has been reported frequently for a variety of lizards, so, contrary to many statements in the literature, lizards do not always possess a single pair of parathyroids. Snakes typically have two pairs of parathyroids, of which one pair is between the anterior and posterior lobes of the thymus and the other is at the bifurcation of the carotid artery, more anterior than in other reptiles. Epithelial proliferations from all five pairs of pharyngeal pouches have been noted for a variety of snakes, but some of those described may have been accessory parathyroids. The two pairs that develop into primary parathyroids seem, at least in the plains garter snake (*Thamnophis radix*, family Colubridae), to be derived from the fourth and fifth pouches rather than from the third and fourth as in other reptiles. However, the embryological derivation of ophidian parathyroids is still confused (Clark, 1967). In crocodilians, although more may develop embryonically, there is only a single pair of parathyroids after hatching; these appear conclusively to be derived from the third pharyngeal pouches. Turtles and the tuatara have two pairs of parathyroids which are derived from the third and fourth pouches.

The physiological effects of parathormone are poorly studied in reptiles but, as in mammals, reptilian parathyroids seem to be involved in the regulation of blood calcium and phosphate levels. Clark (1965) found that the urinary excretion of phosphate decreased in parathyroidectomized turtles (*Chrysemys picta* and *Pseudemys scripta*, family Testudinidae) and that the injection of mammalian parathyroid extract into normal or parathyroidectomized turtles resulted in a significant rise in urinary excretion of phosphate. However, she was unable to induce tetany, paralysis, or significant alterations in total concentrations of calcium or inorganic phosphate by parathyroidectomy of the turtles. She suggested that turtles might be able to obtain sufficient calcium and phosphate by physicochemical exchange between the large bony surface of the shell and the body fluids or that they might have active transport systems for calcium in their pharyngeal or cloacal tissues and, thus, are unresponsive to parathyroidectomy.

There have been no studies of the parathyroid function in crocodilians or the tuatara. Parathyroidectomy of snakes (*Thamnophis sirtalis*, family Colubridae) results in a 30 per cent decrease in blood calcium level after 10 days, indicating the parathyroids are important to calcium metabolism in snakes (Clark and Srivastava, 1970). Similarly, parathyroidectomy of lizards results in at least a 50 per cent reduction in blood calcium and a 70 per cent increase in the concentration of blood phosphate (Clark *et al.*, 1969). The administration of parathyroid extract induces hypercalcemia, hypercalciuria, and hyperphosphaturia in desert iguanas (*Dipsosaurus dorsalis*) and mesquite lizards (*Sceloporus grammicus*), both in the family Iguanidae, but has no effect on the serum phosphate level (McWhinnie and Cortelyou, 1968). It also has no significant effect on serum concentrations of calcium or phosphate in the American anole (*Anolis carolinensis*) (Clark, 1970). Because of their role in the regulation of blood calcium and phosphate levels, the parathyroids may be particularly important to reptiles having calcareous eggshells.

ADRENAL GLANDS

All reptiles have a pair of adrenal glands. Those of turtles are flattened dorsoventrally and lie against the kidneys, as is generally true in other vertebrates. In contrast, those of other reptiles are closely associated with the gonads and the urogenital ducts. The adrenal glands of the tuatara and of Squamata are incorporated in the mesorchium of the male or the mesovarium of the female, while that of crocodilians lies dorsal to the gonads, ventral to the kidneys, and partially insinuated into the parietal peritoneum. In all reptiles other than turtles, the right adrenal is positioned anterior to the left and, particularly in snakes, is larger than the left. As in amphibians, birds, and mammals, each adrenal is composed of a combination of chromaffin cells and interrenal cells. The chromaffin cells, which are homologous with the adrenal medulla of mammals, tend either to occur in groups which are scattered about throughout the interrenal tissue or to be in layers. In some reptiles, such as the American anole (*Anolis carolinensis*), the chromaffin tissue lies mostly on the dorsal side of the interrenal tissue. In the European wall lizard (*Lacerta viridis,* family Lacertidae), the chromaffin tissue is organized in such a way that there is a peripheral layer of noradrenaline cells extending as tongues and islands into the mass of adrenaline cells. Small masses of accessory adrenal tissue are commonly found lying posterior to the main adrenals. In turtles and crocodilians, the kidneys may contain islets of adrenal tissue (Gabe, 1970).

It is generally assumed that the adrenal hormones are similar in nature and effects in all vertebrates, but very little is known about them in reptiles. Adrenaline and noradrenaline have both been found in lizards, snakes, and turtles. Studies of adrenal slices from lizards (*Lacerta viridis*) and snakes (*Natrix natrix*) have demonstrated the synthesis from progesterone of both aldosterone and corticosterone; these appear to be in reptiles the major interrenal tissue secretions, just as they are in amphibians. The control of electrolyte content and water balance is fundamental to the success of reptiles as fully terrestrial animals. This control is presumably similar to that of other kinds of vertebrates and is believed to be effected through hormones secreted by the interrenal tissue of the adrenals and the posterior lobe of the pituitary. All adrenocorticoids studied produce similar effects

in kidney tubules, skin, gut, bladder, and a variety of glands in which transcellular water and salt movements take place.

ISLETS OF LANGERHANS

The islets of Langerhans in reptiles are generally scattered throughout the pancreas, just as they are in amphibians, birds, and mammals. However, those of snakes are concentrated in the anterior region of the pancreas adjacent to the spleen. Both alpha and beta cells have been reported to be present in reptilian islets. In Squamata, rather than being segregated, these two cell types lie adjacent to one another in an alternating pattern along vascular spaces (Miller and Lagios, 1970). In turtles and crocodilians, however, there is a marked segregation of alpha and beta cells such as that occurring in fish, some birds, and mammals. In both of these kinds of reptiles, the tendency is for alpha cells to be concentrated peripherally and beta cells to be aggregated in the central portion of the islet. Alpha cells are much more abundant in the islets of lizards and snakes than they are in mammalian pancreatic islets; they seem to be less abundant in turtles.

Pancreatectomy and treatment with alloxan, a product of the oxidation of uric acid which causes degenerative changes in beta cells, have had differing effects on various reptiles. In turtles, total pancreatectomy and alloxan treatment both produce a diabetes that is very intense in tortoises; partial pancreatectomy seems to induce diabetes more easily in male turtles than in females (see Young, 1963, for references). Total pancreatectomy of the snake *Xenodon merremii* (family Colubridae) produces a slight lowering of the blood sugar level for about three days then a rise to five or six times its normal level (Houssay and Penhos, 1960). Treatment of this snake with alloxan induces an initial hyperglycemia after which there is a serious and lasting hypoglycemia. Treatment of several other kinds of snakes with alloxan has induced hyperglycemia. Lizards treated with alloxan exhibit an initial fall in blood sugar level and then a rise; pancreatectomy produces only a decrease in blood sugar (Miller and Wurster, 1958). It is possible that the tendency in snakes and lizards for the blood sugar level to fall after the removal of the pancreas or damage to the beta cells is related to the greater importance of the secretion of glucagon by the very numerous alpha cells. Turtles, on the other hand, have relatively few alpha cells and immediately become hyperglycemic. In any event, insulin and glucagon seem to play the same role in regulating blood sugar levels in reptiles as in other vertebrates.

PITUITARY GLAND

The reptilian pituitary gland (hypophysis) is basically the same as that in other classes of vertebrates, being formed from an adenohypophysis and a neurohypophysis. The three oldest groups of living reptiles are the tuatara, turtles, and crocodilians, so it is not surprising that their pituitaries are more primitive than those of lizards and snakes. In the former three groups, the pituitary consists of a posterior lobe (*neurohypophysis*), a well-developed intermediate lobe (*pars intermedia*), a *pars tuberalis*, and an anterior lobe (*pars distalis*) that, as in birds, is subdivided into fairly distinct cephalic and caudal lobes. The pars tuberalis is very reduced or absent in lizards, and it fails to even develop embryonically in snakes. In both lizards and snakes, the portal vessels passing to the anterior lobe are accompanied by connective tissue, called the *pars terminalis*, rather than by the pars tuberalis, as in other reptiles. The neurohypophysis of some lizards and of all examined snakes is much more compact than in other reptiles. Probably as a result of cranial adaptations for swallowing large objects, the pituitary of adult snakes is typically flattened dorsoventrally and is very asymmetrical, with the neurohypophysis and intermediate lobe thrust to one side and the anterior lobe to the other. What is interesting is that the right and left relations vary individually and are not constant within a species. They seem to be due to chance, for some individuals have the anterior lobe on the right side and others have it on the left. The pituitary is symmetrical in young embryos and becomes asymmetrical late in development, just prior to hatching. Detailed descriptions and illustrations of the pituitaries of all four living orders of

reptiles may be found in Wingstrand (1966) and Girons (1970).

As in other vertebrates, the adenohypophyseal hormones of reptiles are concerned with regulating growth, reproductive functions, thyroid gland function, and adrenal gland function. Atrophy of testes and degeneration of spermatocytes occur in both male garter snakes (*Thamnophis*, family Colubridae) and lizards (*Agama*, family Agamidae) following hypophysectomy (Schaefer, 1933; Wright and Chester Jones, 1957). Hypophysectomy of female garter snakes (*Thamnophis sirtalis*) causes degeneration of primary Graafian follicles (Bragdon, 1953). The implantation of hypophyses into hypophysectomized snakes (*Thamnophis*) stimulates spermatogenesis, increases the size of the testicular tubules and interstitial cells, and causes an increase in testicular weight (Schaefer, 1933; Cieslak, 1945). Hypophyses implanted into female snakes (*Xenodon merremii*, family Colubridae) causes oviposition to occur (Houssay, 1931). Injections of human chorionic gonadotropin (HCG) stimulate spermatogenesis in male lizards and ovulation and oviposition in female lizards during the winter, when they are normally sexually inactive (Knobil and Sandler, 1963).

The dependence of normal thyroid gland function on the pituitary gland has been clearly established for reptiles within certain temperature ranges. Hypophysectomy of garter snakes (*Thamnophis radix* and *Thamnophis sirtalis*) results in degeneration of the secretory epithelium of thyroid follicles, while injections of thyroid stimulating hormone (TSH) into hypophysectomized snakes results in an increased height of the follicular epithelium (Schaefer, 1933; Hellbaum, 1936). Thyroidal uptake of radioiodine can be increased in turtles by injections of TSH if given at a warm temperature, and either TSH or thyroxine injections cause an increase in oxygen consumption in American anoles (*Anolis carolinensis*) at temperatures of 30 degrees centigrade (Maher and Levedahl, 1959).

The pituitary-adrenal relationship is indicated by the reduction in size of the adrenals and the degeneration of some of the groups of interrenal cells in snakes (*Thamnophis* and *Natrix*) and lizards

(*Xantusia* and *Agama*) following hypophysectomy. Implantation of pituitary tissue into hypophysectomized snakes restores their adrenals to a normal condition. Reptiles do respond to injections of mammalian adrenocorticotropic hormone (ACTH) with changes in their adrenals, but the action of ACTH on the secretory function of reptilian adrenals is not known.

There is very little information regarding the pituitary's regulation of growth in reptiles. Hypophysectomy has caused weight loss in snakes (*Natrix*, family Colubridae) when performed in the summer, but winter operations seem to have no effect on either these snakes or lizards (*Agama*, family Agamidae) (Wright and Chester Jones, 1957). However, hypophysectomy of diabetic turtles (*Chrysemys* and *Phrynops*, family Testudinidae and family Chelyidae, respectively) and otherwise normal lizards (*Eumeces*, family Scincidae) resulted in a lowered blood sugar level, which could be interpreted as being due to the elimination of the anti-insulin action of the pituitary growth hormone (Knobil and Sandler, 1963).

Pituitary hormones play a role in color changes in some reptiles. For example, the American anole (*Anolis carolinensis*) remains permanently pale after hypophysectomy. Furthermore, hypophysectomized individuals can be darkened by injections of intermediate lobes of amphibian or reptilian pituitary glands. This color change, as in amphibians, is produced by the effect of intermedin, the hormone of the pars intermedia, which causes dispersion of melanin granules in the melanophores.

Reptilian neurohypophyseal tissues have been shown to contain substances similar to those of amphibians, birds, and teleost fishes, with vasopressor, oxytocic, and antidiuretic characteristics. A hormone that appears to be arginine vasotocin has been found in tortoises (*Testudo graeca*, family Testudinidae), green turtles (*Chelonia mydas*, family Chelonidae), caimans (*Caiman*, family Crocodylidae), *Iguana* lizards (family Iguanidae), and grass snakes (*Natrix natrix*, family Colubridae). This hormone is also found in teleost fish, amphibians, and birds and has antidiuretic potencies which are similar but not identical to those of the mammalian hormone vasopressin.

Pharmacological and chromatographic studies have shown that pituitaries from tortoises, green turtles, *Iguana* lizards, and grass snakes also contain an oxytocin-like substance, but this has not been positively identified (Heller, 1963).

GONADS

A few studies have been made of the action of testicular hormones in lizards, alligators, and turtles on characters that are sexually dimorphic. It has been shown that a number of male characteristics of lizards, including skin coloration, dorsal crests and spines, femoral and preanal pores, the epididymides, and sex segments of the kidneys, become feminine after castration (literature reviewed by Van Oordt, 1963). It can be demonstrated that the male sex segments of the kidneys are dependent upon testosterone for expression. Testosterone will masculinize female reproductive ducts in the herbivorous lizard *Uromastix* (family Agamidae) and androsterone will masculinize female urinary sex segments (Kehl and Combescot, 1955). Testosterone treatments of juvenile alligators (*Alligator mississippiensis*, family Crocodylidae) cause hypertrophy of the genital tubercles of both sexes and of the epididymides and ductus deferens of young males, and, perplexingly, stimulation of the oviducts in young females (Forbes, 1938, 1939). When young female turtles are treated with this hormone, masculinization of the reproductive ducts and copulatory organ occurs. It is also known that the tail and claws of male turtles are dependent upon testosterone for their development. It is not clear which part of each testis secretes the male hormones in reptiles. However, the interstitial cells of at least lizards change seasonally in size, number, and structure, and these changes are correlated with seasonal changes in sexual characters (Kehl and Combescot, 1955). However, Miller (1948) found neither quantitative nor qualitative changes occurring in interstitial cells of the night lizard (*Xantusia vigilis*, family Xantusiidae) at the time that sex characters developed.

Relatively little is known about the female gonadal hormones. The corpora lutea of ovoviviparous and viviparous lizards and snakes seem to have an endocrine function, for they persist much longer than in oviparous forms. However, as mentioned previously, their hormone, progesterone, is not always necessary for the maintenance of pregnancy and its significance is not known. It is known that progesterone will stimulate the growth of oviducts in viviparous lizards and that the plasma level of this hormone increases during pregnancy in ovoviviparous snakes (Velle, 1963, lists references).

NERVOUS SYSTEM

The brains of reptiles (Fig. 4–46) are organized along the basic vertebrate plan, with three major divisions: the forebrain, consisting of paired olfactory lobes and cerebral hemispheres plus the diencephalon, the region behind and between the hemispheres; the midbrain, consisting of the optic lobes and the cerebral peduncles, tracts of nerve fibers connecting the forebrain and the hindbrain; and the hindbrain, consisting of the cerebellum

and the medulla oblongata, which is continuous with the spinal cord. The brain of pterydactyls fitted closely to the skull, but in most reptiles it lies rather loosely within the cranial cavity and, compared to the brains of mammals, is small. It probably never represents more than about one per cent of the mass of the reptile. For example, some of the largest dinosaurs had bodies that must have weighed over 20 tons but brains that are estimated to have weighed only a few ounces. Despite its small size, the reptilian brain is significantly advanced over the brains of amphibians and fishes, particularly in regard to the cerebral hemispheres and the cerebellum.

In reptiles other than turtles, the cerebral hemispheres are continued anteriorly as olfactory tracts that terminate in the olfactory lobes. Thus, although small, these lobes are distinct, whereas in amphibians and turtles they are barely distinguishable from the cerebral hemispheres. In many reptiles, accessory olfactory lobes occur just posterior to the

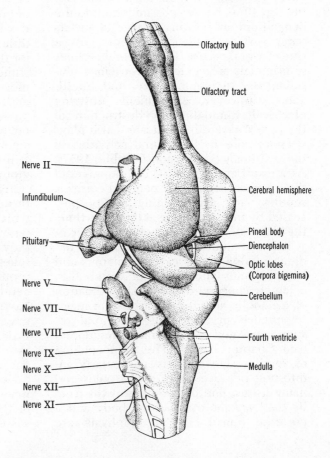

Figure 4–46   The brain of an alligator. (From Torrey, T. W.: *Morphogenesis of the Vertebrates.* 2nd Ed. John Wiley and Sons, 1967.)

primary ones. These receive impulses from well-developed vomeronasal nerves which innervate the paired Jacobson's organ. The cerebral hemispheres are considerably enlarged over their condition in amphibians. Much of the bulk of each is formed by a corpus striatum, a solid mass of tissue bulging into the hemisphere's hollow cavity (lateral ventricle) from its ventrolateral wall. The corpora striata, best developed in birds, are centers for instinctive, as opposed to learned, behavior. Furthermore, the amount of gray matter on the surface of the hemispheres of reptiles is increased so that these thickened layers of nerve cell bodies, their dendrites, and the proximal portions of their axons conceal the corpora striata beneath them. In reptiles other than the tuatara, a new area called the neopallium has formed on the dorsoanterior end of each hemisphere. This receives sensory fibers from other parts of the brain and gives rise to motor fibers which conduct impulses back to the spinal cord. The neopallium is the portion that is very enlarged in the mammalian brains and accounts for much of their bulk. In crocodilians, for the first time, nerve cells migrate into the neopallium and form a layer along its outer surface, which now represents a true cerebral cortex. The cerebral cortices of mammals are centers of conscious sensation and action, memory, and intelligence; however, experiments utilizing electrical stimulation or destruction of the cortex in reptiles indicate that it plays a minor role in the neural activities of these animals (Golby and Gamble, 1957). As a result of the relative development of their corpora striata and cerebral cortices, reptiles lack the learning ability possessed by mammals and, at the same time, the elaborate instinctive behavior of birds.

The remainder of the forebrain, the diencephalon, includes the thalamus and hypothalamus and, as in all vertebrates, is largely concerned with the coordination of metabolic activities. It is also of major importance in triggering behavioral responses related to temperature regulation (see Chapter Eight). The diencephalon of reptiles, however, is particularly interesting because, as in cyclostomes and many fishes, one or two outgrowths from its roof, an anterior parietal body and a posterior pineal body or epiphysis, are frequently present. These are discussed in the following section on receptor organs.

The keen sense of vision possessed by most reptiles is reflected in a very well-developed midbrain (*mesencephalon*). The optic lobes are very conspicuous, more so than in mammals, and receive nearly all of the fibers from the optic nerve. Like those of mammals, the optic lobes of snakes are divided by a transverse fissure into four masses, the corpora quadrigemina. The anterior pair of the corpora quadrigemina, called the superior colliculi, function as centers for vision whereas the posterior pair, called the inferior colliculi, serve as centers for the integration of auditory impulses. In all reptiles, the roof of the midbrain, called the tectum, is an important coordination center with functions paralleling those of the cerebral cortices of mammals.

Although it varies considerably among reptiles, the cerebellum is generally larger than in amphibians but not as enlarged as in mammals, birds, or even some fishes. It is only slightly larger in lizards and the tuatara than in amphibians but definitely larger in snakes, and it is best developed and most prominent in crocodilians and turtles. Since it is a center for the correlation of internal and external stimuli and the coordination of muscular movement, the size of the cerebellum generally varies directly with the animal's powers of locomotion. However, this does not adequately account for the differences among reptiles, for lizards, whose cerebellum is small among reptiles, are as a group very active and would seem to require at least as much coordination as turtles. The wings of the cerebellum, which receive nerve fibers concerned with balance, are primitive in reptiles; they are represented by small floccular lobes on the ventral side. However, the reptilian cerebellum is advanced over the cerebella of fishes and amphibians in having a layer of white matter inside the gray like that occurring in birds and mammals. There is nothing on the ventral surface of the medulla oblongata of reptiles corresponding to the mammalian pons.

Reptiles are the lowest vertebrates to have 12 pairs of cranial nerves in addition to the terminal nerve. The 10 pairs possessed by fishes and amphibians are present and, in addition, both the spinal accessory (XI) and the hypoglossal (XII) are developed as separate nerves. The bases of the vagus (X) and the spinal accessory (XI) may be fused together in some reptiles.

The spinal nerves of reptiles are advanced over those of amphibians and fishes only in that the dorsal roots are composed solely of sensory fibers; visceral motor and somatic motor fibers are restricted to the ventral roots. In at least some snakes and limbless lizards, a distinct lumbosacral plexus is present, providing evidence that these forms evolved from ancestors possessing limbs.

The autonomic nervous system of most reptiles includes the usual double trunk of nerves running the length of the animal. Crocodilians, however, are exceptional in having a second accessory trunk of sympathetic nerves on each side running parallel with arteries through the vertebraarterial canal, just as occurs in the mudpuppy *Necturus* (family Proteidae).

RECEPTOR ORGANS

With few exceptions, the receptor organs possessed by reptiles are the same kinds present in the more terrestrial amphibians. In nearly every instance, however, improvements have evolved so that reptiles have greater perception of their environment than do amphibians.

Ear

As in most higher vertebrates, the reptilian ear (Fig. 4–47) generally serves as an organ of both equilibrium and hearing. The structures in the inner ear concerned with the equilibratory function, the semicircular canals and utriculus, are essentially unchanged from their condition in amphibians. However, modifications have occurred in the sacculus and its derivatives that are concerned with hearing. The lagena, which in amphibians is only a slight bulge in the ventral wall of the sacculus, is elongated in reptiles to form the cochlear duct, and this contains both the macula lagenae and the papilla basilaris. Thus, the sensory patches are concentrated in one outpocketing of the sacculus. The papilla basilaris appears to be the essential receptor organ for hear-

ing, but the function of the macula lagenae is not clear. The cochlear duct is straight in all living reptiles other than the crocodilians; in the latter (as in birds) it forms a simple spiral which is a forerunner of the complicated coil present in mammals. Just as occurred in anuran amphibians, the endolymphatic ducts of some lizards and snakes seem to have become specialized as calcium-secreting glands. Those of geckos (family Gekkonidae), especially, are enlarged and filled with calcium carbonate. In some species they even pass through openings in the skull and extend posteriorly among the muscles of the neck. Apparently these function on a seasonal basis, releasing calcium salts into the blood stream for use by the shell glands during the formation of calcareous eggshells.

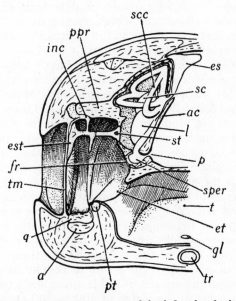

Figure 4–47 Posterior view of the left side of a lizard head to show the auditory apparatus. A shallow external depression leads to the tympanic membrane (*tm*). Internal to the tympanic membrane, the stapes is seen, divided into two parts, the "extracolumella" (*est*), and columella or stapes proper (*st*); processes from the former articulate above with the skull (*inc*) and below with the quadrate anterior to the middle ear cavity. This cavity opens by a broad eustachian tube (*et*) to the throat. The internal ear is shown in diagrammatic fashion. Other abbreviations: *a*, articular; *ac*, inner wall of auditory capsule; *es*, endolymphatic sac; *fr*, fenestra rotunda; *gl*, glottis; *l*, lagena; *p*, perilymphatic duct connecting inner ear with brain cavity; *ppr*, paroccipital process of otic region; *pt*, pterygoid; *q*, quadrate; *sc*, sacculus; *ssc*, semicircular canals; *sper*, position of perilymphatic sac; *t*, tongue; *tr*, trachea. (From Goodrich, E. S.: *Studies on the Structure and Development of Vertebrates.* Volume 1, 1958. Dover Publications, New York.)

There is considerable variation in the middle ear among living reptiles. In snakes, the tympanic membrane (eardrum), the cavity of the middle ear, and the Eustachian tube are missing entirely. A typical reptilian inner ear is present, consisting of three semicircular canals, an utricle, a saccule, a lagena, and a simple cochlea. The proximal end of the columella is in its usual position, resting in the fenestra ovalis, but the distal end (which would ordinarily articulate with the tympanic membrane) articulates with the quadrate bone. The quadrate of snakes is relatively free, being only loosely attached by ligaments to the lower jaw and to the dorsolateral skull. Because of this arrangement, snakes generally have been considered unable to perceive air-borne vibrations (sound waves) and have been thought capable of detecting only those vibrations which travel from the ground to the jaws to the columella to the fenestra ovalis to the inner ear. However, Wever and Vernon (1960) have demonstrated clearly that colubrid snakes (*Pituiphis melanoleucus, Thamnophis sirtalis,* and *Natrix sipedon*) can hear aerial sounds and are moderately sensitive to frequencies in the range of 100 to 700 cycles per second. They have concluded that the quadrate bone acts as a receiving surface for the aerial sounds and have noted that removal of this bone produces a moderate, but not great, reduction in inner ear response to sound. This is apparently because muscle and fiber layers over the distal end of the columella continue to transmit sound pressures to it. The hearing of the snakes tested is superior to that of a cat in the range of 100 to 200 cycles per second; this sensitivity is undoubtedly adaptive in allowing these reptiles to perceive movements of large animals.

In most reptiles other than snakes, the tympanic membrane is visible on the surface of the head, but in some lizards and crocodilians it is found at the bottom of a small depression. In aquatic turtles it is thin and almost transparent, but in terrestrial species it is thick and covered with skin. Studies of turtles (Wever and Vernon, 1956a, 1956b) indicate that their hearing is similar in range and sensitivity to that of snakes.

The tympanic membrane of some iguanid lizards, including earless lizards (*Holbrookia*) and some horned lizards (*Phrynosoma*), is covered with scaly skin. In general, degeneration of the middle ear has occurred in most burrowing lizards, and they tend to have a tympanic membrane which is either vestigial or absent; in some, the cavity of the middle ear has also been lost. Presumably, these animals are insensitive to air-borne sounds, but they probably can detect ground vibrations.

The middle ear of crocodilians is well-developed. The two Eustachian tubes are unusual in that they join just before connecting to the pharynx and open into it by a single median orifice.

The tympanic membrane of the tuatara is covered with scaly skin, but these primitive reptiles appear to have fairly sensitive hearing.

The lateral line system, which is well-developed in fishes and aquatic amphibians, does not occur in any living reptile.

Eyes

Except for burrowing forms, which characteristically have degenerate eyes, vision is a dominant sense in reptiles. Their eyes have undergone considerable adaptive radiation with the evolution of specialized habits. Those of snakes have probably evolved from rudimentary eyes of burrowing ancestors and differ in many ways from those of other reptiles (Fig. 4–48). In reptiles other than snakes, near-vision-focusing is accomplished by contraction of the ciliary muscle that presses the ciliary body against the peripheral edge of the elliptical lens. The resulting pressure causes the lens, especially its anterior surface, to bulge into a more spherical shape; some deformation of the cornea also occurs. What is particularly unusual about this mechanism is that this ciliary muscle (and also that controlling the iris) is striated, thus, under voluntary control. Accommodation in snakes occurs, as in amphibians, with the forward movement of the lens for near vision. However, whereas in amphibians the lens is moved forward by the contraction of the protractor muscle, in snakes it is accomplished by the contraction of the iris, which puts pressure on the vitreous body (the mass of jellylike material which fills the eyeball behind the lens), which in turn pushes the lens forward.

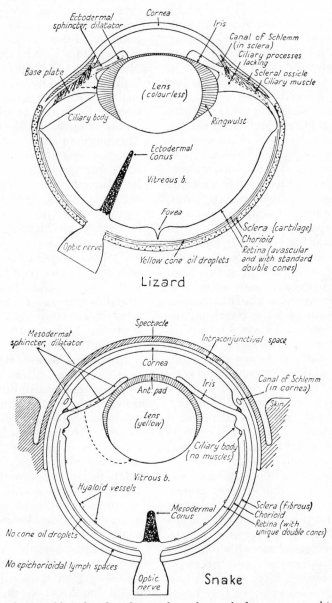

Figure 4-48 Diagrams of eyes of lizard and snake, to show the marked contrasts resulting from presumed loss during underground life and later acquisition by the snake of features paralleling those present in their ancestors. The dotted arrows show the direction of application of force during accommodation. (From Walls, G. L. 1942. *The Vertebrate Eye.* The Cranbrook Press, 1942.)

Both rods and cones are always present in the retinas of tuataras, crocodilians, and turtles. These reptiles usually have a greater number of cones than amphibians, and many have a fovea centralis, a small depression in the retina which lacks rods; this is a point of especially acute vision and is functionally important in the resolution of fine detail. Cones are chiefly concerned with color vision and visual acuity and are most abundant in reptiles, such as the majority of turtles, that are active in the daytime. Because of differences in the relative number of cones present, some reptiles, including turtles, have what appears to be good color vision and acute discrimination, whereas others, including the tuatara, do not. Both red and yellow oil-droplets, distributed in a mosaic pattern, are present in turtle retinal cones. The yellow droplets act as filters to screen out shorter wavelengths, particularly in the blue portion of the spectrum, and both red and yellow droplets reduce chromatic aberration. The red droplets, which have a greater effect upon chromatic aberration than the yellow, are particularly important to a turtle looking over the dazzling surface of water on bright days. On cloudy days, the yellow (or colorless) droplets become functionally the most important.

The relative number of cones and rods varies among Squamata, especially between nocturnal and diurnal species. Rods are primarily used to distinguish different intensities of light and are most abundant in nocturnal lizards. The retinal cells of geckos (family Gekkonidae) all appear to be rods, some of which may have arisen from the transmutation of cones, and only a few species (e.g., *Sphaerodactylus parkeri*) possess a fovea (Underwood, 1951). The slitlike pupil of geckos has a series of serrations which form a line of pinhole openings when the pupil is completely closed (Fig. 4–49). Light passing through each of these tiny holes forms a sharp image, and the images formed by all of them are superimposed on the retina, giving the animal extremely acute vision even in poor light. Among snakes, cones have been lost in burrowing forms such as the blind snakes (*Typhlops*, family Typhlopidae), and only tiny rods are present.

In contrast, many diurnal lizards and snakes in the family Colubridae have lost the rods and have retinas containing only cones, so that the animals are virtually blind at night. Whereas nocturnal species have a vertical pupil and lack the fovea centralis, diurnal lizards have a round pupil and possess a fovea centralis. A fovea is positively known to occur in only two genera of snakes, the East Indian long-nosed tree snake (*Dryophis mycterizans*, family Colubridae) and the African bird snake (*Thelotornis kirtlandi*, family Colubridae). The fovea in these two snakes, rather than being central as in lizards, is on the outer rim of the retina on the temporal (posterior) side of the eye. The pupil in these snakes is horizontally elongate and shaped like a keyhole, and there are grooves on the sides of the lengthened snout along which the two eyes can sight. The fovea, pupil, and groove are aligned so that a line from the fovea through the center of the lens passes through the slot in the keyhole pupil, along the groove in front of the eye, and straight forward, parallel to the axis of the body (Walls, 1942; Fig. 4–50). The combined mechanism gives these two species what is generally agreed to be the most acute vision and greatest depth perception of all ophidians.

In both diurnal lizards and diurnal snakes, yellow filters are present in the eye to screen out shorter wavelengths and reduce chromatic aberration. However, the mechanism is not the same in the two groups. Diurnal lizards have droplets of yellow oil in their retinal cones, whereas diurnal snakes have a yellow lens which filters the light before it reaches the cones. No snake or crocodilian has oil-droplets. A unique double cone, not found in lizards, often occurs in the retinas of diurnal snakes.

Reptilian eyelids, except in species having a spectacle, are generally more movable than those of amphibians, but, as in amphibians, the lower lid is the largest and its upward movement primarily closes the eye. A third more or less transparent eyelid, the true nictitating membrane, is also present at the front of the eye. It functions to cleanse and lubricate the cornea by very rapidly sweeping back over its surface. The irritation of a foreign object on the cornea causes a blood sinus at the base of the nictitating membrane to become swollen with blood.

Figure 4–49   Head of the East Indian tokay (*Gekko gecko*), illustrating the pupil of geckos, which forms a series of pinhole openings when completely closed. (From Mertens, R.: *The World of Amphibians and Reptiles*, George C. Harrap and Co., 1960.)

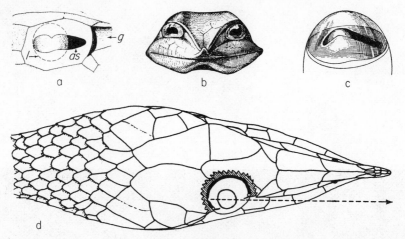

Figure 4–50 Diagrammatic views of the head of the East Indian long-nosed tree-snake (*Dryophis mycteri-zans*). *a*, lateral view of the right eye *in situ*, showing aphakic portion of pupil and cheek groove which permits straight-forward vision. *as*, aphakic space; *1*, lens; *g*, groove. *b*, anterior view of face, showing provisions for binocular vision. *c*, anterior segment of right eye, showing form of iris, lens, and aphakic space. *d*, dorsal view of head, cut away to reveal eye in section, showing line of sight from temporal fovea through lens and aphakic portion of pupil and along cheek groove. (From Walls, G. L., 1942. *The Vertebrate Eye*. Michigan.)

The swelling of this sinus helps force the object to the corner of the eye, where it may be rubbed out by a foot. Horned lizards (*Phyrnosoma*, family Iguanidae) are capable of squirting blood from their eyes by suddenly rupturing the sinus in the nictitating membrane through muscular contraction.

Well-developed Harderian glands, first evolved in amphibians, occur in association with the nictitating membrane, and a posterior lachrymal gland is generally present. As mentioned previously, the posterior lachrymal glands of marine turtles play an important role in osmotic regulation by secreting sodium chloride and are very enlarged. The lachrymal (tear) duct usually originates as two little canals at the anterior edge of the lower lid and in front of the nictitating membrane. The duct drains into the nasal passage in crocodilians, but in lizards, snakes, and tuataras it opens into the duct of Jacobson's organ or into a groove on the palate adjacent to this organ. Turtles apparently lack lachrymal ducts. In snakes, true lachrymal glands are missing, and the oily secretion of the Harderian gland flows into the space between the spectacle and the cornea and from there into the lachrymal duct to the vicinity of Jacobson's

organ. The Harderian gland is always large in snakes, and it appears that its secretion may play a role in the functioning of Jacobson's organ.

In snakes and a variety of lizards, including geckos, the nictitating membrane is absent. The upper and lower eyelids fuse together during embryonic development and are transparent, forming a fixed window of clear skin, called the spectacle, over the movable eye. Several stages in the development of the spectacle exist among skinks (family Scincidae). Some species possess a fully-developed spectacle, whereas others, including *Mabuya vittata*, possess eyelids with a transparent window in the lower. Spectacles also occur in at least two turtles, *Emys granosa* (Family Testudinidae) and *Chelodina longicollis* (family Chelyidae) (Wall, 1942). Both the transparent lower lid and the spectacle function to protect the eye from abrasive dirt, sand, or vegetation. They are important adaptations to living on or beneath the surface of the ground or, in the case of the turtles, probably function to protect the eyes when the animal is feeding or burrowing in bottom mud. Being modified skin, the spectacle and lower lid window are shed during ecdysis. The onset of ecdysis is generally indicated by the formation of a milky film under the layer being shed. This makes the animal temporarily blind and often, as a result, rather pugnacious. The presence of the spectacle in such a variety of reptiles is an excellent example of parallel evolution and the independent development of the same sort of a structure in unrelated forms.

In all reptiles other than snakes, the sclerotic coat covering the eyeball is cartilaginous, and often there is a ring of small overlapping bony plates (scleral ossicles) at the junction of the sclerotic coat and the cornea. Scleral ossicles were also present in many extinct reptiles but their significance is questionable, since many modern reptiles, including crocodilians, lack them. Snakes lack both scleral bones and the cartilage in the sclerotic coat.

Reptiles have the normal vertebrate set of six eye muscles plus a *retractor bulbi,* which operates the nictitating membrane, and a *levator bulbi,* which in part operates the lower eyelid. Normally, reptiles have binocular vision at a distance and in some, including the monitor lizards (family Varanidae), the face is concave in front of each eye so that the two fields of vision overlap to a greater degree than would otherwise be possible. The grooves on the snout of the East Indian long-nosed tree snake, mentioned previously, give this species a binocular field of vision which is 46 degrees wide. Crocodilians have about 25 degrees of binocular field, turtles 18 (*Testudo*) to 38 (*Chelydra*) degrees, lizards other than Varanidae 10 to 20 degrees, Varanidae about 30 degrees, and most snakes 30 to 40 degrees.

True chameleons (family Chamaeleontidae) have what must be the most specialized eyes of any vertebrate. Their eyeballs protrude from the head and can be swiveled independently of one another in a turretlike fashion, giving the animal a 360-degree field of vision from the combined monocular vision of both eyes. When an insect is spotted both eyes converge on it, and its position is watched with binocular vision until the lizard has crept close enough to grasp it with its tongue. It is not known how the chameleon's brain is able to interpret the two different images seen with monocular vision. The eyelids of chameleons are also specialized and form a complete covering over the eye with a small circular opening in the center exposing the cornea.

Median Eye

The diencephalon of reptiles is particularly interesting because, as in lampreys and many fishes, one or two outgrowths from its roof are frequently present, an anterior parietal body (parapineal organ) and a posterior pineal body (epiphysis). The pineal body is the dominant median eye of lampreys but in higher vertebrates it persists as a glandular structure. The parietal body, however, is what is very well-developed in the tuatara and many lizards and has the structure of a degenerate eye with a lens, retina, and a nerve connecting it to the brain (Fig. 4–51). However, there are neither eye muscles nor any apparent mechanism for accommodation. Stebbins and Eakin (1958) studied the effect of removing the parietal eye from North American lizards in the genera *Sceloporus, Uta,* and *Uma* (all in the family Iguanidae) and concluded that it may act as a kind of light meter, regulating the amount of time during which an animal exposes itself to sunlight. It does not seem to be capable of forming images but can probably distinguish light from dark. Therefore, in lizards its function seems to be importantly related to the regulation of body temperature. The parietal eye of the tuatara apparently has a different function, for it either has a greater delay in its response or does not respond in the same manner as that of lizards (Stebbins, 1958).

Chemical Sense Organs

During the evolution of reptiles there has been a tendency for the nasal region to become more complicated than it is in amphibians. As a result of the evolution of the secondary palate, developed to different degrees in various reptiles, the nasal passages are longer than in amphibians, and each includes two more or less distinct membranous sacs. The external nostril opens into an anterior sac (vestibule); this leads into the main olfactory chamber, which is sometimes curved sideways on itself so as to be U-

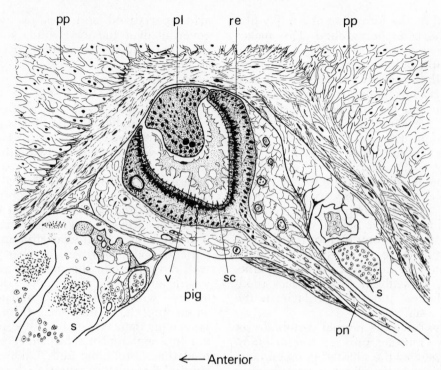

← Anterior

Figure 4–51 Longitudinal section through the parietal (median) eye of a tuatara (*Sphenodon*). *pig*, pigment layer of retina; *pl*, pineal eye lens; *pn*, pineal nerve; *pp*, pineal plug; *re*, retina; *s*, venous sinus; *sc*, sensory cells; *v*, vitreous body. (From Bellairs: *The Life of Reptiles*.)

shaped. Posteriorly, the main olfactory chamber constricts into a nasopharyngeal duct that opens through the internal nostril (choana) into the buccal cavity. Only a limited portion of the lining of the nasal passages is sensory epithelium (containing bodies of the sensory cells of the olfactory (I) cranial nerve), and this is generally found in the dorsal part of the main olfactory chamber. However, there are typically one or more shelflike folds (conchae) projecting inward from the lateral wall of each olfactory chamber; these increase the surface area of the portion lined with sensory epithelium. The olfactory chamber of crocodilians is particularly elongate and contains three conchae; one or two usually occur in other reptiles.

Paired Jacobson's organs (the vomeronasal organs) are lost in adult crocodilians but occur in all other living reptiles. In turtles, as in amphibians, they are confluent with the nasal passages and are not well-developed. In the tuatara, they are relatively large and are located in club-shaped outpocketings from the nasopharyngeal ducts near the internal nostrils. Their greatest development is in lizards and snakes where, during embryonic development, they are completely separated from the nasal passages to become separate blind pouches opening independently through the palate and into the mouth cavity well forward from the internal nostrils (Fig. 4–52). Their function has also been modified in lizards and snakes. In amphibians and presumably in the tuatara and turtles, the Jacobson's organs are primarily used to taste objects being held in the mouth, but, as described earlier in this chapter, in lizards and snakes with protrusible bifid tongues they are accessory olfactory organs used to detect odors in the air. True chameleons (family Chamaeleontidae) are exceptionally specialized for feeding on insects and their tongue, used to capture prey, ends in a sticky bulblike tip rather than being forked. Associated with this, Jacobson's organ is vestigial in these lizards.

The sense of taste is relatively poor in reptiles compared to amphibians and fish. Reptilian taste buds are found primarily in the lining of the pharynx with very few, if any, on the tongue.

Sensory Pits

Rattlesnakes and other pit vipers (subfamily Crotalinae of the family Viperidae) have a pit on each side of the face between the eye and the nostril which experimentation has shown to be heat-sensitive (Noble and Schmidt, 1937). Pythons and boas (family Boidae) have rows of similar pits in the scales along the margins of their jaws which are heat-sensitive also (Bullock and Barrett, 1968). Since some species of boids which lack pits are still sensitive to radiant heat (Bullock and Barrett, 1968), the function of the pits is probably to increase directional sensitivity to the stimuli (Barrett, 1970). These pits each consist of a depression lined with cornified epithelium and partitioned into an outer and an inner chamber by a highly innervated membrane (Fig. 4–53), the innervation being from the trigeminal (V)

cranial nerve. There is a continual transmission of impulses from the pits to the brain and Bullock and Cowles (1952) have shown that the rate of this continual transmission is independent of the animal's body temperature and is dependent upon the average radiation from all objects in the receptive field. The pit organs are highly sensitive to infrared wavelengths between 15,000 and 40,000 Ångströms, and any warm or any cold object causes a temporary change in the rate of impulse transmission, the response being greatest to a sudden temperature change. Within a few seconds following the initial response, the impulse rate adapts to the new level. Objects which are cold relative to the average of the receptive field cause a depression in the nerve activity even if they are warmer than the snake's body. Thus, the snake is able to perceive both warm and cold objects with temperatures differing from the average of surrounding objects.

Based on the study of four species of rattlesnakes (*Crotalus*), Bullock and Cowles (1952) concluded that the receptive fields are irregular cones extending

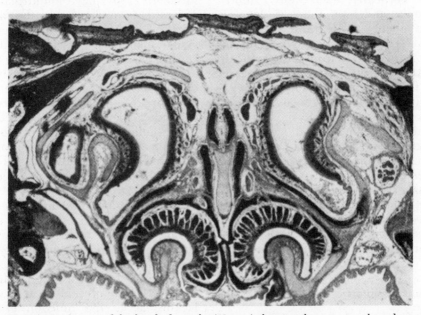

Figure 4–52 Transverse section of the head of a snake (*Vipera*) showing the crescent-shaped organs of Jacobson; the Jacobson's organ on the left has been sectioned so as to show the connection to the mouth cavity, which is at the bottom of the picture. The larger openings above the paired Jacobson's organs are the nasal passages, which are completely independent of Jacobson's organs. (From Cordier, R.: Sensory cells, pp. 313–386 in *The Cell*, vol. VI, Academic Press, 1964.)

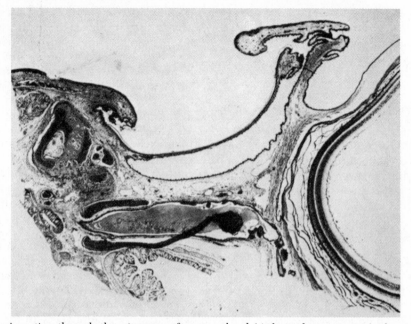

Figure 4-53 A section through the pit organ of a copperhead (*Agkistrodon contortrix*), showing the outer and inner chambers and the sensory membrane. The snout of the animal is to the right. (From Cordier, R.: Sensory cells, pp. 313–386 in *The Cell*, vol. VI, Academic Press, 1964.)

in a horizontal plane about 10 degrees across the midline in front and almost at right angles to the body laterally from the pits. Vertically, each field extends forward only, and from about 35 degrees below to about 45 degrees above the horizontal. Thus, the receptive fields of the two pits overlap in front of the animal, and together the pits survey a 180-degree field in front of them. Their sensitivity varies with the wavelength and is generally greater to infrared in the range of 2 to 3 microns than to shorter or longer wavelengths. They seem to be able to detect temperature differences of 0.2 degrees centigrade or less.

The snakes possessing pit organs are at least partly nocturnal; they undoubtedly use these organs to locate and strike prey such as small mammals whose temperature differs from that of their surroundings.

Cutaneous Receptors

Almost nothing is known about cutaneous receptor organs in reptiles, but undoubtedly thermoreceptors, tangoreceptors, and pain receptors are present.

Because of their dry and cornified epithelium, it is doubtful that the common chemical sense is important. Miller and Kasahara (1967) have found that the skin of lizards is innervated by the same basic types of nerve endings that occur in the skin of mammals. They also found specialized "sensory" areas in a variety of lizard skins. In such areas, the living epidermal cells are taller than in unspecialized areas and are well-innervated by expanded-tip nerve terminals. The overlying keratinized layers of specialized areas may be either thinner or thicker than in unspecialized areas. "Hair-like" organs have been found in some species belonging to the families Agamidae, Gekkonidae, and Iguanidae. Such structures are characterized by high columnar living epidermal cells, which are probably well-innervated with expanded-tip terminals, and an overlying keratinized epithelium that forms a "hair-like" projection. While physiological evidence is lacking as to the function of areas of specialized skin, Miller and Kasahara (1967) have tentatively postulated that such areas might be sensitive to some type of radiant energy.

References

Angel, F. 1950. Vie et Moeurs des Serpents. Paris, Payot 319 pp.

Auffenberg, W. 1962. A review of the trunk musculature in the limbless land vertebrates. Am. Zool. 2:183–190.

Barrett, R. 1970. The pit organs of snakes. In: C. Gans (ed.) 1970. Biology of the Reptilia, Volume 2, Morphology B. New York, Academic Press. Pp. 277–300.

Bellairs, A. 1970. The Life of Reptiles. New York, Universe Books. 590 pp.

Bogert, C. M. 1947. Rectilinear locomotion in snakes. Copeia 1947:253–254.

Bragdon, D. E. 1953. A contribution to the surgical anatomy of the water snake, Natrix sipedon sipedon, the location of the visceral endocrine organs with reference to ventral scutellation. Anat. Rec. 117:145–161.

Bullock, T. H., and R. Barrett. 1968. Radiant heat reception in snakes. Commun. Behav. Biol. (A) 1:19–29.

Bullock, T. H., and R. B. Cowles. 1952. Physiology of an infrared receptor: the facial pit of pit vipers. Science 115:541–543.

Carter, G. S. 1967. Structure And Habitat in Vertebrate Evolution. Seattle, Univ. Washington Press. 520 pp.

Chiu, K. W., J. G. Phillips, and P. F. A. Maderson. 1967. The role of the thyroid in the control of the sloughing cycle in the tokay (Gekko gecko, Lacertilia). J. Endocrinol. 39:463–472.

Cieslak, E. S. 1945. Relations between the reproductive cycle and the pituitary gland in the snake Thamnophis radix. Physiol. Zool. 18:299–329.

Clark, N. B. 1965. Experimental and histological studies of the parathyroid glands of fresh-water turtles. Gen. Comp. Endocrinol. 5:297–312.

_____ 1967. Parathyroid glands in reptiles. Am. Zool. 7:869–881.

_____ 1970. The parathyroid. In: C. Gans and T. S. Parsons (eds.) 1970. Biology of the Reptilia, Volume 3. New York, Academic Press. Pp. 235–262.

_____, P. K. T. Pand, and M. W. Dix. 1969. Parathyroid glands and calcium and phosphate regulation in the lizard, Anolis carolinensis. Gen. Comp. Endocrinol. 12:614–618.

_____, and A. K. Srivastava. 1970. Parathyroidectomy of the garter snake, Thamnophis sirtalis. Am. Zool. 10:298 (Abstr.).

Colbert, E. H. 1967. Adaptation for gliding in the lizard Draco. Am. Mus. Novitates 2283:1–20.

Cole, C. J. 1966a. Femoral glands of the lizard, Crotaphytus collaris. J. Morphol. 118:119–135.

_____ 1966b. Femoral glands in lizards: a review. Herpetologica 22:199–206.

Dimond, M. T., Sr. 1954. The reactions of developing snapping turtles, Chelydra serpentina serpentina (Linné), to thiourea. J. Exp. Zool. 127:93–115.

Dowling, H. G., and J. M. Savage. 1960. A guide to the snake hemipenis: A survey of basic structure and systematic characteristics. Zoologica 45:17–28.

Drexler, E., and B. von Issekutz. 1935. Die Wirkung des Thyroxins auf den Stoffwechsel Kaltblütiger Wirbeltiere. Naunyn-Schmiedebergs Arch. f. exp. Path. Pharm. 177:435–441.

Drummond, F. H. 1946. Pharyngeo-oesophageal respiration in the lizard Trachysaurus rugosus. Proc. Zool. Soc. London 116:225–228.

Drzewicki, S. 1929. Uber den Einfluss der Schilddrusenexstirpation auf die Zauneideschse; (Hemmung des Hautungsprozesses, Veränderungen in der Haut, in den Augen und in den innersekretorischen Drüsen, Wachstumschemmung). Arch. Entw. Mech. Org. 114:155–176.

Dunson, W. A., and A. M. Taub. 1967. Extrarenal salt excretion in sea snakes (Laticauda). Am. J. Physiol. 213:975–982.

Edmund, A. G. 1969. Dentition. In: C. Gans, A. d'A. Bellairs, and T. S. Parsons (eds.) 1969. Biology of the Reptilia, Volume 1. New York, Academic Press, Pp. 117–200.

Eggert, B. 1933. Über die histologischen und physiologischen Beziehungen zwischen Schilddrüse und Hautung bei den einheimischen Eidechsen. Zool. Anz. 105:1–9.

_____ 1936. Zur Morphologie und Physiologie der Eidechsen-Schilddrüse. III. Über die nach Entfernung der Schilddrüse auftretenden allgemeinen Ausfallserscheinungen und über die Bedeutung der Schilddrüse für die Hautung and für die Kaltstarre. Z. Wiss. Zool. 148:221–260.

Etheridge, R. 1967. Lizard caudal vertebrae. Copeia 1967:699–721.

Forbes, T. R. 1938. Studies on the reproductive system of the alligator. III. The action of testosterone on the accessory sex structures of recently hatched female alligators. Anat. Rec. 72:87–95.

_____ 1939. Studies on the reproductive system of the alligator. V. The effects of injections of testosterone propionate in immature alligators. Anat. Rec. 75:51–57.

Gabe, M. 1970. The adrenal. In: C. Gans and T. S. Parsons (eds.) 1970. Biology of the Reptilia, Volume 3. New York, Academic Press. Pp. 263–318.

Gans, C. 1962. Terrestrial locomotion without limbs. Am. Zool. 2:167–182.

Girgis, S. 1961. Aquatic respiration in the common Nile turtle, Trionyx triunguis (Forskal). Comp. Biochem. Physiol. 3:206–217.

Girons, H. S. 1970. The pituitary gland. In: C. Gans and T. S. Parsons (eds.) 1970. Biology of the Reptilia, Volume 3. New York, Academic Press. Pp. 135–199.

Giusti, L. 1931. La tiroidectomia en una tartaruga Clemmys leprosa (Schweig). Revue Elev. Med. Vet. Pays Tropic. 13:16–19.

Goldby, F., and H. J. Gamble. 1957. The reptilian cerebral hemispheres. Biol. Rev. Cambridge Philos. Soc. 32:383–420.

Greep, R. O. 1963. Parathyroid glands. In: U. S. von Euler and H. Heller (eds.), Comparative Endocrinology, Volume I. New York, Academic Press. Pp. 325–370.

Haarmann, W. 1936. Über den Einfluss von Thyroxin auf den Sauerstoffverbrauch überlebender Gewebe. Naunyn-Schmiedebergs, Arch. exp. Path. Pharm. 180:167–182.

Hellbaum, H. W. 1936. The cytology of snake thyroids following hypophysectomy, activation and ultra-centrifuging. Anat. Rec. 67:53–67.

Heller, H. 1963. Neurohypophyseal hormones. In: U. S. von Euler and H. Heller (eds.), Comparative Endocrinology, Volume I. New York, Academic Press. Pp. 25–80.

Houssay, B. A. 1931. Compt. Rend. Soc. Biol. 106:337–378. (Vide Knobil and Sandler, 1963.)

———, and J. C. Penhos. 1960. Pancreatic diabetes and hypophysectomy in the snake *Xenodon merremii*. Acta Endocrinol. 35:313–323.

Jankowsky, H. D. 1964. Die Bedeutung der Hormone für die Temperaturanpassung im normalen Temperaturbereich. Helg. Wiss. Meeresunt. 9: 412–419.

Kehl, R., and C. Combescot. 1955. Reproduction in the Reptilia. Mem. Soc. Endocrinol. 4:57–74.

Klauber, L. M. 1939. A statistical study of the rattlesnakes. VI. Fangs. Occ. Papers San Diego Soc. Nat. Hist. 5:1–61.

——— 1956. Rattlesnakes; Their Habits, Life Histories, and Influence on Mankind. Berkeley, Univ. California Press. 1476 pp.

Knobil, E., and R. Sandler. 1963. The physiology of the adenohypophyseal hormones. *In:* U. S. von Euler and H. Heller (eds.), Comparative Endocrinology, Volume I. New York, Academic Press. Pp. 447–491.

Lynn, W. G. 1970. The thyroid. *In:* C. Gans and T. S. Parsons (eds.) 1970. Biology of the Reptilia, Volume 3. New York, Academic Press. Pp. 201–234.

Maher, M. J., and B. H. Levedahl. 1959. The effect of the thyroid gland on the oxidative metabolism of the lizard, *Anolis carolinensis*. J. Exp. Zool. 140:169–189.

McWhinnie, D. J., and J. R. Cortelyou. 1968. Influence of parathyroid extract on blood and urine mineral levels in iguanid lizards. Gen. Comp. Endocrinol. 11:78–87.

Miller, M. R. 1948. The seasonal histological changes occurring in the ovary, corpus luteum, and testis of the viviparous lizard, *Xantusia vigilis*. Univ. California Publ. Zool. 48:197–224.

——— 1958. The endocrine basis for reproductive adaptation in reptiles. *In:* A. Gorbman (ed.), Textbook of Comparative Endocrinology. New York, Wiley and Sons. Pp. 499–516.

———, and M. Kasahara. 1967. Studies on the cutaneous innervation of lizards. Proc. California Acad. Sci. 34:549–568.

———, and M. D. Lagios. 1970. The pancreas. *In:* C. Gans and T. S. Parsons (eds.) 1970. Biology of the Reptilia, Volume 3. New York, Academic Press. Pp. 319–346.

———, and D. H. Wurster. 1958. The morphology and physiology of the pancreatic islets in urodele amphibians and lizards. *In:* A. Gorbman (ed.), Textbook of Comparative Endocrinology. New York, Wiley and Sons. Pp. 668–680.

Noble, G. K., and H. T. Bradley. 1933. The relation of the thyroid and the hypophysis to the molting process in the lizard, *Hemidactylus brookii*. Biol. Bull. (Mar. Biol. Lab.) 64:289–298.

Noble, G. K., and A. Schmidt. 1937. The structure and function of the facial and labial pits of snakes. Proc. Am. Phil. Soc. 77:263–288.

Parker, G. H. 1948. Animal Colour Changes And Their Neurohumours; A Survey Of Investigations, 1910–1943. Cambridge, Cambridge Univ. Press. 377 pp.

Ratzersdorfer, C., A. S. Gordon, and H. A. Charipper. 1949. The effects of thiourea on the thyroid gland and molting behavior of the lizard, *Anolis carolinensis*. J. Exp. Zool. 112:13–27.

Romer, A. S. 1956. Osteology of the Reptiles. Chicago, Univ. Chicago Press. 772 pp.

Schaefer, W. H. 1933. Hypophysectomy and thyroidectomy of snakes. Proc. Soc. Exp. Biol. Med. 30:1363–1365.

Schmidt-Nielsen, K., and R. Fange. 1958. Salt glands in marine reptiles. Nature 182:783–785.

Scott, A. H. 1935. Thyroxin and tissue metabolism. Am. J. Physiol. 111:107–117.

Shellabarger, C. J., A. Gorbman, F. C. Schatzlein, and D. McGill. 1956. Some quantitative and qualitative aspects of I^{131} metabolism in turtles. Endocrinology 59:331–339.

Smith, H. M., and L. F. James. 1958. The taxonomic significance of cloacal bursae in turtles. Trans. Kansas Acad. Sci. 61:86–96.

Snyder, R. C. 1954. The anatomy and function of the pelvic girdle and hindlimb in lizard locomotion. Am. J. Anat. 95:1–46.

——— 1962. Adaptations for bipedal locomotion of lizards. Am. Zool. 2:191–203.

Stebbins, R. C. 1958. An experimental study of the "third eye" of the Tuatara. Copeia 1958:183–190.

———, and R. M. Eakin. 1958. The role of the "third eye" in reptilian behavior. American Mus. Nov. 1870:1–40.

Templeton, J. R. 1966. Responses of the lizard nasal salt gland to chronic hypersalemia. Comp. Biochem. Physiol. 18:563–572.

Underwood, G. 1951. Reptilian retinas. Nature 167: 183–185.

Van Oordt, G. J. 1963. Male gonadal hormones. *In:* U. S. von Euler and H. Heller (eds.), Comparative Endocrinology, Volume I. New York, Academic Press. Pp. 154–207.

Velle, W. 1963. Female gonadal hormones. *In:* U. S. von Euler and H. Heller (eds.), Comparative Endocrinology, Volume I. New York, Academic Press. Pp. 111–153.

Walls, G. L. 1942. The Vertebrate Eye and Its Adaptive Radiation. Cranbrook Inst. Sci. (Michigan) Bull. 19:1–785.

Waring, H. 1963. Color Change Mechanisms of Cold-blooded Vertebrates. New York, Academic Press. 266 pp.

Wever, E. G., and J. A. Vernon. 1956a. The sensitivity of the turtle's ear as shown by the cochlear potentials. Proc. Natl. Acad. Sci. Washington 42: 213–220.

——— 1956b. Auditory responses in the common box turtle. Proc. Natl. Acad. Sci. Washington 42: 962–965.

——— 1960. The problem of hearing in snakes. J. Aud. Res. 1:77–83.

Wingstrand, K. G. 1966. Comparative anatomy and evolution of the hypophysis. *In:* G. W. Harris and B. T. Donovan (eds.), The Pituitary Gland. Volume 1. Anterior Pituitary. New York, Academic Press, Pp. 58–126.

Wright, A., and I. Chester Jones. 1957. The adrenal gland in lizards and snakes. J. Endocrinol. 15: 83–99.

Young, F. G. 1963. Pancreatic hormones: Insulin. *In:* U. S. von Euler and H. Heller (eds.), Comparative Endocrinology, Volume I. New York, Academic Press. Pp. 371–409.

CHAPTER 5

THE ORIGIN AND PHYLOGENETIC RELATIONSHIPS OF REPTILIA

As is true for the class Amphibia, the majority of the class Reptilia is extinct. Living reptiles represent only four of the 17 orders recognized by Romer (1966). Despite the fact that the fossil record for reptiles is generally much more complete than that for amphibians, many questions regarding relationships remain unanswered. The fundamental difference between primitive reptiles and their amphibian ancestors lies in the types of egg: amphibians produce anamniotic eggs; reptiles produce amniotic eggs. Since there is seldom direct evidence regarding the type of egg possessed by extinct forms, there is confusion about the classification of fossil remains that are literally intermediate between those that definitely represent amphibians and those that definitely represent reptiles. The most primitive unquestionable reptiles known are from the early Pennsylvanian, but already during this period there existed at least two or three different phylogenetic lines, and the relationships of these to one another is uncertain. Some workers have even suggested that the class Reptilia is diphyletic or polyphyletic in origin.

In contrast to the gap separating living amphibians from extinct ones, the phylogenetic lines of living reptiles can all be traced back to the Triassic period (Fig. 5–1). However, prior to that period origins and relationships become subject to speculation. Some families of living reptiles have been studied intensively, but others are badly in need of revision. Therefore, even the taxonomy of the modern reptiles varies in its validity from one family to the next. Authorities do not agree on the number of genera or, in some instances, even the number of families which should be recognized. Through the discovery of new species and the synonymizing of previously recognized ones, the number of species is constantly changing. Thus, it must be assumed that changes in the taxonomy of reptiles will continue to be made for some time as more and more information is gathered.

It is certain that reptiles evolved from amphibians, but there is insufficient evidence to indicate conclusively from which particular group or groups of amphibians they arose. The uncertainty stems basically from a lack of both reptiles and reptilelike amphibians from deposits older than the lower Pennsylvanian. Related to this, most deposits known from the lowermost Pennsylvanian and from the Mississippian represent aquatic habitats, and there is a great lack of information about inhabitants of terrestrial environments during the critical time when reptiles were making their initial appearance. Romer (1946, 1957) and Tihen

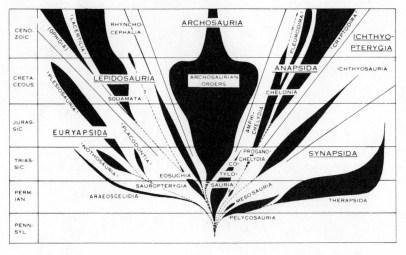

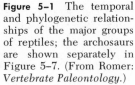

Figure 5–1 The temporal and phylogenetic relationships of the major groups of reptiles; the archosaurs are shown separately in Figure 5–7. (From Romer: *Vertebrate Paleontology*.)

(1960), among others, have proposed that the evolution of the amniotic egg preceded the evolution of fully terrestrial reptiles. In other words, they suggest that the original reptiles, like many modern turtles, were aquatic animals that only deposited their eggs on land. Such a theory seems plausible for several reasons. Environmental conditions were such at the time that, no matter whether the adult reptile was fresh-water aquatic, amphibious, or terrestrial, natural selection would have favored the amniotic egg. Most of the predators of the time were aquatic, so there would have been a great selective advantage in depositing eggs on land where they would be relatively safe. Geological evidence indicates that climatic conditions alternated seasonally between severe drought and flooding; therefore, the elimination of aquatic development and free-swimming larval stages would reduce the hazard of being washed away or of dying from dehydration. Finally, many of the better-known early reptiles, such as *Limnoscelis* (family Limnoscelidae, suborder Captorhinomorpha, order Cotylosauria), were obviously at least amphibious or semiaquatic.

Recent discoveries in early Pennsylvanian deposits at Joggins, Nova Scotia, tend to support the theory that the earliest reptiles were primarily terrestrial animals. Joggins is the site of an old lake bed which became rapidly filled with sediments during the Carboniferous time. These sedimentary deposits, comprising a layer about 5000 feet thick, contain the fossil remains of a variety of fish and embolomere amphibians. What is significant, however, is that the sediments buried giant treelike club mosses, which were common in Carboniferous swamps, in an upright position. Later, as the centers of these trees decomposed, deep pits were formed that acted as traps for animals walking on the surface. Among other terrestrial animals caught and preserved inside hollow stumps of the fossilized trees were some very early reptiles, including *Hylonomus* (family Romeriidae, suborder Captorhinomorpha, order Cotylosauria), with highly ossified skeletons and well-developed limbs reflecting adaptation to terrestrial life (Fig. 5–2). The remains of other terrestrial reptiles, preserved in erect trees as at Joggins, have been found at Sydney, Nova Scotia, in upper Pennsylvanian deposits. Because the remains found at Joggins and Sydney represent not only very early reptiles but forms with terrestrial adaptations, Carroll (1969) has suggested that reptiles first appeared in a terrestrial environment and that their absence from lowermost Pennsylvanian and Mississippian deposits reflects the paucity of information about terrestrial habitats during that time. Reptiles occurred in a variety of habitats by the end of the Pennsylvanian and, according to Carroll's theory, the aquatic characteristics of such forms as *Limnoscelis* (Fig. 5–3), rather than being primitive characteristics, were the result of adaptive radiation and invasion of aquatic or amphibious niches.

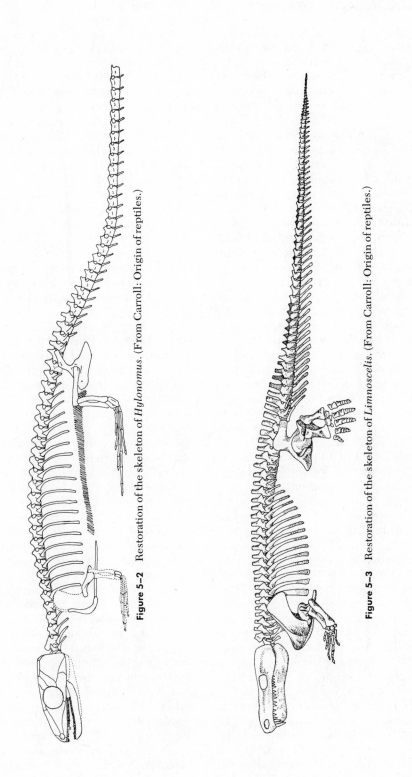

Figure 5–2 Restoration of the skeleton of *Hylonomus*. (From Carroll: Origin of reptiles.)

Figure 5–3 Restoration of the skeleton of *Limnoscelis*. (From Carroll: Origin of reptiles.)

THE EVIDENCE FOR AN ANTHRACOSAURIAN ORIGIN

Again, there were at least two major phylogenetic lines of what were distinctly reptiles present by the end of the Pennsylvanian. These were the pelycosaurs (order Pelycosauria), primitive members of the line of descent leading to mammals, and the captorhinomorphs (suborder Captorhinomorpha of the order Cotylosauria), generally unspecialized reptiles which were probably ancestral to all other reptilian groups. The oldest known pelycosaur, *Protoclepsydrops*, and the oldest known captorhinomorph, *Hylonomus*, have been found in the same fauna at Joggins; this indicates that the two lines of descent were separate in the lower Pennsylvanian. Although *Protoclepsydrops* is relatively poorly known, Carroll (1969) believes that the two genera are so similar that they could have evolved from a common ancestor within the lower Pennsylvanian. He suggests, furthermore, that if they were classified on the basis of their morphological similarity rather than on the evolutionary divergence that occurred subsequently in their lines of descent, *Protoclepsydrops* and *Hylonomus* could be placed in a single family. The important point is that, although two different lines of descent are apparent in the lower Pennsylvanian and the two genera just mentioned are classified in separate subclasses, they are not so different as to require separate origins from amphibians. Romer (1966) has proposed that the line of descent leading to mammals (subclass Synapsida, including pelycosaurs) may have branched directly from typical captorhinomorphs. Thus, except for the problematical diadectids, classified by some as amphibians and others as reptiles but definitely not on the main reptilian line of descent (see page 99), there is no strong evidence indicating a polyphyletic origin for Reptilia. The developmental pattern associated with the amniotic egg is so complex and yet so uniform among reptiles that it is not at all likely that it could have evolved independently in two or more different groups of amphibians (Romer, 1966). At any rate, the general consensus of opinion at the present time is that the class Reptilia is monophyletic and that the captorhinomorphs were the most primitive of known reptilian groups. Their structure, therefore, should most closely resemble that of the amphibian ancestor.

The amphibian group most closely resembling primitive captorhinomorphs is the order Anthracosauria of the subclass Labyrinthodontia. As noted in Chapter Three, the seymouriamorphs (suborder Seymouriamorpha) possessed a combination of amphibian and reptilian characteristics, and *Seymouria*, the best known of them, has often been pictured as intermediate between amphibians and reptiles. However, *Seymouria*, known from the lower Permian, is too recent to have been ancestral to reptiles; it has been generally concluded that it and the other known seymouriamorphs were remnants of a much older group which must have given rise to reptiles. Carroll (1969) considers *Tseajaia* (in Romer's anthracosaurian family Tseajaiidae and, therefore, amphibian), *Solenodonsaurus* (in Romer's captorhinomorph family Romeriidae and, therefore, reptilian), *Diplovertebron* (in Romer's anthracosaurian suborder Diplomeri and, accordingly, amphibian), and *Gephyrostegus* (in synonymy with *Diplovertebron* in Romer's classification) to be almost ideal morphological intermediates between primitive anthracosaurs and captorhinomorph reptiles and suggested that they (his family Solenodonsauridae) are relics of a group ancestral to most, if not all, reptiles. *Tseajaia*, described by Vaughn (1964), really combines in one animal characteristics of captorhinomorphs, diadectids, and seymouriamorphs.

Obviously, what we are dealing with here in part centers around the definition of an amphibian or reptile in the absence of information about the kind of egg produced and around the classification of those forms which are intermediate between what are obviously amphibians on one hand and obviously reptiles on the other. All cotylosaurs, captorhinomorphs especially, are close to the body plan from which all later reptiles probably evolved, and they are also close to the structure of the more terrestrial anthracosaurian amphibians. This, in summary, is the argument favoring an anthracosaurian derivation of reptiles. However, there is no known anthracosaurian which both structurally and temporally fulfills the prerequisites for an ancestor of reptiles.

Some workers, including Parrington (1958) and Gregory (1965), oppose the theoretical derivation of reptiles from anthracosaurian amphibians because it necessitates an unlikely sequence in the evolution of the middle ear. The hyomandibular bone (stapes) is directed ventrolaterally in both rhipidistian crossopterygian fish and captorhinomorph reptiles but is oriented dorsolaterally in all known labyrinthodont amphibians, including the anthracosaurians. Parrington (1958) believes that the differences between labyrinthodonts and captorhinomorphs are sufficient to show that labyrinthodonts could not be ancestral to reptiles. Gregory (1965) has interpreted the same and other features as indicating "the early divergence of the ancestors of labyrinthodonts and captorhinomorphs;" the implication here is that the labyrinthodonts and the line leading to captorhinomorphs had a common ancestor but reptiles were not directly descended from known labyrinthodonts. Because of complications involved in the derivation of reptiles from known seymouriamorphs, it has been suggested from time to time that the microsaurs (subclass Lepospondyli) might have been ancestral to reptiles. The Microsauria do have a number of reptile-like characteristics, but they are no longer considered to reflect any relationship (see page 101; Gregory, 1965; and Carroll and Baird, 1968).

Whether the diadectids should be classified as amphibians or reptiles is not agreed, but there is agreement that they represent a specialized dead-end branch from amphibians that was not ancestral to other reptilian groups. They are considered to be somewhat distantly related to all known seymouriamorphs and captorhinomorphs. Thus, if one considers them to be reptiles, the class Reptilia may be considered diphyletic, with the diadectids coming from a different amphibian ancestor than the captorhinomorphs, or monophyletic, with a very early branching of the diadectids from the main line leading to reptiles.

CLASSIFICATION OF REPTILES

As mentioned in the previous chapter, the nature of the temporal region of the skull has played an important role in the classification of reptiles; for many years, the subclasses of Reptilia have been defined by the number and position of the temporal fenestrae. Emphasizing these temporal openings, Williston (1925) divided the class Reptilia into five subclasses. Williston's five subclasses are defined here because his classification has served as a basis for subsequent arrangements.

1. Subclass Anapsida—reptiles whose skulls lack openings in the temporal region and are completely roofed or secondarily emarginate:

Order Cotylosauria—the extinct "stem reptiles;"

Order Eunotosauria—*Eunotosaurus* only, extinct primitive turtlelike animals with broadly expanded ribs;

Order Testudinata (Chelonia)—extinct and living turtles.

2. Subclass Synapsida—reptiles whose skulls possess a single lateral temporal opening on each side bounded *above* by the postorbital and squamosal bones:

Order Theromorpha (=Pelycosauria)—extinct primitive mammal-like reptiles;

Order Therapsida—advanced mammal-like reptiles, extinct.

3. Subclass Synaptosauria—reptiles whose skulls possess a single superior temporal opening on each side bounded *below* by the postorbital and squamosal bones:

Order Sauropterygia—extinct amphibious and marine reptiles;

Order Placodontia—extinct aquatic mollusc-feeding reptiles.

4. Subclass Parapsida—reptiles with a single superior temporal opening on each side of the skull bounded below by the postfrontal and supratemporal bones:

Order Proganosauria (=Mesosauria) —small extinct slender-bodied and amphibious reptiles (presumed by Williston to be possible ancestors of the ichthyosaurs);

Order Ichthyosauria—extinct marine fishlike reptiles;

Order Protorosauria—small extinct

reptiles with long slender limbs (presumed by Williston to be possible ancestors of the lizards);

Order Squamata—living and extinct lizards and snakes.

5. Subclass Diapsida—reptiles whose skulls have two temporal openings on each side separated by the postorbital and squamosal bones:

? Order Proterosuchia—extinct primitive archosaurian reptiles;

? Order Eosuchia—extinct lepidosaurian lizardlike reptiles;

Superorder Diaptosauria—unspecialized diapsid reptiles—

Order Rhynchocephalia—the living tuatara (*Sphenodon punctatus*) plus extinct rhynchocephalians;

Superorder Archosauria—diapsid reptiles tending toward bipedal adaptations—

Order Parasuchia (included Pseudosuchia and equivalent to what is now Thecodontia)—extinct Triassic ancestors of dinosaurs;

Order Crocodilia—living and extinct crocodilians;

Order Saurischia—extinct saurischian bipedal dinosaurs;

Order Ornithischia—extinct ornithischian dinosaurs;

Order Pterosauria—extinct flying reptiles.

Subsequent modifications of the above classification have included major revisions of Williston's subclass Parapsida, the shifting of some orders and combining of others, and substitutions of names. The origin of turtles is still not clear (Olson, 1965) but Williston's subclass Anapsida has withstood the test of time and is still accepted as a natural grouping; the only major change proposed here has been Romer's (1966) inclusion of the order Mesosauria in this subclass. The temporal region of mesosaurs has been poorly preserved or missing in specimens found to date, but there are indications of a single lateral opening similar to that of Synapsida but farther down on the cheek and apparently independently evolved; *Mesosaurus* also differs greatly from the synapsids in other characters. Williston's presumption that mesosaurs were possible ancestors of the ichthyosaurs is no longer accepted. Instead, the similarities between the two groups are interpreted as

parallel adaptations to aquatic life. This current opinion is supported by the fact that mesosaurs had enlarged hind legs, whereas those of ichthyosaurs were reduced. Again, the temporal region of ichthyosaurs differs from that indicated for mesosaurs. Some workers (Goin and Goin, 1962) have proposed a separate but unnamed subclass for the order Mesosauria. Romer's (1966) placement of these in the subclass Anapsida is based on the belief that they were an independent and early side branch from the cotylosaurs and reflects the current tendency for placing less importance on the temporal region.

The subclass Synapsida is unquestionably a natural grouping containing the early branch that led to mammals. The only change which has occurred here at the order level has been the substitution of the name Pelycosauria for Theromorpha.

The sauropterygians and placodonts, grouped in Williston's subclass Synaptosauria, are still considered to have been closely related; the only difference of opinion regarding these is as to whether the placodonts should be treated as a separate order or should be included in the order Sauropterygia. The small slender lizardlike forms constituting the order Protorosauria make a puzzling group with uncertain relationship to other reptiles. Williston's opinion that they were possibly ancestral to lizards is no longer accepted because, among other reasons, the type of vertebral attachment of the ribs in lizards is not likely to have evolved from the condition in protorosaurs (details may be found in Romer, 1956). Protorosaurian skulls are not known well enough that proper comparisons with those of other groups can be made. However, the temporal fenestra of protorosaurs is now considered to have been bounded by the same bones as that of sauropterygians, and most modern classifications have removed the order Protorosauria from the subclass Parapsida and placed it with the sauropterygians and placodonts in the subclass Synaptosauria. Williston's name Protorosauria, derived from *Protorosaurus*, was originally vaguely applied to an assortment of primitive reptiles that were generally presumed to be diapsids. The grouping was later revived by Williston, but he then used *Araeoscelis*, rather than

Protorosaurus, as the central type. Consequently, in some recent classifications the name of the order has been changed from Protorosauria to Araeoscelidia. Finally, in order that the ending of the name would conform with those of other subclasses, Colbert (1945) suggested substituting the name Euryapsida for Williston's subclass Synaptosauria; this suggestion has been followed in most subsequent classifications. Thus, the subclass Euryapsida presently contains the order Araeoscelidia, the order Sauropterygia, and the order Placodontia.

The changes just described reduced Williston's subclass Parapsida to two orders, Ichthyosauria and Squamata. Williston associated these two orders because both have a similar upper temporal opening which he assumed to be a primitive condition in both. It is now believed, as proposed by Broom (1935), that the temporal pattern in the order Squamata has been derived from a diapsid condition such as occurs in the tuatara through the loss of one (in lizards) or both (in snakes) temporal arches. This conclusion is supported by the fossil record, which indicates that the ichthyosaurs and squamates have been independent lines of descent. The two lines do not converge when traced back through time. Accordingly, the order Ichthyosauria is now placed in its own subclass, the name of which has become Ichthyopterygia.

Williston recognized that the diapsid reptiles formed two natural groups, which he designated as superorders. More recent classifications have elevated these to the subclasses Lepidosauria and Archosauria. Thus, there are six subclasses presently recognized where Williston's scheme included only five. The subclass Lepidosauria includes those reptiles having a diapsid temporal region (one or both temporal arches may be lost in advanced forms) and lacking bipedal specializations of limbs and girdles. This grouping of the more primitive diapsids includes the orders Eosuchia, Rhynchocephalia, and Squamata. The subclass Archosauria includes those reptiles with a diapsid temporal region (without the loss of temporal arches but occasionally with a secondary closure of the upper temporal opening) and with bipedal specializations of limbs and girdles. Williston's order Protero-

suchia has been relegated to a suborder within the order Thecodontia. Thus, the subclass Archosauria includes five currently recognized orders: Thecodontia, Crocodilia, Saurischia, Ornithischia, and Pterosauria.

The division of the class Reptilia into six subclasses as just outlined is almost universally accepted, and most classifications are in agreement down through the level of orders. The following compendium of Reptilia is based on Romer's (1966) classification. Other variations exist because of differences of opinion at family and lower levels.

CLASS REPTILIA

Subclass Anapsida

Order Cotylosauria—Temporal region of skull typically lacking fenestrae and emargination; skull relatively high and narrow; otic notch lacking; small postparietals present; supratemporal small or absent; marginal teeth normally in a single row; palatal teeth often present; vertebrae as in advanced anthracosaurian amphibians, functional centra formed by pleurocentra, intercentra crescent-shaped and moderately large; zygapophyses widely spaced laterally; neural arches swollen; ribs sometimes present on all trunk and anterior caudal vertebrae, anterior ribs with two heads.

Suborder Captorhinomorpha—Limbs short and stout with proximal segments extending parallel to walking surface and perpendicular to body axis; pectoral girdle massive and including a small splintlike cleithrum. This suborder includes three or four extinct families of the most primitive "stem reptiles." The oldest remains known are from the Pennsylvanian, and the group appears to have died out during the Permian. Most captorhinomorphs were relatively small (*Limnoscelis* reached lengths of five feet but was considerably larger than the other known genera) and clumsy lizardlike animals; most were apparently semiaquatic. The families included here are Romeriidae (*Hylonomus,* the oldest known reptile, and seven to nine additional genera from lower and middle Pennsylvanian deposits of North America, the late Upper Carboniferous of Europe, and the lower Permian of North America and Europe), Limnoscelidae (*Limnoscelis* and the questionable *Limnosceloides* from the lower Permian of North America), Captorhinidae (10 to 12 genera from lower and middle Permian deposits of North America and Europe), and, questionably, Bolosauridae (*Bolosaurus* only, from the lower Permian of Texas). *Bolosaurus* was a tiny reptile with unique bulb-shaped cheek teeth, each bearing a single marginal cusp. This reptile also possessed a small lateral temporal fenestra on the lower part of the cheek somewhat like that of synapsids; however, its age and the lack of an otic notch indicate that this genus was probably a specialized side branch from the main line of captorhinomorphs and was not related to the synapsids.

Suborder Procolophonia — This suborder includes three superfamilies of advanced cotylosaurs. The superfamily Procolophonoidea contains two families of small (one to two feet in length) forms with slender limbs, transversely expanded cheek teeth resembling those of diadectids, and a concavity in the cheek region which was probably the forerunner of the otic notch of more advanced reptiles. The two families placed here are Nyctiphruretidae (three or four genera from the middle Permian of South Africa and eastern Europe and the upper Permian of Madagascar and South Africa) and Procolophonidae (about 17 genera from the upper Permian and Triassic of Europe, South Africa, northern and eastern Asia, South America, and North America). The procolophonoids were a very widely distributed and abundant group during the Triassic and were the last surviving members of the order Cotylosauria.

The superfamily Pareiasauroidea includes the family Rhipaeosauridae (three genera from the middle Permian of eastern Europe) and the family Pareiasauridae (eight genera from the middle and upper Permian of Europe and Africa). These were large sluggish herbivores that reached lengths of up to 10 feet. As mentioned previously, the structure of the girdles and limbs changed in this group so that the elbow rotated backward and the knee forward, resulting in the vertical support of the weight on the limbs. The feet were also modified, with the fifth digit frequently being reduced or absent.

The superfamily Millerosauroidea includes only the family Millerettidae, with four to six genera from the middle and upper Permian of South Africa. The millerettids resembled procolophonids in having short jaws and a concavity in the cheek region corresponding to the otic notch of higher forms. The different genera in this family form a sequence in the evolution of lateral temporal fenestrae: *Nannopareia* lacked temporal openings; *Milleretta* had a small unossified space on each side between the jugal and squamosal bones; *Millerosaurus* had a well-developed fenestra on each side. The location of this fenestra is comparable to that characteristic of synapsids, but Romer (1966) believes this to have been the result of parallel evolution rather than close relationship, since millerettids bear no other resemblance to synapsids. He suggests, however, that millerettids may have been primitive diapsids, since their temporal region could be transformed into a diapsid pattern merely by adding the upper opening. They are tentatively grouped with procolophonoids because of the similarity of the jaws and otic region. This, again, reflects the decreased emphasis placed on the temporal region by modern workers.

The remaining reptilian groups are all believed to have been derived directly or indirectly from the order Cotylosauria.

Order Mesosauria — Skull with low lateral temporal fenestrae; jaws very long and slender with numerous needlelike thecodont teeth; palate toothed; body slender with long laterally compressed tail; hind legs enlarged, front legs smaller. This order includes only the family Mesosauridae, containing *Mesosaurus* from the lower Permian of South Africa and South America. The mesosaurs were fresh-water fish-eaters that apparently branched very early from the cotylosaurs and became extinct before the end of the Permian.

Order Chelonia (=Testudines, Testudinata) — Skull with or without emargination but lacking temporal fenestrae; teeth generally absent in adults, functionally replaced by horny bill; quadrate immovable; body short and broad, typically with eight cervical and 10 dorsal vertebrae; cervical ribs generally reduced to processes on vertebrae, ribs of 10 dorsal vertebrae expanded and usually fused with a bony carapace, no sternal ribs or sternum; ventral bony plastron generally present, typically joined to carapace by a bony bridge on each side; pectoral girdle internal to ribs, clavicles and interclavicles incorporated into the plastron; limbs typically primitive with horizontal proximal segments, distal segments modified into flippers in many aquatic species.

The turtles and tortoises (Fig. 5–4) are a very distinctive group which makes its first sudden appearance in late Triassic deposits; it has persisted ever since. Even the oldest fossil turtles had well-developed shells, and their relationships to other reptilian groups have been the subject of much debate. An early theory was that turtles evolved from the diadectids, but Olson (1965) has presented considerable evidence refuting this. He has suggested that the Chelonia may have arisen instead from within the procolophonoids or pareiasaurs.

Suborder Proganochelydia—Nasals, lacri-

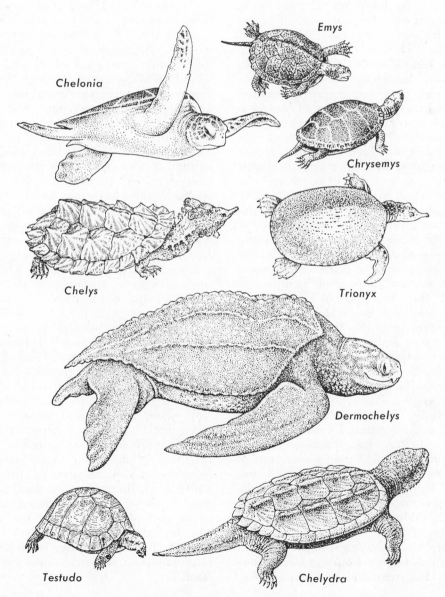

Figure 5–4   Various kinds of turtles, not all drawn to the same scale. (From Young, J. Z.: *The Life of Vertebrates.*)

mals, and prefrontals present in skull; pelvis generally fused to both carapace and plastron; palatal teeth present together with rudimentary teeth on margins of jaws; several pairs of mesoplastra present (intermediate elements of plastron that are not present on most modern turtles); neck and tail covered with bony tubercles, not capable of being withdrawn. This suborder includes only the superfamily Proganochelyoidea and the family Proganochelyidae, with one to three genera typified by *Proganochelys*, from the Triassic of Europe.

Suborder Amphichelydia — Nasals, lacrimals, and prefrontals absent; pelvis may or may not be fused to plastron; teeth absent; no more than one pair of mesoplastra present; neck may be partially retractable. This suborder includes the common turtles of the Jurassic and early Cretaceous. As a group, they survived into the Eocene, and one genus, *Meiolania*, persisted into the Pleistocene of Australia. There are two superfamilies recognized. The superfamily Pleurosternoidea includes the family Pleurosternidae (about 11 genera from the Upper Triassic, Jurassic, and Lower Cretaceous periods of Europe, eastern Asia, and North America); the family Plesiochelyidae (*Craspedochelys* and *Plesiochelys* from the Upper Jurassic of Europe and eastern Asia and the Lower Cretaceous of Europe); the family Thalasemyidae (nine to 12 genera from the Upper Jurassic of Europe and Asia and the Cretaceous of Europe); the family Sinemydidae (*Manchurochelys* and *Sinemys* from the Upper Jurassic of eastern Asia); and the family Apertotemporalidae (*Apertotemporalis* from the Upper Cretaceous of North Africa and *Chitracephalus* from the Lower Cretaceous of Europe). The superfamily Baenoidea includes four small families: Neurankylidae (four genera from the Cretaceous of North America); Baenidae (four to six genera from the Upper Cretaceous of eastern Asia and Cretaceous North America, the Upper Jurassic of Eastern Asia, and the Eocene of eastern Asia and North America); Meiolaniidae (*Crossochelys* from the Eocene of South America, *Meiolania* from the Pleistocene of Australia, and *Niolamia* from the Cretaceous of South America); and Eubaenidae (*Eu-*

baena from the Cretaceous of North America).

Suborder Pleurodira — Nasals and lacrimals absent; pelvis fused to the plastron and sutured to the carapace; mesoplastra present or absent; neck capable of being withdrawn by lateral bending. This suborder includes two families of aquatic "side-necked" turtles that have survived only in the southern hemisphere; it is an aberrant but structurally conservative side branch which sprang from the Amphichelydia (Romer, 1968). The family Pelomedusidae includes about 23 extinct genera known from the Lower Cretaceous through the Pleistocene; some of these have been found in Europe, Asia, and North America in addition to South America and Africa. There are only three living genera: *Pelomedusa* (one species) is found only in tropical Africa and on the island of Madagascar; *Pelusios* (five species) occurs in tropical Africa and on Madagascar and also on some surrounding islands; *Podocnemis* (eight species) is found in South America and on Madagascar. The family Chelydae includes two extinct genera (*Parahydraspis* and *Pelocomastes*) and 10 living genera, four of which are also known as fossils. The oldest fossils assigned to this family are from the Tertiary (Miocene?) of South America. Living members occur in South America, Australia, and New Guinea.

Suborder Cryptodira — Nasals and lacrimals present or absent; pelvis not fused to either the plastron or the carapace; teeth absent; mesoplastra absent; neck capable of being withdrawn by means of a vertical S-shaped bend. This suborder contains the most successful and advanced turtles and includes, by far, the majority of living turtles. Adaptive radiation has resulted in fully terrestrial, amphibious, fresh-water aquatic, and marine species. The oldest cryptodires appeared during the Upper Jurassic as a progressive continuation of the main evolutionary line from the Amphichelydia. By the Upper Cretaceous, they were the dominant turtles in the Northern Hemisphere, and some had already invaded the oceans. The terrestrial tortoises existed throughout the Tertiary and persist today. This large suborder is divided into five superfamilies.

The superfamily Testudinoidea includes the most primitive and generalized of the modern turtles and contains three families. The family Dermatemydidae includes the living *Dermatemys* from Central America plus about 20 extinct genera from the Cretaceous through the Miocene of North America, Europe, and Asia. The family Chelydridae includes the living *Chelydra* (North America), *Claudius* (South America), *Kinosternon* (North and South America), *Macroclemys* (North America), *Staurotypus* (Middle America), and *Sternothaerus* (North America). The former three genera are also known as fossils, and, in addition, there are about five extinct genera dating back to the Miocene of North America and the Eocene of North Africa. The fossil genera plus *Chelydra* and *Macroclemys* are placed in the subfamily Chelydrinae, *Claudius* and *Staurotypus* compose the subfamily Staurotypinae, and *Kinosternon* and *Sternothaerus* constitute the subfamily Kinosterninae. Some authorities (e.g., Carr, 1952) have recognized these three groupings as separate families (Chelydridae, Staurotypidae, and Kinosternidae). The large family Testudinidae (Emydidae), the third family in Testudinoidea, includes about 33 living genera plus 15 extinct genera dating back to the Eocene of Europe, Asia, and North America. These are the common fresh-water, amphibious, and terrestrial turtles of the Northern Hemisphere; *Malaclemys* is a marine littoral form. Three subfamilies of Testudinidae are recognized. The subfamily Platysterninae includes only *Platysternon* of Southeast Asia. Twenty-six living genera plus about nine extinct genera compose the subfamily Emydinae. These are the common terrapins, pond turtles, and box turtles, and the subfamily occurs on all inhabitable continents except Australia and Africa south of the Sahara. Among members of Emydinae are painted terrapins (*Chrysemys picta*), Blanding's turtle (*Emydoidea blandingi*), the European pond terrapin (*Emys orbicularis*), several species of pond turtles (*Clemmys*), map terrapins (*Graptemys*), sliders and cooters (*Pseudemys*), and box turtles (*Terrapene*). The third subfamily, Testudininae, includes six living genera (*Gopherus, Homopus, Kinixys, Malacochersus, Pyxis,* and *Testudo*) and about

four extinct genera of land tortoises. Testudininae is cosmopolitan except for Australia. Some authorities consider these subfamilies to be separate families (Platysternidae, Emydidae, and Testudinidae), while others (e.g., Bellairs, 1970) group Platysterninae and Emydinae in the family Emydidae and place the land tortoises alone in the family Testudinidae.

The superfamily Chelonioidea includes three families of extinct and living marine turtles. The family Toxochelyidae is known only from the Upper Cretaceous of North America, Europe, and eastern Asia and includes nine or 10 genera; this family is thought to have been ancestral to the family Cheloniidae. The family Protostegidae, also extinct, is known from the Upper Cretaceous of North America and of eastern Asia and from the Oligocene of Europe; it includes five or six genera and possibly was the group from which the family Dermochelyidae evolved. The family Cheloniidae contains about 16 extinct genera and four living genera of marine turtles which come onto land only to lay their eggs. Fossil genera date back to the Lower Cretaceous of Europe and the Upper Cretaceous of North America. The living genera (*Caretta, Chelonia, Eretmochelys,* and *Lepidochelys*) occur worldwide in warm seas.

The superfamily Dermocheloidea contains only the family Dermochelyidae, the leatherback marine turtles. *Dermochelys coriacea* is the only living member of the family, but *Dermochelys* is known as a fossil back to the Miocene of Europe. Three genera (*Cosmochelys, Eosphargis,* and *Psephophorus*) are extinct and date back to the Eocene of Europe, West Africa, and North America.

Finally, the superfamily Trionychoidea contains the highly aquatic soft-shelled turtles, all in the family Trionychidae, and the pitted-shelled turtles of the family Carettochelyidae. There are seven living genera in the family Trionychidae, and these are distributed in North America, Africa, and southern and southeastern Asia. No extinct genera are known, but *Chitria, Cycloderma, Lissemys,* and *Trionyx* are known as fossils. *Trionyx* dates back to the Lower Cretaceous of Asia and the Upper Cretaceous of North America, *Cycloderma* to the Miocene of Africa, and

Chitria and *Lissemys* to the Pleistocene of southern Asia. The only living member of the family Carettochelyidae is *Carettochelys,* which occurs in several river systems on the southern coast of New Guinea and is also in northern Australia. Six extinct genera of Carettochelyidae are named from the Paleocene to Oligocene of Europe and the Eocene of Asia and North America.

Suborder Eunotosauria—Teeth present; second through ninth ribs broadly expanded laterally so as to nearly touch one another; reduced number of vertebrae. This is a provisional suborder which includes only the small reptile *Eunotosaurus* (family Eunotosauridae) from the middle Permian of South Africa. The roof of the skull is unknown for this form and, consequently, the condition of the temporal region is unknown. Romer's provisional placement of *Eunotosaurus* in the order Chelonia is qualified with the opinion that it was neither a true turtle nor a typical cotylosaur.

Subclass Lepidosauria

Order Eosuchia—Skull typically like that of an advanced cotylosaur except for a diapsid temporal region; lower temporal arch always at least partially present; quadratojugal present; quadrate fixed in position; lacrimals present; premaxillae not beaklike; palatal teeth well developed; marginal teeth generally subthecodont but thecodont in advanced forms (set in sockets in jaw margins); pterygoids (palate) movable on braincase; posterior margin of cheek concave, but otic notch not present. This order was probably derived from the millerettids. It, in turn, appears definitely to have been ancestral to the lizards and snakes and may also have given rise to the archosaurian line of diapsids. Two poorly known families of eosuchians are recognized: Younginiidae contains about a dozen genera from the upper Permian of Europe and South Africa; Tangasauridae includes only *Hovasaurus* and *Tangasaurus* from the upper Permian of, respectively, Madagascar and East Africa. The tangasaurids had a number of aquatic adaptations and have been classified at various times as mesosaurians and protorosaurians.

Order Squamata—Skull basically diapsid, but lower temporal arch reduced or absent in lizards and both arches missing in snakes; quadrate movable or secondarily fixed; quadratojugal absent; lacrimals small or absent; premaxillae not beaklike; palatal teeth often present; dentition generally pleurodont (attached to sides of jaw without sockets) but may be acrodont (attached to biting edge of jaw without sockets) or subthecodont; pterygoids either movable on braincase, fused to braincase, or not connected to braincase; abdominal ribs rudimentary or absent; male copulatory organ paired; anal opening transverse; vertebrae usually procoelous. This order, which includes the lizards and snakes, is the most successful and diversified living group of reptiles (Fig. 5–5). Except for Antarctica and extreme altitudes and latitudes, the order is cosmopolitan at the present time. Depending upon the author, it is divided into two or three suborders. Some workers consider the worm lizards (family Amphisbaenidae) to be a separate suborder (Amphisbaenia). Others, including Romer (1966), group the amphisbaenids with the lizards in the suborder Lacertilia. All of the snakes are grouped together in the suborder Ophidia.

Suborder Lacertilia (Sauria)—Upper temporal arch generally present; epipterygoid rodlike or rudimentary; mandibles rigidly attached at their symphysis; sternum and pectoral girdle generally present; limbs well developed, reduced, or absent. This suborder includes about 30 families, which are grouped into seven infraorders.

INFRAORDER EOLACERTILIA—Primitive lizards with movable quadrate, but otherwise similar to eosuchians; palatal teeth present; lacrimal well developed; parietal and premaxilla paired. This infraorder includes only the extinct family Kuehneosauridae and four genera from the Upper Triassic of Europe and North America. *Kuehneosaurus* is of particular interest, for it had long trunk ribs that extended horizontally from the body and seem certain to have supported a gliding membrane similar to that of the living *Draco.* This grouping of lizards is transitional between the eosuchians and later lizards and has been arbitrarily distinguished from the order Eosuchia on the basis of a movable quadrate.

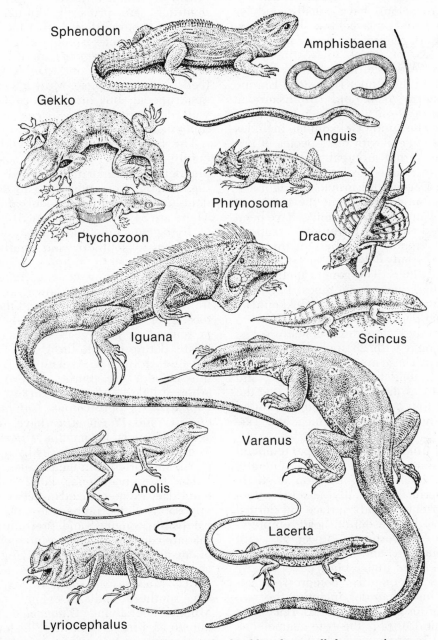

Figure 5–5 The Tuatara (*Sphenodon*) and various kinds of lizards, not all drawn to the same scale. (From Young, J. Z.: *The Life of Vertebrates.*)

INFRAORDER GEKKOTA — Upper temporal arch absent; postorbital, squamosal, and lacrimal bones missing; osteoderms usually absent; epipterygoid present; vertebrae generally amphicoelous, but when procoelous have small condyles and persistent intercentra; limbs generally short. There are four families of geckos, two of which are extinct. The latter are Ardeosauridae (three or four genera from the Upper Jurassic of Europe and eastern Asia) and Broilisauridae (*Broilisaurus* from the Upper Jurassic of Europe). The family Gekkonidae includes 79 living genera with 582 species of primarily nocturnal arboreal and terrestrial insectivores; three fossil genera are known from the Eocene of Europe. Living members are found throughout the tropics of the world and, like house mice, have been distributed among tropical seaports through shipping. Members of this family differ from Pygopodidae in having functional limbs. The genera included in Gekkonidae have sometimes been divided into three families (Gekkonidae, Eublepharidae, Sphaerodactylidae).

The family Pygopodidae contains seven living genera with 23 species; no fossils are known. These are snakelike in appearance with long tails, no forelimbs, and hind limbs as flaps of skin containing the bones of four toes. Pygopodids are found only in Australia, New Guinea, and Tasmania.

INFRAORDER IGUANIA — Upper temporal arch present, temporal fenestra generally open; parietals fused to form a single bone, lacrimal generally absent; osteoderms absent on body surface but dermal armor may be present on head; epipterygoid present; vertebrae procoelous with large condyles, and intercentra absent; well-developed limbs; body never greatly elongated, frequently compressed laterally. This infraorder includes the two huge families of typical lizards, Iguanidae and Agamidae, plus the true chameleons (family Chamaeleontidae). Two extinct families from the Upper Jurassic of Europe, Bavarisauridae (*Bavarisaurus*) and Euposauridae (*Euposaurus*) are also questionably assigned to this grouping. The family Iguanidae includes 50 living genera and about 560 species; it is the dominant family of lizards in the New World, but two genera occur on Mada-

gascar and one on the Fiji Islands. Six genera are known only as fossils, and living genera are also known as fossils; these date from the Eocene onward. The iguanids, characterized by pleurodont dentition, are primarily diurnal terrestrial and arboreal lizards that feed on insects. However, some (*Dipsosaurus*, *Ctenosaura*, and *Iguana*) are at least partially herbivorous, and *Amblyrhynchus* feeds on marine algae. Most iguanids are quadrupedal, but bipedal locomotion is well developed in *Basiliscus* and *Crotaphytus*.

The family Agamidae, the Old World family which is ecologically and morphologically similar to Iguanidae, is distinguishable from Iguanidae by its acrodont (rather than pleurodont) dentition. There are 34 genera and about 315 species of living agamids, and these are distributed over southern and southeastern Asia, southeastern Europe, Africa, and Australia. Of living genera, *Agama* and *Chlamydosaurus* are known as fossils from the Eocene of Europe and the Pleistocene of Australia, respectively. Three or four other genera are known only as fossils dating back to the Upper Cretaceous of eastern Asia. As with the iguanids, the agamids are diurnal terrestrial and arboreal lizards. Most feed on insects, but *Uromastix* is herbivorous. *Calotes*, *Otocryptis*, and *Physignatus* have well-developed bipedal locomotion. *Draco* is able to glide significant distances by means of its rib-supported skin membranes.

The true chameleons, family Chamaeleontidae, are specialized insectivores with opposable digits, prehensile tails, greatly extensible tongues, and protruding eyes that are capable of moving independently of one another. There are six living genera containing 109 species; these occur in Africa, Madagascar, southern Spain, Arabia, southern India, Ceylon, and Asia Minor. The living *Chamaeleo* is also known as a fossil from the Pleistocene of southwestern Asia. The extinct *Mimeosaurus* has been found in Upper Cretaceous deposits in eastern Asia.

INFRAORDER SCINCOMORPHA — Parietals typically expanded laterally; temporal fenestra reduced to a longitudinal slit or closed completely; lacrimal absent, fused to prefrontal, or small; epipterygoid present; osteoderms typically present on head and often present on body; vertebrae procoelous with typically large condyles; numerous species with elongate bodies and reduced limbs, but limbs and body proportions are normal in many. Six living families are recognized; there are no extinct families. The family Xantusidae occurs only in the New World and contains four living genera (*Cricosaura* from Cuba, *Klauberina* from islands off California, *Lepidophyma* from Mexico and Central America, and *Xantusia* from southwestern United States and Baja California) and the extinct genus *Palaeocantusia* from the Paleocene and Eocene of North America. Xantusids are small nocturnal terrestrial insectivores.

The family Teiidae is a large grouping of New World lizards which contains 38 living genera (about 210 species) and eight or nine fossil genera dating back to the Cretaceous of North and South America. *Cnemidophorus* is widely distributed throughout North, Middle, and South America. *Ameiva* extends out of South America through the West Indies and north into Mexico, and *Gymnophthalmus* reaches as far north as Mexico. The rest of the family is primarily South American. Teiids are generally diurnal terrestrial carnivores, but some are herbivorous and *Dracaena* and *Crocodilurus* are semiaquatic.

The family Lacertidae is the Old World family which is similar morphologically and ecologically to the family Teiidae. There are 20 genera and 164 species of living lacertids, and they are distributed in Europe, Africa, Asia, and the East Indies. Four extinct genera are known from the Eocene and Oligocene of Europe and *Palaeolacerta*, questionably assigned to this family, is from the Upper Jurassic of Europe. The living *Lacerta* is known as a fossil back to the Miocene of Europe. Lacertids are diurnal terrestrial carnivores.

The skinks, family Scincidae, form a cosmopolitan group which is absent only from Antarctica and high latitudes and altitudes. Greer (1970) recognized about 73 living genera including about 701 species. There are three known extinct genera (*Capitolacerta* from the Eocene of Europe, *Didosaurus* from the Pleistocene of the Mascarene Islands, and *Sauriscus* from the Cretaceous of North

America). Some authors (e.g., Goin and Goin, 1962) recognize *Anelytropsis papillosus* as a separate family, Anelytropsidae, and the four African species of *Feylinia* as a separate family, Feyliniidae. *Anelytropsis* and *Feylinia* are small burrowing forms lacking temporal arches. Most skinks are arboreal, and others are fossorial.

The family Dibamidae, including only *Dibamus* with three species, is questionably assigned to this infraorder. This family has no fossil history. The living species, fossorial limbless lizards that differ from burrowing skinks in lacking osteoderms, occur in southern Indochina and on various islands in the Malay Archipelago. Goin and Goin (1962) include the family Dibamidae in the infraorder Gekkota.

Finally, the family Cordylidae (*Gerrhosauridae*) contains 10 living genera (about 47 species) of diurnal terrestrial carnivores that are restricted to Africa south of the Sahara Desert and to the island of Madagascar. The living *Gerrhosaurus* is known as a fossil dating back to the Miocene of Africa, and the extinct *Macellodus* and *Pseudolacerta* have been found in European Upper Jurassic and Eocene deposits, respectively. The plated lizards (*Cordylosaurus*, *Gerrhosaurus*, *Paratetradactylus*, *Tetradactylus*, *Tracheloptychus*, and *Zonosaurus*), formerly considered to constitute a distinct family, are included in the subfamily Gerrhosaurinae. The remaining genera (*Chamaesaura*, *Cordylus*, *Platysaurus*, *Pseudocordylus*) compose the subfamily Cordylinae.

INFRAORDER ANGUINOMORPHA (DIPLOGLOSSA)—Skull arches usually present; lacrimal bone present, postfrontal and postorbital frequently fused; epipterygoid reduced; osteoderms present or absent; limbs usually present but may be reduced or modified; vertebrae procoelous with well-developed condyles; teeth subpleurodont, frequently recurved and pointed but sometimes blunt. This is, perhaps, not a natural grouping. It includes a broad spectrum of lizards ranging from the slowworms (*Anguis*) to the monitor lizards (*Varanus*), but the morphological extremes are bridged by intermediates.

Superfamily Anguoidea—Skull relatively short; postorbital and postfrontal generally separate; temporal fenestra may be covered by osteoderms; osteoderms generally covering both the head and body; no joint in mandible. This super-family, which is largely extinct, is com-posed of three families. Following Mes-zoely (1970), the family Anguidae con-tains six living genera with about 88 spe-cies plus six extinct genera dating back to the Cretaceous of North America and the Pliocene of Europe. The subfamily Anguinae includes *Pancelosaurus,* the most primitive of fossil anguids from the late Cretaceous to Oligocene of Wyoming, plus the living *Anguis* and *Ophisaurus.* These are characterized by fused frontals and well-separated frontoparietal epi-dermal scales. The subfamily Gerrhono-tinae includes three living genera (*Ger-rhonotus, Abronia,* and possibly *Colopty-chon*) with fused, hourglass-shaped fron-tals and frontoparietal scales that are almost or barely in contact. The subfamily Glyptosaurinae includes five extinct gen-era (*Xestops, Peltosaurus, Melanosaurus, Arpadosaurus,* and *Glyptosaurus*) with tuberculated osteoscutes and toothed palates. The subfamily Diploglossinae includes only *Diploglossus* of the West Indies (*Celestus* and *Sauresia* have been treated as synonyms of *Diploglossus*), characterized by cycloid scales with peaked gliding surfaces, separate frontals, and separated frontoparietal scales. Most anguids are terrestrial or fossorial species.

The family Anniellidae includes only *Anniella* (two species) and has no fossil history. These are the legless lizards of California and Baja California, burrowing forms which seldom come to the surface. They differ from the Anguidae in lacking temporal arches but are sometimes class-ified as a subfamily (Anniellinae) in the family Anguidae.

The family Xenosauridae includes two known genera (*Xenosaurus* of Mexico and Guatemala and *Shinisaurus* of south-ern China) plus the extinct *Exostinus* from the Cretaceous to Oligocene of North America. *Shinisaurus,* the amphibious Chinese crocodile lizard, is reported to feed partially on tadpoles and fish. King and Thompson (1968) have analyzed the morphological variation in *Xenosaurus* and concluded that there are three spe-cies (*Xenosaurus grandis, Xenosaurus newmanorum,* and *Xenosaurus platy-ceps*), but very little is known about them.

Superfamily Varanoidea (Platynota)— Skull moderately long and slender; post-orbital and postfrontal usually fused; osteoderms sometimes absent; mandibles generally jointed. These are moderate to large carnivorous terrestrial and marine lizards which apparently descended from the anguids and, in turn, are generally considered to be close to the ancestry of snakes. Ten families of varanoids are recognized.

The family Necrosauridae, questionably placed here, includes only *Necrosaurus* from the Paleocene to Oligocene of Eur-ope.

Parasaniwidae is another small family; it contains three extinct North American genera (*Paraderma* and *Parasaniwa* from the Cretaceous and *Provaranosaurus* from the Paleocene) which were closely related to the monitor lizards.

The family Helodermatidae includes the genus *Heloderma* with two living species, the Gila monster (*Heloderma suspectum*) and the Mexican beaded liz-ard (*Heloderma horridum*) of southwest-ern United States and Mexico, and the extinct *Eurheloderma* from the Eocene and Oligocene of western Europe. The latter genus indicates that the family was once much more widely distributed than at the present time. The Gila monster and beaded lizard are the only lizards known to be poisonous.

The family Varanidae, monitor lizards, is a very old group which dates back to the Upper Cretaceous and includes the larg-est living lizards. There are about 24 living species, all in the genus *Varanus,* and these range in length from about 200 millimeters (*Varanus brevicauda* of Aus-tralia) to about 300 centimeters (*Varanus komodoensis,* the "Komodo dragon" of the Indonesian islands of Komodo, Rintja, Padar, and the western extremity of Flores). All are carnivorous, and the larger species fill the ecological niche of large mammalian carnivores, feeding on such prey as pigs and deer. Living species occur in Africa, Arabia, southern Asia, the East Indies, Australia, and on the Marshall, Caroline, and Fiji Islands. Fossils assigned to this genus are known from the Miocene to Pleistocene of Eur-ope, the Pliocene and Pleistocene of southern Asia, and the Pleistocene of Australia, Africa, and the East Indies.

Extinct genera include *Chilingosaurus* from the Upper Cretaceous of eastern Asia, *Megalania* from the Pleistocene of southern Asia and Australia, *Paleosaniwa* from the Cretaceous of North America, *Saniwa* from the Eocene to Oligocene of North America and Europe, and *Telmasaurus* from the Upper Cretaceous of eastern Asia. *Pachyvaranus* from the Upper Cretaceous of North Africa is questionably assigned to this family. *Megalania* was probably the largest lizard that has ever lived and reached a size almost double that of the Komodo dragon.

Lanthanotus boreensis, a poorly known species found only in Borneo, is in a family by itself, Lanthanotidae. This lizard appears much like a miniature crocodile but is structurally similar to both the Gila monster and a monitor lizard. In the past, it was sometimes placed in the genus *Heloderma,* but it is now considered to be an advanced form possibly close to the aquatic varanoid line of descent.

The family Aigialosauridae is extinct; it includes three or four genera of subaquatic lizards with long compressed tails and paddlelike limbs. Aigialosaurids are known from the middle Cretaceous and the Upper Jurassic of Europe and seem to be intermediate between the more primitive varanoids and the marine mosasaurs.

The family Dolichoesauridae is another extinct grouping of four genera (*Acteosaurus, Coniasaurus, Carsosaurus,* and *Proaigialosaurus*) from the Cretaceous of Europe. Dolichosaurids were small lizards with aquatic adaptations and greatly elongated bodies. Apparently descended from the aigialosaurids, they were an evolutionary dead end.

The family Mosasauridae, descended from the aigialosaurids, was a very successful group of Upper Cretaceous marine lizards (Fig. 5-6). These were large animals, ranging in length from about 15 to 30 feet, with a long pointed head, a short neck, and a long body and tail. The tail was used to propel the animal, and the limbs, which had short proximal bones but long spreading digits, functioned only in steering. The skull was so similar to those of varanoids that a close relationship is certain. About 20 genera, worldwide in distribution but known only from the Upper Cretaceous, are recognized, and these may be grouped into surface-swimming, diving, and deep-water categories. Most of the mosasaurids were fish-eaters that competed with plesiosaurs and ecologically replaced the ichthyosaurs, but *Globidens* was a specialized mollusk-eater. Despite their abundance in the Upper Cretaceous, mosasaurs suddenly became extinct at the end of that period and left no descendants.

The family Paleophidae (Cholophidae, Cholophidia) contains three genera (*Anomalopsis, Palaeophis,* and *Pterosphenus*) of snakelike Eocene reptiles which have sometimes been classified as snakes but which have varanoid vertebrae and, from limited evidence, varanoid skulls. Except for the fact that these appear to have been marine forms, they would seem to qualify as intermediates between snakes and lizards. Paleophids have been found in Eocene deposits of Europe, North America, South America, and Africa.

Another group of snakelike reptiles is the family Simoliophidae, containing *Lapparentophis* from the Lower Cretaceous of North Africa, *Mesophis* and *Pachyophis* from the Lower Cretaceous of Europe, and *Simoliophis* from the Upper Cretaceous of Europe, North Africa, and Madagascar. These poorly known fossils have also been classified by some workers as snakes.

INFRAORDER ANNULATA (AMPHISBAENIA)—Skull highly specialized for burrowing, skull arches generally lost; postfrontal and squamosal absent, postorbital generally lost; lacrimal absent or fused to prefrontal; quadrate slanting forward and fixed; epipterygoid reduced or absent; teeth pleurodont; osteoderms absent; hind limbs absent; forelimbs generally absent; girdles reduced; body elongate, with 85 to 156 vertebrae. This infraorder, sometimes considered to be a separate suborder, includes only the family Amphisbaenidae, the worm lizards. There are 16 living genera and about 87 species of these burrowing forms. They are distributed in Florida, Mexico, South America, the West Indies, Africa, southern Europe, and southwestern Asia. Eight other genera are extinct and date back to the Paleocene of North America and the Eocene of Europe and southwestern Asia.

Suborder Ophidia (Serpentes) — Upper

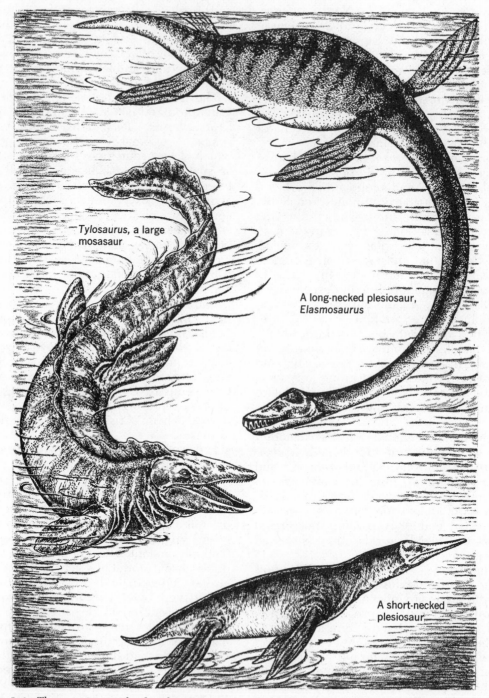

Figure 5–6   Three marine reptiles found in Cretaceous rocks of Kansas. (From *Tales Told By Fossils*, by Carroll Lane Fenton. Copyright, © 1966 by Carroll Lane Fenton. Reproduced by permission of Doubleday & Company, Inc.)

temporal arch missing; epipterygoid absent; mandibles ligamentously joined at their symphysis, proximal elements of each mandible (articular, prearticular, surangular) completely fused; anterior portion of braincase completely enclosed by dermal bone; maxillae, palatines, and pterygoids moveable; vertebrae procoelous and numbering between 141 and 435; pectoral girdle absent; forelimbs absent; sternum and parasternum absent; hind limbs and pelvic girdle generally absent; teeth acrodont and recurved. This suborder, the snakes, is undoubtedly the most recently evolved reptilian group and clearly descended from the lizards, probably from the varanoids. It is now generally agreed that the loss of limbs, elongation of body, development of spectacle, and other characteristics of snakes originated as a result of adaptation to burrowing. As a result of subsequent radiation, the majority of living snakes are surface dwellers, with different groups specialized for arboreal, aquatic, and terrestrial life. Snakes, because of their lightly built skeletons, do not preserve well, so their fossil record is poor compared to that of other reptilian groups. However, it seems likely that they first appeared in the late Mesozoic, underwent their greatest diversification during the Tertiary, and remain, in contrast to other reptiles, a very successful and possibly expanding group.

The classification of snakes, particularly in regard to the bulk of the species in the family Colubridae, is in a state of flux. The only modern original classification of all snakes that gives consideration to comparative anatomy, histology, cytology, biochemistry, and genetics is that of Underwood (1967), but his scheme has not been universally accepted. As emphasized by Hoffstetter (1968), there is a great need for information about extinct snakes and, consequently, much speculation about relationships of recent groupings.

Superfamily Scolecophidia (Typhlopoidea)—Skull bones generally joined solidly together; premaxillary and maxillary not in contact; premaxillary, palatine, and pterygoid lacking teeth; maxillary and dentary may or may not have teeth; jaws shortened; coronoid present; nasals and prefrontals in contact; supratemporal absent; ventral scales same size as dorsals; vestiges of pelvic girdle possibly present; short tail terminating in a spine. This superfamily includes three families of small tropical wormlike snakes that are highly specialized insectivorous burrowers.

The family Anomalepidae includes four genera (*Anomalepis, Liotyphlops, Helminthophis,* and *Typhlophis*) that occur in South America. In the past these have been considered part of the family Typhlopidae, but List (1966) has shown sufficient osteological differences to merit their recognition as a separate family.

The family Typhlopidae includes only the genus *Typhlops,* which is widely distributed in Europe, southern Asia, Africa, South America, the East and West Indies, and Australia.

The family Leptotyphlopidae includes the South American *Rhinoleptus* and the widely distributed *Leptotyphlops* (about 50 species), which occurs in the New World tropics, the West Indies, Africa, and western Asia.

The three scolecophidian families possess a combination of primitive lizardlike characters and highly specialized features and have been classified both as lizards and as snakes. They are now considered to be properly placed in Ophidia (see List, 1966; McDowell, 1967; Underwood, 1967) by most, but not all, workers. In the past there has been a question as to whether the three families were closely related or had merely undergone parallel evolution due to identical habits. Underwood (1967) has presented considerable evidence that there is a real relationship between them.

Superfamily Henophidia (Boidea)—Premaxilla possibly in contact with maxilla; nasals and prefrontals in contact; maxilla, dentary, palatine, and pterygoid generally with well-developed teeth; premaxilla usually toothless; coronoid and supratemporal generally present; vestiges of hind limbs generally present. This grouping includes five living and one extinct family of primitive snakes. The oldest known fossil of what was definitely a snake is *Dinilysia,* a large (six feet in length) boalike form which has been found in later Cretaceous deposits of Patagonia. *Dinilysia,* because of peculiarities of the posterior part of its skull, is placed in a family by itself, Dinilysiidae.

The family Aniliidae (Ilysiidae) includes three living genera (*Anilius* of South America, *Anomochilus* of Sumatra, and

Cylindrophis of Southeast Asia) plus *Anilioides* from the Miocene of North America and *Coniophis* from the Cretaceous to Eocene of North America. These are burrowing stout-bodied cylindrical snakes with short tails. They are commonly referred to as pipe snakes.

The family Uropeltidae includes seven living genera (*Melanophidium, Platyplecturus, Plecturus, Pseudotyphlops, Rhinophis, Teretrurus, Uropeltis*) with about 44 species; no fossils are known. These are the rough-tailed snakes of southern India and Ceylon and are forest-dwelling burrowers with rigid cylindrical bodies and very short tails. There is a very enlarged scale on the tip of their tail which is rugose, spiny, or double-pointed and may be used in digging. The skull elements are joined together much more solidly in rough-tails than in any other family of snakes. This family and the following are considered subfamilies in the family Aniliidae by Bellairs (1970).

The family Xenopeltidae includes only the sunbeam snake of Southeast Asia (*Xenopeltus unicolor*), a nocturnal burrowing snake that feeds on small vertebrates. The family has no fossil record.

The family Acrochordidae includes only two living genera, *Acrochordus* and *Chersydrus,* the wart snakes, which are aquatic fish-eaters found in estuaries and marine habitats in India, Indochina, and the Indo-Australian Archipelago. These specialized aquatic snakes have very stout bodies, granular scales, small ventral scales, and short compressed tails. Hypapophyses are well developed on all the trunk vertebrae.

The family Boidae, the boas and pythons, includes the largest living snake, *Python reticulatus,* which reaches lengths up to 32 feet, but also includes a variety of small burrowing snakes. There are 23 living genera, including about 58 species, and the family is well represented in the fossil record from the Upper Cretaceous on. Nineteen genera are extinct. Fossils have been found on all of the major land masses except Antarctica. Living boids are found primarily in tropical and subtropical regions throughout the world but also occur in western North America. They are a diversified group, some being arboreal, others terrestrial, and many fossorial; the anaconda (*Eunectes murinus*) is semiaquatic. This diversity is reflected in the fact that as many as six subfamilies have been recognized. The subfamily Loxoceminae, including only *Loxocemus,* the dwarf Mexican python, has many primitive features and is regarded as a relict. Two genera, *Bolyeria* and *Casarea,* which occur only on Round Island off Mauritius (near Madagascar) comprise the subfamily Bolyeriinae. These are small semifossorial snakes which differ from all other boids in completely lacking rudiments of the pelvic girdle and hind limbs and in having the maxilla divided into anterior and posterior parts. The subfamily Erycinae, consisting of *Charina* and *Lichanura* from western North America, *Eryx* from Africa and Asia, and *Engyrus* from the East Indies, is another distinct but less divergent grouping of boids. The Erycinae are relatively small boids with short tails, small eyes, and fossorial habits. The remaining boid genera are relatively homogenous but have traditionally been divided into the subfamily Boinae (10 genera distributed in the New World and Madagascar) and the subfamily Pythoninae (six genera distributed mostly in the Old World and Australia). Although these two groupings differ in such features as the supraorbital bone (present in Pythoninae and absent in Boinae) and the premaxillae (usually toothed in Pythoninae and lacking teeth in Boinae), Hoffstetter (1968) argues that the Boinae and Pythoninae should be grouped into a single subfamily, the Boinae. Finally, some authorities (e.g., Goin and Goin, 1962) recognize the genera *Trachyboa* and *Tropidophis,* small snakes of the West Indies and northern South America, as a distinct subfamily, Tropidophinae, characterized by such features as only one true lung (the right) plus a well-developed tracheal lung on the dorsal wall of the trachea.

Superfamily Caenophidia (Colubroidea) —Premaxilla not in contact with maxilla, premaxilla toothless; well-developed teeth generally on maxilla, dentary, palatine, and pterygoid; nasals and prefrontals not in contact; coronoid absent; no vestiges of pelvic girdle or hind limbs; in adaptation to ingestion of large prey, many skull elements concerned with swallowing articulate loosely. This superfamily includes the "advanced" snakes and the majority of living species. Four living and one questionable extinct family are recognized. The majority of the known fossil remains of snakes are fragmentary, making their identification and phylogenetic relationships difficult to determine. Consequently, the family Archaeophidae, including *Archaeophis* and *Onomalophis* from the Eocene of Europe, is questionably included. These fossil forms were very long (*Archaeophis* had about 565 vertebrae!), but other morphological details are poorly known. McDowell and Bogert (1954) have even suggested that *Archaeophis* may have been an eel rather than a snake. At any rate, if these were in fact snakes, they are not assignable to other families. Their jaw structure, especially the position of the quadrate, indicates they belong to this superfamily.

The family Colubridae is huge. It includes the majority of living snakes, about 302 genera and more than 1400 species, and is worldwide in distribution except for Antarctica and a few oceanic islands. At least one colubrid, *Natrix natrix*, extends into the Arctic (67° north) in Sweden. Colubrids occupy every kind of habitat—terrestrial, fresh water, and marine. Nearly all are harmless, but a few are opisthoglyphous. Correlated with the variety of habitats occupied is a diversity of feeding habits. The smaller species tend to feed on insects and other invertebrates. Larger forms frequently prey entirely on vertebrates, especially birds and mammals. Aquatic species feed on amphibians and fish. Some species are generalized predators that feed on almost any prey they can ingest, but others are specialized feeders with restricted diets of such items as eggs, slugs, or snakes. Many schemes have been proposed for dividing the family into subfamilies, but none has been entirely satisfactory. Underwood (1967) has split this vast number of snakes into many smaller groups, but, as noted by Hoffstetter (1968), he has not accounted for all the genera in his various families. In the following scheme, the genera which are sufficiently distinct to be separated from the rest are grouped into subfamilies other than Colubrinae. The subfamily Colubrinae is a composite of all the remaining genera.

Subfamily   Dasypeltinae — Hypapophyses on anterior vertebrae projecting into esophagus; supratemporal and quadrate solidly fused; teeth minute, none grooved; no mental groove; no facial pit. This subfamily has no fossil record and contains only the genus *Dasypeltis* with five species. These are egg-eating snakes found in Africa and southwestern Arabia.

Subfamily Elachistodontinae — Hypapophyses on anterior vertebrae piercing esophagus; no mental groove; one or two enlarged and grooved teeth on rear of maxilla; facial pit present. This subfamily includes only the living *Elachistodon westermanni* of northeastern India, another egg-eater. Some workers include this species in the subfamily Dasypeltinae.

Subfamily Dipsadinae — Anterior vertebrae lack hypapophyses; mental groove sometimes absent; pterygoids widely separated from quadrates; no grooved teeth; sulcus spermaticus divided. This grouping lacks a fossil record and includes three living genera of specialized slug-eating snakes: *Dipsas* (32 species) and *Sibon* (9 species) of Middle America and northern South America, and *Sibynomorphus* (6 species) of northern South America.

Subfamily   Homalopsinae — Hypapophyses present on all vertebrae; opisthoglyphous, maxillary teeth increase in size posteriorly, followed by diastema and two or three enlarged grooved teeth; small valvular nostrils; body stout, tail short. No fossil record exists for this grouping of 10 genera (seven of which are monotypic) and 34 species (Gyi, 1970). All are aquatic fish- and frog-eaters. Most species occur in fresh water, some tolerate brackish water, and some are marine. They are only occasionally found inland in large river systems. The Homalopsinae range from the Indus River basin in West Pakistan to China as far north as Nanking in Kiangsu Province and through the Indo-Australian Archipelago to northern Australia.

Subfamily   Pareinae — Hypapophyses only on cervical vertebrae; teeth not present on anterior part of maxillary; sulcus spermaticus divided; no mental groove; nasal gland greatly enlarged; head short and wide; body slender, neck slim. Pareinae, the bluntheads, has no fossil history and includes only the living *Aplopeltura* (one species) and *Pareas* (15 species) of southern Asia and the East Indies. These are terrestrial and arboreal snakes that feed on small invertebrates.

Subfamily Xenoderminae — Vertebral neural spines elongate and bent so as to form a flat dorsal surface; occipital condyles reduced; scales reduced, attached to dermis, and separated from one another by bare skin. This is a poorly known grouping which includes only four living genera: *Xenodermus* (one species) and *Stoliczkia* (two species) of the East Indies and Southeast Asia, *Cercaspis* (one species) of Ceylon, and *Xenopholis* (one species) of central South America. These are sometimes merely placed in the composite Colubrinae; however, because of their scales and vertebrae, they are similar to one another and distinguishable from other genera.

Subfamily Colubrinae — Composite grouping in which vertebral hypapophyses, teeth, hemipenes, and pupil are variable. However, members of this subfamily tend to be unspecialized, and, accordingly, the ventral scales are well-developed, the nostrils are lateral in position, and the head shields are well developed and symmetrically arranged. The only poisonous species are opisthoglyphous with two or three posterior grooved teeth. Most of the living snakes belong to this subfamily that includes 292 genera and about 1350 species. Of these, 10 genera are known only as fossils which date back to the Miocene of North America, Europe, and eastern Asia. Twenty-five living genera are also known as fossils, most of which only date back to either the Pleistocene or Pliocene. In view of the young age of most fossils, the very large number of living genera and species reflects the great success the group is encountering at the present time.

The family Elapidae is another modern family which includes many of the most poisonous snakes, the cobras, corals, mambas, and kraits. All members of this family are proteroglyphs, each anterior maxillary tooth being grooved or perforated so as to form a fixed fang. Elapids occur on all continental tropical and subtropical land masses, and they are the dominant ophidians of Australia. They also occur throughout the East Indies. This family contains 40 living genera (181 species). Of these, two are also known as fossils from the Pliocene to Recent of North America (*Micrurus*) and the Pleisto-

cene of Europe (*Naja*). The only extinct genus assigned to this family is *Palaeonaja* from the Miocene to Pliocene of Europe.

The sea snakes, family Hydrophidae, are probably closely related to the elapids. This family contains 15 genera (59 species) of proteroglyphous marine fish-eaters. No fossil history exists for the family, but living members are widely distributed in warm seas. *Pelamis platurus*, which has been found hundreds of miles from land, occurs from southeastern Africa and Madagascar across the Indian and Pacific oceans to the western coast of tropical America; other genera are more limited in distribution, around the East Indies and northern coast of Australia and along the coast of Asia from the Gulf of Persia to southern Japan.

The family Viperidae, the vipers and pit vipers, occurs worldwide except for extreme latitudes and altitudes, Antarctica, and Australia. All members of this family are solenoglyphs, having rotatable maxillae bearing one large fang each. Two subfamilies of viperids are recognized. The subfamily Viperinae, which includes the adders and other true vipers, is found only in the Old World. These lack facial pits and tend to be short-bodied and stout. Of the 10 living genera (47 species), *Bitis* and *Vipera* are also known as fossils from the Miocene of Europe. *Provipera*, also from the Miocene of Europe, is the only known extinct genus. The subfamily Crotalinae, pit vipers, contains six living genera (100 species) and is more widely distributed. Crotalinae occur from eastern Europe through Asia to Japan and the East Indies, and they are abundant in North, Middle, and South America. Pit vipers, including the various kinds of rattlesnakes and the water moccasin, copperhead, bushmaster, and fer-de-lance, among others, are the most dangerous New World snakes. All are characterized by the presence of heat-sensitive facial pits; they tend to be nocturnal predators of small mammals which they detect through sensitivity to changes in infrared wavelengths. *Laophis*, questionably a member of this subfamily, is the only extinct genus; it is known from the Miocene of Europe. The living *Crotalus, Agkistrodon,* and *Sistrurus* are also known as fossils from Pliocene to Recent deposits of North America.

Order Rhynchocephalia—Primitive lepido-saurians typically having diapsid skulls with at least the upper arch present; quadrate fixed; dentition acrodont, teeth present on premaxillary, maxillary, pala-tine, and dentary and, vestigially, on vomer; premaxilla beaklike; skull short and broad with expanded cheek region; no tabulars or postparietal, supratemporal generally absent; quadratojugal present; body and limb of normal proportions; vertebrae amphicoelous, intercentra well developed; caudal chevrons present; 23 to 25 presacral vertebrae, two sacral ver-tebrae; anterior ribs with both tubercular and capitular attachments; clavicles slen-der, interclavicle T-shaped; no osteoderms or other dermal armor. This order includes five families, four of which are extinct.

The family Sphenodontidae includes one living species, *Sphenodon punctatus*, the tuatara (see Figure 5–5). Ten or 11 extinct genera were widely distributed in North America, Europe, eastern Asia, and South Africa during the Mesozoic (Upper Triassic to Upper Cretaceous). Fossils assigned to *Sphenodon* are known from throughout the range of the family, but the tuatara survives only on the Chicken Islands, Stephen Island, and Karewa Island, all off the main islands of New Zealand. Sphenodontids were the least specialized of the rhynchocephalians and, obviously, persisted the longest. Their premaxillary beak is relatively poorly developed and teeth are present on the premaxillae, maxillae, and dentar-ies.

The extinct family Rhynchosauridae contains 10 or 11 genera of relatively large stout Triassic forms with highly developed parrotlike premaxillary beaks, large tooth plates on each maxillary, and narrow lower jaws which formed a chopping device by fitting into the groove between the tooth plates. These were herbivorous animals which apparently fed on hard-shelled fruit and used their chopping mechanism to break the shell. Rhynchosaurids were widely distributed during the middle Triassic and are known from all continents except Australia and North America. As a group, they appear suddenly in deposits from early in the Triassic and disappear before the close of that period.

The family Sapheosauridae (=Saurodon-tidae), also extinct, includes only *Sapheo-saurus* from the Upper Jurassic of Europe.

This is a poorly known form which had toothless jaws with sharp cutting edges but, otherwise, was very similar to the sphenodontids.

The family Claraziidae is questionably included in Rhynchocephalia. This fam-ily contains two genera, *Clarazia* and *Hescheleria*, from the Alpine marine Triassic deposits of Europe. These were aquatic animals with moderately elongate bodies and elongate snouts. The pre-maxillae were more or less recurved and formed the shell-cracking mechanism.

Pleurosaurus, another elongate aquatic reptile from the Upper Jurassic of Europe, constitutes the family Pleurosauridae and is another questionable rhynchocephalian. These may have been closely related to the claraziids but lacked the recurved tip on the premaxilla. There was an upper fenestra in the temporal region of the skull, but, rather than a lower opening, there was an emargination from below. Thus, the skull resembled the euryapsid type. However, the remainder of the skeleton and the acrodont dentition appear to be similar to the rhynchocephalian. Thus, at the present time Huene's (1952) suggestion that *Pleurosaurus* is an aber-rant offshoot of the rhynchocephalians is being followed.

Subclass Archosauria (Fig. 5–7)

Order Thecodontia—Skull never flattened posteriorly; snout slender; antorbital vacuity well developed; quadrate vertical or slanting forward, elongated; otic notch present posterior to quadrate and below squamosal; pineal foramen generally absent; postparietal and postfrontal may be present; epipterygoid frequently pres-ent; jaw generally slender with large ex-ternal and internal fenestrae; teeth con-ical, usually pointed, generally homodont, but incisor type or caninelike tusks some-times present; generally 25 to 26 presacral vertebrae, including seven to eight cer-vicals; generally two sacral vertebrae; vertebral centra amphicoelous to platy-coelous (flat anteriorly and concave pos-teriorly); interclavicle always present, clavicles generally present; hind limbs longer than forelimbs; no loss of digits, but outer digits sometimes reduced; meta-tarsals generally elongated; dorsal armor generally present along back and tail. The order Thecodontia includes a variety of

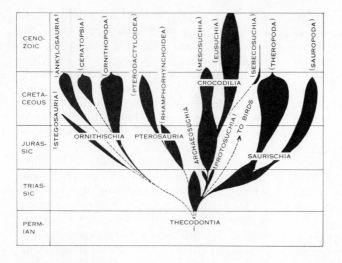

Figure 5-7 The temporal and phylogenetic relationships of the archosaurian reptiles. (From Romer: *Vertebrate Paleontology.*)

basically primitive reptiles in which archosaurian specializations were initiated but not well developed. First appearing in the upper Permian, thecodonts were already extinct by the end of the Triassic. Despite their very short duration, thecodonts played a key role in vertebrate evolution, for they gave rise to dinosaurs and other archosaurian groups and were also ancestral to birds. Four suborders of Thecodontia are generally recognized.

Suborder Proterosuchia—Skull primitive; postparietal and postfrontal present; pineal foramen present; palatal teeth present; posterior surface of quadrate not concave; hind limbs only slightly elongated; quadrupedal locomotion. This suborder includes two families of the most primitive thecodonts, Chasmatosauridae with four or five genera and Erythrosuchidae with eight genera.

Suborder Pseudosuchia—Postfrontals are sometimes absent; few or, generally, no palatal teeth; pubis always turned downward; hind limbs considerably longer than forelimbs; bipedal locomotion; double row of armor plates generally on back. This suborder, typified by *Euparkeria,* contains the following families: Euparkeriidae (*Euparkeria* only), Erpetosuchidae (10 genera), Teleocrateridae (*Teleocrater* only), Elastichosuchidae (*Elastichosuchus* only), and Prestosuchidae (four genera). These "advanced bipedals" were middle and late Triassic forms which positively gave rise to the dinosaur order Saurischia. Pseudosuchians must also have given rise to the pterosaurs, ornithischian dinosaurs, and birds,

but known fossils do not provide any direct lines of descent to these three groups. Pseudosuchians occurred in North and South America, Africa, Europe, and eastern Asia.

Suborder Aetosauria (family Aetosauridae with nine genera)—Hind limbs distinctly longer than forelimbs but locomotion quadrupedal; body covered with a solid layer of armor plating (Fig. 5–8). These late Triassic thecodonts undoubtedly evolved from bipedal ancestors but reverted to quadrupedal locomotion. They have been found in North American and European deposits.

Suborder Phytosauria (family Phytosauridae only with about six genera)—Crocodile-like thecodonts (Fig. 5–8) with armor like that of aetosaurians; skull very elongate; nostrils on top of head and posterior in position, almost between eyes; jaws with numerous pointed teeth; limbs short, hind limbs longer than forelimbs. The adaptations of these common late Triassic reptiles indicate that they were aquatic fish-eaters. They were a very successful group until they faced competition from crocodiles. Phytosaurs were widely distributed in the Northern Hemisphere, and abundant remains have been found in North America, Europe, and India.

Order Crocodilia—Skull usually sculptured, broad, and flattened posteriorly; snout generally elongate; upper temporal opening often reduced and sometimes secondarily roofed; secondary palate highly developed; thecodont dentition, teeth numerous, conical, and pointed; 23 to 24 presacral vertebrae; interclavicle present, clavicles absent; forelimbs shorter than hind limbs; dermal armor generally present. The crocodilians evolved from thecodonts during the Triassic and are the only archosaurians to survive to the present time; all others became extinct before the close of the Mesozoic, the Age of Reptiles. The long survival of crocodilians, as that of turtles, may be related to the fact that they have been a very conservative group of reptiles: the tendency in evolution is for specialization to lead to extinction. Five suborders of crocodilians are recognized, four of which are extinct.

Suborder Protosuchia—Primitive crocodilians; snout moderately long and slender; quadrate vertical; vertebrae amphicoelous; coracoid short; well-developed armor. This suborder includes the family Protosuchidae (*Protosuchus*) from the Upper Triassic or Lower Jurassic of North America plus, questionably, the family Sphenosuchidae (*Sphenosuchus* from the Upper Triassic of South Africa and possibly *Pedeticosaurus* and *Platyognathus* from the Upper Triassic of eastern Asia). These were small reptiles that, aside from crocodilian modifications of girdles and a few other features, were very close to the thecodonts.

Suborder Archaeosuchia—Intermediate crocodilians; nares fused, opening on top of snout; secondary palate partially formed by maxillae. This group, including the family Notochampsidae (*Erythrochampsa* and *Notochampsa* from the Upper Triassic of South Africa and *Microchampsa*

from the Upper Triassic of eastern Asia) and the family Proterochampsidae (*Proterochampsa* from the middle Triassic of South America), had skulls with many more crocodilian features than the protosuchians but were still more primitive than typical crocodilians.

Suborder Mesosuchia—The central stock of typical crocodilians; vertebrae platycoelous; pterygoids not entering secondary palate; armor typically well developed; antorbital openings present; upper temporal openings not greatly reduced. While more primitive than recent forms, members of this suborder were typical crocodilians and were well adapted to aquatic and amphibious habitats. This relatively large grouping includes six families. They are Teleosauridae (11 genera from Jurassic deposits of Europe, Madagascar, North Africa, East Asia, and South America); Pholidosauridae (12 to 14 genera from Upper Jurassic, Cretaceous, and Eocene deposits of Europe, East Africa, and North America); Atoposauridae (five genera from the Upper Jurassic of Europe, North America, and East Asia), Goniopholidae (about 18 genera from Upper Jurassic and Cretaceous deposits of Europe, North America, South America, East Asia, and North Africa); Notosuchidae (four genera from the Upper Cretaceous of North Africa and South America); and Metriorhynchidae (five genera including *Geosaurus*, seen in Figure 5–9, from middle Jurassic to Lower Cretaceous deposits of Europe and South America). Many of these were long-snouted fish-eaters. The teleosaurids and metriorhynchids were

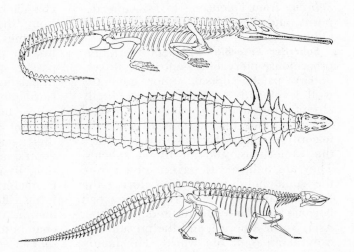

Figure 5–8 Top, *Mystriosuchus*, a Triassic phytosaur, original about 11.5 feet long. Center, dorsal view of the armor of the aetosaur *Desmatosuchus*, portion illustrated about 12 feet long. Bottom, lateral view of the skeleton of the aetosaur *Stagonolepis*, armor omitted; length about 9.5 feet. (From Romer: *Vertebrate Paleontology*.)

Figure 5–9 *Geosaurus*, a smooth-skinned marine crocodile, had webbed feet and a fin on its tail. (From *Tales Told By Fossils*, by Carroll Lane Fenton. Copyright, © 1966 by Carroll Lane Fenton. Reprinted by permission of Doubleday & Company, Inc.)

sea crocodiles, some of which had fish-like tails and paddle-shaped limbs. Pholidosaurids were fresh-water forms. The atoposaurids were very small crocodilians (less than a foot in length) with short skulls. The notosuchids were similarly short-skulled but considerably larger. The goniopholids were most similar to and appear to have been ancestral to the eusuchian crocodiles of the Cenozoic.

Suborder Sebecosuchia—Skull high and narrow; secondary palate short, not involving pterygoids; nares separated by nasals; vertebrae platycoelous. The postcranial skeletons of these South American crocodilians are very poorly known, but their skulls are like those of neither mesosuchians nor eusuchians. Sebecosuchians apparently diverged from the other crocodilians while geographically isolated in South America during the Mesozoic. The two families included here are Baurusuchidae (*Baurusuchus* and *Cynodontosuchus* from the Cretaceous) and Sebecidae (*Peirosaurus* from the Cretaceous, *Sebecus* from Paleocene to Miocene deposits, and, possibly, *Ilchunaia* from the Oligocene).

Suborder Eusuchia—Presacral vertebrae procoelous; pterygoids contribute to secondary palate; dorsal armor generally well-developed, ventral armor generally absent. This suborder includes the typical Cenozoic and Recent crocodilians, the majority of those that have existed since the Cretaceous, and all living species. They are believed to have evolved from the goniopholids some time during the Jurassic, and by the Tertiary they had become abundantly distributed in Europe, Asia, North America, and the present-day tropical areas of the world. The family

Hylaeochampsidae includes the oldest eusuchians, *Hylaeochampsa* from the Upper Jurassic to Lower Cretaceous of Europe and, possibly, *Bernissartia* from the Lower Cretaceous of Europe. These small crocodilians had the eusuchian type of secondary palate but otherwise were not typical eusuchians.

The family Stomatosuchidae includes two to four genera which lacked lower teeth and had small upper teeth; their jaws were long and slender and they lacked dermal armor. Stomatosuchids are known from the Upper Cretaceous of North Africa and East Asia.

The remaining eusuchians fall into three general groupings: the gavials, alligators, and crocodiles. Differences of opinion exist as to whether these should be treated as separate families or as subfamilies. Following Romer (1966), the gavials are treated as a separate family, Gavialidae, while the alligators and crocodiles are all included in the same family, Crocodylidae. Both families have the following characteristics, which distinguish them from the other eusuchians: maxillae, palatines, and pterygoids meeting in midline of palate and enclosing respiratory passages; internal nares opening far back in palate; fixed quadrate; thecodont teeth; cartilaginous sternum; abdominal ribs; pubis not contributing to acetabulum; forefoot with five digits, hind foot with four digits; dorsal dermal armor always present, ventral armor sometimes present. In addition, living members have the following characteristics (not known for extinct forms): fleshy transverse fold on posterior margin of tongue and on palate; four-chambered heart; diaphragm separating thoracic and abdominal cavities; longitudinal slit as anal opening; single male copulatory organ; two pairs of scent glands, one on throat and one within anal lips; oviparity. The family Gavialidae, including the living *Gavialis gangeticus* from the East Indies and southern Asia, has the following characteristics differentiating it from Crocodylidae: extremely long and slender snout; fourth tooth of lower jaw not fitting into a groove or pit in upper jaw; teeth of both jaws angling outward and meshing so as to bite between one another; dorsal (but no ventral) dermal armor. The genus *Gavialis* is also known as a fossil from the Eocene of Europe, the Oligocene of South America, Miocene of Africa, Miocene to Recent of the East Indies and southern Asia, and the Pleistocene of East Asia. In addition, *Rhamphostomopsis* is known from the Pliocene of South America and *Rhamphosuchus*, questionably a gavialid, from the Pliocene of southern Asia.

The family Crocodylidae includes seven living genera with 20 species plus 46 extinct genera from Upper Cretaceous to Recent deposits around the world. The characteristics of this family distinguishing it from Gavialidae include the following: only moderately elongate snout; fourth tooth of lower jaw fitting into a pit or groove in upper jaw; teeth of lower jaw biting inward from those of upper jaw or

in an even vertical plane between those of upper jaw; dorsal dermal armor, ventral armor sometimes present. The Crocodylidae were particularly abundant during the Tertiary and have decreased since. Most living species are in danger of extinction because of man's activities. Extinct crocodylids varied considerably in size but, otherwise, resembled living species. The largest known crocodile was *Deinosuchus* from the Upper Cretaceous of North America and Europe. This giant had a skull six feet long and an estimated total length of 40 to 50 feet.

Order Pterosauria—Flying reptiles with many adaptations for flight; skeleton pneumatic; skull diapsid but thin with many fused elements; lateral temporal opening narrow and diagonal; post-temporal openings slitlike; jaws long and slender; dentition thecodont, teeth conical and sharply pointed; dermal elements of pectoral girdle absent; scapula and coracoid long and slender; sternum very large; forelimbs modified into elongate wings with greatly elongated fourth digit supporting a membrane of skin; 17 to 24 procoelous presacral vertebrae; tail slender and variable in length. The pterosaurs (Fig. 5–10) unquestionably evolved from some member of the order Thecodontia during the Triassic. In order that the forelimbs would be

Pterodactylus

Ramphorhynchus

Figure 5–10 A long-tailed and a short-tailed pterosaur, both of which lived in Europe during the Jurassic. (From *Tales Told By Fossils*, by Carroll Lane Fenton. Copyright, © 1966 by Carroll Lane Fenton. Reprinted by permission of Doubleday & Company, Inc.)

free for modification into wings, the ancestral thecodont must have already possessed bipedal locomotion. From this ancestor, the evolution among pterosaurs involved a progressive development of wings and a general reduction of the hind limbs until, in many, it appears that terrestrial locomotion must have been impossible. There was also a reduction in the length of the tail in more advanced forms. The end result was a gross similarity between the postcranial skeleton of pterosaurs and that of bats; the pterosaur skull was more like that of birds. Nearly all pterosaur remains have been found in salt-water deposits and, as their wings were better adapted for gliding than for true flight, these archosaurians must have ecologically paralleled terns and other coastal and marine birds. Pterosaurs reached their peak in abundance during the late Jurassic but, probably as a result of competition from birds, became extinct before the end of the Cretaceous. They were not ancestral to any other group of vertebrates.

Suborder Rhamphorhynchoidea — Primitive pterosaurs, all from the Jurassic; skull short with relatively little fusion of elements; jaws with well-developed teeth; tail long with up to 40 caudal vertebrae. This suborder includes the family Dimorphodontidae (five genera from Eu-

rope), the family Rhamphorhynchidae (four genera from North America, East Africa, and Europe), and the family Anurognathidae (two genera from Europe and Asia).

Suborder Pterydactyloidea — Advanced pterosaurs (Fig. 5–11); skull very elongate; teeth reduced, on anterior parts of jaws only, or absent; tail shortened, with about 12 vertebrae. The two families included here are Pterodactylidae (five genera from middle Jurassic to late Cretaceous deposits of Europe and East Africa) and Ornithocheiridae (eight genera from Upper Jurassic to late Cretaceous deposits of North and South America, Europe, Africa, and Asia).

Order Saurischia — Dinosaurs with triradiate pelvis (Fig. 5–12); skull relatively short with high postorbital region; external nares separate; postfrontal and postparietal absent; temporal openings well developed; otic notch present, prominently roofed by overhanging squamosal; external fenestra on lower jaw; quadrate typically vertical; dermal elements of pectoral girdle generally absent; forelimbs generally shorter than hind limbs; at least some reduction in digits of forefeet and hind feet; generally 23 to 25 presacral vertebrae; vertebrae platycoelous, amphiplatyan, amphicoelous, or opistho-

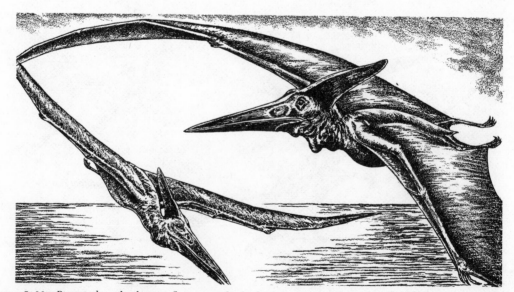

Figure 5–11 *Pteranodon,* the largest flying reptile, had a long, sharp bill and a bony crest on its head. (From *Tales Told By Fossils,* by Carroll Lane Fenton. Copyright, © 1966 by Carroll Lane Fenton. Reprinted by permission of Doubleday & Company, Inc.)

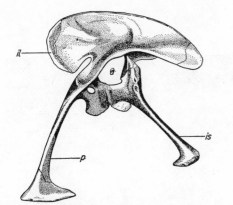

Figure 5–12 Triradiate pelvis of the bipedal saurischian *Ceratosaurus*, viewed from the left side and approximately 1/16 natural size. *a*, acetabulum; *il*, ilium; *is*, ischium; *p*, pubis. (From Swinton, W. E.: *The Dinosaurs*.)

coelous. The saurischians were one of two major groups of "dinosaurs" that evolved from the order Thecodontia during the Triassic. The order Saurischia includes four distinct groups (infraorders) which Romer (1966) has arranged into two suborders.

Suborder Sauropodomorpha — Giant Jurassic and Cretaceous dinosaurs, generally quadrupedal or amphibious herbivores, and their ancestors, primitive semi-bipedal sauropodomorphs of the Triassic.

INFRAORDER PROSAUROPODA — Small to moderate forms with heavy bones; semi-bipedal locomotion; scapula relatively broad; generally a broad and plate-like pubis; forelimbs 40 to 60 per cent of length of hind limbs; vertebrae amphicoelous, platycoelous, or amphiplatyan; neck shorter than trunk; generally 25 presacral vertebrae, including 9 to 10 cervicals. All known prosauropods are from the middle to late Triassic and they are the most commonly found dinosaurs of that time. The three families included here are Thecodontosauridae (=Gryponychidae) with nine genera from North and South America, Africa, Asia, Australia, and Europe; Plateosauridae with four genera from Europe, East Asia, and South America; and Melanorosauridae with six genera from South Africa, East Asia, and Europe. The prosauropods became extinct by the end of the Triassic but by then had given rise to the Sauropoda.

INFRAORDER SAUROPODA (=OPISTHOCOELIA, CETIOSAURIA) — Giant dinosaurs with generally quadrupedal locomotion; scapula broad; pubis broad; forelimbs at least 66 per cent of length of hind limbs; vertebrae opisthocoelous, platycoelous, or rarely procoelous; neck variable but longer than trunk; 25 to 27 presacral vertebrae, including 12 to 17 cervicals; skull relatively small in proportion to body size; tail long. These gigantic herbivores undoubtedly evolved from the prosauropods during the Triassic, but the oldest fossils known are from the Lower Jurassic. They became extinct by the end of the Cretaceous and were not ancestral to any other group. This infraorder, which includes the family Brachiosauridae (17 genera from North and South America, Europe, Africa, Madagascar, and East Asia) and the family Titanosauridae (22 genera from North and South America, Europe, Africa, Madagascar, and Asia), contains the largest qudrupeds that ever existed. *Brachiosaurus*, of the family Brachiosauridae, was an amphibious form (Fig. 5–13) that has been found in both East Africa and North America; it reached a length of about 80 feet, could hold its head above the level of a three-story building, and is estimated to have weighed 50 tons! Other well-known North American giants in this infraorder were *Brontosaurus* (family Titanosauridae), *Camarasaurus* (family Brachiosauridae), and *Diplodocus* (family Titanosauridae, Fig. 5–14). In Europe, one of the larger forms was *Cetiosaurus* (family Brachiosauridae).

Suborder Theropoda — Bipedal, primarily carnivorous dinosaurs; forelimbs reduced and used for grasping; skull variable in size but generally moderate to large; vertebrae platycoelous, amphiplatyan, or opisthocoelous; neck always shorter than trunk; scapula and pubis narrow; tibia elongate, sometimes longer than femur. The theropods originated from thecodonts independently of the sauropodomorphs.

INFRAORDER COELUROSAURIA — Relatively small and slender primitive theropods with lightly built skeletons; skull small, orbits relatively large; teeth pointed, recurved, and tending to be heterodont with an enlarged canine series; neck slender and generally about half the length of trunk; generally 23 presacral vertebrae, including nine cervicals. The coelurosaurs were common predators by the late Triassic and were nearly world-

Skull, 30 inches long

Figure 5–13 *Brachiosaurus*, the largest Late Jurassic dinosaur, ranged from eastern Africa to southern Colorado. (From *Tales Told By Fossils*, by Carroll Lane Fenton. Copyright, © 1966 by Carroll Lane Fenton. Reprinted by Permission of Doubleday & Company, Inc.)

Figure 5–14 *Diplodocus* (1) was the largest dinosaur known, measuring 80 to 87 feet in length. *Brontosaurus* (2), the thunder lizard, was shorter than *Diplodocus* but heavier. Both of these lived on Morrison lowlands. (From *Tales Told By Fossils*, by Carroll Lane Fenton. Copyright, © 1966 by Carroll Lane Fenton. Reprinted by permission of Doubleday & Company, Inc.)

wide in distribution when they became extinct in the Upper Cretaceous. This infraorder includes five families: Procompsognathidae, with 12 genera from the Upper Jurassic of Europe and North America; Segisauridae, with *Segisaurus* from the Upper Triassic of North America; Coeluridae (=Coelurosauridae, Compsognathidae), with 22 genera from Upper Jurassic and Cretaceous deposits of Australia, Europe, Asia, Africa, North and South America; Ornithomimidae, with six genera from the Cretaceous of Europe, Asia, and North America; and Caenagnathidae, with *Caenagnathus* from the Upper Cretaceous of North America. The more primitive procompsognathids were frequently only about three feet long, especially small for dinosaurs. They gave rise to the larger ornithomimids, which were birdlike in proportions and ostrichlike in size. The ornithomimids were typically toothless and are believed to have preyed upon the eggs of other dinosaurs by grasping them with opposable digits on their forefeet and cutting the shell with beaklike jaws.

INFRAORDER CARNOSAURIA—Relatively large flesh-eating theropods with heavily built skeletons; skull generally large, orbits relatively small; teeth sharply pointed, recurved, with much-enlarged canine tooth series on maxillary; neck heavy, about half the length of trunk; generally 25 presacral vertebrae, including 9 to 10 cervicals. This infraorder was already clearly differentiated from the coelurosaurs in the Triassic and by the Jurassic contained the major terrestrial carnivores of the world. The five families in this grouping are Ornithosuchidae (*Ornithosuchus* from the Upper Triassic of Scotland and possibly *Clarenceia* from the Upper Triassic of South Africa); Poposauridae (*Poposaurus* from the Upper Triassic of North America); Megalosauridae (17 genera from Jurassic and Cretaceous deposits of North and South America, Europe, Africa, and East Asia); Spinosauridae (three genera from the Cretaceous of North America, Europe, and North Africa); and Tyrannosauridae (=Deinodontidae, with 10 genera from the Cretaceous of North and South America, Asia, and Madagascar). Of these, *Ornithosuchus* was the most primitive and had so many thecodont characteristics (in-

cluding a double row of armor plates down the back, dermal elements in the pectoral girdle, a broad pubis, three sacral vertebrae, and so forth) that it has sometimes been classified as an advanced thecodont rather than a primitive carnosaur. However, its skull closely resembles that of more advanced carnosaurs. *Ornithosuchus* was relatively small, reaching a maximum length of about 12 feet, but Jurassic carnosaurs such as *Allosaurus* and *Tyrannosaurus* (in the families Megalosauridae and Tyrannosauridae, respectively), were the largest terrestrial carnivores that have ever existed. *Allosaurus* (Fig. 5–15) reached a length of about 34 feet, and *Tyrannosaurus* (Fig. 5–20) measured about 47 feet. These giant carnivores preyed upon the contemporaneous herbivorous dinosaurs and became extinct, along with their prey, at the end of the Cretaceous.

Order Ornithischia—Dinosaurs with tetraradiate pelvis (Fig. 5–16); skull variable but with generally longer facial region than in saurischians; postfrontal and postparietal absent; well-developed upper temporal opening, lower opening often reduced; otic notch with prominent roof formed by overhanging squamosal; jaw generally lacking lateral fenestra; quadrate tall, ventrally slanting anteriorly, concave on posterior margin; dermal elements of pectoral girdle seldom present; limbs generally adapted for bipedal locomotion but reversion to quadrupedal locomotion in some; humerus generally much shorter than femur; fifth digit of both forefoot and hind foot reduced, with unreduced digits of both forefoot and hind foot generally ending in hooflike structure; vertebrae platycoelous, amphiplatyan, or opisthocoelous; generally numerous (24 to 34) presacrals. The order Ornithischia is the second great group of dinosaurs that evolved from the Thecodontia. It includes only herbivorous forms and all those with a birdlike pelvis. The order Ornithischia is a variable group and has been divided into four suborders.

Suborder Ornithopoda—Bipedal, most primitive of ornithischians. These were the duck-billed dinosaurs and their relatives. Six families of ornithopods are recognized: Heterodontosauridae (*Heterodontosaurus* from the upper Triassic of

Horned dinosaur

Figure 5–15 *Allosaurus*, a carnivore of the Morrison lowlands, was 34 feet long and had a 3-foot head. The horned dinosaur was not quite so large. (From *Tales Told By Fossils*, by Carroll Lane Fenton. Copyright, © 1966 by Carroll Lane Fenton. Reprinted by permission of Doubleday & Company, Inc.)

South Africa); Hypsilophodontidae (about 10 genera from the Lower Cretaceous of Europe, the Upper Jurassic to Lower Cretaceous of Europe, North America, and Africa); Iguanodontidae (about nine genera from the Jurassic and Cretaceous of Africa, East Asia, Europe, and North America); Hadrosauridae (=Trachodontidae, 27 genera from the Upper Cretaceous of North America, Europe, and Asia and, questionably, the Paleocene of South America), Psittacosauridae (*Protiguanodon* and *Psittacosaurus* from the Upper Cretaceous of East Asia); and Pachycephalosauridae (*Pachycephalosaurus*, *Stegoceras*, and possibly *Polyodontosaurus* from the Upper Cretaceous of North America). The ornithopods evolved during the Triassic and became common by the late Jurassic and early Cretaceous. Triassic forms, generally speaking, are very poorly known, and the order cannot be traced directly to any particular thecodont ancestral grouping. Primitive ornithopods, such as *Hypsilophodon* (family Hypsilophodontidae), were relatively small, reaching about three feet in length; later forms, such as *Iguanodon* (family Iguanodontidae), reached lengths of up to 40 feet. *Iguanodon* (Fig. 5–17) is of interest because it was the first dinosaur to be described scientifically. The hadrosaurids, duck-billed dinosaurs (Fig. 5–18 and 5–20), were very common, widely distributed amphibious ornithopods with webbed feet; many had peculiar head crests. Psittacosaurids, possibly ancestral to the horned dinosaurs that appeared in

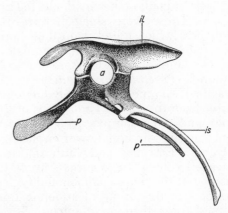

Figure 5–16 Tetraradiate pelvis of bipedal ornithischian *Iguanodon*, approximately 1/30 natural size. *a*, acetabulum; *il*, ilium; *is*, ischium; *p*, pubis; *p¹*, post-pubis. (From Swinton, W. E.: *The Dinosaurs*.)

Figure 5–17 *Camptosaurus*, a primitive ornithischian dinosaur of the American west, and *Iguanodon* of Europe. Both lived during the Cretaceous period. (From *Tales Told By Fossils*, by Carroll Lane Fenton. Copyright, © 1966 by Carroll Lane Fenton. Reprinted by permission of Doubleday & Company, Inc.)

the Upper Cretaceous, were small animals with strong beaks. Pachycephalosaurids had enormous domelike skulls capped with rugosities and spikes.

Suborder Stegosauria—Quadrupedal; two rows of protective plates and spines down the back and tail; bodies moderate to large in size. These were sluggish terrestrial herbivores which relied upon their horns and armor for protection from predators. The two families included here are Scelidosauridae (*Scelidosaurus* from the Lower Jurassic of Europe and possibly *Lusitanosaurus* from the Lower Cretaceous of Europe) and Stegosauridae (eight genera from the Jurassic and Lower Cretaceous of Europe, East Asia, and North America). The well-known *Stegosaurus* (Fig. 5–19) was one of the more common and typical members of this group of armored dinosaurs. Larger than an elephant, *Stegosaurus* had a brain about the size of a golf ball and, interestingly, an enlargement of the spinal cord in the pelvic region that

was about 20 times as large as its brain. This posterior nerve center controlled the movements of the huge hind limbs and tail and gave rise to the popular story that this dinosaur had two sets of brains, one in its head and one in its tail!

Suborder Ankylosauria—Quadrupedal; turtlelike with heavy armor. These "reptilian tanks" were the dinosaurs of the Cretaceous with overlapping layers of bony plates that encased the entire body, head, and tail. Much like that of the armadillo, this was the most efficient armor possessed by any reptile. The suborder Ankylosauria includes the family Acanthopholidae, with six genera from the Cretaceous of Europe and South America, and the family Nodosauridae (=Ankylosauridae), with 25 genera from the Cretaceous of Europe, Asia, South Africa, and North America. Ankylosaurs (Fig. 5–20) were squat, short-legged animals with broad flattened bodies; *Euoplocephalus*, *Palaeoscincus*, and *Nodosaurus* (all in the

Figure 5–18 Numbers 1 and 2, two duck-billed dinosaurs of the Late Cretaceous of Alberta. Number 3 is a saurischian that resembled an ostrich except for its long tail. (From *Tales Told By Fossils*, by Carroll Lane Fenton. Copyright, © 1966 by Carroll Lane Fenton. Reprinted by permission of Doubleday & Company, Inc.)

Figure 5–19 *Stegosaurus*, an armored ornithischian of the Morrison lowlands. At the left are the hip nerve ganglion and brain, drawn to the same scale. (From *Tales Told By Fossils*, by Carroll Lane Fenton. Copyright, © 1966 by Carroll Lane Fenton. Reprinted by permission of Doubleday & Company, Inc.)

Figure 5–20 Typical dinosaurs that lived in Montana and the nearby West near the end of the Cretaceous. Number *1* is a duckbill, *Anatosaurus* (*Trachodon*). Number *2*, *Triceratops* warding off an attack. Number *3* (opposite page), the armor-plated *Ankylosaurus*, which defended itself by crouching on the ground. (From *Tales Told By Fossils*, by Carroll Lane Fenton. Copyright, © 1966 by Carroll Lane Fenton. Reprinted by permission of Doubleday & Company, Inc.)

Figure 5–20 Continued.   Number 4, *Tyrannosaurus*, a carnivore that reached a length of 47 feet. (From *Tales Told By Fossils*, by Carroll Lane Fenton. Copyright, © 1966 by Carroll Lane Fenton. Reprinted by permission of Doubleday & Company, Inc.)

family Nodosauridae) are among the better-known ankylosaurs.

Suborder Ceratopsia — Quadrupedal, horned dinosaurs. All of the known horned dinosaurs are from the Upper Cretaceous, and all are characterized by extremely large heads with great extensions of the parietals and squamosals forming a frill of bone over the neck and nearly to the shoulders. The three families included here are Protoceratopsidae (four genera from East Asia and North and South America), Pachyrhinosauridae (*Pachyrhinosaurus* from North America), and Ceratopsidae (13 genera from East Asia and North America). The protoceratopsids, typified by *Protoceratops*, were small primitive ceratopsians with well-developed bony frills but no horns. The ceratopsids, *Triceratops* (Fig. 5–20), for example, were larger forms reaching lengths of up to 20 feet and having well-developed horns and frills. *Pachyrhinosaurus*, rather than having horns, had an enormous concave structure on its forehead which might have housed a horny cushion.

Subclass Euryapsida (Synaptosauria)

Order Araeoscelidia (= Protorosauria) — Generally primitive, slender reptiles with moderately to greatly elongated cervical vertebrae; vertebrae platycoelous, amphicoelous, or procoelous, 24 to 27 presacrals, intercentra generally present; skull thin; palate toothed and primitive. This order contains five families: Araeoscelidae (*Araeoscelis* from the lower Permian of North America, *Kadaliosaurus* from the lower Permian of Europe, and possibly *Aenigmasaurus* from the Lower Triassic of South Africa); Protorosauridae (four genera from the late Permian and Lower Triassic of Europe); Tanystropheidae (*Tanystropheus* from the Middle Triassic of Europe and southwestern Asia); Weigeltisauridae (*Coelurosauravus* from the upper Permian of Madagascar and *Weigeltisaurus* from the upper Permian of Europe), and Trilophosauridae (six genera from the Triassic and Cretaceous of North America and the Triassic of Europe). These are of uncertain relationship to one another and to other reptilian groups. The temporal region of all araeoscelidian skulls appears to be of the euryapsid type, but even this is not positive;

except for the temporal region, the skulls closely resemble those of captorhinomorph cotylosaurs. The postcranial skeleton is, except for the elongated cervical vertebrae, generally unspecialized. A current theory is that *Araeoscelis* is representative of a group which was ancestral to the nothosaurs, plesiosaurs, and placodonts. Accordingly, the araeoscelidians are provisionally placed in the order Euryapsida.

Order Sauropterygia — Amphibious to marine reptiles with aquatic adaptations; skull low, broad posteriorly; external nares located posteriorly; pineal eye present; interpterygoid vacuities nearly or completely closed; vertebrae amphicoelous or amphiplatyan; trunk ribs articulated to transverse process of neural arch by a single head; ventral ribs well developed; dorsal elements of pectoral and pelvic girdles reduced, ventral elements enlarged; swimming accomplished primarily by paddle-shaped limbs. This order includes the nothosaurs and the plesiosaurs.

Suborder Nothosauria — Relatively primitive amphibious sauropterygians with broad flat skulls; postorbital portion of skull relatively elongate; nasals present but reduced; anterior portion of parietal ending about the level of temporal openings; squamosals not extending inward behind temporal fenestrae; humerus and femur relatively slender but moderately broad distally; up to 49 rows of gastralia. The three families of nothosaurs are Nothosauridae (about 14 genera from the Triassic of Europe, Asia, and North Africa and the upper Permian of Europe); Pachypleurosauridae (*Neusticosaurus* and *Pachypleurosaurus* from the middle Triassic of Europe); and Simosauridae (*Corosaurus* from the Upper Triassic of North America and *Simosaurus* from the Triassic of Europe). The nothosaurs were coastal marine animals that probably spent time both in and out of water, much as sea lions do today. They appear to be close to the ancestral stock from which plesiosaurs evolved.

Suborder Plesiosauria — Predatory marine sauropterygians with higher skulls than nothosaurs and more advanced aquatic adaptations; postorbital portion of skull relatively short; nasals absent; anterior portion of parietal ends well forward of temporal fenestrae; humerus and femur stout, greatly broadened, and flattened distally; generally six to nine rows of gastralia. The plesiosaurs were the most successful of the sauropterygians and persisted the longest. They were highly adapted marine animals that competed primarily with the ichthyosaurs. Most were relatively large, reaching lengths of up to 50 feet. Two or three superfamilies of plesiosaurs are recognized.

Superfamily Pistosauroidea — This grouping, including the family Pistosauridae with *Pistosaurus* from the middle Triassic of Europe and the family Cymatosauridae with *Cymatosaurus* and possibly two other genera from the middle and late Triassic of Europe, is intermediate between the nothosaurs and typical plesiosaurs. The postcranial skeleton of *Pistosaurus* is unknown. This genus has sometimes been assigned to Nothosauria, but it is not typical of that suborder either. *Cymatosaurus* has also been considered to be a nothosaur but has plesiosaur characteristics. It seems likely that the pistosauroideans were ancestral to typical plesiosaurs.

Superfamily Plesiosauroidea (=Dolichodeira) — Typical plesiosaurs with small, short heads, long necks, and short humeri and femora (Fig. 5–6). This relatively large grouping includes the family Plesiosauridae (seven genera from Upper Triassic and Jurassic deposits of Europe, North Africa, and the West Indies), the family Thaurmatosauridae (five genera from the Jusassic of Europe and North Africa), and the family Elasmosauridae (about 13 genera from the Cretaceous of Europe, North and South America, North Africa, New Zealand, and Australia).

Superfamily Pliosauroidea (=Brachydeira) — Typical plesiosaurs with long heads, short necks, and long humeri and femora. The three families included here are Pliosauridae (eight genera from the Jurassic and Cretaceous of North and South America, Europe, Asia, and Australia), Polycotylidae (about 11 genera from the Cretaceous of North and South America, Europe, and Australia), and Leptocleididae (five genera from the Upper Jurassic and Cretaceous of Europe, Africa, and Australia).

Order Placodontia — A small group of specialized mollusk-eating euryapsids adapted for crushing shells but, otherwise, similar to the nothosaurs; jaw heavily built; marginal teeth reduced in number, flattened, blunt; palatine teeth typically forming crushing plates, contacting similar plates formed by teeth on lower jaw; 19 to 46 presacral vertebrae; dermal armor forming a carapace and plastron in some advanced forms. Four small families of placodonts are recognized. They are Helveticosauridae (*Helveticosaurus* from the middle Triassic of Europe), Placodontidae (*Paraplacodus* and *Placodus*, illustrated in Figure 5–21, from the lower to the middle of the Triassic of Europe), Placochelyidae (=Cymodontidae, with five genera from the middle to the Upper Triassic of Europe, North Africa, and southwestern Asia), and Henodontidae (*Henodus*, shown in Figure 5–22, from the Upper Triassic of Europe).

Figure 5–21 *Placodus*, a marine reptile of early Triassic age, was 5 to 8½ feet long and inhabited coastal habitats where it fed on clams and other mollusks. (From *Tales Told By Fossils*, by Carroll Lane Fenton. Copyright, © 1966 by Carroll Lane Fenton. Reprinted by permission of Doubleday & Company, Inc.)

Figure 5–22 *Henodus,* 3 to 4 feet long, had a broad turtle-like shell made of bony plates and lived in swamps, lakes, and broad sluggish streams. (From *Tales Told By Fossils,* by Carroll Lane Fenton. Copyright, © 1966 by Carroll Lane Fenton. Reprinted by permission of Doubleday & Company, Inc.)

Subclass Ichthyopterygia

Order Ichthyosauria — Highly specialized marine reptiles with fishlike bodies and "fins" (Fig. 5–23); skull having long snout; external nares posterior and lateral; eyes relatively very large; little ossification of braincase or girdles; teeth generally numerous (up to 200 in a single series), conical, sharply pointed; vertebrae amphicoelous, 40 to 65 presacrals; fishlike tail with vertebral support in lower lobe; ribs present on all vertebrae anterior to base of tail; limbs modified into flattened paddles, forelimbs larger than hind limbs; humerus and femur relatively short; gastralia present; sharklike dorsal fin; reproduction, at least in some, viviparous. Five families of ichthyosaurs are recognized. They are Mixosauridae (*Mixosaurus* from the Middle Triassic of Europe, southwestern Asia, North America, and the East Indies); Omphalosauridae (*Grippia* and *Omphalosaurus* from the Middle Triassic of Europe and North America); Shastasauridae (six genera from the Middle and Upper Triassic of North America, Europe, and the East Indies); Ichthyosauridae (six genera from Jurassic and Cretaceous deposits of Europe, North and South America, and New Zealand); and Stenopterygiidae (=Longipinnati, with five genera from Jurassic and Cretaceous deposits of Europe and South America). The oldest known ichthyosaurs, the mixosaurids from the Triassic, were somewhat more primitive than subsequent groups but, nevertheless, were highly specialized marine animals. Consequently, their derivation is unknown. They appear most similar to the pelycosaurs, and it is possible that the two groups had a common ancestor. The omphalosaurids were a specialized side branch of mollusk-eaters with mouth adaptations for crushing shells. The shastasaurids were relatively advanced Triassic ichthyosaurs and may have been ancestral to the stenopterygiids. The ichthyosaurids were the typical Jurassic ichthyosaurs, with broad paddlelike appendages; they clearly appear to have descended from the mixosaurids. The stenopterygiids were advanced ichthyosaurs with long and slender paddlelike forelimbs.

Subclass Synapsida

Order Pelycosauria (=Theromorpha) — Primitive Upper Carboniferous and Permian mammal-like reptiles of North America and Europe; small temporal openings; external nares lateral and widely spaced; septomaxilla small; no secondary palate;

vomers separate; quadrate large, fixed in position; articulation of jaw posterior to level of occipital condyle; stapes long; teeth simple, conical, little differentiation of upper canine, no differentiation of lower canine; typically 27 presacral vertebrae; tail long and stout; limbs sprawling like those of cotylosaurs; humerus with screw-shaped articular head. The pelycosaurs were the dominant group of reptiles during the early Permian. Three major groups, designated as suborders, are distinguishable.

Suborder Ophiacodontia — Primitive pelycosaurs with lizardlike proportions; dentition homodont (or nearly so) except for pair of enlarged canine-type teeth on anterior end of each maxilla; snout relatively long; vertebrae primitively short, never sharply keeled nor greatly compressed, transverse processes short; limbs short, feet broad; gastralia well developed. The ophiacodonts were typically long-snouted fish-eaters; they were amphibious and spent considerable periods of time in water. They obviously evolved from the cotylosaurs and shared many characteristics with them. Ophiacodonts, in turn, gave rise to the two relatively terrestrial groups, one herbivorous and the other carnivorous, that make up the remaining

two suborders of Pelycosauria. Two families of ophiacodonts are recognized: Ophiacodontidae (five genera from the Pennsylvanian and Lower Permian of North America, of which *Varanosaurus*, in Figure 3–8, is best known) and Eothyrididae (six genera from the late Upper Carboniferous of Europe and the Lower Permian of North America).

Suborder Sphenacodontoidea — Advanced carnivorous pelycosaurs with moderately to greatly elongated facial regions; teeth sharply pointed, recurved, moderately to highly developed maxillary canines; anterior presacral vertebrae with strongly compressed centra and ventral keels, transverse processes relatively long; limbs relatively elongate, feet relatively long and slender; gastralia not well developed. The sphenacodonts were the dominant carnivores of the early Permian and were ancestral to the order Therapsida. The two families recognized are Varanopsidae, with five genera from the Middle and Upper Permian of South Africa and the Lower Permian of North America, and Sphenacodontidae, with about 12 genera from the Upper Carboniferous, Permian, and Triassic of Europe and the Permian of North America. The most common reptile in Lower Permian deposits of

Figure 5–23   Number *1*, a Triassic ichthyosaur, or "fish-lizard", that swam in seas of California and Nevada. Number *2*, *Ophthalmosaurus*, from Late Cretaceous rocks of western Kansas, resembled Jurassic ichthyosaurs that occurred in European seas. (From *Tales Told By Fossils*, by Carroll Lane Fenton. Copyright, © 1966 by Carroll Lane Fenton. Reprinted by permission of Doubleday & Company, Inc.)

North America is *Dimetrodon* (family Sphenacodontidae), characterized by a well-developed "sail" that was possibly used as a radiator for temperature regulation (Fig. 5–24).

Suborder Edaphosauria—Advanced herbivorous pelycosaurs with low short skulls; teeth generally reduced in number, stout, blunt, bulbous, with little differentiation of maxillary canines; vertebral centra lacking keels, transverse processes moderately long; femur and humerus relatively long, but forelimbs short; gastralia poorly developed. There are four families in this suborder. The family Nitosauridae (about seven genera from the Upper Pennsylvanian and Lower Permian of North America) includes small and primitive edaphosaurians with relatively sharp teeth and some development of the canine series. *Lupeosaurus* (of the family Lupeosauridae) from the Lower Permian of Texas was a large animal whose skull is not known. It may, in fact, not belong to this suborder.

Edaphosaurus (of the family Edaphosauridae) was a marsh-dwelling form distributed over the northern continents during the late Upper Carboniferous and early Permian; like *Dimetrodon*, *Edaphosaurus* had a sail-like structure down its back. Finally, the family Caseidae includes eight genera from Upper Carboniferous and Lower to Middle Permian deposits of North America and eastern Europe.

Order Therapsida—Advanced mammal-like reptiles (Fig. 5–25) descended from sphenacodont pelycosaurs and known primarily from Middle Permian to Middle Triassic deposits of Russia, South Africa, and South America; temporal opening often enlarged and, in advanced forms, extended dorsally to parietal; external nares dorsal and near tip of snout; septomaxilla relatively large; secondary palate present in advanced forms; vomers sometimes fused; quadrat small, supported primarily by squamosal; articulation of

Figure 5–24   Number *1*, the carnivorous fin-backed reptile *Dimetrodon* which reached lengths of 9 to 11 feet. Number 2 is the herbivorous *Edaphosaurus*. (From *Tales Told By Fossils*, by Carroll Lane Fenton. Copyright, © 1966 by Carroll Lane Fenton. Reprinted by permission of Doubleday & Company, Inc.)

Lycaenops, Late Permian

Cynognathus, of
Early Triassic age

Figure 5–25   Two therapsid reptiles, *Lycaenops* of the late Permian of South Africa and *Cynognathus* of the early Triassic of South Africa. (From *Tales Told By Fossils*, by Carroll Lane Fenton. Copyright, © 1966 by Carroll Lane Fenton. Reprinted by permission of Doubleday & Company, Inc.)

jaw opposite or anterior to level of occipital condyle; stapes short; dentition variable, generally distinct canines in both upper and lower jaws; typically 26 presacral vertebrae; tail variable, but slender if long; limbs of advanced quadrupedal pattern with proximal segments parallel to body axis; humerus with oval articular head. The therapsids underwent a great deal of adaptive radiation and, as a result, constitute an extremely diversified grouping which includes all of the carnivorous reptiles of the Upper Permian and a great variety of herbivores.

Suborder Phthinosuchia (=Eotitanosuchia) —Primitive therapsids with skulls resembling those of sphenacodont pelycosaurs but having larger temporal openings, distinct pair of both upper and lower canines, a more anterior jaw articulation. This suborder includes only the family Phthinosuchidae with nine genera from the Middle Permian of Europe and North America. The Phthinosuchidae gave rise to two distinct evolutionary lines, represented by the following suborders.

Suborder Theriodontia — Small to medium-sized carnivorous therapsids descended from Phthinosuchia.

INFRAORDER GORGONOPSIA — Primitive theriodonts with temporal opening bordered above by postorbital and squamosal; postfrontal present; vomers fused; no secondary palate. The 18 families included here are the following: Galesuchidae (six

genera from the Middle and Upper Permian of South Africa); Hipposauridae (=Ictidorhinidae, with five genera from the Middle and Upper Permian of South Africa); Cynariopsidae (*Cynarioides* and *Cynariops* from the Upper Permian of South Africa); Rubidgeidae (eight or nine genera from the Upper Permian of Africa); Gorgonopsidae (*Gorgonops* and *Leptotrachelus* from the Upper Permian of South and East Africa); Scymnognathidae (five genera from the Upper Permian of South and East Africa); Aelurosauridae (*Aelurosaurus* from the Upper Permian of South Africa); Galerhinidae (*Galerhinus* from the Upper Permian of South and East Africa); Gorgonognathidae (three genera from East and South Africa); Arctognathoididae (three genera from the Upper Permian of South Africa); Scylacopsidae (six genera from the Upper Permian of South Africa); Sycosauridae (*Sycosaurus* from the Upper Permian of South Africa); Arctognathidae (*Arctognathus* and *Lycaenodontoides* from the Upper Permian of Europe and South and East Africa); Aelurosauropsidae (*Aelurosauropsis* from the Upper Permian of South Africa); Scylacocephalidae (*Scylacocephalus* from the Upper Permian of South Africa); Broomisauridae (*Broomisaurus* from the Middle to Upper Permian of South Africa); Inostranceviidae (three genera from the Upper Permian of East Europe); and, questionably, Burnetiidae (=Burnetiamorpha, with *Burnetia* and *Styracocephalus* from the Upper and Middle Permian of South Africa). The gorgonopsians were apparently an evolutionary dead end.

INFRAORDER CYNODONTIA—Advanced theriodonts with secondary palates, a gap between the squamosal and postorbital above the temporal opening, no postfrontal. Cynodontia replaced Gorgonopsia as the dominant carnivorous group during the Triassic and comprises the following five families: Procynosuchidae (nine genera from the Upper Permian of Europe and Africa), Thrinaxodontidae (=Galesauridae, with nine genera from the late Permian and Lower Triassic of South Africa and the Lower Triassic of East Asia), Cynognathidae (six genera from the Middle Triassic of South America and the late Lower Triassic of South Africa), Diademodontidae (six genera from the Lower Triassic and Middle Triassic of Africa), and Traversodontidae (14 genera from the late Lower Triassic of Africa, South America, and East Africa).

INFRAORDER TRITYLODONTOIDEA— Dentition rodentlike with enlarged incisors and cheek teeth separated by diastema; skull mammal-like with secondary palate; articular and quadrate bones articulated. The tritylodonts, all in the family Tritylodontidae (with eight genera from the Upper Triassic to Middle Jurassic of East Asia, South Africa, and Europe), were highly specialized therapsids that probably evolved from the cynodonts and were the last survivors of the order Therapsida. They were in many ways intermediate between reptiles and mammals but were too specialized to be directly ancestral to mammals.

INFRAORDER THEROCEPHALIA—Relatively primitive, heavily built carnivores; secondary palate absent; mammal-like digits present. This moderately large grouping represents an early line of theriodonts that was independent of, but paralleled in many ways, the gorgonopsians during the Middle to Upper Permian. It includes the family Pristerognathidae (approximately 26 genera from South Africa and eastern Europe), the family Alopecodontidae (five genera from the Middle Permian of South Africa and the Upper Permian of eastern Asia), the family Trochosauridae (=Lycosuchidae, with four genera from the Middle Permian of South Africa), the family Whaitsiidae (about 16 genera from the Upper Permian of South Africa, East Africa, and eastern Europe), and the family Euchambersiidae (*Euchambersia* from the Upper Permian of South Africa). The therocephalians gave rise to the following infraorder.

INFRAORDER BAURIAMORPHA—Advanced carnivores descended from the Therocephalia; secondary palates present; postorbital arch lost; primitive lower jaw. The bauriamorphs, which may not all represent a natural grouping, paralleled the cynodonts. They possessed many mammalian characteristics and some, such as *Bauria,* may have been close to the direct line of descent leading to mammals; they probably gave rise to the Ictidosauria. The families included here are Lycideopsidae (*Arnognathus* and *Lycideops* from the Upper Permian of South Africa); Ictidosuchidae (*Ictidosuchoides* from the Upper Permian of South and East Africa); Nanictidopsidae (three genera from the Upper Permian of South Africa); Silpholestidae (five genera from the Upper Permian of South and East Africa); Ericiolacertidae (four genera from the Lower Triassic of South Africa and eastern Europe); Bauriidae (six genera from the Lower Triassic of South Africa); and Rubidginidae (three genera from the Upper Permian of South Africa).

INFRAORDER ICTIDOSAURIA—Small, relatively unspecialized carnivores similar to the tritylodonts but with a squamosal-dentary contact supplementing the quadrate-articular joint. This infraorder includes *Diarthrognathus* (family Diarthrognathidae) from the Upper Triassic of South Africa and possibly the family Haramiyidae (=Microcleptidae), with *Haramiya* and *Thomasia* from the Upper Triassic to Lower Jurassic of Europe; the latter are known only by molar teeth and may represent mammals rather than therapsid reptiles. *Diarthrognathus* is of particular interest since it is truly intermediate between reptiles and mammals, and, as Romer (1966) suggests, the Ictidosauria may prove to be the true ancestors of at least some mammals.

Suborder Anomodontia—Small to large therapsids, descended from the Phthinosuchia; incisors rodentlike; progressive changes from carnivore to herbivore specializations; coronoid absent; postorbital and squamosal in contact above temporal opening.

INFRAORDER DINOCEPHALIA — Large-bodied herbivores; skull roof usually thickened; temporal openings dorsal and vertically elongate; squamosal not expanded laterally; limbs massive; scapula lacking acromion; iliac blade not expanded; incisors often modified for crushing, canines generally corresponding in size to adjacent maxillary teeth; anterior maxillary teeth with crushing surface, posterior maxillary teeth generally simple cones. These were the first anomodonts and were common during the Middle Permian in both Russia and South Africa. Two relatively distinct groups of dinocephalians, classified here as superfamilies, existed; neither apparently was ancestral to subsequent anomodonts.

Superfamily Titanosuchoidea — Primitive dinocephalians, primarily carnivorous and generally possessing enlarged canines but chisel-shaped incisors. This superfamily includes the following four families: Brithopodidae (=Titanophoneidae, with eight genera from the Middle Permian of eastern Europe and North America); Estemmenosuchidae (*Estemmenosuchus* and *Molybdopugus* from the Middle Permian of South Africa); Anteosauridae (four genera from Middle Permian deposits of South Africa); and Titanosuchidae (=Jonkeriidae, with three genera from the Middle Permian of South Africa).

Superfamily Tapinocephaloidea — Specialized herbivorous dinocephalians, generally large-bodied and tending to have a massive dome of bone on top of the skull; canines reduced to size of other teeth; teeth all tending to be chisel-shaped. This superfamily includes the family Deuterosauridae (*Deuterosaurus* from the Middle Permian of Europe), the family Tapinocephalidae (22 genera from the Middle Permian of South Africa and *Ulemosaurus* from the Middle Permian of eastern Europe), and questionably, the families Driveriidae (*Driveria* from the Middle Permian of North America) and Mastersoniidae (*Mastersonia* from the Middle Permian of North America).

INFRAORDER VENYUKOVIAMORPHA — Anomodonts without thickened skull roofs; heterodont dentition with large incisors, short stout upper canines, and small irregularly spaced cheek teeth. This infraorder includes only the family Venyukoviidae (*Venyukovia* and one or two other genera from the Middle Permian of Russia, Europe, and North America), which is intermediate between the dinocephalians and the dicynodonts; the Venyukoviidae appear definitely to have been ancestral to the dicynodonts.

INFRAORDER DROMASAURIA — Small-bodied anomodonts with homodont or reduced dentition; slender limbs and digits; long tails. This poorly known grouping includes only the family Galeopsidae (four Middle and Upper Permian genera from South Africa) and appears to have been a side branch in the evolution of anomodonts that gave rise to no other group.

INFRAORDER DICYNODONTIA — Specialized herbivores with beaklike upper and lower jaws; temporal openings broad; squamosal expanded laterally; upper canines generally well developed in at least one sex (males?); scapula with acromion; ilium expanded; digits short. Dicynodontia was nearly cosmopolitan during the Triassic and included a variety of terrestrial and aquatic forms. The family Endothiodontidae includes about 30 genera from the Middle and Upper Permian and the Lower Triassic of South Africa; the family Dicynodontidae includes 21 genera from Middle Permian to Lower Triassic deposits of South Africa, Europe, and eastern Asia; the family Lystrosauridae includes only the aquatic *Lystrosaurus* from the Lower Triassic of Africa and Asia; the family Kannemeyeriidae includes six genera from the Triassic of North and South America, East Asia, Africa, and eastern Europe; the family Stahleckeriidae includes *Dinodontosaurus* and *Stahleckeria* from the Middle Triassic of South America; and the family Shansiodontidae includes *Shansiodon* and *Tetragonias* from the Triassic of East Asia and Africa. The dicynodonts apparently became extinct near the end of the Triassic.

Using the preceding classification as a basis, the following is a tally of the approximate numbers of extinct and living families and genera plus the approximate number of living species for each order of Reptilia.

| Order | Families | | Genera | | Species |
	Extinct	Living	Extinct	Living	Living
Cotylosauria	8	0	57	0	0
Mesosauria	1	0	1	0	0
Chelonia	13	8	147	56	219
Eosuchia	2	0	14	0	0
Squamata					
Lacertilia	12	19	104	360	2839
Ophidia	2	9	37	416	2005
Rhynchocephalia	4	1	26	1	1
Thecodontia	9	0	45	0	0
Crocodilia	14	2	124	8	21
Pterosauria	5	0	24	0	0
Saurischia	28	0	244	0	0
Ornithischia	13	0	111	0	0
Araeoscelidia	5	0	16	0	0
Sauropterygia	11	0	71	0	0
Placodontia	4	0	9	0	0
Ichthyosauria	5	0	20	0	0
Pelycosauria	8	0	45	0	0
Therapsida	55	0	298	0	0

As with Amphibia, these exact numbers for Reptilia are not too meaningful, but they do reflect the morphological diversity within each order, the great variety of extinct reptiles, and the success presently enjoyed by the order Squamata.

References

Bellairs, A. d'A. 1970. The Life of Reptiles. Universe Natural History Series. New York, Universe Books. 590 pp.

Broom, R. 1935. On the structure of the temporal region in lizard skulls. Ann. Transvaal Mus. 18: 13–22.

Carr, A. 1952. Handbook of the Turtles of the United States, Canada, and Baja California. Ithaca, N.Y., Comstock. 542 pp.

Carroll, R. L. 1969. Origin of Reptiles. In: C. Gans, A. d'A. Bellairs, and T. S. Parsons (eds.) 1969, Biology of the Reptilia, Volume I, New York, Academic Press. Pp. 1–44.

———, and D. Baird. 1968. Tuditanus (Eosauravus) and the distinctions between microsaurs and reptiles. Am. Mus. Nov. 2337:1–50.

Colbert, E. H. 1945. The Dinosaur Book. New York, McGraw-Hill. 156 pp.

Goin, C. J., and O. B. Goin. 1962. Introduction to Herpetology. San Francisco, W. H. Freeman. 341 pp.

Greer, A. E. 1970. A subfamilial classification of scincid lizards. Bull. Mus. Comp. Zool. 139(3): 151–184.

Gregory, J. T. 1965. Microsaurs and the origin of captorhinomorph reptiles. Am. Zool. 5:277–286.

Gyi, K. K. 1970. A revision of colubrid snakes of the Subfamily Homalopsinae. Univ. Kansas Publ. Mus. Nat. Hist. 20(2):47–223.

Hoffstetter, R. 1968. A contribution to the classification of snakes. (Review.) Copeia 1968:201–213.

Huene, F. 1952. Revision der Gattung Pleurosaurus

auf Grund neuer und alter Funde. Palaeontographica (A) 101:167–200.

List, J. C. 1966. Comparative osteology of the snake families Typhlopidae and Leptotyphlopidae. Illinois Biol. Monographs 36:1–112.

King, W. and F. G. Thompson. 1968. A review of the American lizards of the genus Xenosaurus Peters. Bull. Florida State Mus. Biol. Sci. 12(2):93–123.

McDowell, S. B. 1967. Osteology of the Typhlopidae and Leptotyphlopidae: A critical review. Copeia 1967:686–692.

McDowell, S. B., and C. M. Bogert. 1954. The systematic position of Lanthanotus and the affinities of the anguinomorphan lizards. Bull. Am. Mus. Nat. Hist. 105:1–142.

Meszoely, C. A. M. 1970. North American fossil anguid lizards. Bull. Mus. Comp. Zool. 139(2): 87–150.

Olson, E. C. 1965. Relationships of Seymouria, Diadectes, and Chelonia. Am. Zool. 5:295–307.

Parrington, F. R. 1958. The problem of the classification of reptiles. J. Linn. Soc. (Zool.) 56:99–115.

Romer, A. S. 1946. The primitive reptile Limnoscelis restudied. Am. J. Science, 244:149–188.

——— 1956. Osteology of the Reptiles. Chicago, Univ. Chicago Press. 772 pp.

——— 1957. Origin of the amniote egg. Scient. Monthly 85:57–63.

——— 1966. Vertebrate Paleontology, 3rd ed. Chicago, Univ. Chicago Press. 468 pp.

——— 1968. Notes and Comments on Vertebrate Paleontology. Chicago, Univ. Chicago Press. 304 pp.

Tihen, J. A. 1960. Comments on the origin of the amniote egg. Evolution 14:528–531.

Underwood, G. 1967. A Contribution to the Classification of Snakes. London, British Museum of Natural History. Publ. 653.

Vaughn, P. P. 1964. Vertebrates from the organ rock shale of the Cutler Group, Permian of Monument Valley and vicinity, Utah and Arizona. J. Paleontol. 38:567–583.

Williston, S. W. 1925. The Osteology of the Reptiles. Cambridge, Harvard Univ. Press. 300 pp.

CHAPTER 6

THE GEOGRAPHICAL DISTRIBUTION OF AMPHIBIANS AND REPTILES

In studying the geographical distribution of any group of organisms, certain relevant facts concerning the Earth's structure, history, and atmospheric features must be taken into consideration. These, while obvious, are very important and sometimes overlooked.

A basic point to keep in mind, especially when considering east-west distances in polar regions, is that the Earth is a sphere. Although lines of latitude are parallel, lines of longitude are not. Lines of longitude converge at the poles, so the distances between them are continually decreasing as one goes north or south from the Equator. Too often we use flat maps which exaggerate polar distances. For example, we become accustomed to seeing Alaska and Greenland pictured about the same distance apart as the Hawaiian Islands and the Azores. In reality, the distance from the Azores to New York is approximately equivalent to that from Greenland to Alaska.

A second fundamental point is that water covers about 71 per cent of the Earth's surface, and the relatively small amount of land is not distributed uniformly. At the present time, there is more than twice as much land in the Northern Hemisphere (north of the Equator) as in the Southern Hemisphere. Similarly, there is approximately twice as much land in the Eastern Hemisphere (Europe, Asia,

Africa, Australia, and associated islands) as in the Western Hemisphere (North, Middle, and South America). The relatively small amount of unevenly distributed land, however, forms an almost continuous system of continents. All of the major land masses except Antarctica are either connected or separated by water studded with islands.

The Earth's crust is dynamically unstable, as is evidenced by the earthquakes, fault movements, volcanic activity, and so forth which occur almost daily somewhere on Earth, and this has had a continuing effect on the local, regional, and global distributions of organisms. The reality of continental drift can no longer be questioned; it appears that during the Triassic, no more than 200 million years ago, all the continents were joined in a single universal continent, called Pangaea (Fig. 6–1). Based on their reconstruction of Pangaea from 1000-fathom isobath measurements, Dietz and Holden (1970) have provided a fascinating description of its breakup and have extrapolated present-day plate movements to indicate how the continents will have drifted in about 60 million years from now. The reader is referred to their article and to other recent publications for evidence and possible mechanisms of continental drift. However, it is pertinent at this point to note that the recent finding of the therapsid

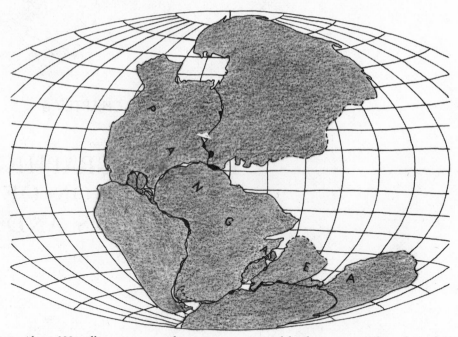

Figure 6-1   About 200 million years ago, the various continental land masses may have formed a universal continent, Pangaea. Panthalassa was the ancestral Pacific Ocean surrounding this continent. The bay separating Africa and Eurasia was the Tethys Sea, the ancestral Mediterranean. Except for India, the relative positions of the continents are based on best fits made by computer, using the 1000-fathom isobath to define continental boundaries. (After Dietz, R. S., and J. C. Holden: The breakup of Pangaea.)

reptile *Lystrosaurus* in Antarctica is truly great evidence that Antarctica was once part of a larger land mass, for this Lower Triassic form was previously known only from southern and eastern Africa and from southern and eastern Asia.

Following the chronological sequence proposed by Dietz and Holden, by the end of the Triassic period, about 180 million years ago, the northern group of continents, known as Laurasia, split away from the southern group, Gondwana, and the latter began to break up (Fig. 6–2). India separated from Gondwana and began drifting north, and the Africa-South America land mass began to separate from the Antarctica-Australia land mass. By the end of the Jurassic period, about 135 million years ago, South America had begun to split away from Africa, so that by the end of the Cretaceous period, about 65 million years ago, the South Atlantic had widened into a major ocean. Australia was still attached to Antarctica, but Madagascar had separated from Africa (Fig. 6–3). Within the past 65 million years the continents have drifted to their present

positions and three important developments have occurred: North and South America were joined together by Middle America, which was formed by volcanism and the arching upward of the Earth's mantle; India collided with the southern edge of Asia, uniting it to that continent and creating the Himalayas; and Australia split away from Antarctica and drifted northward to its present position (Fig. 6–4). At the present time, the distribution of continents is such that there is a large amount of land north of the tropics that is interconnected; within the tropics are large areas of land separated from one another by wide expanses of ocean; and south of the tropics are relatively small land masses that are very widely separated from one another by ocean.

The distribution of all organisms is materially affected by climate, the principal components of which are temperature and precipitation. The temperature of any locality on the Earth's surface is determined primarily by the amount of radiant energy it receives from the sun.

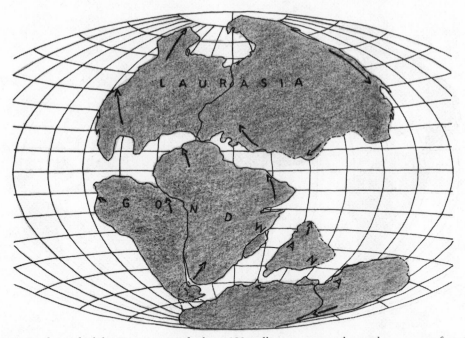

Figure 6–2 By the end of the Triassic period, about 180 million years ago, the northern group of continents, known as Laurasia, split away from the southern group, known as Gondwana, and the latter had started to break up. India had been set free and was drifting north as a result of a Y-shaped rift which also began to separate the Africa-South American land mass from the Antarctica-Australian land mass. (After Dietz, R. S., and J. C. Holden: The breakup of Pangaea.)

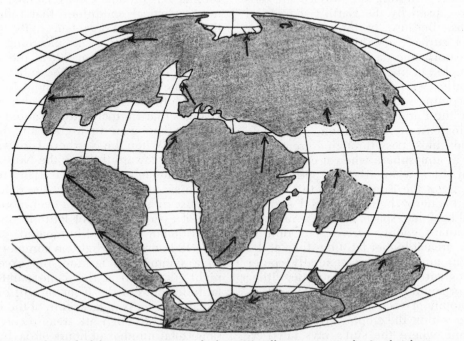

Figure 6–3 By the end of the Cretaceous period, about 65 million years ago, the South Atlantic was a major ocean, Madagascar had split away from Africa, the Mediterranean Sea was recognizable, and a rift developed on the eastern side of Greenland. Australia and Antarctica were still joined together. (After Dietz, R. S., and J. C. Holden: The breakup of Pangaea.)

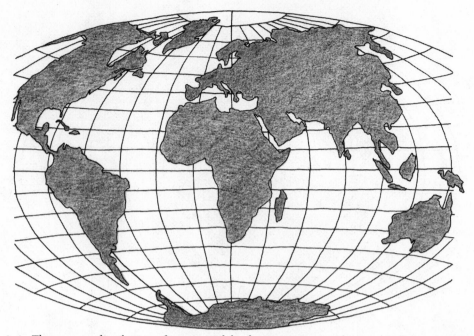

Figure 6–4 The present distribution of continental land masses. During the course of the past 65 million years, nearly half of the oceanic floor was created, India collided with Asia and ended its northward drift, Australia separated from Antarctica, North America separated from the rest of Laurasia, and Central America joined South America to North America.

Geographical variations in the amount of this energy striking the surface are primarily caused by the Earth's shape and motions. Because of the curve in the Earth's surface, the higher the latitude the greater is the slant (the more obtuse is the angle) of incoming radiation. Consequently, with an increase in latitude the incoming solar energy covers a larger area of the Earth's surface, and its dispersion increases. In addition, the sun's radiation must pass through a greater expanse of atmosphere when it comes in at a slant, than when it strikes perpendicularly, as at the Equator, and this results in more energy being scattered and reflected back into space. For these reasons, less radiant energy arrives per unit of surface area at higher latitudes than at the Equator, and there are generally cooler climates with increases in latitude, either north or south of the Equator. The greatest amount of solar radiation strikes the surface within the tropics—by definition, the zone where the sun's rays fall perpendicularly—bounded by 23°27' north (the tropic of Cancer) and south (the tropic of Capricorn).

The tipping of the Earth on its longitudinal axis produces seasonal changes in meteorological conditions that influence the distribution of organisms in its various regions. The angle of incoming solar radiation changes from 90° at the tropic of Capricorn on about December 21 to 90° at the tropic of Cancer about June 21. This shift causes the seasonal climatic changes that are increasingly more noticeable as one goes to higher and higher latitudes. When the Northern Hemisphere is pointed toward the sun, the North Pole is continuously illuminated. This results in the six-month summer at the pole and, because an area extending over more than half of every parallel of latitude north of the Equator will be illuminated, greater than 12 hours of daylight on summer days. In a symmetrical fashion, the excess of night (darkness) at the same parallel of latitude south of the Equator will exactly equal the excess of day north of the Equator. Thus, although all areas receive the same total number of hours of daylight in a year, the poles receive their six months of daylight in one photoperiod, the Equator has 12 hours of daylight alternated with twelve hours of night all year long, and intermediate latitudes have longer days in the summer and shorter days in winter.

Because air expands as it rises and expansion is a cooling process, the average temperature decreases with increasing altitude. Thus, a decrease in temperature comparable to that encountered in moving toward the poles is felt as one makes a trip up a mountain. The vegetation as well as the fauna of a tropical mountain may change from rain forest at the base to tundra or ice at the top if the mountain is sufficiently high. However, the effect of altitude on temperature is different than the effect of latitude on temperature, since altitude does not produce the seasonal variations encountered with changes in latitude.

Precipitation depends upon circulation of air to transport water vapor. The general pattern of atmospheric circulation results from unequal heating of the Earth's surface and rotation of the Earth. Incoming solar radiation is composed of relatively short waves, which are not absorbed readily by the atmospheric gases. Therefore, this energy tends either to be reflected back into space (approximately 54 per cent of the solar constant is reflected back by the Earth's outer atmosphere) or to pass through the atmosphere largely unabsorbed. At the Earth's surface, it is either absorbed or is reflected back to outer space without a change in wavelength. Energy absorbed causes the surface to heat, and it starts to radiate long infrared wavelengths. Infrared radiation is absorbed by water and water vapor in the atmosphere, so the atmosphere is heated from below by radiation from the Earth's surface. As the air next to the surface is heated, it expands, becomes less dense, and rises. The rising air produces a low-pressure zone which draws in air from adjacent regions. Because the maximum concentration of solar energy strikes the surface within the tropical zone, causing that belt of air to be constantly heated from below, the tropics are a permanent low-pressure zone. As the tropical air rises, it is replaced by air drawn in from the subtropics to the north and south. When the rising tropical air reaches high altitudes, it spills away to the north and south, cools, and eventually sinks back to the surface. Thus, a permanent meridional

pattern of circulation exists wherein air rises from the tropics, moves away from the tropical zone at high altitudes, sinks, and moves back into the tropics at the surface. Consistent with the principles of Coriolis force, two additional patterns result: the air leaving the tropics does not travel all the way to the poles of the rotating Earth before sinking, but rather tends to cool and sink at latitudes about 30° north and south, making these latitudes permanent high-pressure zones; and winds veer to the right in the Northern Hemisphere and to the left in the Southern Hemisphere, which accounts for the easterly component of the "trade winds." The northeast and southeast trade winds represent the meridional circulation moving toward the Equator between 30° north and south. The two temperate zones are relatively low-pressure zones and the polar regions are high-pressure zones. However, temperate and subpolar latitudes tend to be dominated by moving air masses which respond to variable pressures and winds.

Precipitation systems significant in the distribution of organisms are produced by the pattern of atmospheric circulation. As the belt of air rises from the tropics, it cools. Since cool air holds less moisture than warm air, some of the water vapor condenses as droplets, forming clouds and eventually rain. Having lost much of its moisture, the air flowing at high altitudes toward the subtropics is dry. As it falls, it warms and picks up moisture from the surface. Therefore, most of the major deserts of the world, in North Africa, southwestern Africa, southwestern Asia, Australia, southwestern North America, and southwestern South America, lie on either side of the tropics in high-pressure zones near 30° latitude. The deserts of eastern Asia (the Gobi, for example) are present owing to great distance from oceans and are exceptions to the rule. Because of variable pressures and winds, precipitation in temperate and subpolar latitudes tends to be seasonal and variable.

Oceans have an effect on both the temperature and precipitation of terrestrial climates and the degree of dispersal possible for certain organisms. The general effect of oceans is to raise humidity and increase precipitation. Because of the thermal properties of water, oceans also tend to moderate any extremes of temperature on adjacent land masses. Warm (the Gulf Stream, for example) and cold (the Humboldt Current, for example) water currents extend warm and cold climatic conditions far beyond their normal limits, thus influencing the occurrence of different life forms in affected regions. In addition, oceans (and land masses) may act as barriers to distribution: for example, terrestrial amphibians and reptiles have trouble crossing major bodies of water, and aquatic amphibians and reptiles have trouble crossing land.

The distribution of vegetation has a significant effect upon the distribution of animals. Vegetation tends to be generally distributed according to climate and, therefore, to be zonal. However, it is locally affected by historical and topographical features.

Time plays an important role in biogeography, and it is to be expected that the distribution of an organism will reflect the age of the phylogenetic group. Very young, newly emerging groups, regardless of the existence of favorable and accessible geographical areas, will simply not have had the time to become widely distributed and, therefore, will tend to be endemic, relatively restricted in distribution. If such a group is flourishing, its geographic range may be expected to expand through time; consequently, the most widely distributed group is expected to be one that has been both flourishing and around long enough to have spread out from its place of origin. Through competition from more advanced forms and through long-term environmental changes, old phylogenetic lines tend to become extinct locally and to survive as small relict populations. Thus, old groups generally have fragmented distributions, the distances between populations giving indications of how widespread the group was formerly or, in some cases, indications of past distributions of continents.

Consideration must be given here to the fact that the Amphibia and Reptilia are ancient groups and are largely extinct. Their fossil records are not complete enough to provide clear historical patterns of distribution; this means that discussions of centers of origin and patterns of dispersal, particularly of orders, families, and genera, tend to be speculative and are frequently based on opinions rather than facts. In the absence of good fossil records, the temptation to fit distributional patterns into preconceived ideas of what might have happened is seldom resisted. It must be continuously borne in mind that geographical ranges fluctuate through time in response to the instability of the Earth's crust, long-term climatic changes, and biological evolution.

Amphibians and reptiles, being ectothermic, are particularly sensitive to their environment, and this, together with their relatively low vagility, narrows their distributional ranges. On the other hand, large geographical ranges tend to encompass varied environmental conditions and to widen and become occupied through adaptive radiation. Thus, there is almost always a correlation among the factors of relative numbers, taxonomic and ecological diversity, and extent and diversity of the geographical area occupied for any phylogenetic grouping.

THE WORLDWIDE DISTRIBUTION OF MODERN AMPHIBIANS

Three features of amphibians are important in analyzing their distributions. First, amphibians have little tolerance for desiccation, although they do vary in this somewhat, so they require habitats providing them with a constant source of fresh water. The variation from species to species in ability to withstand desiccation (see Chapter Seven) makes arid areas absolute barriers to some groups but mere filters to other groups, preventing the expansion of only certain species. A second important feature of amphibians is that they have very little tolerance for salt water: this means that oceans are substantial barriers to dispersal. Some species can withstand limited contact with salt water, however, and some forms are more easily rafted than others. Thus, moderate gaps of salt water also tend to act as filters, particularly for anurans, and the distribution of amphibians indicates that moderate bodies of water have frequently been crossed by some groups. Thirdly, consideration must be given to the probability that amphibians may be introduced into an area under more or less artificial circumstances. Several species, including bullfrogs (*Rana catesbeiana*), edible frogs (*Rana esculenta*), and giant salamanders (*Andrias japonicus*), have been introduced into various areas by man for food. Other species, such as the marine toad (*Bufo marinus*), have been introduced into agricultural areas to control insects. Children everywhere carry frogs about for pets. Tree frogs and some salamanders may be transported in shipments of fruit.

As noted in Chapter Three, the order Anura has been the most successful order of Lissamphibia and exhibits the greatest taxonomic and ecologic diversity. Extant forms make up 16 families, 218 genera, and about 2600 species; they occupy a great variety of habitats, ranging from fully aquatic to amphibious to terrestrial. Some of the terrestrial anurans are adapted to burrowing, some to arboreal existences, and some to surface-dwelling. Living anurans range farther north and south than other amphibians and occur in all climatic zones from tropical to polar and from very wet to arid. Anurans exist on all continents except Antarctica and occur on most continental islands, including the British Isles, Japan, the Riu Kius, Formosa, Ceylon, Sumatra, Java, Borneo, the Philippines, Solomons, Fiji Islands, Madagascar, and the Seychelles. The only amphibian occurring on the island of Newfoundland is an anuran, the green frog (*Rana clamitans*), which was apparently introduced in loads of hay from Nova Scotia. By far the majority of anurans occur in the tropics (approximately 80 per cent of the named species), but there are anurans from within the Arctic Circle (two ranids: the common frog, *Rana temporaria*, in Europe; the wood frog, *Rana sylvatica*, in Alaska) to the southern tip of South Africa, South America and Tierra del Fuego, New Zealand, Australia, and Tasmania.

The order Urodela is more limited in numbers and taxonomic diversity and is more restricted ecologically and geographically than Anura. Living salamanders comprise eight families, 54 genera, and about 316 species. Most adult salamanders live on the surface in terrestrial or aquatic habitats, but a few are arboreal and a few inhabit subterranean water. The Urodela are primitively a group of the North Temperate Zone and are generally confined to wet areas. However, the family Plethodontidae has entered the New World tropics and greatly diversified there. Neotropical plethodontids, in the tribe Bolitoglossini, comprise 132 of 194 species in their family and over 40 per cent of all living salamander species (Wake, 1970). *Hynobius keyserlingii* (family Hynobiidae), in contrast, ranges as far north as Verkhoyansk, Siberia, one of the coldest places in the world and within the Arctic Circle. Despite the extremes in latitude for the order as a whole (from within the Arctic Circle to about South 20°) salamanders are generally much more restricted latitudinally than anurans, and no species other than *Hynobius keyserlingii* is known to even approach the Arctic Circle. Their general distribution indicates that salamanders have difficulty crossing salt water, for they are confined to continental land masses and a few islands, such as the British Isles, Japan, and Formosa, which have only recently become isolated from adjacent continents. Climatically, the more primitive salamanders are adapted to survive within a narrow range of cool temperatures, and very few can survive in arid regions. The tiger salamander complex (*Ambystoma tigrinum*, family Ambystomatidae) is exceptional in its wide distribution throughout arid regions of central and western United States and the Mexican plateau. The more primitive tropical plethodontids live in upland terrestrial habitats where climatic conditions are almost temperate. Lowland tropical plethodontids almost invariably occupy either burrowing or arboreal niches (Wake, 1970). No salamanders occur in the Old World tropics or in the South Temperate Zone.

The order Apoda is the most restricted of living amphibian orders. There are only four families (one according to many authors), 20 genera, and about 150 species of caecilians; these are all aquatic or burrowing as adults. Caecilians are almost entirely confined to the Torrid Zone at the present time, but the fact that they occur in the widely separated tropics of Africa, southern Asia, and the New World indicates that they either were more widely distributed in the past or existed, as an order, before these land masses became so widely separated. Like salamanders, caecilians do occur on some continental islands that have only recently been separated from adjacent mainlands. They appear to have very limited, if any, ability to cross salt water; this is probably directly related to their burrowing and freshwater aquatic modes of life.

The present distribution of land masses is such that the greatest potential for an exchange of Old and New World terrestrial faunas exists in the Northern Hemisphere. Such an exchange of amphibians (and reptiles) is now prevented by the climatic barrier of the Arctic, which keeps groups from approaching even those places where only narrow gaps of water separate the continents. However, this climatic barrier has not always existed, and exchanges of northern amphibians have occurred so that some primarily north-temperate families have distributions overlapping the Old and New Worlds. The family Pelobatidae has long been distributed in both the Old and New Worlds, fossils being known back to the Eocene in both. The greatest diversification of pelobatids is in the Old World, and the suggestion that they are descendents of a discoglossid-like ancestor (Estes, 1970) implies that the family originated there. Although modern subfamilies were represented in North America by the time of the Tertiary (Taylor, 1941, 1942; Zweifel, 1956; Hecht, 1959), there is no fossil evidence that they have ever been more successful or more widely distributed than they are now. This is possibly due to their ecological isolation as fossorial and aquatic animals (Griffiths, 1963). The family Salamandridae, the newts, is widely distributed in Europe, Asia, and North America. It is known back to the Paleocene and possibly the Cretaceous of Europe, whereas the oldest North American fossils are from the Miocene or Oligocene. This, together with the fact the family has its greatest diversification in Europe and Asia, indicates that it originated in the Old World and dispersed from there into North America. Plethodontidae (lungless salamanders) is primarily a North American group, but the family includes two European species of a genus, *Hydromantes*, that is present in California. Living mole salamanders (family Ambystomatidae) occur only in North America, but fossils are known from Europe.

Some amphibian families are primarily tropical and have their greatest diversity in either the Old World or the New World but occur in both because of particularly successful phylogenetic lines. The family Ranidae (true frogs) has its greatest diversity in the tropics of Africa and seems to have originated there, but it is nearly cosmopolitan because of the genus *Rana*. *Rana* is widely distributed in temperate parts of Europe, Asia, North America, Middle America, and the northern half of South America. The presence of ranids in Australia is significant, for it indicates that the family was already dispersed over much of Eurasia by the Cretaceous. Similarly, the occurrence of ranids on Madagascar and their radiation in Africa indicate that the family spread into the Ethiopian region by at least the beginning of the Tertiary (Griffiths, 1963). The family Bufonidae (true toads), which probably originated in South America, is nearly cosmopolitan because the genus *Bufo* has spread through all habitable continents except Australia. Its absence from Australia, which was isolated from Asia in the Cretaceous, indicates that bufonids radiated later than ranids. The family Microhylidae contains seven subfamilies, all of which occur in the Old World tropics, and has its greatest diversification in Madagascar, tropical Asia, and the Indo-Australian Archipelago exclusive of Australia (only six species are present in northern Australia). Microhylinae, with 16 genera, is the only microhylid subfamily to leave the Old World tropics, and it now occurs in tropical and temperate Asia and in New World tropical areas, overlapping into subtropical North America and South America. The presence of microhylids in Australia and on Madagascar suggests that this family is at least as old as the ranids. Tree frogs (family Hylidae) have their greatest center of diversification in Middle America and tropical South America, but the genus *Hyla* is almost cosmopolitan outside of the Australopapuan region and Africa. The only other hylid genera occurring outside of the New World are *Nyctimystes* and *Litoria*, both confined to the Australopapuan area. The presence of hylids in Australia and the divergence of these at the generic level suggest that this family also radiated at least as early as the ranids. *Rana, Bufo,* and *Hyla* are unquestionably the most widespread and successful genera of amphibians at this time.

The only amphibian family which is primarily distributed in the Southern Hemisphere but occurs in both the Old World and the New World is Leptodactylidae. Only reaching as far north as southern United States, leptodactylids are very diversified in both South America and the Australian region and occur as well in South Africa. As noted by Griffiths (1963), Bufonidae and Leptodactylidae have long competed ecologically, and the present distribution of leptodactylids may represent a withdrawal before the more recent and successful bufonids. This conclusion is supported by the fact that primitive and unspecialized leptodactylids are enjoying success in Australia where bufonids are absent. In contrast, recent leptodactylids in the New World have diversified ecologically in the presence of conservatively terrestrial bufonids (Griffiths, 1963).

Three anuran and two salamander families that occur in both the Old World and the New World represent old phylogenetic groups which have few living species and disjunct ranges. Anuran families in this category are Ascaphidae, with *Ascaphus* (tailed frogs) in the Pacific Coast highlands of northwestern United States and *Leiopelma* in New Zealand, Pipidae (aquatic toads), with *Pipa* in South America and three genera in Africa, and Pelodytidae, with the living European *Pelodytes* and the North American fossil *Miopelodytes*. The salamander families are Proteidae, with *Proteus* (the olm) in Europe and *Necturus* (mudpuppies) in eastern United States, and Cryptobranchidae, with *Andrias* (giant salamanders) in China and Japan and *Cryptobranchus* (hellbenders) in eastern United States.

Finally, the apodan families Caeciliidae and Ichthyophiidae occur in both the Old World and New World tropics. Caecilidae may not represent a natural group, but Ichthyophiidae appears to be representative of a single phylogenetic line which occurs throughout the tropics either because of continental drift or because it was more widely distributed in the past and has since become restricted to the tropics.

All other living families of amphibians are endemic to either the Old World or the New World. The family Rhacophoridae occurs in Africa, Madagascar, the East Indies, the Philippines, Japan, and tropical Asia and in these areas fills the niches occupied by hylids elsewhere. Competition between hylids and rhacophorids likely prevents their ranges from overlapping. The family Hyperoliidae is distributed only in Africa south of the Sahara, on Madagascar, and on the Seychelles Islands and is another hylidlike grouping. The rather primitive and presumably old family Discoglossidae occurs only in Europe, Asia, and the Philippines. Time would not seem to have been a factor preventing discoglossids from getting into the New World, but there is no evidence that they have ever been more successful or widely distributed than they are now. The only endemic Old World salamander family is Hynobiidae, a widely distributed group in eastern Asia that is believed to be primitive. It, too, would seem to have had ample time to reach the New World but, despite the fact that it has been geographically in a favorable position to do so, apparently has never made the crossing. However, as mentioned in Chapter Three, Tihen (1958) has suggested that the family Ambystomatidae evolved from Asiatic hynobiids.

The two remaining families of living salamanders are the aquatic Amphiumidae and Sirenidae, both having few species, both restricted ecologically, and both geographically confined to southeastern United States. Similarly, the only families of anurans endemic to the New World are Brachycephalidae, Centrolenidae, Pseudidae, and Rhinophrynidae, all of which are small taxonomically, restricted ecologically, and confined geographically to Mexico and Central and South America.

When the above distributions are analyzed, certain patterns are revealed. One obvious pattern is that the distributions of amphibians are significantly affected by climate and, as a result, tend to be zonal. Apodans are almost entirely tropical, overlapping to a very limited extent into the edges of temperate zones only in eastern Asia and northern South America. Salamanders are primitively north-temperate animals, but the Bolitoglossini plethodontids have greatly proliferated in the New World tropics. Anurans are primarily tropical, but three of the more primitive families (Ascaphidae, Discoglossidae, and Pelobatidae) are almost entirely confined to the temperate zones. Other anuran families are primarily tropical but overlap into the South Temperate Zone and North Temperate Zone. Finally, the south-temperate amphibian fauna is relatively limited compared to the north-temperate fauna and is dominated by the anuran families Leptodactylidae and Hylidae, both of which are primarily tropical.

Another pattern, one that reflects the distribution of land masses, is that there are more families common to both the Old World and the New World in the Northern Hemisphere than in the Southern Hemisphere. However, the overlapping of Leptodactylidae from South America into Australia and Africa is noteworthy. Leptodactylidae is a relatively old group of amphibians and its disjunct distribution may have resulted from continental drift or simply from rafting, the probability of which increases with time.

A third pattern which is well illustrated by amphibian distributions is that reflecting the rise and decline of phylogenetic groups through time. Very old groups (Ascaphidae, for example) tend to have very limited disjunct relict populations. Moderately old groups (Pelobatidae, for example) tend to be more numerous and diversified but are not generally cosmopolitan. Modern advanced groups (*Rana*, for example) that are flourishing are very numerous and diversified and have nearly cosmopolitan distributions.

Finally, the overall pattern of distribution for amphibians is such that there are usually no sharply-defined regional faunas. Rather, there tend to be broad transitional areas with progressive changes in faunal composition; this is because there are very few absolute barriers to the dispersal of amphibians. In Africa and Asia, there are broad areas of overlap between the tropical and north-temperate amphibians. Similarly, Middle America is a transitional area between North American and South American faunas and contains many amphibians from both continents, plus many endemics. In contrast, very little movement across Wallace's Line (the supposed boundary between the Oriental and Australian faunas and running across the Indo-Australian Archi-

pelago south of Mindanao, between Borneo and Celebes, and between Bali and Lombok) and Weber's Line (the line of faunal balance of Oriental and Australian forms) has occurred by amphibians and, consequently, there is only limited overlap of Australian and Asian faunas. The rich hylid fauna in Australia and rhacophorid fauna in Asia almost completely replace one another, as do the subfamilies of microhylids. Bufonids and pelobatids are on the Asian side of the lines, while the leptodactylids are on the Australian side. Only the ranids seem to have successfully done some island-hopping since the Cretaceous isolation of Australia. The aquatic barrier between Africa and Madagascar and that surrounding the Seychelles Islands in the Indian Ocean have been formidable to amphibian movement. Other than water, the most absolute barrier to amphibians at the present time is extreme cold, but because long-range climatic changes occur, such a barrier is apt to only be temporary. Since amphibians have an ability to hibernate under water or in burrows, thereby avoiding winter climatic extremes, summer temperatures are the most important in limiting amphibian dispersal wherever there is refuge available for hibernation. The ability to aestivate in burrows similarly allows some anurans to exist in what would otherwise be prohibitively hot and arid regions. Suitable habitats for aquatic larvae sharply restrict the geographic distribution of many amphibians, and certainly an important factor in the success of the tropical plethodontid salamanders has been their abandonment of the aquatic larval stage and the acquisition of direct terrestrial development (Wake, 1966).

THE WORLDWIDE DISTRIBUTION OF MODERN REPTILES

Most of the general principles influencing the distributions of amphibians also apply to reptiles, but there are significant differences of importance to biogeographers. Like amphibians, reptiles are an ancient and largely extinct group, but their fossil record is better than for amphibians and, therefore, historical patterns of distribution are more evident. Reptiles, unlike amphibians, are truly terrestrial and do not require environments rich in water. Consequently, reptiles are found in a greater variety of terrestrial habitats than amphibians and are not as restricted by arid regions. Because the reptilian skin is relatively impervious to it, salt water forms less of a barrier to the dispersal of reptiles than to the dispersal of amphibians. Reptiles, like amphibians, are ectothermic, and their distributions are also affected by environmental temperatures. There are, however, important differences between amphibians and reptiles regarding the restriction of distributions by temperature. Amphibians, by utilizing cutaneous respiration, are able to avoid cold temperatures by seeking refuge or hibernating under water. Reptiles, generally lacking cutaneous respiration, are unable to do this. Thus, the cold Arctic climatic barrier is even more effective in restricting the dispersal of reptiles than the dispersal of amphibians. Because the amphibian skin must be kept moist and because evaporative rates increase with temperature, amphibians are unable to cope with high environmental temperatures. Reptiles have dry skin, and high temperatures are tolerable, not greatly upsetting their water balance. Consequently, although amphibians have an advantage over reptiles in cold climatic regions, reptiles have the advantage in hot regions.

Man has influenced the abundance and distribution of reptiles even more than he has affected amphibians. Geckos (family Gekkonidae) have been introduced into most of the tropical seaports of the world, as have Old World rats and mice, by shipping, and some species of these lizards, like house mice, occur only around human habitations. The carrying of flower pots from place to place has resulted in the passive introduction of small burrowing worm snakes (*Typhlops*) into many areas of the world. Snakes, lizards, alligators, and turtles are often kept as pets, and the ranges of these reptiles have undoubtedly been affected accordingly. Reptiles, especially turtles and lizards, are often used as food and probably have been introduced into areas as a result. Man has eliminated many of the larger reptiles from some parts of the world either by exploiting them for food or be-cause, as in the case of many snakes, they are considered dangerous. Finally, the hunting of reptiles such as crocodiles and alligators for their skins has greatly reduced their numbers and altered their ranges.

The expected correlation among relative abundance, taxonomic and ecological diversity, and diversity of areas occupied certainly exists for reptilian orders. Living lizards and snakes (order Squamata) comprise 28 families, 776 genera, and about 4800 species; this order is, by far, the most numerous and widely distributed at the present time. Both lizards (suborder Lacertilia) and snakes (suborder Ophidia) are nearly cosmopolitan, and both occur in a great variety of habitats and climatic conditions, ranging from deserts to tropical rain forests. Various extinct lizards, including the mosasaurids, were aquatic; some living species are amphibious, but none is fully aquatic. In contrast, several groups of living snakes are highly aquatic, including the sea snakes (family Hydrophidae). The relative abundance and diversity of snakes and lizards varies from one terrestrial region to another. There are more species of lizards than snakes in Europe, New Guinea, Australia, and probably Africa, but more species of snakes than lizards in southern and eastern Asia (Darlington, 1957). Lizards occupy a greater variety of terrestrial habitats than snakes and are very widely distributed on islands The greater occurrence of lizards than snakes on islands is probably more a reflection of different food requirements than it is of ability to cross water. Lizards tend to be insectivorous and, as insects occur nearly everywhere, can readily find food even in such restricted areas as islands. Snakes, on the other hand, tend to feed on larger prey that may not always be available in some places. Thus, while snakes and lizards may have nearly equal probabilities of crossing water and getting to islands, lizards are more apt to be able to establish and maintain a population once they arrive. Both lizards and snakes are ecologically diversified into burrowing, surface-dwelling, and arboreal species, but there is a tendency for snakes to inhabit densely-vegetated wet areas and lizards drier and open areas. Lizards and snakes are primarily tropical groups, but they have wide latitudinal distributions. In the Old World, both groups range from within the Arctic Circle (*Lacerta vivipara* and *Vipera berus*, both in Europe) to the southern limit of Africa and Tasmania (several species of each in both cases). In the New World, the northernmost reptile is the common garter snake (*Thamnophis sirtalis*), which reaches to about 60° north in Northwest Territories, Canada. Several species of lizards and snakes occur up to a latitude of about 50° north. Several lizards are present in Tierra del Fuego, but the southernmost snake in the New World (*Bothrops ammodytoides*, a pit viper) is found only as far south as Santa Cruz, Argentina.

The second largest living order is Chelonia, with eight families, 56 genera, and 219 species of existing turtles and tortoises. The majority of these are aquatic or amphibious, but the tortoises (in the family Testudinidae) are highly terrestrial; both marine and fresh-water aquatic species exist. Thus, this order is fairly diversified ecologically. It is also widely distributed. Aquatic and terrestrial chelonians occur throughout the tropics, where turtles are most abundant and diversified, and they extend regionally into both the North Temperate and South Temperate Zone. However, no turtle approaches the Arctic Circle. In Europe, *Emys orbicularis* (family Testudinidae) reaches to about 57° north, but other turtles are south of 50° north. Soft-shelled turtles (*Trionyx* in family Trionychidae) extend to about 48° north in eastern Asia, but other Asian turtles are south of 40° north. In North America, the painted turtle (*Chrysemys picta*, family Testudinidae) goes the farthest north, with a range extending from about 51° north in British Columbia to about 48° north along the Atlantic Coast. Terrestrial turtles reach the southern limit of Africa, southern Australia (but not Tasmania), and northern Argentina. Marine turtles are most abundant in tropical seas but extend north and south into subtropical waters in warm currents such as the Gulf Stream, and they appear occasionally on the coast of Tasmania.

The order Crocodilia is much more restricted in diversity, ecology, and distribution than either Chelonia or Squamata. Its living members make up two families, eight genera, and 21 species. They are all amphibious, and their distribution is limited to the tropics and a few warm temperate areas. Crocodilians extend north of the tropics to the Carolinas in eastern United States, to the Yangtze River in China, and into temperate areas of India and Burma. They barely extend south of the tropics in Africa and South America.

The order Rhynchocephalia is, of course, an extremely limited group in all respects. *Sphenodon punctatus* (the tuatara) is the only living species, and it is surviving only on remote islands off the main islands of New Zealand, near the southern limits of reptiles in general The existence of closely related fossils in North America, Europe, eastern Asia, and South Africa clearly establishes the fact that the tuatara is a relict of what was once a very widely distributed, almost cosmopolitan, group. It is now nearing extinction.

The flourishing families of snakes, headed by the Colubridae, have broad distributions in both the Old World and New World. Colubridae, with its 302 genera and over 1400 species, occupies virtually all terrestrial habitats, has spread to all major tropical and temperate land masses, and is abundant everywhere except Australia, where it apparently only recently arrived and is represented by only two or three species. The dominant reptilian group in Australia is the family Elapidae, occurring elsewhere throughout the tropics and subtropics but nowhere else so diversified as in Australia. Worm snakes (family Typhlopidae), aided by man, are found throughout the tropics. Viperidae, Leptotyphlopidae, and Boidae all have wide distributions in both the Old World and New World. The family Aniliidae (burrowing snakes) occurs in both the New World and Asia but is restricted to tropical areas. Sea snakes (family Hydrophidae) are widely distributed in the South Pacific and the Indian Ocean and are particularly abundant around the Australian region and in the southeastern Indian Ocean. The taxonomically very small families Xenopeltidae (sunbeam snakes) and Uropeltidae (rough-tailed snakes) have comparably small geographical distributions in southern Asia.

Lizards belonging to the family Scincidae are distributed worldwide in the tropics and warm temperate regions. Geckos (family Gekkonidae), widely introduced by man, are found in Middle and South America, the West Indies, Africa, southern Asia, the East Indies, Australia, New Zealand, and most Pacific islands. Other flourishing families of lizards, however, tend to occur primarily in either the Old World or the New World (rather than to abound in each) and to have extensive north-south patterns of distribution. In the New World, the family Iguanidae ranges from southern British Columbia and Alberta southward to Tierra del Fuego. This is the largest New World family of lizards, with more than 50 genera, but it does not occur on Old World continents. Significantly, two iguanid genera (*Chalarodon* and *Oplurus*) occur on Madagascar, and one (*Brachylophys*) exists on the Fiji and Tonga Islands. In the Old World, one of the dominant lizard families is Lacertidae, which ranges from the Arctic in Europe to South Africa and covers much of temperate Asia but does not enter the New World. A parallel situation involves the family Agamidae in the Old World and the family Teiidae in the New World: both are particularly abundant and diversified in the tropics yet have extensive longitudinal distributions. The significance of these longitudinal distributions is that the Old and New World lizard faunas differ more sharply than most other groups of animals and certainly more than any amphibian group. It should be noted that the Old World family Agamidae is similar ecologically and morphologically to the New World Iguanidae, and the Old World family Lacertidae is equally similar to the New World family Teiidae. These similarities may be the result of convergent evolution or each pair of families may represent close relationships. If the latter is true, the four families represent two phylogenetic lines, both of which are worldwide in distribution.

The highly specialized worm lizards (family Amphisbaenidae) are climatically restricted to tropics and subtropics but do occur in both the Old World and the New World. The family Anguidae is primarily a New World tropical group but includes *Ophisaurus* (the glass snakes), which occurs in both the Old World and the New World, plus *Anguis* (the slowworm), which occurs only in the Old World. The very old family Varanidae is declining and, although once worldwide in distribution, is now restricted to the Old World tropics and representation by one living genus, *Varanus*. The remaining families of lizards are restricted taxonomically and geographically: Xantusiidae, Anniellidae, Xenosauridae, and Helodermatidae are confined to limited areas of the New World; Pygopodidae, Dibamidae, Cordylidae, and Lanthanotidae are equally restricted in the Old World.

Turtles are a very old, persistent, and conservative group, and only three families have living members restricted to either the Old World or the New World. The two families presently limited to the New World are Dermatemydidae and Chelydridae. Dermatemydidae is nearly extinct and includes but one living species (*Dermatemys mawi*) from southeastern Mexico, Guatemala, and Honduras. Twenty extinct genera of dermatemydids have been found in North America, Europe, and Asia. Chelydridae (snapping turtles) is also a very old family but is still widespread in North, Middle, and South America. All fossil chelydrids are from the New World. The only family of turtles currently limited to the Old World is Carettochelyidae, and this, too, is a very old group that is nearing extinction. Carettochelyidae includes only one living species (*Carettochelys insculpta*), in New Guinea and northern Australia, but six extinct genera are known from Europe, Asia, and North America. The remaining four families of freshwater and terrestrial turtles (Pelomedusidae, Chelydae, Testudinidae, and Trionychidae) occur in both the Old World and the New World. Testudinidae is probably the most successful of living families and is cosmopolitan except for Australia. It is also the most diversified taxonomically and ecologically, the 33 genera including highly aquatic species, amphibious species, and truly terrestrial land tortoises. Again, as in other widely distributed and successful groups, the broad range of this family is a reflection of its adaptability to a variety of habitats. The pleurodirans (Pelomedusidae and Chelydae) are very old families that have survived only in the Southern Hemisphere. Pelomedusidae was formerly cosmopolitan but is now limited to Africa and South America. Chelydae, also primitive and presumably very old, is known only from South America, New Guinea, and Australia. Because the southern continents, especially South America and Australia, have a history of isolation and because turtles do not rapidly disperse when opportunities exist, the survival of the pleurodirans in the southern hemisphere is undoubtedly due to a lack of competition from more progressive forms which spread over the Northern Hemisphere. The highly aquatic soft-shelled turtles (family Trionychidae) date back to the Upper Cretaceous but include no known extinct genera. The family is now distributed over North America, Africa, southern Asia, and the Indo-Australian Archipelago and was formerly even more widely distributed as indicated by fossils found in Europe. The distribution of marine turtles (family Cheloniidae and family Dermochelyidae) is essentially limited only to the presence of warm seas; both families are widely distributed.

The family Crocodylidae (alligators and crocodiles) is another very old grouping which, since the Tertiary, has experienced continual reduction in diversity and numbers. These strictly amphibious animals live in fresh-water lakes, swamps, and rivers and occasionally in brackish and salt water and, of course, have distributions limited to the presence of these aquatic habitats. Living crocodiles (subfamily Crocodylinae) exist in both the Old World and New World tropical and subtropical areas. They are found in Africa, Madagascar, southern and southeastern Asia, the Indo-Australian Archipelago, northern Australia, and tropical and subtropical America. The alligators (subfamily Alligatorinae) have a disjunct distribution at the present time: *Alligator*, a warm-temperate genus and the only truly non-tropical group of crocodilians, occurs in southeastern United States and China (one species in each place); *Caiman* occurs only in Middle and South America. Fossils assigned to Alligatorinae are also known from Europe. The family Gavialidae was formerly distributed over Europe, Africa, southern and southeastern Asia, and South America, but it is now restricted to a single living species (*Gavialis gangeticus*) and a range that has been sharply reduced to tropical India and Burma.

As with amphibians, certain genera of reptiles have experienced great success and are widely distributed in both the Old World and the New World. Among snake genera, water snakes in the genus *Natrix* (as restricted by Malnate, 1960) occur in Southeast Asia from China through the Malay region to Java, in North America from southern Canada to northern Cuba and southern Mexico, in much of Europe and east to Lake Baikal, Sinkiang and Afganistan, and in western and northern Africa. *Elaphe* (rat snakes) and *Coluber* (racers) occur throughout inhabitable portions of the Northern Hemisphere and into the tropics. All three of these genera are in the family Colubridae. Of the viperid genera, *Agkistrodon* (copperheads and cottonmouths) and *Bothrops* (also pit vipers) occur in Asia and North and Middle America. Among burrowing snakes, *Leptotyphlops* (blind snakes, family Leptotyphlopidae) occurs in North and South America and also Africa, while *Typhlops* (worm snakes, family Typhlopidae) is cosmopolitan in the tropics and warm temperate areas. *Phyllodactylus* (leaf-toed geckos), certainly one of the most widely distributed gekkonid lizard genera, has a disjunct distribution which includes Madagascar, tropical Africa, tropical Asia, Australia, North, Middle, and South America. *Mabuya* is the only genus of skinks (family Scincidae) to occur in South America, and it is also found in Africa, Madagascar, southern Asia, Middle America, and the West Indies. *Eumeces* and *Lygosoma* are also large genera of skinks and occur in both the Old World and the New World, but neither has radiated into South America. (It should be noted that *Lygosoma* is a catchall for many valid genera, but no two authorities currently agree on how many.) As mentioned, the anguid genus *Ophisaurus* occurs in both the Old World and the New World. The turtle genus *Clemmys* (family Testudinidae) is known from Europe, Asia, North Africa, and North America and is the only turtle genus extending completely around the North Temperate Zone. *Testudo*, a genus of tortoises in the family Testudinidae, is distributed in Africa, Madagascar, tropical and subtropical Europe, Asia, and Middle America. Soft-shelled turtles in the genus *Trionyx* (family Trionychidae) are widely distributed in southern and southeastern

Asia, tropical Africa, and eastern North America. Finally, the pelomedusid (side-necked turtles) genus *Podocnemis* occurs in both South America and Madagascar but is declining, since as recently as the Pleistocene it also existed in Africa.

Analysis of reptilian patterns of distribution shows that all living groups except Rhynchocephalia, isolated in the South Temperate Zone, are primarily tropical, and there are no well-defined north-temperate or south-temperate reptilian faunas. Crocodilians barely extend beyond the tropics. The widely distributed members of Squamata and Chelonia belong to families which are primarily tropical, indicating that tropical and temperate climatic differences do not form barriers to the dispersal of the latter two orders. This conclusion is supported by the fact that some families, lizards especially, have north-south patterns of distribution that extend the length of continents and across climatic zones. The greatest barrier to the dispersal of reptiles, however, is cold, and the latitudinal limits of reptiles do not equal those for amphibians in the Northern Hemisphere. Cold also effectively limits the altitudinal distribution of reptiles more than that of amphibians. Arid regions are barriers only to amphibious and aquatic reptiles and act as filters for others. Xeric species tend to be filtered by wet areas. Many reptiles are capable of crossing moderate stretches of salt water and, obviously, nothing has stopped the spread of the currently dominant family of reptiles, Colubridae.

The present restriction of reptiles by cold results in a hiatus between the North American and Asian reptilian faunas, but the number of families overlapping these continents gives evidence that exchanges took place between the two faunas in the past. In addition, there is a broad overlap between the North American reptilian fauna and that in South America. For example, the genus *Cnemidophorus* belongs to the family Teiidae, primarily a South American group of lizards, but extends from South America to Oregon and Wisconsin. Similarly, *Pseudemys*, a genus of aquatic turtles in the family Testudinidae, ranges from the United States to northern Argentina. The Australian reptilian fauna and that in Southeast Asia overlap in the Indo-Australian Archipelago, but in this case most of the dispersal has been in one direction—from the Orient toward Australia.

THE DISTRIBUTION OF EXISTING NORTH AMERICAN AMPHIBIANS AND REPTILES

Climate has been and continues to be a major factor influencing the distribution of North American amphibians and reptiles. During the Tertiary period, when it appears that much geographical radiation of modern amphibians and reptiles took place, climatic conditions were much different than they are now. Although there was a gradual decrease in temperature throughout the Tertiary, in general the climate of North America was both warmer and moister than it is today. Evidence of this lies in the presence of early Tertiary palms northward to a latitude of 62° north in Alaska while, at the same time, crocodilians infested waterways as far north as New Jersey (Schwarzbach, 1963). Using Mollusca as indicators of temperature conditions, Durham (1950) deduced that in the early Tertiary the 20° centigrade February isotherm on the west coast of North America ran north of the 49th parallel, whereas now it lies at a latitude of 24° north. Such favorable climatic conditions allowed most existing amphibians and reptiles to disperse over wide areas, and they were probably limited primarily by physical rather than climatic barriers. The Tertiary climatic conditions also promoted exchanges of faunas between Asia and North America during the several times that the continents were connected by the Bering land bridge.

In contrast to the favorable climate which prevailed over much of North America for Tertiary amphibians and reptiles, the climate of North America is now much cooler and tends to be zoned latitudinally. The cold moist belt in northern Canada is a barrier to amphibians and reptiles, and they tend to be zoned in a north-south direction. There is a general decrease in both number and kinds of amphibians and reptiles as one moves from southern North America (southern Mexico) northward. In addition, there are east-west zones in some groups of amphibians and reptiles because of topographical changes and associated local climatic conditions. In the following table are given the approximate numbers of species of amphibians and reptiles in different regions of North America to illustrate the changes occurring with latitude and topographical differences. "Southern Mexico" includes the Mexican states of Campeche, Chiapas, Oaxaca, Quintana Roo, Tabasco, Veracruz, and Yucatan and generally represents the area adjacent to and southeast of the Isthmus of Tehuantepec. "Eastern U.S." refers to the area covered by Conant (1958) falling within the United States and generally includes all the area east of a line from central North Dakota through central Texas. "Rocky Mountains" includes the central and western portions of Montana, Wyoming, Colorado, and New Mexico and adjacent portions of Idaho, Utah, and Arizona. "Western U.S." refers to the area covered by Stebbins (1966) falling within the United States and generally includes Montana, Wyoming, Colorado, New Mexico, and the states to the west of these; it does not include Alaska.

The amphibian and reptilian faunas of North America have also been influenced to a great degree by intercontinental exchanges. These exchanges have been primarily by way of land bridges. Complete land bridges across the Bering Strait, established several times during the Tertiary, existed until the Pleistocene and allowed relatively large numbers of cold-adapted northern amphibians and reptiles to invade North America from Asia; movement in the reverse direction also occurred. Another important land bridge affecting North American amphibian and

reptilian faunas is Central America. North and South America were separated until late in the Tertiary when the Central American land bridge was formed. Thus, the Bering Strait was most important in affecting the North American amphibian and reptilian faunas during the Tertiary, before they were kept south of it by the cold climatic barrier. The Central American land bridge has been most important in Recent time.

The fossil record suggests that, down to species groups, all existing North American phylogenetic lines of amphibians and reptiles were on this continent prior to the end of the Tertiary. Some present-day distributional patterns, particularly in unglaciated areas, have undoubtedly existed since the Tertiary. However, many current distributions in North America of species and subspecies can be attributed to Pleistocene and Recent climatic changes and have their origins in four geographical areas: extreme southeastern United States, southwestern United States and the central plateau of northern Mexico, the Mexican tropics, and the Appalachian Mountains.

Although much of Florida was under water in the Pleistocene, the southeastern United States has existed as a center for warmth- and mesic-adapted amphibians and reptiles and is a major center of distribution for North American Hylidae, Ranidae, and Scincidae. It is the primary center for the ranids and skinks and the secondary center for the tree frogs.

The southwestern region of the United States and, more significantly, the Mexican plateau to the south have probably constituted an arid area since the late Tertiary. This area has been a major cen-

APPROXIMATE NUMBERS OF AMPHIBIAN AND REPTILIAN SPECIES IN REGIONS OF NORTH AMERICA

	Southern Mexico	Florida	Eastern U. S.	Rocky Mountains	Western U. S.	Canada	Alaska
Urodela	64	19	47	4	26	18	2
Anura	161	26	50	19	34	23	3
Apoda	2+	0	0	0	0	0	0
Ophidia	163	41	74	21	70	22	0
Lacertilia	112	19	42	16	56	5	0
Chelonia	22	19	34	4	11	13	0
Crocodilia	3	2	2	0	0	0	0

ter of distribution for North American warmth- and desert-adapted amphibians and reptiles in the following families: Pelobatidae, Bufonidae, Iguanidae, Teiidae (racerunners, in the genus *Cnemidophorus*), and Viperidae (rattlesnakes, in the genus *Crotalus*).

The Mexican tropics have been especially important as a center of distribution for tropical amphibians that have entered North America. They are a major center for Hylidae and a minor center for the family Leptodactylidae (which has undergone secondary radiation at the specific level in *Eleutherodactylus* in this part of North America) and the subfamily Microhylinae of the family Microhylidae.

The Appalachian region is the major center of distribution for North American salamanders, especially the family Plethodontidae.

The centers in southeastern United States, southwestern United States, and the Appalachians developed as a result of climatic changes brought about by Pleistocene glaciation. During the peak of glaciation, the glacial advances were accompanied by climatic changes which extended far south of the limits of glaciers (see Wright and Frey, 1965, for various kinds of evidence); a latitudinal displacement of both cold- and warmth-adapted amphibians and reptiles accompanied the southward shift of cold climatic conditions. The effect of these climatic changes and range shifts on reptilian distributions was different than for amphibians, and, among the latter, urodeles responded differently than anurans.

As a group, salamanders are north-temperate, adapted to cool moist conditions, and historically have been associated with the old Arcto-Tertiary forest. When this forest shifted its position in response to climatic changes, the salamanders moved with it. Today, North American salamanders have their major center of distribution in the Appalachian Mountains and a secondary center in the Pacific Northwest. In both of these areas remnants of the Arcto-Tertiary forest remain, and the intervening area has been all but wiped clean of salamanders. The tiger salamander (*Ambystoma tigrinum* complex) is the only widely distributed and abundant salamander in the Great Plains region. Because they were a boreal group,

the primary effect of Pleistocene glacial build-up and associated climatic changes was to push the salamanders southward along with the Arcto-Tertiary forest. When the glaciers retreated northward and climatic conditions warmed, salamanders also retreated northward. Thus, although North American salamanders are rather sharply divided into West Coast and East Coast groupings, few members of either group have east-west disjunct distributions but many have north-south disjunct distributions, especially in the eastern group. The salamanders with east-west disjunct distributions have these patterns because, with the retreat of glaciers and cool moist climatic conditions, the Arcto-Tertiary forest was eliminated from the intervening environment. Accordingly, in the West, *Dicamptodon ensatus* (Pacific giant salamander), *Taricha granulosa* (rough-skinned newt), and *Plethodon vandykei* (Van Dyke's salamander) have relict populations in northern Idaho, with the remainder of their ranges in western Washington and Oregon (Fig. 6–5, 6–6, and 6–7). In the East, *Gyrinophilus danielsi* (mountain spring salamander) and *Plethodon richmondi richmondi* (ravine salamander), for example, have relict populations to the west of their main ranges (Fig. 6–8 and 6–9). The dwarf sal-

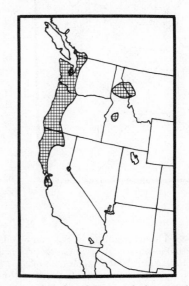

Figure 6–5 The distribution of the Pacific giant salamander, *Dicamptodon ensatus*. (After Stebbins, Robert C.: *A Field Guide to Western Reptiles and Amphibians*, 1966. Reprinted by permission of the publisher, Houghton Mifflin Company.)

Figure 6–6   The distribution of the rough-skinned newt, *Taricha granulosa*. (After Stebbins, Robert C.: *A Field Guide to Western Reptiles and Amphibians*, 1966. Reprinted by permission of the publisher, Houghton Mifflin Company.)

Figure 6–7   The distribution of Van Dyke's salamander, *Plethodon vandykei*. (After Stebbins, Robert C.: *A Field Guide to Western Reptiles and Amphibians*, 1966. Reprinted by permission of the publisher, Houghton Mifflin Company.)

Figure 6–8   The distribution of the mountain spring salamander, *Gyrinophilus danielsi*. (Conant, R.: *A Field Guide to Reptiles and Amphibians of the United States and Canada East of the 100th Meridian*, 1958. Reprinted by permission of the publisher, Houghton Mifflin Company.)

Figure 6–9   The distribution of the ravine salamander, *Plethodon richmondi*. (Conant, R.: *A Field Guide to Reptiles and Amphibians of the United States and Canada East of the 100th Meridian*, 1958. Reprinted by permission of the publisher, Houghton Mifflin Company.)

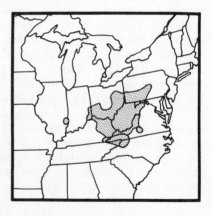

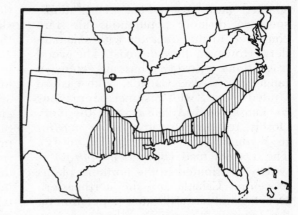

Figure 6-10 The distribution of the dwarf salamander, *Eurycea quadridigitata*. (Conant, R.: *A Field Guide to Reptiles and Amphibians of the United States and Canada East of the 100th Meridian*, 1958. Reprinted by permission of the publisher, Houghton Mifflin Company.)

amander (*Eurycea quadridigitata*) and the mole salamander (*Ambystoma talpoideum*) are unusual in having northern disjunct populations (Fig. 6–10 and 6–11), which are attributed to a northward shift of these southern warmth-adapted salamanders during the post-Wisconsin (Postglacial) warm moist "climatic optimum" (Smith, 1957). The Mexican tropics, together with the area from Costa Rica to northern Argentina, now constitute an important center for the evolution and dispersal of the tropical Bolitoglossini plethodontid salamanders.

Again, the majority of anurans were affected differently than salamanders by Pleistocene glaciation and subsequent climatic changes. This is because, as Blair (1958) has expressed it, "urodeles. . . moved with the invading environment that fragmented anuran ranges." In contrast to salamanders, anurans are primarily a tropical group, adapted to warm conditions, and have centers of distribution (southeastern United States, the Southwest, and the Mexican tropics) which are at more southern latitudes than those of salamanders. During the Pleistocene glacial periods, as the glaciers and cold climatic conditions were shifted southward, the ranges of many of the more southern warmth-adapted amphibians and reptiles were displaced toward the Gulf Coast and then became fragmented as the populations retreated southward toward Florida and Mexico. This fragmentation of ranges is evidenced today by the Floridian fauna, which includes such species as Woodhouse's toad (*Bufo woodhousei*) and the eastern narrow-mouthed toad (*Gastrophryne carolinensis*), fence lizard (*Sceloporus undulatus*), six-lined racerunner (*Cnemidophorus sexlineatus*) and eastern diamondback rattlesnake (*Crotalus adamanteus*); all of these belong to groups with centers of distribution in the Southwest (Blair, 1958).

Because relatively few anurans are cool-

Figure 6-11 The distribution of the mole salamander, *Ambystoma talpoideum*. (Conant, R.: *A Field Guide to Reptiles and Amphibians of the United States and Canada East of the 100th Meridian*, 1958. Reprinted by permission of the publisher, Houghton Mifflin Company.)

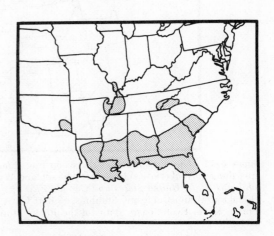

adapted and have boreal distributions, southern disjunct populations are rare compared to the situation with salamanders. However, they do exist. The wood frog (*Rana sylvatica*) has an extensive range in Alaska, Canada, and eastern United States, with disjunct populations in Kansas and Arkansas and the very closely related derivative *Rana maslini* in Wyoming and Colorado (Fig. 6–12). The Dakota toad (*Bufo hemiophrys*) is widely distributed in the northern plains of central Canada and the northeastern Dakotas and has a disjunct population on the Laramie plains in southeastern Wyoming (Fig. 6–13). Both of these cases certainly seem to reflect a southward displacement during the Wisconsin glacial stage and subsequent return northward of the main bodies, the relict populations being left behind where local conditions allowed them to survive. The eastern narrow-mouthed toad, a warmth-adapted species, has a northern disjunct population which is attributable to a northward shift of the southern population during the post-Wisconsin "climatic optimum."

The more general pattern for anurans involves the existence of east-west disjunct populations and pairs of warmth-adapted species. Such a pattern is especially prevalent in the eastern United States and is attributed to a splitting of previous ranges by the invasion of the southward-shifting cool climate during the Pleistocene (Glacial) stages. In some instances the populations involved have remained separated because of an uninhabitable intervening environment, whereas in other cases the isolated populations diverged as a result of evolution, becoming adapted to different habitats. Examples of the former include the eastern spadefoot toad (*Scaphiopus holbrooki*), which was in Florida in the Miocene; at present the subspecies *Scaphiopus holbrooki holbrooki* and *Scaphiopus holbrooki hurteri* are separated

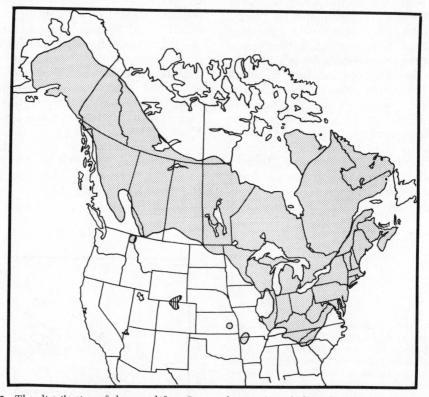

Figure 6–12 The distribution of the wood frog *Rana sylvatica* (stippled), and of the Rocky Mountain wood frog *Rana maslini* (vertical lines, in Colorado and Wyoming). (Conant, R.: *A Field Guide to Reptiles and Amphibians of the United States and Canada East of the 100th Meridian*, 1958, by permission of the publisher, Houghton Mifflin Company; Stebbins, Robert C.: *A Field Guide to Western Reptiles and Amphibians;* and Porter, K. R.: Evolutionary status of the Rocky Mountain population of Wood Frogs, Evolution 23:163–170, 1969.)

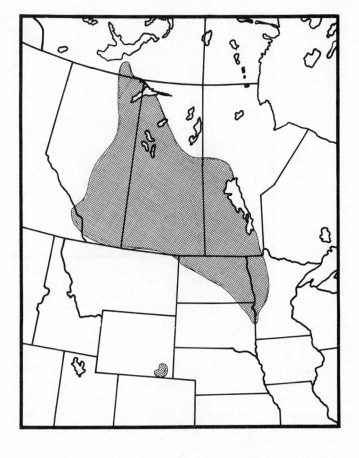

Figure 6–13 The distribution of the Dakota toad, *Bufo hemiophrys*. (From Porter, K. R.: Evolutionary status of a relict population of *Bufo hemiophrys* Cope, Evolution 22:583–594, 1968.)

by a wide alluvial soil corridor straddling the Mississippi River that is uninhabitable by either. The species pair of chorus frogs *Pseudacris ornata* and *Pseudacris streckeri* have a distributional pattern similar to that of the eastern spadefoots but with an even wider gap separating the two (Fig. 6–14). An example of a species pair in which evolutionary divergence has occurred involves the eastern narrow-mouthed toad (*Gastrophryne carolinensis*) and the Great Plains narrow-mouthed toad (*Gastrophryne olivacea*). The former microhylid is adapted to forests and the latter to grasslands, so their ranges are essentially allopatric with a narrow overlap zone where the two habitats meet (Fig. 6–15).

Disjunct anuran distributions in the Southwest generally involve either north-

Figure 6–14 The distribution of Strecker's chorus frog, *Pseudacris streckeri* (horizontal lines), and of the ornate chorus frog, *Pseudacris ornata* (stippled). (Conant, R.: *A Field Guide to Reptiles and Amphibians of the United States and Canada East of the 100th Meridian*, 1958. Reprinted by permission of the publisher, Houghton Mifflin Company.)

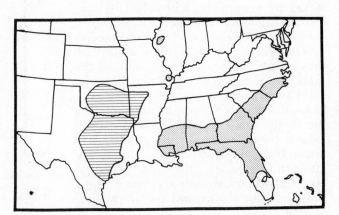

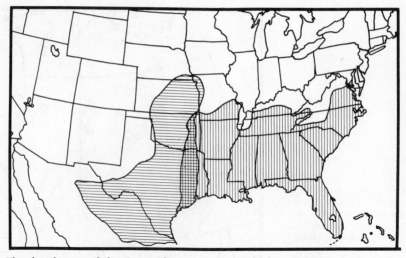

Figure 6–15 The distribution of the Great Plains narrow-mouthed toad *Gastrophryne olivacea* (horizontal lines), and of the eastern narrow-mouthed toad *Gastrophryne carolinensis* (vertical lines). (Conant, R.: *A Field Guide to Reptiles and Amphibians of the United States and Canada East of the 100th Meridian,* 1958. Reprinted by permission of the publisher, Houghton Mifflin Company.)

ern populations of warmth-adapted species made relicts through range changes during the post-Wisconsin "climatic optimum" or relict populations of species that were widely distributed during the water-rich Wisconsin pluvial time and now survive essentially because desert water is still available locally. Examples of distributions established due to the former situation include that of the barking frog (*Hylactophryne augusti*, family Leptodactylidae) and the Arizona tree frog (*Hyla eximia*). The distribution of the southwestern toad (*Bufo microscaphus*) exemplifies the latter pattern where it occurs across Arizona and New Mexico (Fig. 6–16).

Finally, in the Far West, the distributions of amphibians are significantly affected by topography and associated climatic variation. The general effect of mountains is to allow cool-adapted species to extend their ranges farther south than they would otherwise. In addition, the presence of the Sierra Nevada, Cascade, and Pacific Coast ranges results in a longitudinal pattern of precipitation paralleling the Pacific Coast. On the leeward eastern side of these mountains are parallel arid areas. Thus, the distributions of amphibians tend to parallel the coast, with the cool-moist-adapted forms ranging along the mountain chains and the warm-arid-

adapted forms occurring in the lower elevations to the east of the mountains. There are several instances of species pairs with distributions paralleling one another in this manner. In the Northwest, the red-legged frog (*Rana aurora*) occurs in the lowlands to the west of the Cascades,

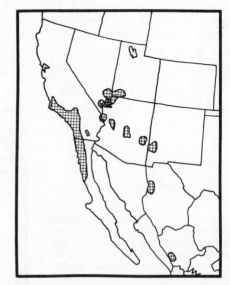

Figure 6–16 The distribution of the southwestern toad, *Bufo microscaphus*. (After Stebbins, Robert C.: *A Field Guide to Western Reptiles and Amphibians,* 1966. Reprinted by permission of the publisher, Houghton Mifflin Company.)

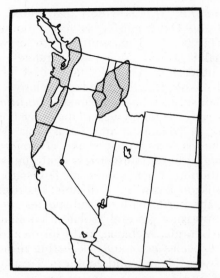

Figure 6–17 The distribution of the tailed frog, *Ascaphus truei*. (After Stebbins, Robert C.: *A Field Guide to Western Reptiles and Amphibians*, 1966. Reprinted by permission of the publisher, Houghton Mifflin Company.)

while the Cascade frog (*Rana cascadae*) occurs parallel to the red-legged frog but in the higher elevations of the mountains. In California, the western toad (*Bufo boreas*) has a range which encircles that of the Yosemite toad (*Bufo canorus*) in the Sierra Nevada because the range of the latter species is consistently higher, there only being one known sympatric area (Karlstrom, 1962). The tailed frog (*Ascaphus truei*), like the urodeles, has been associated with the Arcto-Tertiary forest and has a disjunct distribution resembling those of several salamanders (*Dicamptodon ensatus, Taricha granulosa, Plethodon vandykei*): populations of tailed frogs in the coastal ranges and in the Rocky Mountains of Idaho and Montana are separated by the plateau of eastern Washington and Oregon (see Figure 6–5, 6–6, 6–7, and 6–17).

Like those of amphibians, the distributions of modern North American reptilian species have been shaped by the fluctuating environment of the Pleistocene. North-south displacements occurred with the shifting climates accompanying the different glacial ages and the three Interglacial stages of this epoch; these resulted in northern disjunct populations of a few southern reptiles, including two colubrids, the narrow-headed

garter snake *Thamnophis rufipunctatus* and the sharp-tailed snake *Contia tenuis* (Figs. 6–18 and 6–19), and several southern disjunct populations of more northern species. Among the latter are the painted turtle *Chrysemys picta*, the sagebrush lizard *Sceloporus graciosus*, and the colubrid smooth green snake *Opheodrys vernalis* (Figs. 6–20, 6–21, and 6–22). However, the proportion of reptiles with disjunct distributions is small compared to that of amphibians.

Based on fossil evidence, Auffenberg and Milstead (1965) have concluded that cold climatic conditions shifting north and south during the Pleistocene primarily affected the northern range of reptiles as a whole and that this effect was essentially confined to the periglacial zones and involved relatively few species. This conclusion is supported by the fact of the relatively small number of disjunct distributions, but the greater dispersal abilities and relative independence of water of reptiles compared to amphibians would, at least theoretically, reduce the probability that distributions of reptiles would remain disjunct for long. Blair (1958), in arguing that climatic effects extended well south of periglacial zones and affected southern warmth-adapted species as well as northern cool-adapted species, has cited a number of east-west distributional

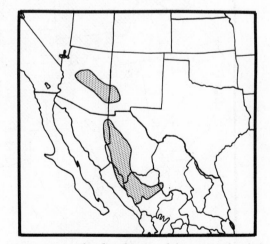

Figure 6–18 The distribution of the narrow-headed garter snake, *Thamnophis rufipunctatus*. (After Stebbins, Robert C.: *A Field Guide to Western Reptiles and Amphibians*, 1966. Reprinted by permission of the publisher, Houghton Mifflin Company.)

Figure 6-19 The distribution of the sharp-tailed snake, *Contia tenuis*. (After Stebbins, Robert C.: *A Field Guide to Western Reptiles and Amphibians*, 1966. Reprinted by permission of the publisher, Houghton Mifflin Company.)

patterns of southern reptiles supporting this thesis. The diamondback rattlesnake *Crotalus adamanteus* occurs east of the Mississippi, and the western diamondback

Crotalus atrox is widely distributed west of the Oklahoma-Texas forest boundary (Fig. 6–23). The southeastern crowned snake (*Tantilla coronata*) ranges over southeastern United States east of the Mississippi, and its close relative, the Plains black-headed snake (*Tantilla nigriceps*), is found west of the forest boundary in Oklahoma and Texas (Fig. 6–24). The eastern indigo snake (*Drymarchon corais*) has a more or less continuous distribution from southern Texas to the Amazon Basin but a disjunct distribution in the north, with the subspecies *couperi* comprising several isolated populations in Georgia, Alabama, and Florida and the subspecies *erebennus* occurring in northeastern Mexico and southern Texas (Fig. 6–25).

As with many similar situations involving amphibians, the preceding examples are allopatric patterns in which one or both populations are limited by the Mississippi embayment and also generally involve sister populations in which one has become adapted to forest habitats and the other to grasslands. Several other reptilian species with similar adaptations

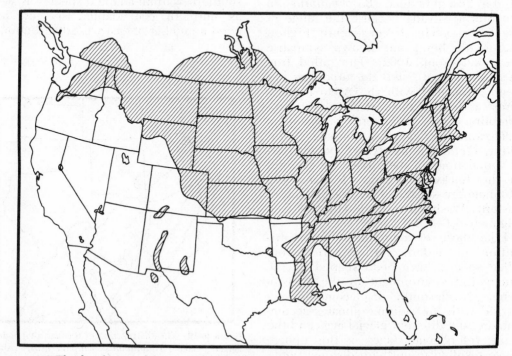

Figure 6-20 The distribution of the painted turtle *Chrysemys picta*. (After Stebbins, Robert C.: *A Field Guide to Western Reptiles and Amphibians*, 1966. Reprinted by permission of the publisher, Houghton Mifflin Company.)

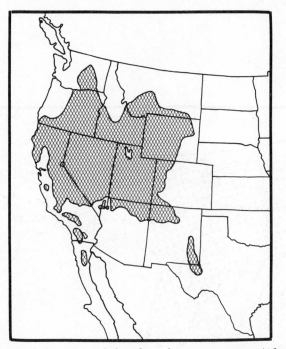

Figure 6-21 The distribution of the sagebrush lizard, *Sceloporus graciosus*. (After Stebbins, Robert C.: *A Field Guide to Western Reptiles and Amphibians*, 1966. Reprinted by permission of the publisher, Houghton Mifflin Company.)

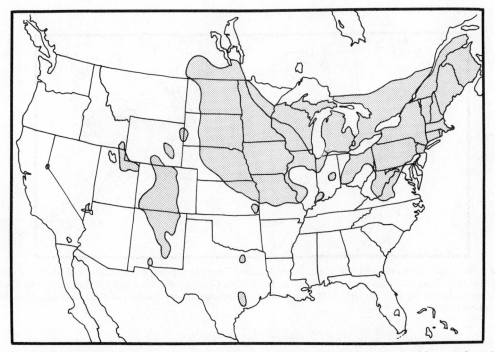

Figure 6-22 The distribution of the smooth green snake, *Opheodrys vernalis*. (After Stebbins, Robert C.: *A Field Guide to Western Reptiles and Amphibians*, 1966. Reprinted by permission of the publisher, Houghton Mifflin Company.)

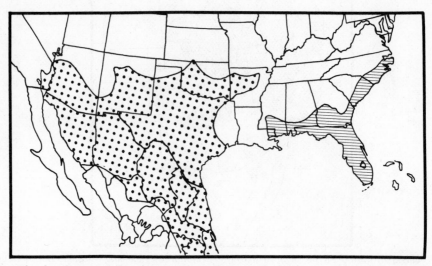

Figure 6–23 The distribution of the western diamondback rattlesnake, *Crotalus atrox* (dotted area), and of the eastern diamondback rattlesnake, *Crotalus adamanteus* (parallel lines). (Conant, R.: *A Field Guide to Reptiles and Amphibians of the United States and Canada East of the 100th Meridian*, 1958. Reprinted by permission of the publisher, Houghton Mifflin Company.)

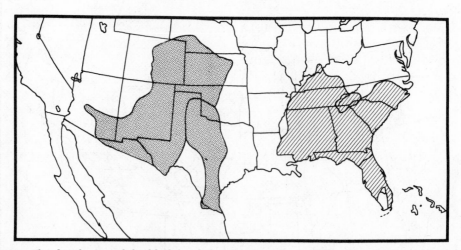

Figure 6–24 The distribution of the black-headed snake, *Tantilla nigriceps* (stippled), and of the crowned snake, *Tantilla coronata* (diagonal lines). (Conant, R.: *A Field Guide to Reptiles and Amphibians of the United States and Canada East of the 100th Meridian*, 1958. Reprinted by permission of the publisher, Houghton Mifflin Company.)

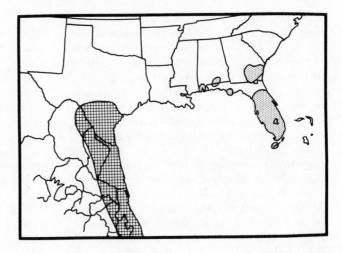

Figure 6–25 The distribution of the Texas indigo snake, *Drymarchon corais erebennus* (squares), and of the eastern indigo snake, *Drymarchon corais couperi* (stippled). (Conant, R.: *A Field Guide to Reptiles and Amphibians of the United States and Canada East of the 100th Meridian*, 1958. Reprinted by permission of the publisher, Houghton Mifflin Company.)

apparently were more widely isolated in the past and have expanded their ranges until they are now contiguous or overlap slightly where the two habitats (forest and grassland) meet. Such species, apparently limited in further dispersal by habitat boundaries, include the eastern box turtle (*Terrapene carolina*), a forest species, and the ornate box turtle (*Terrapene ornata*), a closely related grassland species (Fig. 6–26). There are other examples, but these suffice to indicate that there is considerable evidence supporting the idea that Pleistocene climatic shifts also fragmented southern reptilian ranges. Consequently, this viewpoint and that supported by Auffenberg and Milstead (1965) continue to be debated.

The distributions of amphibians and reptiles inhabiting lowlands along the Gulf coasts of the United States and Mexico and the Pacific Coast of Mexico were affected by sea-level fluctuations during the Pleistocene. Vast quantities of water

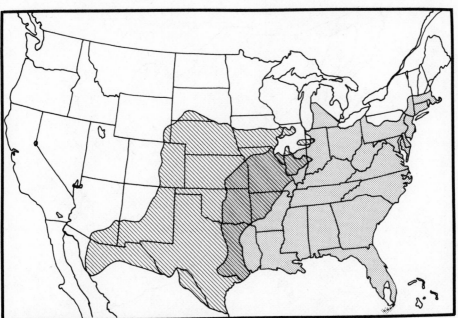

Figure 6–26   The distribution of the ornate box turtle, *Terrapene ornata* (diagonal lines), and of the box turtle, *Terrapene carolina* (stippled). (Conant, R.: *A Field Guide to Reptiles and Amphibians of the United States and Canada East of the 100th Meridian*, 1958. Reprinted by permission of the publisher, Houghton Mifflin Company.)

were stored on land as ice and snow during each glacial age, and this caused the levels of the seas to lower. During each of the three Interglacial stages the melting of this ice and snow resulted in higher sea levels. The lowering and raising of sea levels alternately opened and closed lowland dispersal routes, and in addition the raising of sea levels must have fragmented some distributional ranges by isolating populations on local high terrains. The disjunct distribution of the *Bufo gemmifer–Bufo mazatlanensis* complex of Mexican toads was undoubtedly produced in this manner. As shown in Figure 6–27, *Bufo gemmifer* has a restricted range in the lowlands around Acapulco that is approximately 250 miles from the nearest locality for the very similar *Bufo mazatlanensis* (Porter, 1964). The Pacific lowlands are very narrow in this region since the Sierra Madre del Sur and the Sierra Madre Occidental ranges, separated by the Balsas Basin northwest of Acapulco, rise sharply near the coast. When the sea level rose and the toads were forced to higher ground, those near Acapulco were probably isolated from the remainder of

their kind by the Balsas Basin, giving rise to *Bufo gemmifer*. Several reptiles (rosy boa *Lichanura trivirgata*, western patch-nosed snake *Salvadora hexalepis*, common kingsnake *Lampropeltis getulus*, night snake *Hypsiglena torquata*, sidewinder rattlesnake *Crotalus cerastes*, western whiptail *Cnemidophorus hyperythrus*) exhibit subspecific variation around the Gulf of California which may have originated in the Pleistocene through the isolation of populations by Interglacial high waters or, as suggested by Savage (1960), by climate changes and geofloral shifts (with or without flooding of some terrain).

Between Miocene and Recent time there have been seven important corridors for the dispersal of North American amphibians and reptiles (Fig. 6–28): the Gulf Arc Corridor from the northern Yucatan Peninsula to northeastern Mexico; the Circumferential Gulf Coast Corridor from northeastern Mexico to Florida; the Southern Rocky Mountains Corridor across southern New Mexico and Arizona and the extreme northern end of the Central Plateau of Mexico; the Trans-plateau

Figure 6–27 The distribution of two closely related Mexican toads, *Bufo gemmifer* (two circles in Guerrero) and *Bufo mazatlanensis* (spots) with a hiatus of approximately 250 miles in between. (From Porter, K. R.: Distribution and taxonomic status of seven species of Mexican *Bufo*.)

Figure 6–28 Primary dispersal corridors which have affected the distributions of North American amphibians and reptiles. *1*, the Gulf Arc Corridor from the northern Yucatan Peninsula to northeastern Mexico; *2*, the Circumferential Gulf Coast Corridor from northeastern Mexico to Florida; *3*, the Southern Rocky Mountains Corridor across southern New Mexico and Arizona and the north end of the Central Plateau of Mexico; *4*, the Trans-Plateau Corridor connecting the Sierra Madre Occidental Range and the Sierra Madre Oriental; *5*, the Southern Great Plains Corridor from eastern New Mexico and Texas to southern Canada; *6*, the Prairie Peninsula Corridor from Iowa and Missouri eastward to Ohio; and *7*, the Pacific Coast Range Corridor between California and Washington. (After Auffenberg, W., and W. W. Milstead: Reptiles in the Quaternary of North America, Peabody, F. E., and J. M. Savage: Evolution of a Coast Range corridor in California and its effect on the origin and dispersal of living amphibians and reptiles, and Martin, P. S.: A biogeography of reptiles and amphibians in the Gomez Farias region, Tamaulipas, Mexico.)

Corridor connecting the Sierra Madre Occidental and the Sierra Madre Oriental ranges around a latitude of 25° north; the Southern Great Plains Corridor from eastern New Mexico and Texas to southern Canada; the Prairie Peninsula Corridor extending from Iowa and Missouri through Illinois, Indiana, and Ohio; and the Pacific Coast Range Corridor between California and Washington.

During the Pleistocene, the Gulf Arc appears to have been occupied by a dry lowland tropical forest (evidence summarized by Martin, 1958) which allowed the dispersal of faunas from the northern Yucatan Peninsula to northeastern Mexico. Such a dispersal is indicated by the disjunct ranges of *Hypopachus cuneus, Laemanctus serratus, Sceloporus serrifer, Agkistrodon bilineatus,* and *Terrapene mexicana,* all of which appear in northeastern Mexico and the Yucatan Peninsula but not between.

The Circumferential Gulf Coast Corridor was extremely important in allowing east-west dispersal in southern United States and in promoting the northward dispersal of amphibians and reptiles from the Mexican tropics. It is reasonable to assume that during relatively arid times the grasslands of central United States spread eastward and during more humid times the forests of eastern United States spread westward. As noted by Auffenberg and Milstead (1965) such arid and humid corridors must have existed many times in the past, even before the Pleistocene. The movement of amphibians and rep-

tiles from the humid montane forest in eastern Mexico to those in southeastern United States had occurred by the Lower Miocene, and the eastward movement of some of the reptile genera of arid Florida (*Stilosoma, Neoseps, Rhineura*) occurred during the Pliocene (Auffenberg and Milstead, 1965). Living species of amphibians and reptiles which are representative of phylogenetic lines that probably utilized the Gulf Coast Corridor during Pleistocene to Recent time are listed in the table below.

The Southern Rocky Mountains Corridor was important as a dispersal route over the Continental Divide between the Central Great Plains–Chihuahuan Desert faunas to the east and the Sonoran Desert fauna to the west. At the present time, this region is a semiarid short grass steppe; as with much of southern United States, however, it undoubtedly was more humid during Pleistocene glacial ages, probably allowing Central Great Plains amphibians and reptiles to disperse westward. During the Interglacial stages it possibly was more arid than it is today and would have allowed an exchange between the amphibians and reptiles of the Sonoran Desert and those of the Chihuahuan Desert. The living species of amphibians and reptiles listed in the table on the following page, or their ancestors, probably moved through the Southern Rocky Mountains Corridor in the direction indicated.

There is floral and faunal evidence (summarized by Martin, 1958) for a Pleistocene spread of montane woodland or

LIVING REPRESENTATIVES OF PHYLOGENETIC LINES THAT UTILIZED THE GULF COAST CORRIDOR

Phylogenetic Lines That Moved East	Phylogenetic Lines That Moved West
Amphibia	Amphibia
Scaphiopus holbrooki	Plethodon glutinosus
Gastrophryne carolinensis	
Acris crepitans	
Reptilia	Reptilia
Pseudemys concinna	Kinosternon subrubrum
Pseudemys scripta	Terrapene carolina
Gopherus polyphemus	Deirochelys reticularia
Sceloporus undulatus	Anolis carolinensis
Cnemidophorus sexlineatus	Lygosoma (=Scincella) laterale
Masticophis flagellum	Natrix sipedon
Pituophis melanoleucus	Natrix rigida
Micrurus fulvius	Natrix cyclopion
Sistrurus miliaris	Farancia abacura

LIVING REPRESENTATIVES OF PHYLOGENETIC LINES THAT UTILIZED THE
SOUTHERN ROCKY MOUNTAINS CORRIDOR

Phylogenetic Lines That Moved East	Phylogenetic Lines That Moved West

Amphibia
  Bufo punctatus
	Amphibia
	  Bufo microscaphus
	  Bufo woodhousei
	  Bufo cognatus
	  Scaphiopus hammondi
	  Scaphiopus couchi
	  Rana catesbeiana
	  Pseudacris triseriata triseriata

Reptilia
  Gopherus agassizi
  Crotaphytus collaris
  Sceloporus magister
  Urosaurus ornatus
  Cnemidophorus tigris
  Heloderma suspectum
  Sonora semiannulata
	Reptilia
	  Terrapene nelsoni
	  Trionyx spiniferus
	  Eumeces obsoletus
	  Masticophis flagellum
	  Arizona elegans
	  Lampropeltis getulus
	  Rhinocheilus lecontei
	  Thamnophis marcianus
	  Crotalus atrox
	  Crotalus molossus

oak savanna across the Central Plateau of Mexico, connecting the Sierra Madre Oriental and the Sierra Madre Occidental. Post-Pleistocene increased aridity closed this corridor and resulted in subspecific disjunct distributions of at least four species of reptiles. *Sceloporus scalaris slevini, Sceloporus grammicus disparilis, Thamnophis cyrtopsis,* and *Barisia imbricata ciliaris* all occur in montane woodland or in savannas of the Sierra Madre Oriental and Sierra Madre Occidental, isolated by other subspecies to the south (Martin, 1958). The very closely related *Bufo mazatlanensis* and *Bufo valliceps* provide additional indication of dispersal via the Trans-plateau Corridor: *Bufo mazatlanensis* occurs on the Pacific drainages of northern Mexico, while *Bufo valliceps* occurs on the Gulf Coast and only occurs on the Pacific Coast southeast of the Isthmus of Tehuantepec.

The Southern Great Plains Corridor allowed faunas to disperse parallel to and east of the Rocky Mountains. A great deal of north-south displacement occurred in this region during the Pleistocene, and many, if not all, inhabitant amphibians and reptiles have probably been pushed in both directions at some time by past climatic changes. The upper table on the following page indicates the direction of dispersal required to produce the present distributions in the Central Plains.

The Prairie Peninsula Corridor today consists of discontinuous areas of temperate glasslands surrounded by forests. It is believed that in the post-Pleistocene the North American prairie extended eastward, wedging itself between the northeastern coniferous forest and the southeastern deciduous forest, to form this corridor. If one assumes that competitive exclusion prevented the prairie species entering the corridor from spreading into southeastern United States, then theoretically all faunal elements that have an eastward range north of the Appalachian region but are absent from the Southeast could have had their origin in the western prairie (Schmidt, 1938). The present distributions of the amphibians and reptiles listed in the lower table on the following page, well illustrated by that of the massasaugas (*Sistrurus,* Fig. 6–29) and that of Blanding's turtle (*Emydoidea,* Fig. 6–30), indicate they probably dispersed eastward through the Prairie Peninsula Corridor.

The Pacific Coast Range Corridor consists of parallel mountain ranges (the Coast Ranges and the Sierra Nevada–Cascades range) with the Great Valley of California in between. This corridor has primarily allowed a southward dispersion of Arcto-Tertiary species out of the Northwest, a northward dispersion of Mexican and southern Great Basin species associated with the Madro-Tertiary flora, and a south-

AMPHIBIANS AND REPTILES THAT DISPERSED THROUGH THE SOUTHERN GREAT PLAINS CORRIDOR

Phylogenetic Lines That Moved North	*Phylogenetic Lines That Moved South*
Amphibia	Amphibia
Bufo cognatus	Ambystoma tigrinum
Bufo debilis	Rana maslini
Bufo compactilis	
Bufo woodhousei	
Bufo punctatus	
Gastrophryne olivacea	
Scaphiopus bombifrons	
Rana catesbeiana	
Rana sylvatica	
Reptilia	Reptilia
Kinosternon flavescens	Chrysemys picta
Terrapene ornata	Eumeces septentrionalis
Trionyx spiniferus hartwegi	Thamnophis sirtalis parietalis
Holbrookia maculata	Thamnophis radix
Sceloporus undulatus	
Phrynosoma cornutum	
Eumeces obsoletus	
Ophisaurus attenuatus attenuatus	
Natrix grahami	
Tropidoclonion lineatum	
Heterodon nasicus	
Arizona elegans	
Lampropeltis triangulum	
Crotalus viridis viridis	

AMPHIBIANS AND REPTILES THAT DISPERSED EASTWARD THROUGH THE PRAIRIE PENINSULA CORRIDOR

Pseudacris triseriata triseriata	Rana pipiens pipiens
Rana sylvatica	Ambystoma tigrinum tigrinum
Terrapene ornata	Chrysemys picta marginata
Pseudemys scripta elegans	Emydoidea blandingi
Natrix sipedon sipedon	Natrix kirtlandi
Thamnophis radix radix	Thamnophis butleri
Thamnophis sauritus proximus	Tropidoclonion lineatum lineatum
Opheodrys vernalis blanchardi	Heterodon nasicus nasicus
Coluber constrictor flaviventris	Pituophis melanoleucus sayi
Elaphe vulpina	Lampropeltis triangulum syspila
Agkistrodon contortrix mokeson	Sistrurus catenatus catenatus

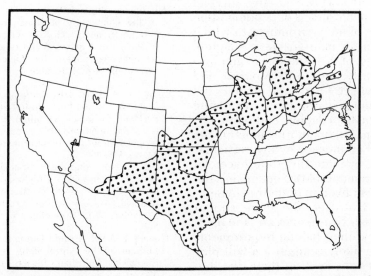

Figure 6–29   The distribution of the massasauga, *Sistrurus catenatus.* (Conant, R.: *A Field Guide to Reptiles and Amphibians of the United States and Canada East of the 100th Meridian,* 1958. Reprinted by permission of the publisher, Houghton Mifflin Company.)

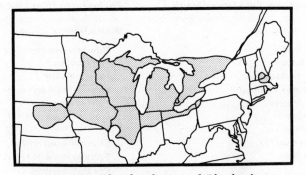

Figure 6–30 The distribution of Blanding's turtle, *Emydoidea blandingi.* (After Conant, R.: *A Field Guide to Reptiles and Amphibians of the United States and Canada East of the 100th Meridian,* 1958. Reprinted by permission of the publisher, Houghton Mifflin Company.)

ern dispersion of northern Great Basin Madro-Tertiary species (Peabody and Savage, 1958). Some of the species that have utilized this corridor are listed in the table below.

Corridors are important in allowing organisms to disperse, and they affect the probability of a given taxon reaching a particular area. However, the survival of the group in that area is dependent upon the existence of an environment to which the organism is more or less adapted. Thus, the present distributions of amphibians and reptiles in North America and elsewhere reflect to a large degree the local and regional distributions of suitable habitats. Climatic, edaphic, and biotic factors all play an important role affecting the presence and abundance of any particular species. The greater the differences in habitat requirements of two organisms, the less likely is an overlap of their geographical ranges. However, the more similar (closely related) two species are, the more similar will be their habitat requirements and the more likely competition will play an important role limiting their distributions where they meet in nature. Species frequently evolve through geographical isolation of sister populations (allopatric speciation), and this process together with the importance of competitive exclusion are reflected in the generally accepted Jordan's law: The most closely related species do not occur together geographically but occupy adjacent ranges.

References

Auffenberg, W., and W. W. Milstead. 1965. Reptiles in the Quaternary of North America. *In:* H. E. Wright, Jr., and D. G. Frey (eds.), The Quaternary of the United States; VII Congress of the International Association for Quaternary Research. Princeton, N.J., Princeton Univ. Press. Pp. 557–568.

Blair, W. F. 1958. Distributional patterns of vertebrates in the southern United States in relation to past and present environments. *In:* C. L. Hubbs (ed.), Zoogeography. Washington, D.C., American Association for the Advancement of Science. Publ. 51, pp. 433–468.

———— 1965. Amphibian speciation. *In:* H. E. Wright, Jr., and D. G. Frey (eds.), The Quaternary of the United States; VII Congress of the International Association for Quaternary Research. Princeton, N.J., Princeton Univ. Press. Pp. 543–556.

Conant, R. 1958. A Field Guide to Reptiles and Amphibians of the United States and Canada East of the 100th Meridian. Boston, Houghton Mifflin. 366 pp.

Darlington, P. J., Jr. 1957. Zoogeography: The Geographical Distribution of Animals. New York, Wiley and Sons. 675 pp.

LIVING REPRESENTATIVES OF PHYLOGENETIC LINES THAT UTILIZED THE PACIFIC COAST RANGE CORRIDOR

Phylogenetic Lines That Moved South	Phylogenetic Lines That Moved North
Arcto-Tertiary species	Madro-Tertiary species
Taricha granulosa	*Bufo microscaphus*
Taricha torosa	*Rana boylii*
Ensatina eschscholtzi	*Rana muscosa*
Batrachoseps attenuatus	*Scaphiopus hammondi*
Aneides ferreus	*Hyla cadaverina*
Bufo boreas	*Rana catesbeiana (?)*
Rana aurora	————
Ascaphus truei (?)	
Hyla regilla	*Clemmys marmorata*
	Crotaphytus wislizenii
Madro-Tertiary species	*Uta stansburiana*
Sceloporus graciosus	*Phrynosoma coronatum*
Sceloporus occidentalis	*Eumeces gilberti (?)*
Eumeces skiltonianus	*Cnemidophorus tigris*
Gerrhonotus coeruleus	*Gerrhonotus multicarinatus*
Charina bottae	*Masticophis lateralis*
Diadophis punctatus	*Tantilla planiceps*
Contia tenuis (?)	*Hypsiglena torquata nuchalata*
Coluber constrictor	
Lampropeltis zonata	
Thamnophis elegans	

Deevey, E. S., Jr. 1949. Biogeography of the Pleistocene. Bull. Geol. Soc. America *60*:1315–1416. (27 figures.)

Dietz, R. S., and J. C. Holden. 1970. The breakup of Pangaea. Scient. Amer. *223(4)*:30–41.

Durham, J. W. 1950. Cenozoic marine climates of the Pacific coast. Bull. Geol. Soc. America *61*:1243–1264.

Estes, R. 1970. New fossil pelobatid frogs and a review of the genus *Eopelobates*. Bull. Mus. Comp. Zool. *139(6)*:293–339.

Griffiths, I. 1963. The phylogeny of the Salientia. Biol. Rev. Cambridge Philos. Soc. *38*:241–292.

Hecht, M. 1959. Amphibians and reptiles. Bull. American Mus. Nat. Hist. *117*:130–146.

Karlstrom, E. L.: 1962. The toad genus *Bufo* in the Sierra Nevada of California. Univ. California Publ. Zool. *62(1)*:1–104.

Malnate, E. V. 1960. Systematic division and evolution of the colubrid snake genus *Natrix*, with comments on the Subfamily Natricinae. Proc. Acad. Nat. Sci. Philadelphia *112*:41–71.

Martin, P. S. 1958. A biogeography of reptiles and amphibians in the Gomez Farias region, Tamaulipas, Mexico. Misc. Publ. Mus. Zool. Univ. Michigan *101*:1–102.

Peabody, F. E., and J. M. Savage. 1958. Evolution of a Coast Range corridor in California and its effect on the origin and dispersal of living amphibians and reptiles. *In:* C. L. Hubbs (ed.), Zoogeography. Washington, D.C., American Association for the Advancement of Science. Publ. 51, pp. 159–186.

Porter, K. R. 1964. Distribution and taxonomic status of seven species of Mexican *Bufo*. Herpetologica *19*:229–247.

Savage, J. M. 1960. Evolution of a peninsular herpetofauna. Systematic Zool. *9*:184–212.

Schmidt, K. P. 1938. Herpetological evidence for the postglacial eastward extension of the steppe in North America. Ecology *19*:396–407.

Schwarzbach, M. 1963. Climates of the Past; An Introduction to Paleoclimatology. London, Van Nostrand Reinhold. 328 pp.

Smith, P. W. 1957. An analysis of post-Wisconsin biogeography of the prairie peninsula region based on distributional phenomena among terrestrial vertebrate populations. Ecology *38*:205–218.

Stebbins, R. C. 1954. Amphibians and Reptiles of Western North America. New York, McGraw-Hill. 536 pp.

_____ 1966. A Field Guide to Western Reptiles and Amphibians. Boston, Houghton Mifflin. 279 pp.

Taylor, E. H. 1941. A new anuran from the middle Miocene of Nevada. Univ. Kansas Sci. Bull. *27*: 61–69.

_____ 1942. Extinct toads and frogs from the upper Pliocene deposits of Meade County, Kansas. Univ. Kansas Sci. Bull. *28*:199–235.

Tihen, J. A. 1958. Comments on the osteology and phylogeny of ambystomatid salamanders. Bull. Florida State Mus. Biol. Ser. *3(1)*:1–50.

Wake, D. B. 1966. Comparative osteology and evolution of the lungless salamanders, family Plethodontidae. Mem. Southern California Acad. Sci. *4*:1–111.

_____ 1970. The abundance and diversity of tropical salamanders. Am. Naturalist *104*:211–213.

Wright, H. E., Jr., and D. G. Frey (eds.). 1965. The Quaternary of the United States; VII Congress of the International Association for Quaternary Research. Princeton, N.J., Princeton Univ. Press. 922 pp.

Zweifel, R. G. 1956. Two pelobatid frogs from the Tertiary of North America and their relationship to fossil and recent forms. American Mus. Nov. *1762*:1–45.

CHAPTER 7

MOISTURE RELATIONS OF AMPHIBIANS AND REPTILES

Water constitutes a large proportion of living material, protoplasm being composed of about 70 to 90 per cent, and one of the major problems of all organisms is the maintenance of a proper water balance. Because they are surrounded by a generally dehydrating atmosphere, terrestrial organisms continually face the danger of desiccation. Regardless of this danger, oxidative metabolism requires a continuous supply of oxygen and constantly produces carbon dioxide which must be dissipated. Therefore, an intimate contact must be maintained between actively respiring cells, with a high water content, and the surrounding atmosphere, with a variable but generally lower water content. Although terrestrial animals have evolved simple moist chambers, branched tracheal systems, or complicated lungs to replace the gills found in many aquatic forms, the fundamental requirement of moist membranes for the exchange of oxygen and carbon dioxide between body fluids, tissues, and the surrounding atmosphere remains the same. Whenever it is necessary to maintain a moist membrane in contact with a generally dry atmosphere, the loss of water by evaporation is unavoidable. The internal placement of respiratory membranes decreases this loss by restricting the volume of atmosphere to which these membranes are exposed, but much water is still removed during respiration since the inspired air, whether saturated or almost dry, is nearly

saturated at the body temperature of the animal by the time it is expired. Additional volumes of water are lost through the elimination of feces and nitrogenous wastes. Thus, the primary water problem of terrestrial animals is one of maintaining an adequate internal supply by replacing the water continuously being lost. This is the fundamental reason why few environmental factors are comparable to the availability of water in limiting the distribution of terrestrial organisms.

Because they are surrounded by a medium which is hypotonic to their body fluids, fresh-water organisms continually face the danger of overhydration through osmosis; their primary water problem is one of removing excess water so as to prevent dilution of their fluids. Marine organisms, on the other hand, are surrounded by a medium which is generally hypertonic to their body fluids and, thus, live in a desiccating environment because of osmotic forces.

The maintenance of a proper water balance is not simply a matter of retaining a particular volume of fluid within the body, since body water is partitioned into extracellular (plasma and interstitial fluid) and intracellular fluids that are separated by semipermeable cell membranes and are subject to diffusional and secretory exchanges with one another. Thus, osmoregulation, the maintenance of a relatively constant osmotic concentration, is an integral part of the problem of water balance. Mechanisms possessed by animals to solve this problem include those which limit permeability to water, limit permeability to salts (or solutes of other kinds), secrete salts (in or out) against a gradient, secrete water (in or out) against a gradient, or store water or solute (Prosser, 1950).

AMPHIBIAN WATER RELATIONS

A complete discussion of the water relations of Amphibia is impossible at this time since many areas of research have been neglected and most studies have been made on a limited variety of Anura, primarily in the genera *Rana* and *Bufo*, and on very few Urodela other than *Necturus* and *Ambystoma*. Virtually nothing is known about the problems of water balance in any of the Apoda. Thus, whenever statements are made about amphibians in general, the reader should bear in mind the fact that relatively few species have been studied in most instances.

Completely aquatic amphibians face the problem of hydration while those that are entirely terrestrial face the problem of dehydration. The many amphibians that spend part of their time in water and part of it on land face the double problem of hydration while in water and dehydration while out of water. It is clear that the exchange of water and solutes between amphibian body fluids and the external environment occurs through all of the respiratory epithelium (skin, buccal and pharyngeal mucosa, and gills and lungs when present), through the kidney tubules, and through the lining of the gastrointestinal tract (stomach, intestines, cloaca, and urinary bladder). However, little is known regarding the importance of gills, lungs, or mucosa to amphibian water balance. Their relatively low metabolic and respiratory rates would presumably minimize the amount of water amphibians lose in breathing; in fact, Mellanby (1941) could measure no change in the rate of water loss after breathing stopped in the common frog (*Rana temporaria*). Many studies have been made on amphibian skin and kidney function, and a limited amount of information is available concerning the role of the gastrointestinal tract in amphibian water balance.

Because of the development of the stratum corneum, the integument of terrestrial animals is generally, at least to some degree, a barrier to the passage of water. Although amphibians maintain a moist skin, this last statement is true even for them, since studies of anurans have shown that the skin of a metamorphosed individual has a greater ability to resist the passage of water than that of a tadpole. However, this is only a relative difference, and all available evidence indicates the adult amphibian skin is a semipermeable membrane that unquestionably prevents the entrance or exit of some solutes but has little effect on the passage of water. Variations in permeability may occur from one part of the skin to another and with sex, changes in blood-hormone titers, ecdysis, temperature, and season (Hevesy *et al.*, 1935; Boyd and Whyte, 1938; Koefoed-Johnsen and Ussing, 1953); Minor species-specific differences have also been reported (Koefoed-Johnsen and Ussing, 1953). However, studies by a variety of workers on different anurans and urodeles have shown that the adult amphibian skin behaves in a similar manner in all species studied.

The skin of xeric species, such as the spadefoot toad (*Scaphiopus*, family Pelobatidae), has no water-retaining features not present in those of mesic-adapted species, such as most ranid frogs (family Ranidae). Therefore, the skin is of primary importance as a major surface through which there is an evaporative loss of water when the amphibian is out of water and through which there is a diffusional gain of water when the animal is submerged. Studies of anurans (primarily of the leopard frog, *Rana pipiens*) have shown that they lose water through evaporation at a rate that is nearly inversely proportional to the relative humidity of the ambient atmosphere. That this evaporative loss is uncontrolled is indicated by experiments which have demonstrated that the rate is unaffected by the functionings of the central nervous system, the circulatory system, or the skin itself. It is significant that no equilibrium of zero evaporation can be established for living frogs. It still goes on from their surface when the atmosphere is saturated with water vapor. Importantly, under no circumstance can a frog absorb water vapor from the atmosphere, not even when the relative humidity is 100 per cent. The skin's continuous evaporation of moisture but inability to absorb it is explained by the fact that the production of heat through metabolism keeps the surface of the animal at least slightly warmer than the atmosphere, raising its dew point above that of the surrounding atmosphere and making it impossible for water vapor to condense on the surface of the animal. Again, *no matter what the atmospheric conditions, an amphibian out of water will continuously lose water by evaporation from its skin, and only the rate of evaporation will change with the relative humidity of the atmosphere.*

Consideration of this point alone makes it obvious that amphibians are poorly adapted for terrestrial life. When an amphibian is in fresh water, water may of course pass through the skin in both directions, but ordinarily diffusion results in a net gain in body water because the fresh-water environment is hypotonic to body fluids. The rate of flow across resting skin has been measured for a variety of toads and frogs and is generally about 4.7 to 9.2 microliters per square centimeter per hour at a temperature of 20° centigrade (Deyrup, 1964). The outer skin of anurans is also slightly permeable to sodium and chloride ions, their permeability coefficient being about 0.02 that of heavy water (Garby and Linderholm, 1953), and sodium uptake through the skin does occur through energy-dependent active transport mechanisms. Thus, under certain conditions, additional water uptake may occur through the skin because of osmosis (water molecules accompanying chloride ions), active transport (water molecules accompanying sodium ions), or because of the electrical potential of the skin, through electro-osmosis.

Amphibian kidney function has been very well studied. Because of pressure differentials, fluid is filtered from the blood plasma through capillary walls in the glomerulus and into the kidney tubule. It is then forced down the tubule by a combination of the pressure and ciliary activity. As it passes down the tubule, cells lining the lumen reabsorb water, sodium, chloride, glucose, and some other solutes, and these pass back into the blood flowing through capillaries surrounding the tubule. Simultaneously, other solutes, including urea, are secreted from the blood into the tubular fluid. Water reabsorption occurs primarily in the proximal convoluted tubule, but in the mesonephros this is not an efficient process: from 14 to 90 per cent of the glomerular filtrate of even normally hydrated anurans may be lost from the body (Deyrup, 1964). The fluid in distal tubules is always hypotonic to the plasma, but dehydrated animals exhibit an increased tubular reabsorption of water and reduced formation of urine. Speaking more generally, urine formation tends to parallel net uptake of water. The reabsorption and secretion of substances occurs throughout the excretory system, including the cloaca and urinary bladder. The primary nitrogenous waste products of amphibians are ammonia and urea, both of which are so toxic that they must be excreted in dilute solutions. This factor and that of the relative inefficiency of the mesonephros in reabsorbing water mean that the amphibian kidney is better adapted to ridding the body of excess water than it is to conserving body water.

The importance of the gastrointestinal tract to water balance in amphibians is not clear. Adult amphibians apparently never drink free water; however, they of course swallow the water contained in their prey. Krogh (1937, 1939) has suggested that intestinal absorption of water may be more important to the water balance of tadpoles than to that of adults, and Rey (1938) estimated that up to 50 per cent of the water intake of tadpoles was through the gastrointestinal tract. The facts that potassium ion transport may occur in both the stomach and the large intestine and that sodium and chloride ions are transported (actively and passively, respectively) in the large intestine indicate that both stomach and large intestine are at least potentially important in osmoregulation. The urinary bladder is particularly important to amphibian water balance, for water, sodium, and other ions can all be absorbed through the bladder wall. Experimentation has shown that bladder fluid is absorbed by a variety of anurans when they are subjected to dehydration, and observations indicate that at least some species utilize the bladder as a water storage reservoir for use during dry seasons or hibernation. It has also been established that in toads and frogs the bladder wall can actively transport sodium ions from the lumen side through to the fluid bathing the serosa. Chloride ions and water may move with the sodium ions from the lumen through the bladder wall, but, at least in the giant toad (*Bufo marinus*), the rate of water movement across the bladder wall may be independent of changes in net sodium transport.

Little is known about the involvement of neural processes in the osmoregulation of amphibians, but hormonal processes have been extensively studied in frogs (primarily *Rana pipiens*). Neurohypophyseal hormones increase the permeability of skin, kidney tubule, and urinary bladder membranes to small molecules (thus accelerating the diffusional uptake of water from ambient fresh water and from hypotonic urine in the kidney tubules and urinary bladder) and decrease glomerular filtration. This results in a decreased urine flow and an increase in body water, most of which is stored in lymphatic sacs and body cavities. The clawed frog (*Xenopus laevis*, family Pipidae) is unique in that it is the only anuran studied so far in which posterior pituitary hormone injections increase the uptake of water by the skin but apparently do not suppress urine formation (Ewer, 1952). Two neurohypophyseal hormones present in anurans are arginine vasotocin (8-arginine oxytocin) and oxytocin; other unidentified substances have also been found. Arginine vasotocin, when applied directly to isolated skin, has been shown to stimulate sodium transport. Morel *et al.* (1961) report that another separate but unidentified substance in the neurohypophysis of the edible frog (*Rana esculenta*) also has the primary effect of enhancing sodium transport by the skin.

Adrenaline has been shown to stimulate active transport of chloride from the inside to the outside of isolated frog skin and to cause a variable effect on glomerular circulation and filtration. Bastian and Zarrow (1954) also report that adrenaline stimulates secretion of an unidentified material from skin glands of the clawed frog (*Xenopus laevis*); this may be related to the active pumping out of chloride (Deyrup, 1964).

There have been many experiments on the effect of interrenal hormones on osmoregulation of anurans. These have shown that in a variety of species aldosterone and cortisol cause an increase in sodium uptake by the skin. Deoxycorticosterone glucoside has also been shown to enhance sodium transport by the skin of toads (*Bufo*, family Bufonidae).

In amphibians, as in mammals, adrenocorticotropic hormone (ACTH) stimulates aldosterone release by adrenal tissue and indirectly, therefore, is involved in the maintenance of active sodium pumps in the skin. Similarly, the activity of the adenohypophysis is regulated by the neurohypophysis, which in turn is regulated by the hypothalamus. Thus, the mechanism of hormonal control of osmoregulation by anurans (Fig. 7–1) appears to involve initially the hypothalamus (preoptic nucleus), which acts when the animal is in fresh water by stimulating the neurohypophysis. Stimulation of the neurohypophysis results in the release of neurohypophyseal hormones that activate sodium pumps in the skin, tubules, and urinary bladder and cause an increase in the diffusional uptake of water as well. In addition, the neurohypophyseal hormones activate the adenohypophysis, causing it to release ACTH, and this stimulates the interrenal bodies to release the aldosterone and cortisol that also stimulate sodium pumps.

To summarize, amphibians (at least the anurans) have an efficient mechanism for absorbing water and salts when they are submerged; they have kidneys which are efficient in ridding the body of excess water but are not efficient in reabsorbing water; they have limited abilities to rid the body of excess salt; and they have no ability to control the loss of water from the body when they are on land. What should be obvious from these statements is that amphibians are adapted to live in environments that are rich in water but poor in salt.

The above conclusion indicates the importance of habitat selection to the survival of amphibians, but it does not explain how some amphibians, such as the pelobatids and bufonids, are better able to survive under xeric conditions than others, such as the salamanders and hylids. Thorson and Svihla (1943) made a com-

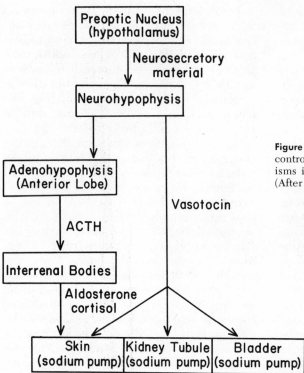

Figure 7–1 Factors in the neuroendocrine control of the ionic regulatory mechanisms in the leopard frog (*Rana pipiens*). (After Scheer, B. T.: *Animal Physiology.*)

parative study of a variety of anurans adapted to different kinds of habitats and, in agreement with the conclusions just outlined, found all species studied to be highly susceptible to desiccation. However, they found that mesic-adapted anurans could tolerate less loss of water than xeric-adapted anurans. Below is a summary of the data of Thorson and Svihla showing vital limits, expressed as percentages of body weight and body

water, of water loss for 10 species of anurans correlated with their habitats.

Littleford, Keller, and Phillips (1947) made a study similar to that of Thorson and Svihla but on plethodontid salamanders. They found the average percentage water loss representing vital limits for *Desmognathus fuscus*, *Eurycea bislineata*, *Plethodon glutinosus*, and *Pseudotriton ruber* to be between 18.04 per cent and 18.98 per cent of the original weight of the

	Highest Loss in Body Weight (Per Cent)	Average Loss in Body Weight (Per Cent)	Highest Loss in Body Water (Per Cent)	Average Loss in Body Water (Per Cent)	Habitat
Scaphiopus holbrooki	48.1	47.9	60.5	60.2	Terrestrial-fossorial
Scaphiopus hammondi	48.4	47.6	61.0	59.5	Terrestrial-fossorial
Bufo boreas	44.6	43.6	55.8	54.8	Terrestrial
Bufo terrestris	43.3	43.0	54.9	54.5	Terrestrial
Hyla regilla	40.0	39.0	50.3	49.1	Terrestrial
Hyla cinerea	39.3	37.3	49.0	46.5	Terrestrial-arboreal
Rana pipiens	36.6	35.5	46.3	44.9	Terrestrial-semiaquatic
Rana aurora	34.3	34.0	43.0	42.6	Semiaquatic
Rana sphenocephalia	32.6	32.4	(no data)		Semiaquatic
Rana grylio	31.2	29.5	40.2	38.0	Aquatic

specimen, but for *Plethodon cinereus* it was about 25.85 per cent of the original weight. This difference is correlated with the environmental distribution of these species, since *Plethodon cinereus* inhabits much drier habitats than do the others. Thus, the same relation between tolerable moisture loss and habitat prevails with the salamanders as with the anurans. Furthermore, in the study of the salamanders, as with the anurans, there was no significant difference from species to species in the rate at which water was lost through the skin. The general conclusion is that *Plethodon cinereus* must possess mechanisms that enable it to withstand a greater percentage of water loss than the other salamanders but that it does not differ in its ability to reduce the loss of water.

Thus, the answer to the question of how some amphibians are able to tolerate more xeric conditions than others lies in how some are able to tolerate greater loss of water than others. As mentioned, body water is partitioned into extracellular and intracellular components, and a recent study by Gordon (1965) indicates that the ability to tolerate desiccation lies in the ability to maintain relatively stable intracellular fluid levels. Gordon's study was a detailed analysis of intracellular osmoregulation in skeletal muscles of two species of toads, *Bufo viridis* and *Bufo boreas*, adapted to fresh water, 20 per cent, 40 per cent, and 50 per cent sea water (the latter having a salinity of 16 per cent). *Bufo viridis*, which in nature lives in environmental salinities as high as 29 per cent, could tolerate 50 per cent sea water. *Bufo boreas*, a more stenohaline species, did well in 40 per cent sea water but died within a week in 50 per cent sea water. In neither species did muscle inulin ("extracellular") spaces differ significantly among individuals adapted to different salinities. Both species maintained plasma osmotic concentrations above that of the medium in which they were submerged until maximum salinity tolerances were reached, and, in both, increases in plasma osmotic concentrations were due primarily to sodium chloride. Plasma osmolarity increased 80 per cent in *Bufo viridis* between fresh water and 50 per cent sea water, and 135 per cent in *Bufo boreas* between fresh water and 40 per cent sea water. In contrast,

muscle weights increased by only 33 per cent in both species, indicating they maintained a relatively constant water content in the face of major changes in extracellular osmotic concentration. It was found that the relative stability of muscle hydration is due to accumulation of intracellular solutes so that intracellular osmotic concentrations probably change with and are equal to simultaneous extracellular osmotic concentrations. The changes in intracellular osmotic concentration were broadly partitioned in both species: 47 per cent inorganic ions (chloride, sodium, and potassium), 33 per cent free amino acids and related compounds, and 20 per cent urea in *Bufo viridis*; 43 per cent inorganic ions, 40 per cent amino acids, and 17 per cent urea in *Bufo boreas*. Free carbohydrates were virtually absent. It was also observed that the increase in intracellular free amino acids and related compounds involved different substances in the two species: *Bufo viridis* accumulated primarily taurine, glycine, and alanine; *Bufo boreas* accumulated primarily taurine, glutamic acid, and carnosine. The intracellular urea concentrations seemed always to be significantly higher than the plasma concentrations. It would seem that the results of this study could be extended to explain the differences in water loss which may be tolerated by different species. Those species which have an ability to maintain a relatively stable muscle hydration, or cell hydration in general, despite extracellular osmotic changes, should theoretically be best able to withstand desiccation. Species which cannot maintain such a stable cell hydration will be unable to withstand such a high degree of desiccation. Interspecific variation in this regard would result from strong selection for an ability to withstand desiccation in water-poor habitats but little or no selection for it in water-rich habitats.

The loss of water from terrestrial amphibians is regulated to a major degree by their habitat selection and behavior, both of which appear to be influenced by rates of evaporation from the skin. The burrowing habit is extremely important to some amphibians since soil moisture is a natural source of water for the animal and, while buried, evaporation from the skin is minimized. Goldsmith (1926) found that the digging behavior of western spadefoots (*Scaphiopus hammondi*) could be induced by evaporation and that the species was sensitive to a 10 per cent change in humidity when at a temperature of 27° centigrade. Shelford (1914) observed that both anurans and salamanders would attempt to avoid evaporative conditions when exposed to them, no matter whether the evaporation was induced by low humidity, higher temperature, or movements of air. He concluded that this avoidance behavior was triggered by changes in body fluids due to the loss of water by evaporation. In nature, the same sort of avoidance is manifested by amphibians seeking shelter during droughts, sometimes in congregations under damp logs, vegetation, or in damp soil. Noble (1931) observed that even though they may have the same climbing mechanisms of large species, many small species of tree frogs will only rarely ascend tall trees, apparently because of the high evaporation rate in such exposed positions. Associated with the avoidance behavior which causes them to seek shelter during droughts is a pattern of behavior which permits most amphibians to remain in an inactive state during dry seasons or periods and then to suddenly become active on rainy nights when the environment is more favorable for their moisture relations. Spadefoot toads (*Scaphiopus*, family Pelobatidae) are particularly well known for being extremely difficult to find until spring or summer rains occur and then to suddenly appear in large numbers. Similarly, many amphibians are only active at night, when the humidity is higher than in the daytime and the temperature lower. As a general rule, the only amphibians active in the daytime in arid or semiarid regions are aquatic or semiaquatic species which are associated with permanent bodies of water and can replenish their body water at will.

REPTILIAN WATER RELATIONS

As with amphibians, the water relations of reptiles are incompletely known. Most studies have been concentrated on a limited variety of lizards; very little is known about the water balance of turtles or crocodilians, and virtually nothing is known about that of the tuatara. As with any animal, reptiles can only maintain their water content if losses occurring through the elimination of feces and urine and through evaporation are balanced by intakes derived from drinking water, from free water in food, and from oxidation water produced by metabolic processes.

Possessing a dry skin which is relatively impermeable to water, reptiles are truly terrestrial animals. Tercafs (1963) has made *in vitro* comparisons of the evaporative loss of water through the skin of two Old World lizards, *Lacerta viridis* and *Uromastix acanthinurus,* and that of the common European frog, *Rana temporaria,* and has found that the rate of water loss through the lizard skins was less than a tenth that for the common frog. However, this loss of water through the skin of a reptile, though small in comparison to that of an amphibian, may represent a significant proportion of the total volume lost when the animal is at a moderate temperature. Benedict (1932) concluded that there was an appreciable loss of water through the shell of tortoises and the skin of pythons. Chew and Dammann (1961) have estimated that a third of the total evaporative loss of water from the western diamondback rattlesnake (*Crotalus atrox*) occurred through the skin at 26° to 27° centigrade. Dawson *et al.* (1966) found that cutaneous losses accounted for 70 per cent of the evaporative loss of water at 20° C in the agamid lizard *Amphibolurus ornatus* and the gecko *Gehyra variegata* and for 40 per cent of that lost in the skink *Sphenomorphus labillardieri.* Because of the greater effect of temperature on pulmonary loss than on cutaneous loss, these percentages dropped to 59, 57, and 27 per cent, respectively, of the total water loss at 30° C. The point here is that, while reptilian skins are relatively waterproof compared to those of amphibians, they are not absolute barriers to the evaporative loss of body water. There appear to be major differences in the permeability of the skin of different reptiles and, aquatic or amphibious reptiles appear to have a more permeable skin than desert-adapted species. For example, Bentley and Schmidt-Nielsen (1966) have found the rate of water loss in a crocodilian (*Caiman sclerops*) to be 19 times higher than that in a desert lizard (*Sauromalus obesus*).

Since it has been shown that water will pass out through the skin of reptiles, the question arises as to whether or not reptiles can take in water through the skin. Bentley and Blumer (1962) have demonstrated that water does not pass through the skin of the Australian moloch (*Moloch horridus*, family Agamidae), as has frequently been stated. However, when the animal is belly deep, water will move by capillarity through the stratum corneum, and when it reaches the mouth the lizard can ingest it by moving its jaws. Pettus (1958) made *in vitro* studies of the permeability of water snake (*Natrix sipedon*) skin and concluded that no water transfer occurred through it. The question of water uptake through the skin is not a dead issue, however, for Bogert and Cowles (1947) noted that the Florida worm lizard (*Rhineura floridana*) loses weight relatively rapidly if it is kept in dry sand but gains weight again if placed in wet sand. They concluded that water must be absorbed through the skin, since their sand was not wet enough to allow the animals to ingest water with their mouths.

Reptilian lungs tend to have a greater internal surface area per unit of lung volume than those of amphibians; this means that the respiratory surfaces are exposed to a relatively small internal atmosphere, minimizing the evaporative loss of water through respiration. One would expect the respiratory loss of water to generally increase with an increase in respiratory rate and ventilation of the lungs, but it appears that this may not always be true. Templeton (1960) has found that although the total evaporative loss of water from the iguanid lizard *Dipsosaurus dorsalis* increased exponentially with temperature, there was no abrupt change at 44° C when panting began.

The loss of water through elimination of body wastes is minimized in reptiles by absorption of water from both urine and feces. Although reptiles are able to excrete ammonia, urea, and uric acid, there is a correlation between the habitat and the predominant nitrogenous waste product. Crocodilians and marine and fresh-water aquatic or amphibious turtles excrete primarily ammonia and urea; terrestrial turtles and Squamata excrete primarily uric acid. Reptiles cannot form urine with a concentration exceeding that of their plasma; therefore, the ability to excrete uric acid is particularly advantageous when there is a need to conserve body water. Because it is relatively insoluble compared to ammonia or urea, uric acid allows a greater reabsorption of water in the kidney tubules than occurs in amphibians: the uric acid precipitates out before the osmotic concentration of the solution reaches any large value, and the "work" of reabsorption in the tubule is reduced. The roles of filtration, secretion, and reabsorption in the kidney function of reptiles are not well known, but some interesting data are available. Marshall (1932) studied kidney secretions in the iguana (*Iguana iguana*) and found that uric acid is not only filtered in the glomerulus of the kidney but is also added to the urine by active secretion as the fluid passes down the tubule; in fact, his data indicated that only 6 per cent of the uric acid entered the urine by filtration, the remaining 94 per cent being added by secretion. Shoemaker *et al.* (1966) studied the effects of temperature on kidney function in the skink *Tiliqua rugosa* and found that both the glomerular filtration rate and the net tubular reabsorption of sodium were temperature dependent. However, the stimulatory effect of temperature on sodium reabsorption was neutralized by the increased urine production, so that the sodium concentration of urine remained similar between temperatures of 17.5° and 37° C.

In reptiles (and birds) additional reabsorption of water may take place in the cloaca so that both urine and feces are frequently voided in a semisolid condition. Theoretically, the removal of water from cloacal contents could be accomplished by either the active transport of water or the reabsorption of osmotically active substances, with the water following passively through osmosis. As noted by Schmidt-Nielsen *et al.* (1963), if water were actively reabsorbed and cations such as sodium and potassium were to remain in the cloaca, the osmotic work involved would be very great and would increase as the cloacal contents became more concentrated. It would seem impossible to produce semisolid excrements in this manner. On the other hand, if the solutes were reabsorbed by active transport from the cloacal contents and the water followed passively, osmoregulation would require another mechanism for the elimination of ions. In a variety of reptiles, as in birds, this mechanism consists of salt-secreting glands located adjacent to nasal passages or the orbits. Salt-secreting glands are found in all marine reptiles, including a variety of turtles, the marine crocodile *Crocodylus porosus*, the marine iguana *Amblyrhynchus cristatus*, and sea snakes in the family Hydrophidae. They are also known to occur in some terrestrial iguanid lizards, including *Iguana iguana*, *Dipsosaurus dorsalis*, and chuckwallas in the genus *Sauromalus*, and in the agamid lizard *Uromastix aegyptius* (Templeton, 1963). Analyses of the secretions from these glands, which empty into the nasal passages of lizards and the orbits of turtles, sea snakes, and the crocodile, have shown the presence of sodium, potassium, chloride, and bicarbonate. The relative amounts of these to be eliminated depend upon the salt content of food ingested and, therefore, vary with the diet. Plant material is high in potassium and bicarbonate, and herbivorous species secrete high proportions of these from their salt glands. Marine species take in major amounts of sodium and chloride and thus secrete these.

The hormonal regulation of water balance in reptiles has not been studied in much detail but it is generally assumed to be much the same as that occurring in amphibians. Arginine vasotocin has been identified in the pituitaries of all reptiles examined and, as in amphibians, is considered to primarily function as an antidiuretic hormone. Dantzler (1967) has been able to demonstrate that arginine vasotocin functions in the water snake (*Natrix sipedon*) to depress tubular permeability to water and also causes a reduction in the glomerular filtration rate, an increase in tubular reabsorption of sodium, and a decrease in tubular secretion of potassium. The results of several studies on a variety of reptiles indicate that reduced glomerular filtration rates in reptiles are due to a reduction in the number of functioning glomeruli rather than to changes in the filtration process. Presumably when reptiles are in a normal condition, arginine vasotocin is released continuously into circulation, causing an osmotic uptake of sodium which is accelerated by an increased pore diameter of the tubule cell membranes, and a minimal elimination of water. When a large volume of water is taken in, the secretion of arginine vasotocin decreases and there is a sharp increase in the elimination of water. Oxytocin has also been found in the neurohypophyseal extracts of turtles, and 8-isoleucine oxytocin has been found in neurohypophyseal extracts of rattlesnakes. Although very high concentrations of these have caused an antidiuretic effect in some reptiles, the function of the oxytocins in reptiles is not clear at this time, since the doses injected are higher than would be expected to ever occur naturally. Bentley (1959) found that adrenalectomy of the lizard *Tiliqua rugosa* resulted in a rise in the plasma concentration of potassium but no change in the concentration of sodium. Otherwise, little is known about the relationship between the adrenocortical hormones and osmoregulation in reptiles. Injections of ACTH caused an increase in plasma sodium concentration in the lizard *Agama agama*. This, again, indicates a hormonal mechanism similar to that of amphibians. The reptilian response to desiccation appears to be a glomerular "shutdown," which would conserve body water at the expense of an alteration in composition of body fluids. Some substances, such as uric acid and probably potassium, can be excreted by tubular secretion in reptiles and, being normally low in concentration in the plasma, would not require much urine for their excretion. However, the concentration of other substances, such as sodium, must rise as a result of continued water loss due to evaporation. Thus, desert-adapted reptiles must be able to either eliminate surplus salts through salt glands or to tolerate hypernatraemia and increases in intracellular concentration.

In recent years, comparative studies have been made of a variety of lizards, primarily desert-adapted iguanids and agamids, which indicate rates of evaporation and ability to tolerate desiccation vary significantly from species to species. Bogert and Cowles (1947) maintained experimental conditions at a temperature of 38° C and a relative humidity of 37 per cent and then studied the ability to resist desiccation in different species of lizards. They found that xeric species were able to withstand the laboratory conditions for 99.5 hours and lost water at the rate of 0.2 per cent of their body weight per hour. Species from mesic habitats were only able to withstand the artificial conditions for 9.5 hours and lost water at the rate of 1.4 per cent of their body weight per hour. Aquatic species lost water at even higher rates. Warburg (1966) has studied the evaporative water loss of several small Australian lizards under different conditions of temperature and humidity. He has found that, in general, agamids (five species) conserved water more effectively than either skinks (six species) or geckos (six species) but that both geckos and skinks from arid habitats lost relatively small quantities of water at temperatures up to 35° C. Although differences between studies in technique and body temperature and differences in weight of species studied tend to cloud interpretation of the adaptive significance of variation in rates of water loss, the general conclusion from these and other studies is that lizards occupying xeric habitats are better able to conserve their body water than species occurring in mesic habitats. In addition, species in xeric habitats seem to be able to tolerate greater losses of body water than those in mesic habitats. At least some xeric-adapted reptiles also have special water-storing abilities which are undoubtedly important to their survival. Chuckwallas (*Sauromalus*) possess accessory lymph spaces along their lateral abdominal folds which allow them to store volumes of extracellular fluid when moisture is available during favorable periods of the year and use it during dry seasons. Norris and Dawson (1964) have found that the volume of these spaces seemed to vary from species to species in correlation with rainfall in their habitats. Like anurans, at least some reptiles seem to use the bladder as a storage reservoir for use during dry seasons or hibernation. Dantzler and Schmidt-Nielsen (1966) have found that normally uric acid precipitates out in the bladder of the desert tortoises (*Gopherus agassizii*) and water is reabsorbed through the bladder wall and that when the animal is hydrated urine does not enter the bladder but runs directly out of the cloaca, probably as a result of closing of the bladder sphincter.

The amount of free water required by reptiles is generally not known, even for desert lizards, which have been studied to the greatest degree. Desert reptiles rarely have access to permanent bodies of water, and it may be assumed that they drink little or nothing for extended periods of time each year. Observations of certain species of lizards bear this out. *Holbrookia maculata* (lesser earless lizard) have been kept under laboratory conditions for five weeks with no water and appeared to have suffered no ill effects; and Schmidt-Nielsen (1964) mentions a captive snake (*Spalerosophis diadema*) that lived for five years without water. Many people who have kept and studied lizards have observed them drinking, indicating that their instinct is to do so, but these observations do not clarify the importance of free water to their water balance. In his study of adaptation to sea water, Pettus (1963) has concluded that *Natrix sipedon clarki*, which inhabits salt-water marshes, is able to survive in this habitat because it doesn't drink the water, whereas the closely related *Natrix sipedon confluens*, which inhabits fresh-water marshes, drinks (and dies as a result) when put in salt water.

Although they generally tolerate lower humidities than amphibians, reptiles respond to evaporative conditions and regulate their water balance to a large degree through habitat selection and behavior. Some reptiles which occur in rather arid regions confine themselves to very specific microhabitats where moisture is available. For example, the many-lined skink (*Eumeces multivirgatus*) lives in a variety of macrohabitats from short-grass prairie to mountains but is most abundant where there is water or moist subsoil. In more arid areas, these skinks are found almost exclusively under surface objects such as cow chips, where both food and moisture are available. Patterns of activity also play an important role in the water relations of reptiles. A variety of lizards may be active throughout the day in desert habitats, but many reptiles are crepuscular or nocturnal. For example, the Gila monster (*Heloderma suspectum*) and desert tortoises (*Gopherus agassizii*) remain in burrows during the heat of the day and venture out only when the humidity is higher and the evaporation rates lower. Burrowing is an extremely effective means of conserving water, since the animal avoids the hot surface temperatures and surrounds itself with a cool moist environment in which, by remaining inactive, it will loose very little, if any, water. Like amphibians, desert reptiles often aestivate in similar burrows to avoid dry seasons. In addition to having behavioral patterns for conserving their water, desert reptiles seem to take advantage of the availability of free water, such as occurs after infrequent heavy rains, by instinctively drinking great quantities of it. For example, Miller (1932) reported that desert tortoises increased their weight by 41 to 43 per cent by drinking! Woodbury and Hardy (1948) concluded that such tortoises are not dependent upon drinking water for their survival but, rather, obtain their water from free water in food and from metabolic water. However, since they both drink instinctively and go weeks or months without voiding their bladder water, it would seem that they store water for future use whenever it is possible to do so.

Obviously, a great deal remains to be learned about water relations of amphibians and reptiles. It is to be expected that there are many differences between xeric and mesic species and that water balance plays an important role in natural selection. For example, selection should favor large individuals, with small ratios of surface to mass, in xeric habitats. Smaller forms might be more successful in moist habitats.

References

Bastian, J. W., and M. X. Zarrow. 1954. Stimulation of the secretory glands of the skin of the South African frog (*Xenopus laevis*). Endocrinology 54: 116–117.

Benedict, F. G. 1932. The physiology of large reptiles with special reference to heat production of snakes, tortoises, lizards and alligators. Washington, D.C., Carnegie Institute. Publ. 425, 539 pp.

Bentley, P. J. 1959. Studies on the water and electrolyte metabolism of the lizard *Trachysaurus rugosus* (Gray). J. Physiol. (London) 145:37–47.

———, and W. F. C. Blumer. 1962. Uptake of water by the lizard *Moloch horridus*. Nature 194:699–700.

———, and K. Schmidt-Nielsen. 1966. Cutaneous water loss in reptiles. Science 151:1547–1549.

Bogert, C. M., and R. B. Cowles. 1947. Results of the Archbold expedition. No. 58. Moisture loss in relation to habitat selection in some Floridian reptiles. American Mus. Nov. 1358:1–34. (12 figures.)

Boyd, E. M., and D. W. Whyte. 1938. The effect of extract of the posterior hypophysis on the loss of water by frogs in a dry environment. Am. J. Physiol. 124:759–766.

Chew, R. M., and A. E. Dammann. 1961. Evaporative water loss of small vertebrates, as measured with an infrared analyzer. Science 133:384–385.

Dantzler, W. H. 1967. Glomerular and tubular effects of arginine vasotocin in water snakes (*Natrix sipedon*). Am. J. Physiol. 212:83–91.

———, and B. Schmidt-Nielsen. 1966. Excretion in fresh-water turtle (*Pseudemys scripta*) and desert tortoise (*Gopherus agassizii*). Am. J. Physiol. 210:198–210.

Dawson, W. R., V. H. Shoemaker, and P. Licht. 1966. Evaporative water losses of some small Australian lizards. Ecology 47:589–594.

Deyrup, I. J. 1964. Water balance and kidney. In: J. A. Moore (ed.), Physiology of the Amphibia. New York, Academic Press. Pp. 251–328.

Ewer, R. F. 1952. The effects of posterior pituitary extracts on water balance in *Bufo carens* and *Xenopus laevis*, together with some general considerations of anuran water economy. J. Exp. Biol. 29:429–439.

Garby, L., and H. Linderholm. 1953. The permeability of the frog skin to heavy water and to ions, with special reference to the effects of some diuretics. Acta Physiol. Scand. 28:336–346.

Goldsmith, G. W. 1925–1926. Habits and reactions of *Scaphiopus hammondi*. Yearbook Carnegie Inst. Washington 25:369–370.

Gordon, M. S. 1965. Intracellular osmoregulation in skeletal muscle during salinity adaptation in two species of toads. Biol. Bull. (Mar. Biol. Lab.) 128: 218–229.

Hevesy, G., E. Hofer, and A. Krogh. 1935. The permeability of the skin of frogs to water as determined by D_2O and H_2O. Skand. Arch. Physiol., 72:199–214.

Koefoed-Johnsen, V., and H. H. Ussing. 1953. The contribution of diffusion and flow to the passage of D_2O through living membranes. Acta Physiol. Scand., 28:60–76.

Krogh, A. 1937. Osmotic regulation in the frog (*Rana esculenta*) by active absorption of chloride ions. Skand. Arch. Physiol. 76:60–74.

——— 1939. Osmotic Regulation In Aquatic Animals. London, Cambridge Univ. Press. 242 pp.

Littleford, R. A., W. F. Keller, and N. E. Phillips. 1947. Studies on the vital limits of water loss in the plethodont salamanders. Ecology 28:440–447.

Maderson, P. F. A. 1964. The skin of lizards and snakes. British J. Herpetol. 3:971–973.

Marshall, E. K., Jr. 1932. Kidney secretion in reptiles. Proc. Soc. Exp. Biol. 29:971–973.

Mellanby, K. 1941. The body temperature of the frog. J. Exp. Biol. 18:55–61.

Miller, L. 1932. Notes on the desert tortoise (*Testudo agassizii*). Trans. San Diego Soc. Nat. Hist. 7:187–208.

Morel, F., J. Maetz, R. Archer, J. Chauvet, and M. T. Lenci. 1961. A "natriferic" principle other than arginine-vasotocin in the frog neurohypophysis. Nature 190:828–829.

Noble, G. K. 1931. The Biology of the Amphibia. New York, McGraw-Hill. 577 pp.

Norris, K. S., and W. R. Dawson. 1964. Observations on the water economy and electrolyte excretion of Chuckwallas (*Lacertilia, Sauromalus*). Copeia *1964*:638–646.

Pettus, D. 1958. Water relationships in *Natrix sipedon*. Copeia *1958*:207–211.

_____ 1963. Salinity and subspeciation in *Natrix sipedon*. Copeia *1963*:499–504.

Prosser, C. L. (ed.) 1950. Comparative Animal Physiology. Philadelphia, W. B. Saunders. 888 pp.

Rey, P. 1938. Récherches expérimentales sur l'économie de l'eau chez les Batraciens. Ann. Physiol. Physico-Chim. Biol. *14(1)*:1–66.

Schmidt-Nielsen, B., and E. Skadhauge. 1967. Function of the excretory system of the crocodile (*Crocodylus acutus*). Am. J. Physiol. *212*:973–980.

Schmidt-Nielsen, K. 1964. Desert Animals; Physiological Problems of Heat and Water. London, Oxford Univ. Press. 277 pp.

_____, A. Borut, P. Lee, and E. Crawford, Jr. 1963. Nasal salt excretion and the possible function of the cloaca in water conservation. Science *142*:1300–1301.

Shelford, V. E. 1914. Modification of the behavior of land animals by contact with air of high evaporation power. J. Animal Behav. *4*:31–49.

Shoemaker, V. H., P. Licht, and W. R. Dawson. 1966. Effects of temperature on kidney function in the lizard *Tiliqua rugosa*. Physiolog. Zool. *39*:244–252.

Templeton, J. R. 1960. Respiration and water loss at the higher temperatures in the desert iguana, *Dipsosaurus dorsalis*. Physiolog. Zool. *33*:136–145.

_____ 1963. Nasal salt secretion in terrestrial iguanids. Am. Zool. *3*:530.

Tercafs, R. R. 1963. Phénomènes de perméabilité au neveau de la peau des reptiles. Arch. Intern. Physiol. Biochem. *71*:318–320.

Thorson, T., and A. Svihla. 1943. Correlation of the habitats of amphibians with their ability to survive the loss of body water. Ecology *24*:374–381.

Warburg, M. R. 1966. On the water economy of several Australian geckos, agamids, and skinks. Copeia *1966*:230–235.

Woodbury, A. M., and R. Hardy. 1948. Studies of the desert tortoise *Gopherus agassizii*. Ecol. Monographs *18*:145–200.

CHAPTER 8

TEMPERATURE RELATIONS OF AMPHIBIANS AND REPTILES

Of all the extrinsic factors affecting organisms, temperature is obviously one of the most important. All chemical processes, including those of living systems, proceed at a rate which is dependent upon the temperature. Metabolism, complex as it is, is no exception. The metabolic rates of amphibians and reptiles and all correlated phenomena increase with temperature, from the lowest level tolerated to a level near the upper lethal limit. The increase in metabolic rate is generally exponential, and a 10° centigrade rise in temperature approximately doubles the metabolic rate. Related to this is the fact that all animals, including amphibians and reptiles, can function normally only within a limited range of body temperatures; the regulation of body temperature, passively or actively, is essential to the well-being of the individual. The ramifications of temperature are not limited, however, to metabolic rate and its effect on behavior and activity. Temperature also has an effect on such things as duration of different stages in life.

Amphibians and reptiles have classically been described as being cold-blooded and poikilothermic because of the general belief that their body temperatures are both low and extremely variable. In contrast, birds and mammals have classically been described as being warm-

blooded and homeothermic because of their generally high and supposedly constant body temperatures. However, the results of various studies have indicated that these terms are not accurate, since the temperatures of "homeothermic" animals are known to fluctuate significantly, both within a 24-hour period as a result of changes in activity (eating, sleeping, etc.) and through life as a result of aging effects. Similarly, studies of "poikilotherms" have indicated that at least some of the animals in this category maintain body temperatures which may be higher than those of so-called "warm-blooded" animals and which may be relatively constant. As noted by Brattstrom (1965), some lizards live at and maintain relatively constant body temperatures above those which are lethal for monotremes, and yet lizards have been categorized as "cold-blooded" and monotremes as "warm-blooded." As a result of the inaccuracy of older terms, modern physiological ecologists have adopted terms which indicate the source(s) of heat primarily affecting organismal temperatures. The terminology followed here is that of Cowles (1962).

Two general sources of heat available to organisms are their environment and metabolism; the terms indicating these sources are, respectively, *ectotherm* and *endotherm*. Ectothermic animals are those which depend primarily upon external (environmental) sources of heat to maintain their body temperature. Endothermic animals primarily utilize internal (metabolically-produced) sources of heat to maintain their body temperatures. There are fundamental differences between ectotherms and endotherms. Because their heat is produced internally, it is advantageous for endotherms to have insulation (hair, feathers, blubber, and so forth) to restrict heat loss through the body surface. A similar layer of insulation surrounding an ectotherm would effectively deprive it of most of its needed thermal energy. The circulatory system of an ectothermic organism has a primary function of heat uptake at the surface and distribution of this heat into the body, whereas in the endotherm the energy-expensive heat generated by metabolic processes in deeper tissues is distributed outward by the circulatory system.

Because they all depend primarily upon environmental sources of heat, amphibians and reptiles are ectotherms. It is important to realize that the terms endotherm and ectotherm imply nothing regarding the relative temperature of the organism. It is also important to realize that the body temperatures of ectothermic organisms are usually, or at least frequently, very different from that of their environment. For example, Pearson (1954) collected an iguanid lizard (*Liolaemus multiformes*) at 12,900 feet in the Andes of Peru which had a cloacal temperature of 31° C when the air temperature in the shade nearby was zero!

Ectotherms may derive their heat either by basking or through conduction from their environment. The term *heliotherm* is used to describe ectotherms, including many lizards and turtles, which rely on periodic basking for their thermoregulation. In contrast, *thigmotherms* are those ectotherms which derive their heat solely by conduction from the medium in which they live, whether air, water, or soil. Thigmotherms have little or no capacity for thermoregulation other than the avoidance of excessive environmental temperatures or thermal changes through their distribution and movements. Burrowing lizards and snakes, all strictly nocturnal amphibians and reptiles, and some aquatic amphibians (larvae and adults) and reptiles are thigmotherms. It should be noted, however, that many aquatic amphibians and reptiles periodically bask and, therefore, are heliotherms.

It is not possible to categorize all animals as being either ectothermic or endothermic. Several small birds (swifts and hummingbirds, for example) are unable to rely solely on either metabolic or external sources of heat but, instead, must regularly utilize both sources. These organisms are termed *heterotherms*. What is more pertinent here is that some ectothermic reptiles are capable of utilizing metabolic heat to supplement external sources in order to maintain a body temperature that is higher than that of their surroundings. Hutchison *et al.* (1966) found that a brooding Indian python (*Python molurus bivittatus*) could regulate her body temperature by physiological means analogous to those in endothermic animals. Brooding and non-brooding pythons were placed in an environmental control chamber, and changes in metabolic rate as measured by oxygen consumption were recorded as temperature was changed. The oxygen consumption of a non-brooding individual followed a pattern which was characteristic of ectothermic animals, increasing with increasing temperature and decreasing with decreasing temperature. However, the oxygen consumption of the same individual during brooding followed a pattern similar to that of an endotherm between temperatures of 25.5° and 33° C, increasing with decreasing temperature and decreasing with increasing temperature. The increased metabolic rate was correlated with increased muscular spasmodic contractions, and the heat produced resulted in an elevation in body temperature of up to 4.7° C above that of the substrate or ambient air. However, below temperatures of 25.5° C the animal was unable to increase its metabolic rate. Thus, within a limited temperature range, the brooding Indian python exhibits what might be termed facultative endothermy. Interestingly, this is not new information, since Lamarre-Picquot read a communication before the French Academy in 1832 in which he stated that the python, after laying eggs, coils about them and produces sensible heat as an aid to incubation. A committee of the French Academy, however, rejected Lamarre-Picquot's statements as being "hazardous and questionable"!

Temperature regulation by internal heat production and by changes in heat transport by the circulatory system have also been described in monitor lizards (family Varanidae). The metabolically-produced heat is appreciably less in the lizards than in the brooding Indian python, however, for the maximum elevation of body temperature above ambient known for the lizards is 2° C, compared to 7.3° C recorded for the brooding python (Bartholomew and Tucker, 1964). Lizards in the families Agamidae, Scincidae, Iguanidae, and Varanidae have also been found capable of exerting physiological control over the rates at which their body temperatures change (Bartholomew and Tucker, 1963, 1964; Bartholomew et al., 1965). This control is by the metabolic production of heat, which accelerates heating and retards cooling, and by cardiovascular adjustments which modify heat transport. The importance of this control is that it increases the duration of time that the lizard can spend at its desired temperature.

AMPHIBIAN TEMPERATURE RELATIONS

Many observations have been made of the effects of environmental temperature on the behavior of amphibians. The behavioral patterns of non-tropical amphibians generally include retraherence, which is temporary retreat from adverse weather. For example, laboratory frogs (Rana pipiens) may generally be induced to seek a retreat under objects in the bottom of an aquarium by lowering the temperature below 8° C. The same, or analogous, behavior is exhibited by most anurans living in temperate areas where they are subjected to seasonal climatic changes. Salamanders also avoid extremely cold temperatures by seeking shelter under objects, either on land or in water. Caecilians, being tropical animals, are not exposed to greatly fluctuating temperatures and do not have comparable behavior patterns. Most amphibians avoid high temperatures by seeking bodies of water

or by burrowing under objects or into the soil. Diurnal patterns of activity also are adaptive to prevailing temperature conditions. Nocturnal behavior is an adaptation which results in exposure to cooler temperatures (and higher humidities) than occur in the daytime, and it is especially noteworthy that the majority of amphibians in warm arid areas are nocturnal. In contrast, amphibians which occur in subpolar regions or at high elevations tend to be diurnally active and, accordingly, are exposed to warmer temperatures than occur at night.

Amphibians outside the tropics frequently hibernate over the winter. The advantages of hibernation are obvious. It allows survival without injury for extended periods of time during which amphibians would surely die if they attempted to winter actively in their usual habitat. Frogs, in general, can survive temperatures of from 0° to 9° C while hibernating, and some European frogs have been credited with surviving temperatures of −4 to −6° C (Müller-Erzbach, 1891). Amphibians typically hibernate under debris, in the soil, or in water. However, the same species may hibernate in different situations in different parts of its range. Amphibians frequently migrate in the fall to particular places for purposes of hibernation and their winter habitat may be different than their summer habitat. The cricket frog (Acris gryllus) has been observed to leave ponds and marshes where it summers and to migrate to rocky ravines in the fall, where it hibernates out of water under stones and logs. Some terrestrial salamanders, such as the dusky salamander (Desmognathus fuscus), hibernate under rocks in mountain streams. Other salamanders, including the European fire salamander (Salamandra salamandra), have been found in aggregations in hibernating dens.

Not only do amphibians avoid lethal temperatures through their behavior, but they also exhibit definite preferences within the temperature range of tolerance. Thus, the temperatures which are preferred by one species may be avoided by another, and consequently the two species will be active at different times or will occur in different habitats. Temperature exerts a strong control over the distribution of amphibians, and those with wide

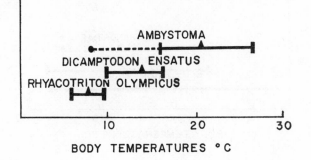

Figure 8–1 Ranges (bars) and means (triangles) of body temperatures of active individuals of three genera of salamanders in the family Ambystomatidae. (From Brattstrom: A preliminary review of the thermal requirements of amphibians. Reproduced by permission of the Duke University Press.)

tolerances for temperature tend to be more widely distributed and occur in a greater variety of habitats than those with narrow temperature tolerances. Ecotypic variation in regard to temperature tolerance undoubtedly also occurs in widely distributed species.

Brattstrom (1963) reviewed the thermal requirements of amphibians on the basis of about 6500 body, air, soil, and water temperatures taken on 99 species from various parts of North and Central America with a Schultheis quick-recording thermometer or a Telethermometer thermister. The following discussion of amphibian temperature relations has been based largely on Brattstrom's preliminary conclusions, and the reader is urged to consult his paper for details and additional references.

SALAMANDERS

Body temperatures of salamanders in nature tend to be relatively low, but in Brattstrom's data they range from −2° to 26.7° C and Zweifel (1957) mentions collecting red-spotted newts (*Notophthalmus v. viridescens*) at Mountain Lake, Virginia, in water at 34° C. Most salamanders appear to have no "preferred" temperature but rather will generally accept the temperatures within their range of tolerance that happen to be available. Despite this general lack of temperature preference, there seem to be specific, generic, and familial differences in temperature tolerance ranges (Figs. 8–1, 8–2, and 8–3), indicating that there may be thermoregulation by habitat selection and distribution. However, such data as are presented in these figures must be used cautiously, since they probably do not encompass the total range of temperatures experienced by each species and one does not know the activities of the individuals at the time their temperatures were recorded.

Aquatic salamanders generally seem to be thigmotherms and usually have a body temperature identical with the tempera-

Figure 8–2 Ranges (bars) and means (triangles) of body temperatures of active individuals of several genera of salamanders in the family Plethodontidae. (From Brattstrom: A preliminary review of the thermal requirements of amphibians. Reproduced by permission of the Duke University Press.)

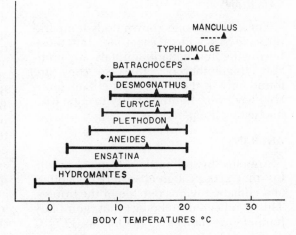

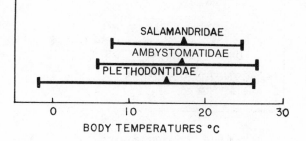

Figure 8–3 Summarized ranges (bars) and means (triangles) of body temperatures of active individuals of three families of salamanders. (From Brattstrom: A preliminary review of the thermal requirements of amphibians. Reproduced by permission of the Duke University Press.)

ture of the water in which they are found. They apparently do not bask at the surface of the water, but there is some indication that in shallow water they may absorb a small amount of heat by conduction from the substrate of the pond.

Terrestrial salamanders appear to be generally thigmothermic, but their temperature is significantly affected by evaporative cooling. When they are found under objects on a moist substrate, their temperature is usually the same as that of the substrate. When terrestrial salamanders are out in the open they may be as cold or colder than the air, owing to the evaporation of water from their skin. Their temperature approximates that of a wet-bulb thermometer: the drier the air, the more rapidly they evaporate water and the colder their temperature will be compared to that of the air.

Salamanders have the ability to alter their temperature tolerance range by acclimation, but otherwise there is no indication that they can control their temperature relations by any means other than behavior.

CAECILIANS

There is only one known body temperature (28.8° C) for a caecilian, but these burrowing and aquatic animals are probably thigmothermic. Because they are found only in the tropics, caecilians probably have very limited ability to cope with fluctuating temperatures.

ANURANS

Anurans are primarily a tropical group, but they occupy a great variety of habitats and climatic regions outside of the tropics. This is reflected in the fact that their body temperature ranges from 3.0° to 35.7° C

in Brattstrom's data. The same caution as with the salamander data must be used here, although anurans, too, seem to exhibit specific, generic, and familial differences in their body temperatures (Figs. 8–4 and 8–5). Some species of anurans (the northern chorus frog *Pseudacris triseriata*, for example) are able to tolerate wide ranges of temperature and appear to have no "preferred" temperature, but others (most hylids) seem to have rather definite "preferred" temperatures and narrow ranges of tolerance. The anurans that are tropical never encounter temperatures lower than 15° or 20° C and probably have little ability to tolerate lower temperatures.

Like those of other amphibians, anuran body temperatures are affected by the evaporation of water from their skin, and many species seem to use this as a cooling mechanism, both in normal temperature regulation and in stress situations. Many anurans, including most ranids, spend much time perched on shore or rocks out of the water and are able to raise their body temperatures above that of the surrounding air by basking and conduction. This, of course, increases the rate of cutaneous evaporation of water and, consequently, imposes a water balance problem on the amphibian. Thus, basking among anurans seems to be restricted to those species, populations, or individuals which live in the vicinity of permanent water. Because of the effects of evaporative cooling, conduction, and basking, the body temperatures of anurans may deviate considerably from that of the ambient air, substrate, or water. If the wind is not blowing and it is cloudy or night, or if the anuran is in a secretive situation, its body temperature may approach that of the substrate, air, or water. Free-swimming an-

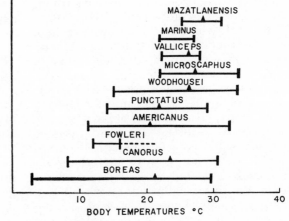

Figure 8–4 Ranges (bars) and means (triangles) of body temperatures of active individuals of several species of toads in the genus *Bufo*. (From Brattstrom: A preliminary review of the thermal requirements of amphibians. Reproduced by permission of the Duke University Press.)

uran larvae, particularly those of species breeding in temporary water, often possess the thermotaxic response of dense aggregation. Such thermal aggregations maintain the temperatures of the individual developing larvae at a higher level than they would otherwise be and, therefore, increase the tadpoles' metabolic and developmental rates, shortening the time required to reach metamorphosis.

There is no evidence that any amphibian can raise its body temperature above that of the environment by utilization of metabolic heat. Whatever heat is generated metabolically is of such a small magnitude (Fromm, 1956) that it must be lost immediately to the environment. Thus, amphibians gain heat primarily by basking in the sun, by conduction from their substrate and water, and by convection from air. They lose heat primarily by conduction and radiation to the surrounding environment (substrate, air, water), by convection to air, and, very importantly, through evaporative cooling. The effectiveness of evaporative cooling in amphibian temperature regulation is indicated by Thorson's (1955) data indicating that a frog maintained a body temperature of 36.8° C while in a desiccator at a temperature of 50.0° C. Mellanby (1941) reported that the body temperature of toads (*Bufo*) was 26.5° C when the environmental air temperature was 27.6° C and the relative humidity was 82 per cent; at the same temperature, with a relative humidity of 27 per cent, the toad's body temperature was 17.5° C. Mellanby also found that at a temperature of 20° C a frog produced CO_2 at a rate equivalent to six calories per hour but simultaneously lost 3.2 grams of water per hour by evaporation and, hence, 2000 calories.

Amphibians have the ability to acclimate to changes in environmental temperatures and, thereby, alter their

Figure 8–5 Summarized ranges (bars) and means (triangles) of body temperatures of active individuals of several families of anurans. (From Brattstrom: A preliminary review of the thermal requirements of amphibians. Reproduced by permission of the Duke University Press.)

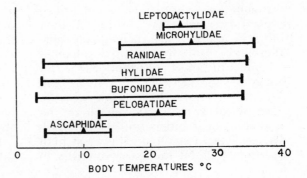

temperature tolerance ranges. This acclimation may occur in a few days (Brattstrom and Lawrence, 1962; Hutchison, 1961) and, while not being rapid enough to allow them to adjust to diurnal temperature fluctuations, is rapid enough to be of survival value in adjusting to seasonal changes in the environment. That this occurs naturally is indicated by Hutchison's (1961) observation that newts (*Notophthalmus viridescens*) had critical thermal maximum–temperature curves which were significantly higher for individuals collected in the fall and winter than for those collected in spring and early summer. The mechanism of acclimation is not completely understood but changes in body water accompanying acclimation have been found in several different kinds of animals. In general, there seem to be an increase in weight in animals acclimated to lower temperatures and a decrease in weight in animals exposed to higher temperatures. This seems to be a factor in amphibian acclimation, for Hutchison (1961) has found that newts desiccated until they had lost 14.0 to 21.5 per cent of their original weight had significantly higher (1.07° C) critical thermal maxima than control animals that were not dehydrated.

Several studies have shown that the rate and range of acclimation are dependent upon the temperature to which the animal is already acclimated and that there are differences between species. Hutchison's (1961) data indicate that the rate of acclimation of newts is slower at lower than at higher temperatures. Interspecific variation in acclimation ability was found by Brattstrom and Lawrence (1962); two ranids (*Rana pipiens* and *Rana clamitans*), *Scaphiopus holbrooki* and *Bufo fowleri* were able to adjust very rapidly to changing environmental temperatures (*Rana pipiens* lowered their critical thermal maximum 3.3° C in two days), but a third ranid, *Rana palustris*, showed a poor ability to adjust. That such interspecific differences result from natural selection and in turn affect distributions is indicated by the fact that, in general, amphibian species which occupy wide geographical areas or occur where the environmental temperature changes rapidly, such as in small streams or at high elevations, exhibit the greatest ability to

acclimate. The ability of an animal to acclimate in an environment where rapid temperature changes occur is advantageous for two reasons: first, by being able to adjust its temperature tolerance range, the animal can survive what would otherwise be adverse temperature extremes; and second, acclimation allows the animal to function efficiently at different temperature levels as they occur seasonally in the environment.

In summary, the regulation of body temperature by amphibians is a complex subject, for it involves distribution and habitat selection, behavior (time of emergence, retreat, basking, hibernation, aestivation, and so forth), evaporative cooling, and physiological acclimation.

REPTILIAN TEMPERATURE RELATIONS

Increasing attention, generally stimulated by the early studies of Cowles and Bogert (1944), is being paid to thermal factors in ecological studies of reptiles. Most of the studies have been concentrated on heliothermic lizards inhabiting temperate arid and semiarid regions, but nevertheless a considerable amount is known about the thermal relations of other lizards, snakes, turtles, crocodilians, and even the tuatara. It is known that many reptiles have "preferred" body temperatures and, while active, are able to maintain their temperatures within a relatively narrow range by behavioral means. Other reptiles appear to have no preference and are active over a wide range of body temperatures. In order for the regulation of body temperature by behavior to be effective, the triggering of a specific behavioral response must be related to a given body temperature. Recent studies have begun to elucidate this relationship. Apparently, a given behavioral pattern occurs only over a narrow range of temperatures, but the problem of determining the relationship between temperature and behavior is complicated by the fact that, at least in some lizards, considerable temperature differences develop among parts of the body, one of which must act as the thermostat.

It is known that lizards may experience differences between head and body temperatures when basking with only part of the animal, usually the head, exposed to the sun. Bogert (1959) proposed that this behavior among *Holbrookia* (earless lizards) results in a warming of the entire body before morning activity is initiated. Stebbins (1954) suggested that this form of basking might warm only the head. Heath (1964) has tested both of these hypotheses and found that the exposure of just the head to radiant energy does not result in more rapid heating of the lizard's body than occurs in animals which are completely buried. This is because heat conducted from ambient sand contributes more to the body temperature of a lizard than heat conducted from the head. However, Heath found head-body temperature differences to be as much as 3° to 5° C and that head temperatures consistently exceeded body temperatures in partially buried animals. Animals which were active upon the substrate surface generally exhibited head-body temperature differences of 2° to 4°C, but the head-body temperature differences diminished as the body temperature approached the normal activity range. Heath proposed that the emergence of a lizard from a buried position during normal activity hours may be dependent upon a particular head temperature and may be independent of body temperature.

The fact that the head-body temperature difference diminishes at higher body temperatures suggests that the differences are regulated. Heath observed a relationship between eye-bulging and diminishing head-body temperature differences. Since eye-bulging results from the constriction of the *jugularis* muscle, a small sphincter surrounding the internal jugular vein near the posterior edge of the skull, a circulatory mechanism was implicated as the regulatory mechanism. The internal jugular vein, which is the only major vessel draining the head in lizards, and the internal carotid artery, a principal vessel bringing blood into the head from the body, are bound together in a connective tissue sheath. Heath proposed that the proximity of these two vessels may cause them to act as a counter-current heat exchange system when blood is flowing freely through both: Heat in the high-temperature blood leaving the head by way of the internal jugular vein is trans-ferred to the cool blood entering the head through the internal carotid artery and, therefore, is retained in the head, maintaining separate head and body temperatures. During eye-bulging, constriction of the *jugularis* causes an interruption of the blood flow in the jugular vein above where it is bound to the internal carotid artery, allowing cooler blood from the body to enter the head through the internal carotid while the hotter blood leaves the head by way of other vessels. The body is, accordingly, used as a sink into which excess cephalic heat is dumped until there is no head-body temperature differential. Heath's findings that the emergence of lizards is dependent upon head temperature and that head temperature seems to be what is regulated are in agreement with those of Rodbard (1948), who found that vasomotor activity in turtles is related to hypothalamic temperature and concluded that the hypothalamic region is the coordination center for metabolic adjustment to internal changes in temperature.

Regardless of an ability, or lack of an ability, to regulate body temperature, all reptiles studied have limitations in regard to maximum and minimum tolerable body temperatures. Lizards react to high temperatures in rather characteristic ways, typically assuming an erect posture with legs extended and tail arched sharply just above the base. This position is advantageous because it results in a reduction in the conduction of heat from the substrate into the animal's body. The fact that desert lizards tend to have long legs is related to this; they therefore have an ability to hold their body off the surface on hot days. Desert lizards are frequently out and active even on the hottest days, whereas snakes, unable to avoid surface high temperatures, are most frequently active at night when the surface is relatively cool. When they are exposed to extremely high temperatures, reptiles characteristically assume a defensive behavior including defecation, biting, hissing, and rolling over. All reptiles seem to depend to some degree upon panting as a cooling mechanism. Because this involves the evaporation of water from the lungs and is more efficient under conditions of low humidity than high humidity, the high temperature which may be tolerated by a reptile is partially determined by the humidity at the time.

In comparison to studies of the high temperature relations of reptiles, low temperature tolerances have been little investigated. Active lizards have been found with body temperatures as low as 4.9° C, but the general effect of such a low temperature is to slow down all bodily processes. Hence, the animal tends to become sluggish and finally, as its temperature drops, to become immobilized completely. Most reptiles studied have the ability to recover from immobilizing temperatures if gradually warmed. In temperate and subpolar regions, reptiles typically hibernate to avoid otherwise lethal cold temperatures. Many species, particularly snakes, characteristically hibernate in dens and form masses of large numbers of individuals. The advantage of this sort of massing is that it reduces the exposed surface area and, therefore, conserves the body heat of the mass as a whole.

As with amphibians, studies of reptiles indicate that many have an ability to acclimate to extreme temperature conditions if they are gradually exposed to increasing or decreasing temperatures over a period of time. As a result of acclimation, at least some species not only become tolerant of severe maximum and minimum temperatures but also undergo an adaptive change such that their metabolic rate at a high temperature will gradually approach that of a lower temperature. The reverse may also be true, in which case an animal at a low temperature will gradually exhibit a metabolic rate similar to those it had previously at higher temperatures. Like that of amphibians, reptilian acclimation is rapid enough to allow them to adjust to seasonal variations in temperature but does not allow them to deal with the normal diurnal temperature fluctuations they encounter. In a study of *Anolis carolinensis,* Licht (1968) observed that, while lizards acclimated to 32° C were more heat-resistant than those acclimated to 20° C (measured as survival time at 42.5° C), there was virtually no change in their preferred body temperature. These results indicate that at least this species will attempt to maintain its preferred temperature under a variety of temperature conditions and that the real

value of acclimation lies in resistance to extreme temperatures when behavioral regulation of body temperature fails.

Brattstrom (1965) has summarized the temperature relations of reptiles on the basis of about 9000 body, air, soil, and water temperatures for 161 species. The following are tentative conclusions and observations based on the data presented by Brattstrom. The reader is referred to his paper for specific data and literature references.

TUATARA

Based on 72 nocturnal and two diurnal body temperature determinations, the active body temperature range for *Sphenodon punctatus* is lower than for many salamanders, ranging from 6.2° to 18.0° C (Bogert, 1953). Tuataras are primarily nocturnal but are known to bask in the sun and to expose their heads above their burrows for periods of time. The two daytime temperatures recorded were 14.0° and 18.0° C and were of basking individuals, indicating that these animals might even prefer temperatures which are higher than they normally encounter at night.

CROCODILIANS

The only temperature data available for crocodilians are for the American alligator (*Alligator mississippiensis*). Colbert, Cowles, and Bogert (1946) found that active body temperatures for this species ranged from 26° to 37° C but that their preferred temperature range seemed to be 32° to 35° C, close to the lethal maximum of 38° to 39° C. Alligators utilize both basking and conduction in regulating their temperatures; there is no evidence that they are able to alter their body temperature with metabolic heat. An alligator exposed to temperatures at or near the lethal maximum of 38° C was able to maintain its body temperature slightly below the ambient temperature by evaporative cooling. When placed in water which was gradually heated, alligators stayed in the water until their body temperatures approached the lethal maximum of 38° C and only then left the water, indicating that they do not instinctively expect high temperatures while in water. Also, alligators placed in water that was gradually cooled left eventually, when their temperature dropped below their preferred range.

TURTLES

Body temperatures of 10 species in the families Chelydridae, Testudinidae, and Trionychidae range from 8.0° to 37.8° C, and most species appear to be active over a wide range of temperatures. Brattstrom concluded that there appear to be specific, generic, and possibly familial differences in the ranges of thermal tolerance and mean body temperatures. However, the recorded temperatures of active turtles from Chelydridae and Testudinidae broadly overlap (Fig. 8–6), and his data indicate the critical thermal maximum is about 41° C (39° to 42.3° C) for members of Trionychidae and Testudinidae (*Trionyx ferox, Testudo hermanni, Gopherus agassizii, Chrysemys picta*, and *Pseudemys scripta*). Some turtles, including painted turtles (*Chrysemys picta*), are heliothermic and bask at certain times of the day, particularly when the wind is not blowing. Judith Erb (*vide* Brattstrom, 1965) found that wind velocity, air temperature, water temperature, solar radiation, and position of the turtle on a log are critical factors affecting basking and rate of heat gain by painted turtles. She also noted that small juveniles of this species do not bask but apparently regulate their temperature by seeking warm water and floating with portions of their shells above water. Other species of turtles, such as musk turtles (*Sternothaerus*), are primarily thigmothermic and regulate their body temperatures by selection of habitat (water) temperatures. No turtle is known to be able to raise its body temperature by the use of metabolic heat.

Behavior plays an important role in the temperature relations of turtles, and activity patterns may change seasonally in response to changes in ambient temperatures. In temperate regions, turtles hibernate where they are protected from freezing temperatures. Fitch (1956) observed that in Kansas, ornate box turtles (*Terrapene ornata*) hibernate in the soil within a few inches of the surface, where the soil remains moist and where leaves and litter provide an insulating blanket. The frequent observation of painted turtles swimming about beneath the ice of lakes and ponds together with Sexton's (1959) observation that they enter deeper water in the autumn as temperatures become cooler indicate that this species hibernates under water, where evidently their cloacal and buccal respiration is sufficient to sustain them through the winter in an inactive state. Turtles also adjust their activity periods seasonally and thereby avoid extreme temperatures. Fitch (1956) noted that the ornate box turtles in Kansas become crepuscular and then nocturnal as hot summer weather develops.

SNAKES

Body temperatures from active snakes representing 38 species in the families Boidae, Colubridae, Viperidae, and Hydrophidae range between 9.0° and 38.0° C. Both the upper and lower extremes in this body temperature range are set by members of the family Colubridae (the rat snake *Elaphe obsoleta* and the common garter snake *Thamnophis sirtalis*, respectively), reflecting the tremendously broad distribution and diversification of this family. As with the amphibian data discussed earlier, caution must be taken in using these data. However, generic and

Figure 8–6 Ranges (bars) and means (triangles) of body temperatures of active turtles of Chelydridae and Testudinidae. (After Brattstrom: Body temperatures of reptiles.)

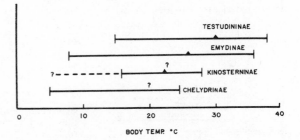

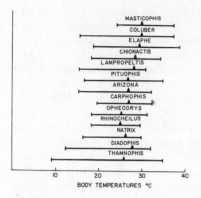

Figure 8-7 Ranges (bars) and means (triangles) of body temperatures of active individuals of several genera of snakes in the family Colubridae. (From Brattstrom: Body temperatures of reptiles.)

familial differences in body temperature are not obvious (Figs. 8–7 and 8–8). Many species of snakes are ecologically very adaptable and seem to have broad temperature tolerances. *Elaphe obsoleta* is distributed from southern Canada to southern United States, and although it provided the highest known active body temperature for a snake it has also had body temperatures as low as 18.2° C recorded. Similarly, *Thamnophis sirtalis*, distributed from British Columbia and Alberta into northern Mexico, provided the lowest body temperature recorded, but others ranged as high as 35.0° C.

Many snakes are heliothermic, at least during cool weather, and through their basking are able to raise their body temperatures above those of the air or soil. Snakes also generally select thermal gradients in their habitats and utilize the conduction of heat (in or out of their body) to regulate body temperature. Because the temperature of blacktop roads is generally warmer than that of adjacent soil, heating up faster in the daytime and cooling off more slowly at night, some nocturnal snakes find preferred thermal conditions on the road surface and are commonly collected by "road hunting." When snakes are unable to maintain their body temperatures within their activity range, they retire to underground retreats. Be-

yond this, they have no means of coping with either maximal or minimal extremes in temperature. The brooding Indian python is the only snake known to be able to raise and maintain its body temperature above environmental conditions by utilization of metabolic heat.

Like turtles, snakes frequently respond to seasonal temperature fluctuations by changing their patterns of activity, being diurnal during cool seasons, especially in the early spring, and nocturnal during hot seasons. Hibernation is the general rule for snakes inhabiting regions with cold winters, and this commonly involves migration to a common hibernaculum by many individuals whose summer ranges are distributed over an extensive area.

LIZARDS

The greatest volume of information on reptilian temperature relations is for lizards; the preferred temperatures are clearly defined for some species. Body temperatures recorded for 95 species representing 12 Old World and New World families range from 11.0° to 46.5° C. The upper temperature range (above 38° C) is higher than that recorded for any other major group of reptiles. There are distinct familial and generic differences in body temperature ranges of lizards

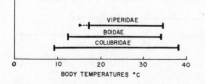

Figure 8-8 Ranges of body temperatures of active individuals of three families of snakes. (After Brattstrom: Body temperatures of reptiles.)

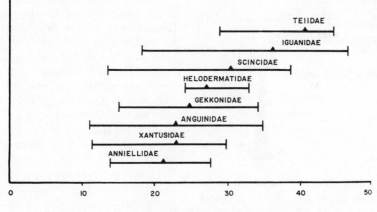

Figure 8–9 Ranges (bars) and means (triangles) of body temperatures of active individuals of eight families of lizards. (From Brattstrom: Body temperatures of reptiles.)

(Figs. 8–9 and 8–10). Some species, however, have no apparent preferred temperature and are active over a broad range of body temperatures, while others are able to regulate their body temperatures effectively and exhibit sharply-defined preferred temperatures and narrow activity ranges (Fig. 8–11). The majority of lizards are diurnally active and are at least partially heliothermic. Because of basking, lizards may have body temperatures which are significantly higher than environmental temperatures. The monitor lizards (family Varanidae) are the only lizards known to be able to raise and maintain their body temperature above that of their environment by utilization of metabolic heat. However, as noted in the introductory comments of this chapter, many, perhaps most, lizards can exert some control over the rate at which their body temperature changes by physiological means. Some burrowing lizards are totally thigmothermic but may select thermal gradients within the soil and regulate their temperature in that manner. At least some lizards which are active throughout the year, including lined tree lizards (*Urosaurus ornatus*) and side-blotched lizards (*Uta stansburiana*), exhibit essentially the same preferred body temperature and activity range throughout the year, indicating that they are efficient in regulating their body tem-

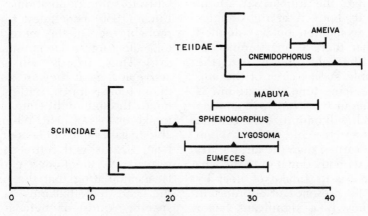

Figure 8–10 Ranges (bars) and means (triangles) of body temperatures of active individuals of four genera of skinks and two genera of teiids. (From Brattstrom: Body temperatures of reptiles.)

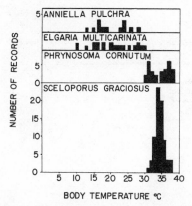

Figure 8–11 Histograms of body temperatures of active individuals of four species of lizards, showing the apparent lack of a preferred temperature in *Anniella* and *Elgaria,* a relatively wide range of preferred temperatures in *Phrynosoma,* and a relatively sharply defined preferred temperature in *Sceloporus graciosus.* (From Brattstrom: Body temperatures of reptiles.)

peratures under a variety of conditions and have no need for physiological acclimation. Acclimation may be of more significance to other species, especially those which are thigmothermic, with more limited temperature regulating abilities.

Active lizards which exhibit narrow body temperature ranges with clearly defined preferred temperatures do so because they have developed very refined behavioral patterns regulating their intake and loss of heat. Rather than merely resting in the sun, lizards orient their bodies so as to regulate the amount of surface area exposed to sunlight. If the lizard's body temperature is below the preference range, the animal will orient itself so that its body is at right angles to the sun's rays, often on an inclined surface, so that there is maximum concentration of the solar energy over the greatest possible surface area of the animal. If, at the same time, the ground is colder than the air the lizard will assume a posture which will minimize the contact between itself and the cold surface. When the ground is warmer than the air, the lizard will press its body down onto the surface and increase its intake of heat by conduction. The presence of appropriate basking perches is a significant factor determining the location of a lizard's territory or home range.

Once basking has raised the lizard's body temperature to the activity range, the animal will regulate its heat intake in such a way as to maintain its temperature near the preferred level. This is done by regulating the time spent in direct sunlight, regulating the area of body surface exposed to sunlight by orienting the body axis so that it is either parallel or at right angles to the sun's rays, regulating the intake or loss of heat through conduction by seeking warmer or cooler surfaces in the habitat, and assuming various postures to minimize or maximize contact with these surfaces. As the environment gets hotter and the lizard's body temperature approaches the upper limit of its range of tolerance, the lizard will do everything it can to reduce the rate of heat gain (seek shade, orient its body, assume various postures) and will utilize panting to cause evaporating cooling. If these actions fail to keep the animal's temperature within the range of tolerance, it will seek a retreat underground. Once underground, it has no further means of coping with high temperatures.

With decreasing environmental temperatures, an active lizard will have an increased problem of maintaining a sufficient intake of heat, and as its temperature approaches the lower limit of the range of tolerance it will generally seek retreat underground where, again, it has no further means of regulating its temperature. Most lizards (and other reptiles) burrow sufficiently deep to avoid freezing, but numerous reports in the literature of frozen reptiles (and amphibians) indicate that this may be a major source of mortality to some populations. (For instance, Fitch (1956) concluded that freezing is probably one of the more important factors affecting snake populations in Kansas.) Thus, lizards will generally stay active as long as they are able to maintain their temperatures within the desired range through deliberate and measured intake or loss of heat. When they are unable to gain sufficient heat or lose sufficient heat, lizards will retire to underground retreats. As mentioned earlier, evidence is accumulating that the temperature of the head (hypothalamus) is what triggers specific kinds of behavior at specific temperatures.

The behavioral regulation of temperature by lizards is so efficient that two species having different temperature preferences may utilize the same sources of heat in the same habitat and at the same time but, through their behavioral differences, maintain body temperatures differing by several degrees. On the other hand, Corn (1971) and Ruibal (1961) have both noted that sympatric species of *Anolis* reflect microhabitat differences in their preferred temperatures and upper thermal limits. In a study of five Cuban anoles, Ruibal found that the two species inhabiting open areas (*Anolis sagrei* and *A. allisoni*) were true heliotherms and had a mean body temperature of 33° C. One species (*Anolis homolechis*) occurred in forest margins and, while at least partially heliothermic, had only filtered sun available to it. The mean body temperature for this species was 31.8° C. The other two species (*Anolis allogus* and *A. lucius*) are forest dwellers and, restricted to shaded areas, are not heliotherms. The average body temperature for individuals of these last two species was found to be only 29° C, and, apparently because of heat lost by respiration and conduction to the substrate, their body temperatures were usually below the air temperature of their environment. Similarly, Corn found that *Anolis carolinensis* and *A. sagrei* both perch less than six feet off of the ground on grass or bushes in open sunlight, whereas the sympatric *Anolis distichus* is arboreal and basks in filtered light on tree trunks. These microhabitat differences were reflected in mean preferred temperatures of 34.0° and 33.3° C, respectively, for *A. carolinensis* and *A. sagrei*, compared to 31.0° C for *A. distichus*.

Recent studies have revealed the importance of proper temperatures to reptilian physiology. Wilhoft (1958) found that if western fence lizards (*Sceloporus occidentalis*) are maintained continuously at their preferred temperature (35° C) for thirteen weeks, some of them die and all show an excessive activity of the thyroid gland. These results indicate alternating cool nocturnal temperatures may be necessary to provide the rest which all living systems require between periods of activity. Licht and Basu (1967) studied the effect of temperature on lizard testes, using *Anolis carolinensis* and

Uma scoparia (the Mohave fringe-toed lizard). In their study, the anoles regulated their body temperature between 28° and 36° C and averaged body temperatures of 32.5°, whereas *Uma* maintained temperatures between 34° and 40° C and averaged 37.5° C. Licht and Basu found that at temperatures of 35° or higher in the anoles and at 40° C and higher in *Uma*, there was spermatogenic arrest and progressive necrosis of all tissues within three to five days. This study certainly documents the fact that lizards are not poikilothermic but rather have narrow ranges of temperature within which they will function normally.

In summary, temperature regulation in reptiles is both behavioral (time of emergence and retreat, selection of temperatures in habitat, regulation of heat gained or lost through basking, orientation, postural changes, and so forth) and physiological (evaporative cooling through panting, vasomotor responses, and limited metabolic heat production in some). Brattstrom (1965) suggests that reptiles may be grouped into the following categories in terms of their thermoregulatory behavior: burrowing forms; aquatic forms not selecting temperatures; aquatic forms selecting temperatures; aquatic forms which bask at the surface or on shore; nocturnal thigmothermic forms; nocturnal thigmothermic forms that occasionally bask; diurnal, primarily thigmothermic forms that occasionally bask and become crepuscular or nocturnal as hot seasons develop; diurnal non-baskers; diurnal limited baskers; and true heliothermic baskers. The importance of pigmentation to thermoregulation will be discussed in the next chapter.

References

Bartholomew, G. A., and V. A. Tucker. 1963. Control of changes in body temperature, metabolism, and circulation by the agamid lizard, *Amphibolurus barbatus*. Physiol. Zool. 36:199–218.

―――― 1964. Size, body temperature, thermal conductance, oxygen consumption, and heart rate in Australian varanid lizards. Physiol. Zool. 37:341–354.

――――, and A. K. Lee. 1965. Oxygen consumption, thermal conductance, and heart rate in the Australian skink *Tiliqua scincoides*. Copeia 1965: 169–173.

Bogert, C. M. 1953. Body temperatures of the Tuatara under natural conditions. Zoologica 38:63–64.

——— 1959. How reptiles regulate their body temperature. Scient. Amer. 200(4):105–120.

Brattstrom, B. H. 1963. A preliminary review of the thermal requirements of amphibians. Ecology 44:238–255.

——— 1965. Body temperatures of reptiles. Am. Midl. Nat. 73:376–422.

———, and P. Lawrence. 1962. The rate of thermal acclimation in anuran amphibians. Physiol. Zool. 35:148–156.

Colbert, E. H., R. B. Cowles, and C. M. Bogert. 1946. Temperature tolerances in the American alligator and their bearing on the habits, evolution, and extinction of the dinosaurs. Bull. American Mus. Nat. Hist. 87:327–374.

Corn, M. J. 1971. Upper thermal limits and thermal preferenda for three sympatric species of Anolis. J. Herpetol. 5:17–21.

Cowles, R. B. 1962. Semantics in biothermal studies. Science 135:670.

———, and C. M. Bogert. 1944. A preliminary study of the thermal requirements of desert reptiles. Bull. American Mus. Nat. Hist. 83:261–296.

Fitch, H. S. 1956. Temperature responses in free-living amphibians and reptiles of northeastern Kansas. Univ. Kansas Publ. Mus. Nat. Hist. 8:417–476.

Fromm, P. O. 1956. Heat production of frogs. Physiol. Zool. 29:234–240.

Heath, J. E. 1964. Head-body temperature differences in horned lizards. Physiol. Zool. 37:273–279.

Hutchison, V. H. 1961. Critical thermal maxima in salamanders. Physiol. Zool. 34:92–125.

———, H. G. Dowling, and A. Vineger. 1966. Thermoregulation in a brooding female Indian python Python molurus bivittatus. Science 151:694–696.

Licht, P. 1968. Response of the thermal preferendum and heat resistance to thermal acclimation under different photoperiods in the lizard Anolis carolinensis. Amer. Midland Naturalist 79:149–158.

———, and S. L. Basu. 1967. Influence of temperature on lizard testes. Nature 213:672–674.

Mellanby, K. 1941. Body temperature of frogs. J. Exp. Biol. 18:55–61.

Müller-Erzbach, W. 1891. Die Widerstandsfähigkeit des Frosches gegen das Einfrieren. Zool. Anz. 14:383–384.

Pearson, O. P. 1954. Habits of the lizard Liolaemus multiformis multiformis at high altitudes in southern Peru. Copeia 1954:111–116.

Rodbard, S. 1948. Body temperature, blood pressure, and hypothalamus. Science 108:413–415.

Ruibal, R. 1961. Thermal relations of five species of tropical lizards. Evolution 15:98–111.

Sexton, O. J. 1959. Spatial and temporal movements of a population of the painted turtle, Chrysemys picta marginata. Ecol. Monographs 29(2):113–140.

Stebbins, R. C. 1954. Amphibians and Reptiles of Western North America. New York, McGraw-Hill. 536 pp.

Thorson, T. B. 1955. The relationship of water economy to terrestrialism in amphibians. Ecology 36:100–116.

Wilhoft, D. C. 1958. The effect of temperature on thyroid histology and survival in the lizard, Sceloporus occidentalis. Copeia 1958:265–276.

Zweifel, R. G. 1957. Studies on the critical thermal maxima of salamanders. Ecology 38:64–69.

CHAPTER 9

COLOR AND COLOR CHANGE ADAPTATIONS OF AMPHIBIANS AND REPTILES

Although some species may possess colors with little or no significance to survival, coloration of animals generally tends to be adaptive to biotic and abiotic environmental factors. Color has four primary functions in amphibians and reptiles. First, it provides protection from predators through coloration which result in obliterative patterns, background color-matching, countershading, pattern-matching, warning colors, and mimicry. Secondly, by affecting the albedo of the animal's surface, it functions in thermoregulation. Thirdly, color functions to provide radiation shields for critical structures and organs. Finally, color is often important in interspecific and intraspecific interactions of individuals because of its involvement in sexual displays and in species and sex recognition.

PROTECTIVE FUNCTIONING OF COLOR

Concealment provides passive protection from predators; particularly where uniform color habitats extend over wide areas, natural selection results in a blending of body coloration with environmental background. Because amphibians and reptiles generally feed on live animals,

their concealment also functions importantly in allowing them to get close enough to their prey to capture it. Concealment of an amphibian or reptile may result either because of the prevailing color and tone or because of the pattern of coloration. The selective advantage of concealment, especially in diurnally active species, is so significant that in widely distributed species one can frequently trace color clines from one region to another. For example, many reptiles exhibit coloration clines along a transect from eastern forest communities in North America westward into the prairie communities. This transect represents a moisture gradient, and the moisture gradient, in turn, results in a gradient in the amount of plant cover and growth. The greatest amount of cover and shade is in the east in the forests, the least is in the west in the prairie. The greater density and growth of plants in the east results in a greater deposition of organic matter on the surface; consequently, there is a gradient of soil color, with dark soils in the east and pale soils in the west. As a result of these interrelated gradients, environmental background coloration tends to be dark in the east and lighter in the west; correlated with this, the coloration of several reptilian species is dark in the east and progressively becomes lighter toward the western limit of their distribution. The coachwhip snake (*Masticophis flagellum*) is dark brown from the eastern forests westward to eastern Texas. It is pale tan in central Texas, still paler in western Texas, and frequently pink in desert regions of the southwestern United States. Blue racers (*Coluber constrictor*) are almost black in the eastern part of their range and exhibit progressively lighter shades of greenish and grayish blue or tan in the west. Duellman and Schwartz (1958) observed that, even within Florida, black racers are found in pine forests and hammocks while light-colored gray-green or gray-tan racers are essentially confined to prairie habitats. In areas where the eastern rim rock breaks into a series of "islands" of pine woods separated by sloughs, Duellman and Schwartz found the racers to be intermediate between the black and bluish gray extremes.

The degree to which an animal's coloration matches its background seems to be related to six factors: (1) the degree of color uniformity of the background in question, (2) the degree of exposure of the color-matched species to predators in its habitat, (3) the illumination levels prevalent in the habitat, (4) the size range of the color-matched species, (5) the degree of ecological restriction of the species in question, and (6) the qualities of the visual apparatus of predators upon the species (Norris and Lowe, 1964). These six factors affect the effectiveness of coloration in concealing the animal, but coloration has various other roles in amphibians and reptiles; these other functions result in coloration that reflects a compromise

dictated by the relative importance of each function to survival. Very little or nothing is known about the visual powers of most predators, but the spectroreflectometric data of Norris (1967), encompassing near-ultraviolet, visible, and near infrared wavelengths (185 to 3500 mμ), indicate that color-matching of desert lizards seldom extends into the invisible spectrum except for short distances into ultraviolet and, in fact, doesn't even encompass the entire human visible spectrum (400 to 700 mμ).

Local variations in color, particularly of reptiles, are frequently the result of changes in natural selection stemming from changes in soil coloration due to localized mineral deposits or parent material. For example, many of the lizards living in the White Sands area of New Mexico, approximately 270 square miles of almost pure gypsum, are very pale in coloration, sometimes almost white, whereas members of the same species living a few miles away on basaltic lava deposits are very dark in coloration. Norris and Lowe (1964) found the flat-tailed horned lizards (*Phrynosoma m'calli*) living on reddish sands in southeastern California and adjacent Sonora to be reddish, while those living 70 miles north on whitish Thousand Palms dunes, Riverside County, are very light. Reflectance curves of individuals from these two areas, obtained with a recording reflectance spectrophotometer by Norris and Lowe, are reproduced in Figure 9–1. Whiteness is expressed by a flattening of the entire curve, whereas the reddish color is expressed by the strong emphasis in the red end of the spectrum. Norris and Lowe

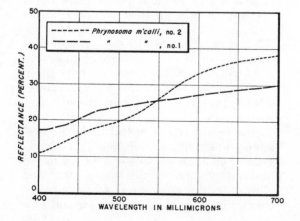

Figure 9–1 Color differences within two populations of a single species: (1) the flat-tailed horned lizard (*Phrynosoma m'calli*) from the light-colored sands at Thousand Palms Junction, Riverside County, California, and (2) the same species from the reddish sands of San Luís, Sonora, Mexico. (After Norris and Lowe: An analysis of backbround color-matching in amphibians and reptiles.)

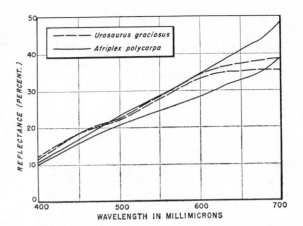

Figure 9–2 The middorsal reflectance of the long-tailed brush lizard, *Urosaurus graciosus* (two different individuals), from Algodones Dunes, Imperial County, California, measured against the bark of the salt bush (*Atriplex polycarpa*) from which it was taken. (From Norris and Lowe: An analysis of background color-matching in amphibians and reptiles.)

noted a comparable variation in the coloration of Cowles' fringe-toed lizards (*Uma notata rufopunctata*) over a very short distance near Puerto Peñasco, Sonora, where pinkish inland dune masses meet whitish beach dunes. Individuals living on the beach dunes are white, whereas those living inland are reddish, the difference in coloration being on the order of that illustrated in Figure 9–1. Similar differences in coloration are evident among populations of rock-dwelling reptiles. Thus, the canyon lizard (*Sceloporus merriami*), which occurs in areas of Texas and Mexico where there are both limestone formations and basalt formations, is pale where limestone formations prevail and blue-black in basaltic areas. Rattlesnakes typically frequent rocky areas, and their color tends to vary locally with changes in the background of their habitat. For example, the coloration of the rock rattlesnake (*Crotalus lepidus*) is dark red where it occurs in Texas on red rhyolitic lava formations and pale white where it occurs on limestone formations. Interestingly, there is evidence that changes in the total amount of melanin present occur in the first few weeks of life in some lizards and that these changes are controlled by vision and albedo (Norris, 1967). Hatchling lizards are sometimes not color-matched with their background but within a month or two become increasingly closer in their match. Once the deposition of a pattern of pigment occurs, it seems to be fixed and cannot be changed in adults.

The degree to which ground color is important in determining such variation in coloration as described above depends upon how much time the animal spends on the surface. Those forms that are ground-dwellers have colorations which are subjected to very strong selective forces and will always tend to blend in with the prevailing soil color. Those which are scansorial and spend the majority of their time on tree trunks or in foliage will have coloration matching the vegetative background and not necessarily the ground. Thus, in the White Sands area, little striped whiptail lizards (*Cnemidophorus inornatus*), which are ground-dwellers, are very pale sky blue with faint stripes in comparison to the darker individuals from areas with darker soils. Eastern fence lizards (*Sceloporus undulatus*), which climb woody vegetation and only spend part of their time on the surface of the ground, are paler in the White Sands area than in dark soil areas, but the reduction in their coloration is not as extreme as in strict sand-dwellers. Tree lizards (*Urosaurus ornatus*) are primarily arboreal, and their color matches that of their vegetative background rather than that of the sand. In their study, Norris and Lowe (1964) found that long-tailed brush lizards (*Urosaurus graciosus*) exhibit a remarkable color- and pattern-match with the branches of desert shrubs and low trees (Fig. 9–2) and observed that, when frightened, these little lizards clasp themselves to a branch of their home bush and become nearly invisible. This observation illustrates the important relationship between behavior and coloration.

In a very interesting aspect of their study, Norris and Lowe (1964) showed how different members of a single community approximate the coloration of their particular microhabitat. To do this, they tested reflectance curves of specimens of the zebra-tailed lizard (*Callisaurus draconoides*), the chuckwalla (*Sauromalus obesus*), the desert spiny lizard (*Sceloporus magister*), and the horned rattlesnake or sidewinder (*Crotalus cerastes*) from a single area against the reflectance curve of their most common background material (Fig. 9–3). The zebra-tailed lizards inhabit sandy flats and washes, and their coloration corresponds very precisely to that of the sand. Chuckwallas are large wary lizards which are rock dwellers and use crevices in their home boulders for protection. The combination of the large size of this species and the mottled background produced by irregular rock surfaces means that chuckwallas would be very conspicuous at moderate and short distances if they had a uniform coloration. Thus, the patterned coloration of chuckwallas probably results from selection for background color-matching when viewed at a distance but disruptive pattern and coloration, enhanced in this species by loose folds of skin on the neck and body, when viewed from closer ranges. To demonstrate this, Norris and Lowe recorded reflectance curves from different portions of the dorsal surface of an individual chuckwalla (curves *A, B, C,* and *D* in Figure 9–3) and found that the average of these (curve *E*) corresponds very closely to the reflectance curve from granite; such an average is what would be perceived by the eye at a distance. At the locality in which Norris and Lowe collected, the desert spiny liz-

ards are found on living Joshua trees, on fallen trunks of these trees, on granite boulders, and occasionally on sandy slopes under bushes. Spending time in such a variety of places, this species cannot possibly always color-match its background, so its coloration probably functions to conceal the animal primarily by being disruptive. Interestingly, its reflectance curve is intermediate between those of the sand and Joshua bark. Although frequently active at night, sidewinders also spend time in daylight basking in the sand, and their reflectance curve closely matches that of the sand in their habitat.

Norris and Lowe (1964) also demonstrated color-matching in amphibians. A variety of salamanders (*Aneides lugubris, Aneides ferreus, Taricha torosa, Taricha rivularis,* and *Dicamptodon ensatus*) all exhibit reflectance curves which correspond closely with those of leaf litter and humus from the forest floor they inhabit. However, it is difficult to evaluate the importance of background color-matching for these animals since they tend to be secretive and nocturnal. The California yellow-legged frog (*Rana muscosa*) provides a clear example of the important relationship between escape behavior and concealing coloration. Norris and Lowe collected this species in the vicinity of a swift stream. The stream bed is characterized by large boulders which above the water line are grayish white but below are covered by a film of yellowish brown algae. The frogs are yellowish brown dorsally and very conspicuous when they sit on the boulders. However, when frightened they jump into the water and swim immediately to the bottom, where their

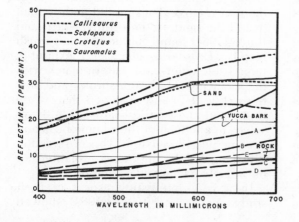

Figure 9–3 Reflectance of four species of reptiles living in the same community and the substrata from which they were taken at Piute Butte, Los Angeles County, California. Curves *A, B, C,* and *D* are from different portions of the dorsal surface of an individual chuckwalla, and curve *E* is the average reflectance from these four curves. (From Norris and Lowe: An analysis of background color-matching in amphibians and reptiles.)

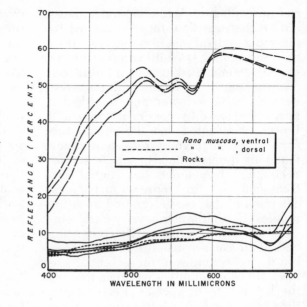

Figure 9–4 Reflectance curves from the California yellow-legged frog, *Rana muscosa* (three different individuals), and of rocks from their habitat. (From Norris and Lowe: An analysis of background color-matching in amphibians and reptiles.)

coloration matches that of the background in saturation, hue, brightness, and pattern (Fig. 9–4). Thus, the frogs are protected out of water by their alertness and their instinct to jump into the water at the slightest disturbance. Once in the water, they are protected by their concealing coloration and the instinct to remain motionless on the background to which they are color-matched.

Because color patterns, as well as color tones, contribute to concealing coloration, the color patterns possessed by amphibians and reptiles will correlate with the patterns of colors in their environments. Terrestrial environmental color patterns are largely produced by the density and physiognomy of vegetation present. Dense forests result in uniformly dark backgrounds, and the amphibians and reptiles inhabiting these are generally a uniform dark color. Shadows in open forests, brushy regions, and contiguous vegetated and barren surfaces tend to be blotchy; accordingly, amphibians and reptiles found in such situations tend to have blotchy patterns. Species inhabiting grassy or marshy areas generally have longitudinal stripes, and this is correlated with the linear shadows in these habitats. Rat snakes (genus *Elaphe* in the family Colubridae) well illustrate each of these variations in color pattern. Those inhabiting heavily forested regions in southeastern United States, where there is little con-

trast in forest floor background color, are a uniform black color. On Edward's Plateau in central Texas, an area characterized by patches of mesquite and limestone outcroppings, rat snakes have blotched patterns. In western Texas, which is primarily a grassy plain, rat snakes are striped and blend in with the shadows cast by the sparse vegetation. In general, patterns with dorsal stripes, lateral stripes, facial markings, and limb bars all tend to disrupt the body pattern and conceal the animal by making perception of the body outline more difficult. The uniqueness of vertebrate eyes makes them visually conspicuous, and many amphibians and reptiles have eye strips which tend to camouflage these sense organs.

Contrasting dorsal and ventral coloration is another common feature of animal coloration. Nearly all amphibians, many lizards and snakes, and numerous turtles have white or light-colored bellies and undersurfaces of limbs and dark-patterned or uniformly colored dorsal surfaces (compare the reflectance curves for dorsal and ventral surfaces in Figure 9–4). At least four hypotheses have been proposed to explain what are frequently gleaming white ventral surfaces: (1) Such white surfaces promote color-matching of the dorsal surface in reflected light by accurately reflecting substratum color. (2) Light reflected from anterior white surfaces into dimly lit areas assists the animal in feeding. (3) White ventral surfaces reduce the heat flow into the animal by reflecting radiant energy reflected from the substratum. (4) White ventral surfaces provide countershading which helps conceal the animal by reducing the contrast between shaded lower and non-shaded upper surfaces. In general, Norris and Lowe (1964) noted that the ventral surfaces with the highest reflectivities are found in desert animals, and lesser reflectivities characterize the ventral surfaces of animals from more mesic habitats. They found that coastal ground-dwelling reptiles in California have ventral surfaces which are often strongly contrasting with their color-matched dorsal surfaces, but the reflectivities of these ventral surfaces are considerably lower than those of the lightest desert species. For example, they found the light ventral surface of the coast horned lizard (*Phrynosoma coronatum blainvilli*) has a maximum reflectance of about 43 per cent, whereas that of the desert iguana (*Dipsosaurus dorsalis*) has up to 72 per cent reflectance. Exceptions to this pattern among lizards sometimes are correlated with differences in background coloration. For example, chuckwallas reflect less light from their ventral surfaces than from their dorsal surfaces (Hutchinson and Larimer, 1960; Norris and Lowe, 1964); and side-blotched lizards (*Uta stansburiana*) exhibit a white-ventered condition and a bluish-black venter in different desert populations (Norris and Lowe, 1964). The chuckwallas normally live on rocks which are frequently dark-colored, while variation in the side-blotched lizards matches that of the habitat; white-ventered individuals occur on light-colored sandy areas and alluvium while dark-ventered individuals are found on lava flows or other dark rocks (Norris and Lowe, 1964). Sometimes, as with male desert spiny lizards, which have two dark bluish ventral lateral blotches, lower-reflectance ventral surfaces occur because of coloration exhibited in sexual displays.

Obviously, in order for countershading to be effective in camouflaging it, the animal must rest in the proper attitude. In addition, rock-climbing and arboreal species may expose their ventral surfaces to view from higher, lower, and lateral angles. Probably because a very light ventral surface would make such animals conspicuous to predators, their ventral surfaces tend to be various shades of gray or other tones which are not highly reflective. Examples of reptiles in this category include the rock lizard (*Streptosaurus mearnsi*), a rock-climbing species which inhabits boulder-filled canyons, and the vine snake (*Oxybelis aeneus*), an arboreal species. In such instances, the tendency of selection seems to be a ventral coloration which blends in with that of the substratum upon which the animals climb (rock and bark, respectively, in the two examples mentioned) because of exposure to predators' view. Similarly, the very reflective ventral surfaces of aquatic turtles such as the marine green turtle (*Chelonia mydas*) and of many anuran amphibians are protective when the animal is in the water and is being viewed from below against the light sky.

Another way in which coloration functions in a protective manner is through flash colors. Flash colors are bright patches of contrasting colors on the caudal side of the femur, lateral surfaces of the groin and axillary regions, or occasionally the ventral surface. These color patches are concealed when the animal is resting but are prominently exposed when the animal moves. Consequently, when the resting animal is flushed by a predator the latter gets a glimpse of the flash color, continues to look for it, and is apt to pass right by the resting animal which has quickly covered the flash color and is, again, camouflaged. Flash colors are particularly common on tree frogs (family Hylidae), which often have red or yellow patches that sharply contrast with the general tone of their body and are exposed to view only when they jump.

A comparable degree of deception is achieved in lizards which practice caudal autotomy and have a tail which contrasts in color with the overall color of their body. For example, juvenile western skinks (*Eumeces skiltonianus*) have a bright blue tail which contrasts with the brown and cream striped body; young Gilbert's skinks (*Eumeces gilberti*) have either blue or red tails. When such a brightly-colored tail is snapped off, it twitches convulsively and distracts the predator by its movement and color while the rest of the animal escapes.

Head mimicry, the resemblance of the tail to the head, is used effectively by a variety of snakes to deceive enemies. By making an enemy believe the harmless tail is the potentially harmful head, head mimicry protects the snake, since the tail can be damaged without jeopardizing the life of the snake. The brunt of the enemy's attack will be received by the tail while the more vulnerable head is ignored and possibly facilitated in its ability to strike the enemy. By deceiving the enemy as to which end of the snake is the head, head mimicry can also be protective to small harmless snakes, since the predator will be uncertain as to which direction the snake will move in attempting to escape. Head mimicry is often associated with a behavioral trait called tail display. In tail display, the snake raises its tail and waves it about, making it appear like a head poised to strike, while the head is kept hidden among body coils or lies inconspicuously on the ground. A few species, including rubber boas (*Charina bottae*), have tail-display patterns of behavior but tails which do not differ in color from the rest of the body. However, nearly all snakes with tail-display behavior have red undersurfaces on their tails; these are exposed when the tail is raised, giving the impression of an open mouth. Myers (1965), who observed that the tail display (tail-coiling) of ringneck snakes (*Diadophis punctatus*) is associated closely with populations of red-tailed individuals, determined that red tail displays have been observed in the following families and species of snakes:

Aniliidae
    Anilius scytale
    Cylindrophis rufus
Colubridae
    Calamaria spp.
    Farancia abacura
    Diadophis amabilis
    Diadophis punctatus
    Diadophis regalis
Elapidae
    Callophis trimaculatus
    Maticora bivirgata
    Maticora intestinalis
    Micrurus affinis
    Micrurus frontalis
    Micrurus mipartitus.

He also notes that yellow tail displays are characteristic of two colubrids, *Drymarchon corais corais* and *Thamnophis eques megalops.*

It is advantageous for an animal that is either dangerous or unpalatable to advertise this fact so that it is easily recognized and not subjected to repeated attacks by ignorant predators. Consequently, some animal species have evolved conspicuous colors and patterns making them easy to identify. Such colors, referred to as warning (aposematic) colors because they advertise the fact that the bearer is obnoxious, unpalatable, or dangerous, are generally shades and combinations of red, yellow, black, and white. Natural selection has favored these colors for warning coloration for the same reason that they are the colors used for highway signs; they stand out conspicuously against natural backgrounds. Certain amphibians and reptiles have unusually bright color patterns which appear to function as warning colors. Among these are the Gila monster and Mexican beaded lizard (*Heloderma suspectum* and *Heloderma horridum*, respectively), the only known poisonous species of lizard, which are colored gaudy patterns of irregular brown or black markings upon a yellow to salmon or orange background (Fig. 9–5). The Neotropical subfamily Dendrobatinae (family Ranidae) includes some extremely poisonous frogs that are vividly colored with bright patterns of metallic green, blue, yellow, and black, which are presumed to be warning colors. The red eft stage of the red-spotted newt (*Notophthalmus viridescens*) has a noxious and toxic secretion that it seems to advertise with its bright red-orange coloration. A number of elapid species in the American genera *Micrurus* and *Micruroides*, the African genus *Aspidelaps*, the Australian genus *Brachyurophis*, and the Asian genera *Maticora* and *Callophis*, all generally referred to as coral snakes, are brilliantly colored with bands of yellow, red, and black that may function to advertise the venomous nature of the snakes.

Rather than being entirely brightly colored, some distasteful or venomous species of amphibians have cryptic coloration on their dorsal surfaces and brightly colored bellies. When approached by a predator, such amphibians roll over on their back and raise their limbs so as to

Figure 9–5 The Gila monster (*Heloderma suspectum*). (From R. Mertens: *The World of Amphibians and Reptiles*.)

exhibit their ventral surface and coloration. Such a behavioral reaction is referred to as an "Unkenreflex". Unkenreflexes and associated brightly colored ventral surfaces are characteristic of a variety of amphibians, including firebelly toads (*Bombina,* family Discoglossidae), which possess brilliant orange or yellow patterns on their ventral surfaces, and rough-skinned newts (*Taricha granulosa,* family Salamandridae), which have an orange belly. These species are particularly venomous. The red tail displays of snakes mentioned previously may function as warnings if the predators associate the display with an undesirable taste or odor, as of the anal sac secretions. In the case of the elapids, the tail display and red color advertise the owner's truly venomous capabilities.

Once warning colors have evolved in some species, palatable species may be provided selective advantages by mimick-

ing the warning coloration of unpalatable species. This sort of mimicry, where there is a resemblance between edible and inedible species, is known as *Batesian mimicry,* and it has been observed here and there throughout the animal kingdom. Howard and Brodie (1970) have shown that the close resemblance of northern red salamanders (*Pseudotriton ruber*), a relatively palatable species, to the noxious red efts of *Notophthalmus viridescens* is functional mimicry. Blue jays, brown thrashers, and domestic chickens ate the *Pseudotriton* before being conditioned and showed no signs of adverse effects. They were then given red efts and rejected them after one or two encounters. Subsequently, they also rejected *Pseudotriton*. In this case, the mimicry of the *Pseudotriton* involves not only color pattern but also a defense posture similar to that of the red eft. Batesian mimicry of coral snakes is well known, and the "false coral snakes" belong to a variety of phylogenetic lines. The scarlet king snake (*Lampropeltus triangulum doliata*) of North America and the harmless *Simophis rhinostoma* of South America are two of a number of colubrid species which look deceptively like coral snakes of the genus *Micrurus*, as does the South American pipe snake *Anilius scytale*. The resemblance of the non-venomous South American leptodactylid frog *Lithodytes lineatus* to highly venomous *Dendrobates* appears to be another case of Batesian mimicry.

COLOR PATTERNS AND SEX RECOGNITION

The function of color as a means of identifying sex is the best known of intraspecific variation. In some amphibians and reptiles this function results in strong sexual dimorphism. In fact, in many species, color variation seems to be the primary means by which a female recognizes a male of her species and, just as importantly, a male recognizes another male. Very striking sexual differences in color are present in the lizard families Iguanidae and Agamidae; as with birds, the male is usually the more brightly colored. During the breeding season, sexually mature male rainbow lizards (*Agama agama*) develop orange-red heads and blue bodies, whereas females and juveniles are brownish yellow with olive-green heads. Males of most species of spiny lizards (*Sceloporous*) have bright blue patches on each side of the belly and on the throat which are exhibited when the individual is defending a territory or otherwise challenging another male and in courtship behavior. Collared and leopard lizards (*Crotaphytus*) characteristically have gular pouches that in males are generally a brighter color than in females. These seem to function in sex recognition, since an aggressive male will flatten its body laterally, stand high on its legs, and extend the gular skin (dewlap) so as to exhibit its coloration. Depending upon the species, the throat coloration may be a deep orange, yellow-orange, deep-green, or greenish blue. Sometimes, as in yellow-headed collared lizards (*Crotaphytus collaris auriceps*), the head and forefeet of the male are also vividly colored, being a bright yellow in this subspecies. These are just some of the examples of sexual dimorphism in coloration of lizards. Some snakes are also sexually dimorphic in coloration. For example, male adders (*Vipera berus*) have more contrasting patterns than females, and, in addition, males have red irises while females have brown irises. Crocodilians do not exhibit sexual differences in coloration, nor do most turtles. However, eastern box turtles (*Terrapene carolina*) are an exception: males have reddish eyes, females greenish or brownish eyes; males have reddish coloration on their feet, whereas females have yellowish colorations. Male tuataras are larger than females but are not colored differently.

Among amphibians and reptiles, the greatest difference in coloration of sexes is present in diurnally active species which depend primarily on their sense of vision for perceiving other individuals. However, for unknown reasons, some nocturnal amphibians exhibit sexual dimorphism in coloration. Males of the plains spadefoot toad (*Scaphiopus couchi*) are uniformly green on their dorsal surfaces, while females are mottled. Males of *Bufo marmoreus*, a Mexican species of toad, have a broad dark green band down their back that females lack. Because both of these species breed at night, it is difficult to imagine that these color differences could operate at night as a means of identifying sex. The throat skin which covers the vocal pouch of male anurans generally contrasts in color with the rest of the ventral surface and is frequently yellow, greenish, bluish, or black, whereas the throat region of females is usually the same color as the remainder of the ventral surface. Although the expanded vocal pouch certainly is a very prominent feature of a calling male, the fact that many anuran species breed at night makes the significance of the colored throat of males questionable as a factor in sex recognition. However, it seems more than coincidental that anuran species which lack a mating call and have poorly developed vocal structures, such as the boreal toad (*Bufo boreas*), also tend to lack the male throat coloration. It is possible that coloration of the vocal pouch, no matter what tone, reduces the albedo of the surface and makes the calling male less visible to predators and that this is why natural selection has favored it.

The Yosemite toad (*Bufo canorus*) exhibits the greatest degree of sexual dimorphism in coloration of all North American amphibians. In this species, the female has numerous dark or rust-colored blotches which are outlined in white and on a green or grayish background. Males, on the other hand, are a uniform yellow-green or dark olive and, although small scattered flecks may be present, blotches are lacking. In contrast to the general situation in anurans, male Yosemite toads have a pale throat like that of the female. Yosemite toads occur at high elevations in the Sierra Nevada of California and, probably as an adaptation to temperature, are chiefly diurnal. Thus, the sexual difference in coloration in this species may be importantly related to mating behavior and sex recognition. The pale throat of the males may be advantageous to them because they call diurnally and, through countershading, they may be less conspicuous than they would be if their throats were colored.

Many other anuran species exhibit varying degrees of sexual dimorphism in their coloration, and the examples mentioned merely indicate some of the differences which may be present.

THE FUNCTION OF COLOR IN THERMOREGULATION

Color patterns and the ability to regulate them are selected for as adaptations to physical factors in the environment as well as to biotic factors. Coloration plays an important role in the thermoregulation of all ectotherms, nocturnal and diurnal. As noted by Cole (1943), when two animals are simultaneously exposed to intense radiant energy, the darker-colored individual will be subject to the greater heating effect. On the other hand, if the two animals are sheltered from the radiation and have higher-than-ambient temperatures, the dark-colored animal will lose heat more rapidly than the light-colored animal. Thus, by darkening or lightening its coloration, a diurnal animal may regulate its temperature by exerting some control over how much heat it ab-

sorbs. A nocturnal animal which can darken or lighten its coloration, on the other hand, regulates its temperature by exerting some control over how much heat it loses.

Sarah Atsatt (1939) made a pioneering investigation which formed the foundation for what is presently known about the relationship between color change and temperature in lizards. She studied 16 species of lizards inhabiting the desert regions of southern California and found that geckos and iguanids respond to high temperatures by assuming their light phase and to low temperatures by assuming their dark phase, just as a diurnal animal should if it is regulating its temperature. She found that the critical temperature for inducing the light phase was generally within the range of 35° to 43° C except for side-blotched lizards (*Uta stansburiana*), where the critical temperature was 25° C, near the minimum voluntary temperature for this species (see Brattstrom, 1965). Atsatt found that when she maintained the temperature at a moderate level and exposed the animals to light or darkness, some of the iguanids assumed their dark phase when illuminated and their light phase in the absence of light. Others (*Dipsosaurus, Callisaurus,* and *Uta*), however, did not respond in this manner. She also found that the nocturnal night lizards (*Xantusia*, family Xantusidae) responded oppositely to the iguanids, assuming their dark phase at high temperatures and when under illumination but not invariably assuming their light phase at low temperatures or when in darkness.

In a follow-up study to that of Atsatt, Caswell (1950) quantitatively measured the degree of darkness of desert night lizards (*Xantusia vigilis*) over a 24-hour period in a dark room where light, temperature, and humidity were constant within narrow limits. He found that this species has an internal rhythm which causes a diurnal cycle in the albedo of the skin. The skin is dark in color (high in absorption and low in albedo) early in the morning, becomes paler in the afternoon, and is palest at sunset (about 7:00 PM in his study); it then darkens at rather a constant rate until morning. Caswell found that this rhythmic cycle could be overridden by an emotional upset of the individual, the effect of stress being variable but generally resulting in the animal blending in with its background. The difference in color response of the nocturnal *Xantusia,* as compared to the iguanids, would seem to reflect the differences in thermoregulation in the dark and in light. The diurnal animal which is exposed to radiant energy responds to high temperatures by increasing its albedo and thereby reducing its intake of radiant energy. The nocturnal animal, on the other hand, has a behavioral pattern which is *not* adapted to exposure to radiant energy. It responds to high temperatures by decreasing its albedo which, *in darkness,* accelerates the rate at which it loses heat.

Field observations by various people have shown that wild lizards belonging to desert iguanid species go through sequential color changes comparable to those described above for laboratory animals. The sequence of reflectivity changes which may occur during any given day has been described by Norris (1967). When the desert iguanids are at a very low temperature and shielded from radiant energy, as in their burrows, they are dark. However, if the temperature in the burrow is more moderate and near the minimum activity level for that species, the lizard will be relatively light and probably will be undergoing a relatively slight diurnal reflectivity fluctuation. If the lizard emerges into sunlight at this temperature it may rapidly darken, or some mechanism such as a diurnal rhythm may cause the lizard to darken before it emerges. Upon emerging, basking begins, and as the body temperature rises the lizard will begin to lighten in coloration. As higher and higher temperatures are reached, the activity thermal range is entered and illumination no longer affects the lizard's coloration. It is now generally color-matched to its environmental background and ready to follow its daily routine.

Two genera, *Uta* and *Sceloporus,* depart from this pattern in that individuals may become active while they are still darkly colored. In at least two other genera, *Callisaurus* and *Dipsosaurus,* coloration continues to lighten as the body temperature rises within the activity range, and the lizards become "superlight," reducing

their heat uptake still more (Norris, 1967). As noted by Norris (1967), the cool lizard often is not color-matched during the brief time before it becomes warm because it is dark; hence, temperature relationships take precedence over concealment coloration at that time. Color-matching is dominant, however, when the lizard is active and within its activity temperature range.

As mentioned earlier, color-matching generally does not extend into the invisible spectra and is especially absent in the infrared spectrum. Norris (1967) has shown that small lizards may have a high reflectivity in the visible spectrum but compensate for this in their temperature relations by having low reflectance in both the ultra-violet and infra-red spectra. He found such a reflectance pattern to be characteristic of small lizards in the size range from that of side-blotched lizards (*Uta stansburiana*) to that of fringe-toed sand lizards (*Uma*) (snout-vent lengths of about 40 to 110 millimeters). Lizards of the size of adult collared lizards (*Crotaphytus collaris*) to that of chuckwallas (*Sauromalus obesus*) (snout-vent lengths of about 110 to 200 millimeters) are color-matched in the visible spectrum and reflective in the infrared spectrum. Thus, infrared absorption, like background color-matching, appears to be size-related.

Light, temperature, and humidity have been shown to be the most important stimuli for color change in amphibians. Although their responses are generally much slower than those of reptiles, many anurans (ranids, bufonids, hylids) become pale when placed in warm, dry situations and dark when in cold, damp conditions. Thus, they react in a manner comparable to that of the diurnal lizards, increasing their albedo when warm and decreasing it when cold. Edgren (1954) determined that a temperature of 3° to 5° C would induce darkening of gray tree frogs (*Hyla versicolor*) but thresholds for color changes in other anurans have not been determined. It is rather common for frogs (ranids and hylids) which are resting with part of their body in sunlight and the remainder in shade or under water to exhibit a sharp color change at the boundary of the two temperature conditions. Green frogs (*Rana clamitans*) in such a situation, for example, will be bright green where they are in the direct sunlight and

a bronze or olive-green where under water or in the shade.

The lack of an ability to change color obviously restricts an animal's capability to regulate its temperature. Cowles (1939) has suggested that the lack of a capacity for color change in snakes is important in restricting many species to nocturnal, crepuscular, or subterranean activuty. Desert reptiles lacking an ability to change their albedo are particularly vulnerable to temperature changes. The aridity of deserts results in wide-spacing of vegetation, a lack of cloud cover, and, consequently, an unavoidable exposure to direct solar radiation of surface-dwellers during daylight hours. Because of the absence of significant quantities of water vapor in the atmosphere, deserts cool off rapidly following sunset. The result of these conditions is that desert temperatures fluctuate over a wide range, particularly at the surface of the ground. A problem facing diurnal desert animals is that of overheating, while that of nocturnal species may be excessive cooling. Thus, the tendency for desert reptiles to be light shades of red, brown, and yellow is advantageous to their temperature relations, since these tones tend to reflect direct solar radiation in the daytime, when overheating is a problem, and tend to promote retention of heat at night, when cooling is a potential problem. Thus, the same colorations which are advantageous for concealment are advantageous for temperature regulation.

The coloration of animals may be influenced by humidity or a combination of humidity and heat. Correlations in this regard are summarized in *Gloger's rule:* Exceptions aside, races of birds or mammals living in cool, dry regions are lighter in color (have less melanin pigment) than races of the same species living in warm, humid areas. Although Gloger's rule was originally applied to birds and mammals, it is also applicable to at least some amphibians and reptiles. As discussed earlier in this chapter, the tendency for amphibians and reptiles living in humid regions to be darkly colored may reflect natural selection for camouflaging in the darker background of a humid habitat. However, dark coloration in warm humid habitats may also be selectively advantageous for temperature regulation. As discussed in

Chapter Eight, evaporative cooling is utilized by both amphibians and reptiles to prevent overheating. However, evaporative cooling decreases in efficiency as the humidity of the surrounding atmosphere increases, since the rate of evaporation from the animal will vary inversely with the relative humidity of the atmosphere. Consequently, the animal living in a habitat which is both hot and humid will have difficulty ridding itself of excess heat by evaporative cooling. If the animal in such a situation can avoid direct solar radiation, which is relatively easy to accomplish in humid habitats because of the density of vegetation, it is advantageous for it to be dark-colored, since the dark pigmentation will not impede the outward flow of heat from the body. Light coloration impedes this loss of heat and is particularly disadvantageous when evaporative cooling is also impeded.

THE FUNCTION OF COLOR AS A RADIATION SHIELD

The vital organs of any animal exposed to radiant energy must be shielded from wavelengths which might, otherwise, penetrate them and cause heating effects or abnormal chemical reactions. Any penetrating radiation which is absorbed by the internal organs may produce deep heating of the tissues and photochemical reactions to an extent dependent upon the intensity of the radiation and the amount of energy absorbed by the tissue and inversely proportional to the wavelength of the radiant energy. It has long been recognized that pigmentation of the skin protects the internal organs of animals from radiant energy. For example, Wallace (1887) observed that diurnal tropical animals have highly pigmented skins or thick layers of hair or feathers while nocturnal forms may lack these. When hair and feathers are lacking, as is the case with amphibians and reptiles, the presence of melanin in the skin is a very effective general filter for solar radiant energy. Accordingly, translucent amphibians and reptiles are typically nocturnal (e.g., the Mediterranean gekkonid, *Hemidactylus turcicus turcicus*), burrowing (e.g., worm lizards, *Rhineura floridana*),

or cave-inhabiting (e.g., the olm, *Proteus anguineus*). It is noteworthy that, if exposed for a long enough time to light, the skin of the olm darkens, first to violet and eventually, over a few months, almost to uniform black (Steward, 1969).

Although melanin in the skin is a very effective general filter for solar radiant energy, under intense light even wavelengths in the visible spectrum, particularly in the yellow range, may penetrate deep within the body of a vertebrate and necessitate the presence of additional reflecting surfaces or increased pigmentation. Watkins-Pitchford (1909) observed that lower vertebrates exposed to solar radiation have pigmented peritoneum if their epidermis is not pigmented and that the peritoneum is black in some colorless fishes while it may be colorless in dark-colored species. Krüger and Kern (1924) also noted that diurnal amphibians and reptiles frequently have central nervous systems and gonads protected by pigmented tissues but that this internal pigmentation is generally absent in closely related nocturnal species. Klauber (1939), Cole (1943), and Collette (1961) correlated the presence of black melanized peritoneum in lizards with diurnal versus crepuscular activity patterns. Collette (1961) was even able to demonstrate that the relative amount of peritoneal pigmentation present in each of six species of anoles (*Anolis*) is directly related to the length of time that each spends in direct sunlight. Pigmented membranes are apparently absent in most desert snakes of southwestern United States, these being largely nocturnal, but the posterior peritoneum is pigmented in species such as coachwhips (*Masticophis flagellum*), which have the largest proportion of daylight activity (Klauber, 1939; Porter, 1967). The tuatara is nocturnal but spends much time basking, and it has black peritoneum (Hunsaker and Johnson, 1959). The implication of all of these observations is that the internal pigmentation of diurnal animals functions to absorb radiant energy in order to protect vital organs or as part of the animal's temperature regulating mechanism. What has been an enigma is that some lizards, such as whiptails (*Cnemidophorus*), are diurnally active and are exposed to very intense solar radiation but lack pigmented peritoneum.

Porter (1967) made a detailed study of the penetration of solar radiation through the living body wall of vertebrates in an attempt to determine the biological significance of black peritoneum. Using light and electron microscopic studies in conjunction with spectrophotometric measurements, he found the quantity and quality of light transmitted through the body wall of all diurnal animals examined were remarkably consistent, whether or not the peritoneum was melanized! His microscopic studies, however, revealed the explanation for what at first might seem to be an incongruity. Animals (e.g., *Cnemidophorus*) lacking black peritoneum have concentrations of melanin in their skin or musculature, lizards with thin body walls have dense intramuscular deposits of melanin, and the surface configuration of keratin on many forms may function as a three-dimensional diffraction grating. Diurnally active species which possess pigmented peritoneum are characterized by the other light-absorbing mechanisms, such as cutaneous and intramuscular melanin deposits, being poorly developed or absent. Thus, the shielding of internal organs from radiation has been accomplished in different ways in different phylogenetic groups. Porter (1967) found that black peritoneum excludes significant amounts of ultraviolet radiation over the spectral interval of 290 to 400 mμ and concluded that this was its functional significance. He considered it to be insignificant in thermoregulation.

Although the emphasis in Porter's study was on desert reptiles, his conclusions probably also apply to amphibians. The pattern and intensity of internal melanophores vary among different species of amphibians (Hunsaker and Johnson, 1959). Some salamanders, including the tiger salamander (*Ambystoma tigrinum*) and the red-spotted newt (*Notophthalmus viridescens*), have melanophores only on the dorsal peritoneum. Pacific giant salamanders (*Dicamptodon ensatus*) have melanophores only on the lateral peritoneum. Dunn's salamander (*Plethodon dunni*) has melanophores scattered over the entire peritoneum, and several species of salamanders (e.g., *Siren intermedia* and *Plethodon glutinosus*) lack peritoneal pigment. Males of some anuran species (such as the cricket frogs, *Acris crepitans*, and chorus frogs, *Pseudacris*) have distinctly pigmented testes while others (such as green tree frogs, *Hyla cinerea*) have non-pigmented testes. Some anurans have anterior-posterior lateral bands of melanophores in the peritoneum, while others lack these. In each of these cases, it might be expected that other radiation shields are present where peritoneal pigmentation is lacking.

Much remains to be learned about adaptations to radiant energy and, as suggested by Cole (1943), it is possible that in the past too much emphasis has been placed on adaptations of color and coloration to biotic factors in the environment.

References

Atsatt, S. R. 1939. Color changes as controlled by temperature and light in the lizards of the desert regions of southern California. Publ. Univ. California Los Angeles Biol. Sci. *1*:237–276.

Bogert, C. M. 1959. How reptiles regulate their body temperature. Scient. Amer. *200(4)*:105–120.

Brattstrom, B. H. 1965. Body temperatures of reptiles. Am. Midland Naturalist *73*:376–422.

Caswell, H. H. 1950. Rhythmic color change in the lizard *Xantusia vigilis*. Copeia *1950*:87–91.

Cole, L. C. 1943. Experiments on toleration of high temperature in lizards with reference to adaptive coloration. Ecology *24*:94–108.

Collette, B. 1961. Correlations between ecology and morphology in anoline lizards from Havana, Cuba and southern Florida. Bull. Mus. Comp. Zool. *125*:137–162.

Cott, H. B. 1957. Adaptive Coloration in Animals. New York, Barnes and Noble. 508 pp.

Cowles, R. B. 1939. Possible implications of reptilian thermal tolerance. Science *90*:465–466.

Duellman, W. E., and A. Schwartz. 1958. Amphibians and reptiles of southern Florida. Bull. Florida State Mus. *3(5)*:181–324.

Edgren, R. A. 1954. Factors controlling color change in the Tree Frog, *Hyla versicolor*. Proc. Soc. Exp. Biol. Med. *87(1)*:20–23.

Howard, R. R., and E. D. Brodie, Jr. 1970. A mimetic relationship in salamanders: *Notophthalmis viridiscens* and *Pseudotriton ruber*. Amer. Zool. *10*: 475.

Hunsaker, D., and C. Johnson. 1959. Internal pigmentation and ultraviolet transmission of the integument in amphibians and reptiles. Copeia *1959*:311–315.

Hutchison, V. H., and J. L. Larimer. 1960. Reflectivity of the integuments of some lizards from different habitats. Ecology *41(1)*:199–209.

Klauber, L. M. 1939. Studies of reptile life in the arid Southwest. I. Night collecting on the desert with ecological statistics. II. Speculation on protective coloration and protective reflectivity. III. Notes on some lizards of the southwestern United States. Bull. Zool. Soc. San Diego *14*:1–100.

Krüger, P., and H. Kern. 1924. Die physikalische und physiologische Bedeutung des Pigmentes bei Amphibien und Reptilien. Pflugers Arch. Physiol. *202*:119–138.

Myers, C. W. 1965. Biology of the Ringneck Snake, *Diadophis punctatus*, in Florida. Bull. Florida State Mus. *10(2)*:43–90.

Norris, K. S. 1967. Color adaptation in desert reptiles and its thermal relationships. *In*: W. W. Milstead (ed.), Lizard Ecology; A Symposium. Columbia, Univ. Missouri Press. Pp. 162–229.

———, and C. H. Lowe. 1964. An analysis of background color-matching in amphibians and reptiles. Ecology *45*:565–580.

Porter, W. P. 1967. Solar radiation through the living body walls of vertebrates with emphasis on desert reptiles. Ecol. Monographs *37(4)*:273–296.

Steward, J. W. 1969. The Tailed Amphibians of Europe. New York, Taplinger. 180 pp.

Wallace, R. 1887. On the colour of the skin of men and animals in India. Proc. Roy. Soc. Edinburgh *15*:64–65.

Watkins-Pitchford, W. 1909. The Etiology of Cancer. Wm. Clowes and Sons, London.

CHAPTER 10

FOOD RELATIONS OF AMPHIBIANS AND REPTILES

A discussion of the food relations of any group of organisms must give consideration to both their feeding role and their role as food in their biotic community. Thus, this chapter will be concerned with what amphibians and reptiles eat and what predators prey upon amphibians and reptiles.

Very few studies of either amphibian or reptilian species have been sufficiently detailed to provide complete data on the animals' feeding habits and food requirements; there are many subjects related to food which need to be studied. A very basic kind of information which is generally lacking is that indicating what types and quantities of food a given species of amphibian or reptile will eat. Ecologically what is even more important than knowing what an animal *will* eat is knowing what it *does* eat. Unfortunately, there is also a paucity of information regarding food preferences of amphibians and reptiles and factors which influence these. In general, quantitative data are needed reflecting the percentage frequency of different items and percentage volume of different items in the diet. Such data are obtained through study of stomach contents, scats, and regurgitations, the latter being particularly useful in studying snakes.

Studies are needed to determine if food is a limiting factor for amphibian and reptilian populations. As a limiting factor, food may determine the number of individuals present in a population and may limit the geographic range of a species. In general, the more specialized a feeder is, the more its geographic range is likely to reflect the range and abundance of its prey.

Studies of the feeding roles of amphibians and reptiles are also needed to clarify interspecific relations within communities by determining if there is interspecific competition for food and, if so, what this competition means to the populations involved. Is competition for food affecting population densities? Is competition for food affecting distributional ranges? Is competition for food causing feeding specializations? These questions, and others, remain unanswered for most existing species of amphibians and reptiles.

THE FEEDING ROLE OF AMPHIBIANS

In view of the fact that they represent a diversiform and very old group of animals, it is rather unusual that all living members of the class Amphibia are primarily carnivorous as adults. Moreover, they usually feed only on live animals, either actively seeking prey or passively waiting for it to come close enough to be caught. Their restriction to live prey seems to reflect the great dependence of amphibians upon the sense of vision and their tendency to react only to moving objects. There are exceptions to this last statement, however, for the marine toad (*Bufo marinus*) will eat dog food from a dish. This species, and other amphibians as well, may also use their olfactory sense and respond to chemical stimuli while feeding. Since most, if not all, species of caecilians are blind, they presumably use their tactile organ instead of vision in locating prey.

Adult salamanders feeding on land will take prey of any appropriate size, including insects, arachnids, annelids, and other amphibians, and in the case of larger species, small mammals may even be eaten. Their protrusible tongue plays an important role in the capturing of prey by terrestrial feeders. They often capture their prey by making a very slow approach which is culminated in a short leap. The tongue is flicked out onto the prey and rapidly withdrawn with the food item adhering to its sticky surface. A small prey will be swallowed immediately; larger organisms will be grasped with the jaws and gradually worked into the esophagus, often with the aid of the front feet. The actual prey eaten by a species undoubtedly varies from one geographical locality to another, with seasonal changes in the relative abundance and availability of prey species, and from individual to individual because of different haunts and habits. Examples of the food eaten by the northern two-lined salamander (*Eurycea bislineata bislineata*) and the red-backed salamander (*Plethodon cinereus cinereus*), expressed in volume of each type of food, are depicted in Figure 10–1 and Figure 10–2, respectively.

Adult salamanders that feed under water generally do so by means of a snapping behavior. The mouth is brought close to

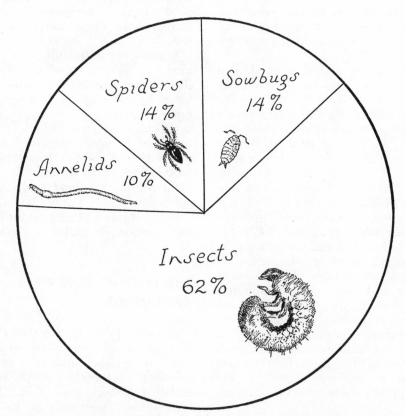

Figure 10–1   Food consumed by the northern two-lined salamander (*Eurycea bislineata bislineata*) expressed in volume of each kind of food. (From Oliver: *The Natural History of North American Amphibians and Reptiles.*)

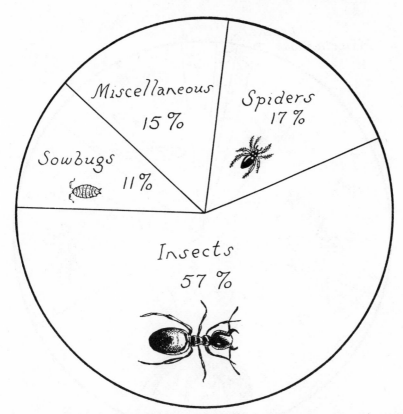

Figure 10–2   Food consumed by the red-backed salamander (*Plethodon cinereus cinereus*) expressed in volume of each kind of food. (From Oliver: *The Natural History of North American Amphibians and Reptiles*.)

the small prey animal and then is very rapidly opened. Water, presumably carrying the prey with it, rushes into the mouth, which is then snapped shut. Larger prey are grasped with the jaws. Small neotenic salamanders such as mudpuppies (*Necturus maculosus*) feed on insects, invertebrates of various kinds, small fish, and fish eggs. Larger aquatic salamanders such as amphiuma (*Amphiuma means*) feed on insects, invertebrates, and aquatic vertebrates such as amphibians, fish, and reptiles. Examples of what these two species have been known to eat are illustrated in Figures 10–3 and 10–4, respectively. Free-swimming larval salamanders will take arthropods and soft-bodied invertebrates. As particulate feeders, they have not evolved the elaborate branchial filter found in anuran tadpoles but, instead, have a well-developed buccal apparatus for ingesting and swallowing prey. The primary factor affecting the selection

of food by aquatic salamanders, just as with terrestrial species, is that the prey must be of the appropriate size.

Many adult anurans, including bufonids and ranids, capture prey on land by a flicking contact with their tongues. Gans (1961) has provided an excellent elucidation of the feeding of a bullfrog (*Rana catesbeiana*). Upon locating the prey, the frog will jump aggressively toward it. When about two or three inches away from the food, the frog lifts its head so that it is bent posteriorly on the neck and drops the lower jaw. The tongue, which is attached anteroventrally just behind the tip of the jaw, is thrust out of the snout by being swung through a vertical arc of 180 degrees. If the frog's aim is proper, the tongue will contact the prey before completing its rotation and, being elongated and limp, will wrap around the prey just as the end of a rope will wrap around a post when swung against it. The tongue

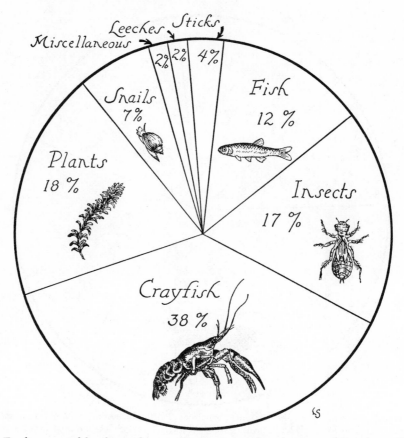

Figure 10–3 Food consumed by the mudpuppy (*Necturus maculosus*) expressed in volume of each kind of food. (From Oliver: *The Natural History of North American Amphibians and Reptiles.*)

is then retracted quickly, carrying the wrapped prey into the mouth. This whole process occurs very rapidly, requiring 0.05 seconds for the tongue to be flicked out and even less time for it to complete the retraction phase. If the prey is relatively large, the frog will use its front feet to help the jaws position the prey for swallowing. When feeding under water, anuran adults may either grasp their prey directly with their jaws or utilize the snapping behavior that adult salamanders use. Like salamanders, anurans feed on almost any moving prey which is of the appropriate size. Thus, the food eaten depends to a large degree upon the size of the anuran and may consist of either relatively small items such as insects or larger animals such as worms, other amphibians, and even small mammals. Examples of food consumed by the green frog (*Rana clamitans*) and the American

toad (*Bufo americanus*) are shown in Figures 10–5 and 10–6, respectively.

Anuran tadpoles, as a group, are omnivorous and gain food in a variety of ways. However, the tadpole diet for any given species is more restricted and is dependent upon their ecological "position". Some species live in the littoral zones of ponds and lakes and are herbivorous, using their horny teeth to graze, and ingest particles from soft plant tissues. Others are bottom or surface filter-feeders, concentrating bacteria, protozoans, small animals, and inorganic detritus by filtration of water that is being pumped through the mouth into the branchial apparatus and out the spiracle during respiration. Some tadpoles are pelagic planktonic feeders and consume the minute plants and animals which aggregate at or near the surface of the water. Cooperative feeding by aggregations of spadefoot toad (*Scaphiopus*) tadpoles have been observed by Bragg (1946, 1965). These aggregations swim about in masses and create whirlpools through their movements; the whirlpools suck up and concentrate plankton and organic material which has settled on the bottom of the pool. The concentrated food is then "inhaled" in the respiratory current of water passing through the tadpole and filtered by the branchial apparatus. As Bragg (1965) notes, feeding aggregations of tadpoles are adaptive because they secure on the average more scarce food for each individual than it could normally find alone. This is particularly important with species such as Hurter's spadefoot (*Scaphiopus holbrooki hurteri*), which deposit their eggs in shallow temporary water and subject the larval stages to a "race" with the evaporation of their aquatic habitat, because the improved diet that results from cooperative feeding hastens development and shortens the time required to reach metamorphosis.

The tadpoles of some anurans are characteristically cannibalistic; other tadpoles have been found to be cannibalistic at some times but not at others. Bragg (1962, 1964) observed that the tadpoles of plains spadefoot toads (*Scaphiopus bombifrons*) in Oklahoma are sometimes cannibalistic and that cannibals also prey on small tadpoles of other species as well. Tadpoles of Couch's spadefoot toad (*Scaphiopus couchi*), which sometimes occur in the same pond with those of the plains spadefoot, do not prey on living tadpoles but will eat dead ones of any kind present. In both cases, the ingestion of other tadpoles accelerates the rate of both growth and development so that less time is required to reach metamorphosis, and there must be strong selection for such a feeding habit in ponds subject to drying. Bragg (1965) observed that the cannibalistic tadpoles of the plains spadefoot have jaw and tooth characteristics which are quite different from those of non-cannibalistic

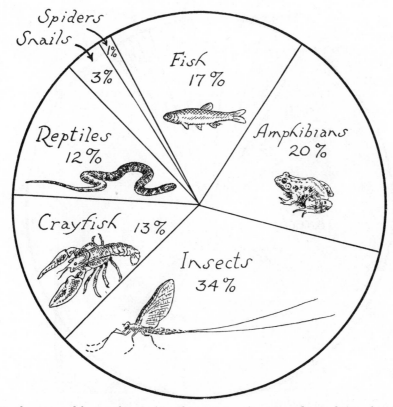

Figure 10–4 Food consumed by amphiuma (*Amphiuma means*) expressed in volume of each kind of food. (From Oliver: *The Natural History of North American Amphibians and Reptiles*.)

tadpoles belonging to the same species. However, it is not clear whether or not these differences are genetically determined.

Tree frogs (family Hylidae) sometimes reproduce in very restricted bodies of water, such as collects at the base of a bromeliad leaf, and have a tadpole diet consisting largely of insect larvae. The food supply for larval stages in such limited aquatic habitats is very limited; consequently, the tadpoles of at least one species (*Anotheca spinosa*) have evolved the habit of eating the eggs of their own or related species laid in the same situation. The result is that the most rapidly developing individuals exist at the expense of the slower individuals. The species as a whole survives because the food supply represented by the total egg mass is concentrated into the growth and development of a few individuals. Without

the cannibalistic habit, the entire group would possibly face starvation, few would be likely to metamorphose successfully, and the species as a whole would be endangered.

Adult caecilians appear to devour almost any prey animal available within their particular habitat as long as it is of the appropriate size. Terrestrial forms probably feed largely on earthworms, burrowing snakes, and other caecilians. Aquatic caecilians probably feed on much the same sort of food as aquatic salamanders. The caecilian tongue is basally attached, as it is in typical mammals, and is capable only of slight protrusion through the contraction of its radial muscles. Thus, caecilians must capture their food primarily by grasping it with their jaws. As previously mentioned, most, if not all, species of caecilians are blind, and they presumably use their tactile organ in locating prey.

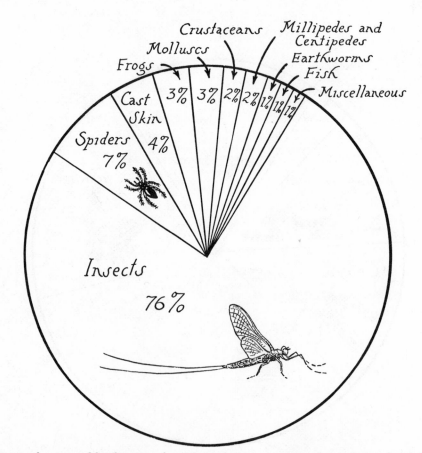

Figure 10–5   Food consumed by the green frog (*Rana clamitans*) expressed in volume of each kind of food. (From Oliver: *The Natural History of North American Amphibians and Reptiles*.)

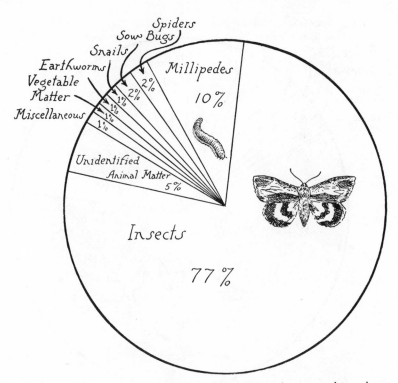

Figure 10-6 Food consumed by the American toad (*Bufo americanus*) expressed in volume of each kind of food. (From Oliver: *The Natural History of North American Amphibians and Reptiles*.)

THE FEEDING ROLE OF REPTILES

The feeding role of reptiles is more complex than that of amphibians. Whereas amphibian adults are all carnivorous and vary in their feeding primarily in regard to the size of prey taken, living reptiles are variable in diet. Turtles, as a group, are omnivorous and will eat almost any form of organic material, but individual species may be very selective in their feeding. Some species, such as painted turtles (*Chrysemys picta*), eat considerable plant material (Fig. 10-7), whereas others, such as the midland soft-shelled turtle (*Trionyx spiniferus*), seem to be strictly carnivorous (Fig. 10-8). Some turtles, including the snapping turtle (*Chelydra serpentina*) consume much carrion in addition to living prey (Fig. 10-9). All crocodilians are strictly carnivorous, feeding on a variety of animal life, from insects and spiders to birds and mammals. Lizards, as a group, are primarily insectivorous but some feed on larger animals,

and occasional lizards are herbivorous. The two large lizard families, Agamidae and Iguanidae, are as diversified in feeding habits as they are in habitats occupied. The marine iguana (*Amblyrhynchus cristatus*) subsists on marine plants which grow along the shores of the Galapagos Islands and dives down fifty feet or more under water to feed. The desert iguana (*Dipsosaurus dorsalis*), as indicated in Figure 10-10, is largely herbivorous but feeds occasionally on a variety of other foods. The majority of other New World herbivorous lizards are also in the family Iguanidae. In the Old World, the mastigure (*Uromastix acanthinurus*, family Agamidae) is largely herbivorous, and the Australian shingleback skink (*Tiliqua rugosa*) feeds on berries and mushrooms. Although seasonal changes in the abundance of prey species undoubtedly influence what insectivorous lizards eat, habitat differences and species-specific preferences in the kinds of food eaten result in the diet of one insectivore being different from that of another (compare

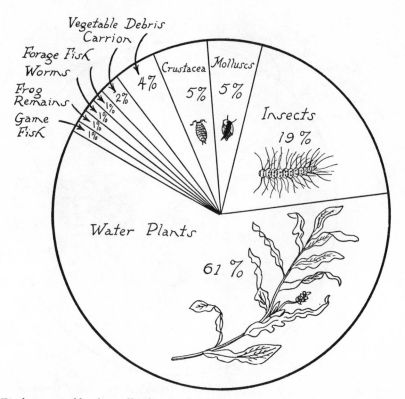

Figure 10–7   Food consumed by the midland painted turtle (*Chrysemys picta marginata*) expressed in volume of each kind of food. (From Oliver: *The Natural History of North American Amphibians and Reptiles.*)

Figures 10–11, 10–12, and 10–13). Most snakes are carnivores and feed on relatively large prey, but small burrowing species are generally insectivorous and some larger species, including racers (*Coluber constrictor*), also feed on grasshoppers and other insects (Fig. 10–14). Many snakes feed primarily on small mammals: none is known to be herbivorous. Finally, tuataras are carnivorous and, being nocturnal feeders, prey upon beetles, crickets, snails, and lizards (especially geckos) which are active in the evening and at night.

Most reptiles, particularly those which are carnivorous, locate their food by a combination of sensory mechanisms. Crocodilians use their sense of vision primarily while hunting; their sense of hearing is also fairly well-developed and may be used in locating prey. Turtles have a keen sense of vision and in discrimination tests have been shown to be able to detect differences between lines varying but 0.1 millimeter in width. Thus, they

probably use this sense and their sense of smell, moderately developed in most species, to detect food. All turtles have ears, but the sense of hearing is generally not well developed, and most turtles respond to mechanical vibrations of the substrate rather than to air-borne sounds. Lizards, with their exceptionally good eyesight, depend upon their vision to find food. As with amphibians, lizards often appear to be unable to detect objects unless they are moving. Snakes find their food by a combination of senses. Eyes are well developed in most species, and burrowing species are generally the only ones to lack keen vision. Jacobson's organ supplements normal olfactory perception by the nasal region in snakes, and their sense of smell is always acute; it is the primary sense in some species. Thus, most snakes find their prey visually and by their sense of smell; they probably use their sense of hearing more for detecting enemies. In addition to visual and olfactory detection of food, pit vipers are able to detect prey with their heat-sensitive facial pits; they are carnivores which eat a variety of animals. Cottonmouths (*Agkistrodon piscivoros*) eat fish and frogs, copperheads (*Agkistrodon contortrix*) may feed on caterpillars, and rattlesnakes (*Crotalus*) eat lizards. However, all of these also feed on small mammals, particularly at night when there is usually a considerable difference between the temperature of their prey and that of the surrounding atmosphere; they are able to locate this kind of prey, both before and after striking it, by heat-sensing. Little is known about the sensory mechanisms of tuataras, but they probably locate their food visually.

All carnivorous reptiles are faced with the problem of catching their prey. As a group, reptiles tend to be stalkers, making slow, deliberate movements until they are within striking range of the prey and then suddenly snapping at it. Crocodilians and monitor lizards use their strong tails as

clubs to overpower their prey and sweep it to where it can be grasped by the jaws. The highly specialized chameleons have tongues that may be extended rapidly to a length several times the length of the head. They also have a prehensile tail and opposing digits, both adaptations for climbing, and their turretlike eyes are capable of moving independently of one another. Thus, chameleons are highly adapted for insectivorous feeding. They merely climb to an advantageous position, sit motionless, and rotate their eyes in search of insects. When an insect comes within range, the chameleon rapidly extends its tongue, which is covered with a sticky mucus and has a bulb on the end, and picks off the prey instantaneously (Fig. 4–36). As described in Chapter Four, many of the cranial features of snakes are adaptations to their habit of feeding on large prey and result in a head that is expandable and flexible throughout, so that snakes may swallow prey which is

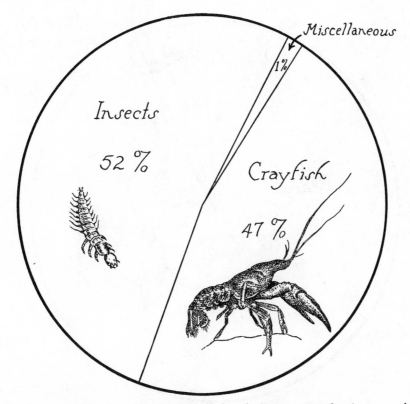

Figure 10–8 Food consumed by the midland soft-shelled turtle (*Trionyx spiniferus*) expressed in volume of each kind of food. (From Oliver: *The Natural History of North American Amphibians and Reptiles*.)

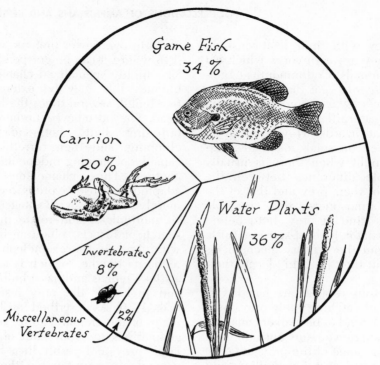

Figure 10–9 Food consumed by the snapping turtle (*Chelydra serpentina*) expressed in volume of each kind of food. (From Oliver: *The Natural History of North American Amphibians and Reptiles.*)

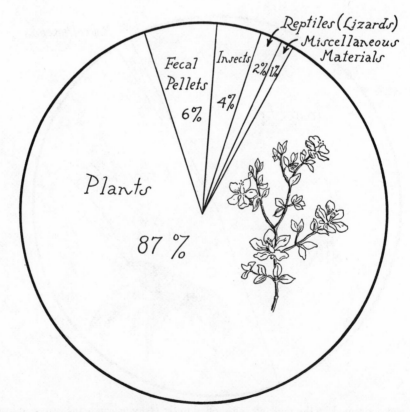

Figure 10–10 Food consumed by the desert iguana (*Dipsosaurus dorsalis dorsalis*) expressed in volume of each kind of food. (From Oliver: *The Natural History of North American Amphibians and Reptiles.*)

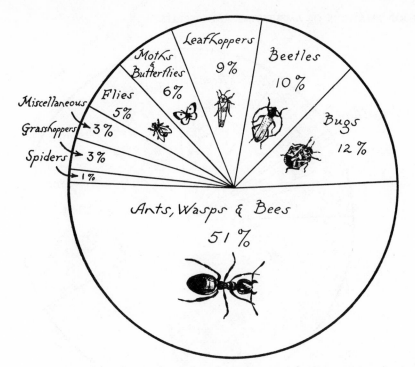

Figure 10–11 Food consumed by the sagebrush lizard (*Sceloporus graciosus graciosus*) expressed in the number of items of food. (From Oliver: *The Natural History of North American Amphibians and Reptiles.*)

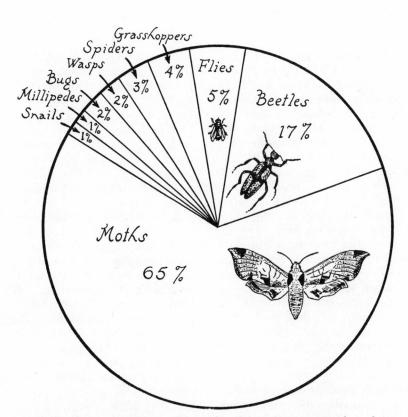

Figure 10–12 Food consumed by the Shasta alligator lizard (*Gerrhonotus coeruleus shastensis*) expressed in the number of items of food. (From Oliver: *The Natural History of North American Amphibians and Reptiles.*)

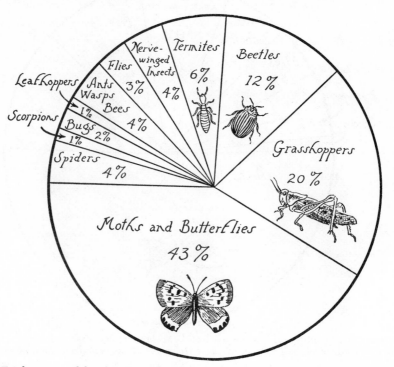

Figure 10–13 Food consumed by the Great Basin whiptail (*Cnemidophorus tigris tigris*) expressed in the number of items of food. (From Oliver: *The Natural History of North American Amphibians and Reptiles.*)

larger in diameter than their heads. The capturing and swallowing of prey is also aided by the arrangement of teeth in snakes. Because the teeth point backward, once the prey is grasped in the mouth of the snake it can never slip forward, and the snake gradually works it back through the mouth and into the throat by alternating movements of the upper and lower jaws. Teeth are frequently broken off by victims' struggles, so the snake's ability to replace teeth is also an important adaptation to feeding. The teeth of venomous snakes are, of course, adapted to various degrees for delivering the poison. The poison is employed advantageously in feeding: when attacking a large prey animal, the venomous snake will strike it, injecting the venom, and then will often wait for the poison to take effect before attempting to swallow the prey; the snake thus avoids the problem of swallowing a large struggling victim.

Associated with the tendency of snakes to feed on large animals is their propensity to feed irregularly. A captive python has been reported to have lived three years

without food (Romer, 1959)! Nevertheless, the food consumption of a large snake may be considerable. Doflein (1914) states that in the Trivandrum Museum in Travancore, India, a python measuring seven meters in length ate 100 hens, a dog, a kangaroo, and four smaller animals during the course of a year. A second snake, five meters long, ate 54 hens, 2 dogs, 2 guinea pigs, and 5 other small animals in the same time.

A few snakes have become very specialized in their diets and are adapted accordingly. The eggshell-cutting apparatus of Dasypeltinae and Elachistodontinae has been discussed in Chapter Four. African snakes in the genus *Lycophidion* (colubrid subfamily Colubrinae) are adapted for feeding upon skinks. The genus *Lycophidion* has evolved enlarged canine-like teeth in the front of both jaws that aid these snakes in holding onto the lizards just as canines aid mammalian carnivores in grasping prey. The shield-tailed snakes (family Uropeltidae) feed almost entirely on earthworms and utilize their spadelike tails to dig out prey. Thirst snakes (*Dipsas* in the colubrid subfamily Dipsadinae) feed primarily on snails, which they extract from shells by means of elongated lower jaws. Although most of the sea snakes feed primarily on fish, members of the genus *Emydocephalus* apparently feed entirely on the eggs of teleost fish. Another genus of marine snakes, *Microcephalophis*, is characterized by a head and anterior part of the body that are much narrower than more posterior portions; this is probably an adaptation for feeding on eels, which seek cover in rocky crevices.

Because large individuals tend to select larger prey than smaller individuals, differences in size tend to reduce competition between juveniles and adults of the same species. This has been well illustrated by Cott's (1961) study of the Nile crocodile (*Crocodylus niloticus*). As shown in Figure 10–15, young crocodiles prey largely on insects, mollusks, crabs, and amphibians. When they are about eight to 10 feet in length, crocodiles of this species feed primarily on fish. As greater lengths are attained, the diet includes more and more mammals and reptiles.

Differences in size also tend to reduce competition between many species of amphibians and reptiles. However, the few studies which have been made indicate that amphibian and reptilian feeding habits overlap those of a wide variety of animals and, therefore, interspecific competition for food may be a factor affecting populational densities and distributions. Aquatic amphibians probably compete most directly with fishes, other amphibians, turtles, and predaceous invertebrates such as beetles and insect larvae. Terrestrial amphibians are generally insectivores and, therefore, compete primarily with other vertebrates, including mammals, birds, snakes, lizards, and other species of amphibians.

Farner (1947) compared the stomach

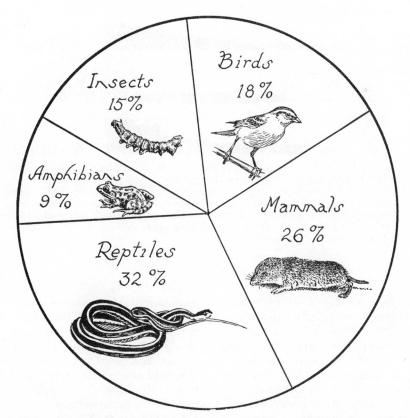

Figure 10–14 Food consumed by the black racer (*Coluber constrictor constrictor*) expressed in volume of each kind of food. (From Oliver: *The Natural History of North American Amphibians and Reptiles.*)

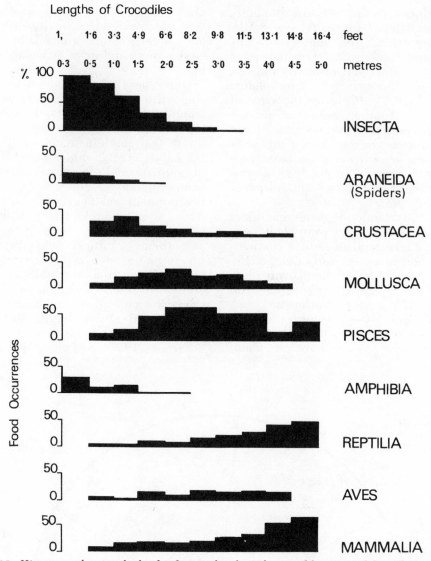

Figure 10–15 Histograms showing the kinds of prey taken by Nile crocodiles (*Crocodylus niloticus*) of different sizes. The lengths of the crocodiles are shown at the top in meters and approximate equivalent number of feet. For each length group, the food occurrences are expressed as a percentage of the stomachs found to contain remains of prey of any kind. For example, over 50 per cent of the stomachs of four- to five-meter crocodiles contained remains of mammals. Based on data from crocodiles in Uganda and Northern Rhodesia. (From Cott: Scientific results of an enquiry into the ecology and economic status of the Nile Crocodile (*Crocodilus niloticus*) in Uganda and Northern Rhodesia.)

contents of two kinds of salamanders, the rough-skinned newt (*Taricha granulosa*) and the long-toed salamander (*Ambystoma macrodactylum*), occurring around Crater Lake, Oregon. He found that there was an overlap in diets but that little competition for food occurred between these species. The long-toed salamanders were scavengers, primarily feeding on terrestrial arthropods but oc-

casionally taking insect larvae. The rough-skinned newts, on the other hand, fed primarily on fresh-water amphipods and snails but occasionally took aquatic insect larvae and terrestrial arthropods. Thus, it would seem that niche segregation is a factor allowing these two species to co-exist in the same habitat.

In contrast to Farner's results, Zweifel (1949) found that two sympatric plethodontid salamanders, *Aneides lugubris* and *Ensatina eschescholtzii,* compete for all food items eaten (sow bugs, centipedes, spiders, springtails, beetles, caterpillars, and ants). *Aneides* was less abundant than *Ensatina* in the area studied, and Zweifel concluded that competition for food might explain the relative abundance of the two species. Clearly, no general statement can be made regarding the role of interspecific competition for food in the niche segregation of salamanders. Its role in anuran ecology is generally unknown, and nothing is known about its importance to caecilians.

Comprehensive studies of a variety of lizards (eastern fence lizards, rusty lizards, five-lined skinks, collared lizards, and whiptails) indicate that they potentially compete primarily with other vertebrates for food. Among mammals which are possible or known users of the foods eaten by rusty lizards (*Sceloporus olivaceus*) are opossums, armadillos, skunks, gray foxes, feral house cats, and pocket mice (Blair, 1960). All insectivorous birds eat the same sort of food as lizards, and in the southwestern part of North America, where lizards are most abundant, mockingbirds and roadrunners may be among lizards' greatest competitors. Snakes may sometimes be food competitors of lizards. Blue racers (*Coluber constrictor*) and green snakes (*Opheodrys aestivus*) feed heavily on grasshoppers and, thus, overlap adult rusty lizards in their diet (Blair, 1960).

Among sympatric species which feed on the same sorts of food there is generally a reduction or complete elimination of potential competition because of habitat and behavioral differences. For example, sympatric anuran amphibian populations and lizard populations may all eat insects. However, most anurans are nocturnal and most lizards are diurnal, and they probably feed on entirely different insect populations because the insects that are active at night are generally different from those that are active in the daytime. Even the sympatric occurrence of different species of diurnally active insectivorous lizards does not mean that all prey on the same populations because there may be habitat differences and resulting niche differerences. For example, whiptails (*Cnemidophorus*), horned lizards (*Phrynosoma*),

and anoles (*Anolis*) sometimes occur sympatrically and are of the same general size. However, horned lizards are terrestrial and feed largely on ants. Whiptails forage about for insects on the ground in open areas. Anoles are climbers that spend a good portion of their time in vegetation above the ground and feed on insects in a microhabitat which is very different from that of the whiptails.

The majority of snakes feed on small vertebrates, including mammals, birds, other reptiles, and amphibians. In addition, many snakes feed extensively on large insects such as grasshoppers and crickets and on earthworms. Thus, snakes potentially may compete primarily with avian and mammalian carnivores, with other snakes, and, to a lesser degree, with lizards.

Because they tend to be omnivores, many turtles potentially compete with a wide variety of consumer organisms. However, turtles are often capable of shifting their diet in response to the relative abundance of different foods and are probably not severely affected by competition for food. For example, green turtles (*Chelonia mydas*), hawksbill turtles (*Eretmochelys imbricata*), and loggerhead turtles (*Caretta caretta*) are all marine species that, respectively, are primarily herbivorous, omnivorus, and primarily carnivorous. However, all three at least occasionally eat crustaceans, jellyfish, mollusks, fish, and marine grasses.

Crocodilians are generally without food competitors because they tend to feed on the next largest vertebrates in their communities. Because of the geographical isolation of their habitat, tuataras probably face little potential competition for food other than from insectivorous birds and geckos.

THE ROLE OF AMPHIBIANS AND REPTILES AS FOOD

As is true of the feeding role of amphibians and reptiles, their role in biotic communities as food has received very little attention. In general, amphibians and reptiles are carnivores and, therefore, represent either secondary or tertiary trophic levels. Their predators represent tertiary or quaternary levels. Scattered observations indicate that both amphibians and reptiles are preyed upon by all other classes of vertebrates, and, of course, some amphibians or reptiles prey on other amphibians or reptiles.

Anurans and salamanders are preyed upon by such birds as herons and storks. Most mammalian carnivores and omnivores, including raccoons, opossums, armadillos, weasels, and shrews, prey upon amphibians. Opossums and armadillos are, accordingly, both potential competitors and predators to amphibians. In the Old World tropics, large-winged bats (*Megaderma*) are a major predator on amphibians. Reptiles, especially snakes, prey upon amphibians, and snakes frequently seem to be the most important source of predation on anuran populations. Other amphibians eat amphibians; large species prey upon small species and adults eat juveniles. Wherever they occur together, fish prey upon amphibians. Predaceous invertebrates such as diving beetles, fresh-water hydra, and aquatic insect larvae prey upon amphibian larval stages and sometimes even on adults. It appears that almost anything will eat an amphibian!

Lizards are known to be preyed upon by many mammals, including armadillos, skunks, feral house cats, opossums, domestic dogs, coyotes, tarsiers, badgers, foxes, and, in the Old World tropics, large-winged bats. Lizards are also eaten by a variety of birds, including roadrunners, sparrow hawks, red-tailed hawks, jays, crows, mockingbirds, and shrikes. Snakes which prey upon lizards include blue racers, coachwhips, rat snakes, gopher snakes, and patch-nosed snakes. Several of these predators are also potentially food competitors for lizards. The Texas patch-nosed snake (*Salvadora grahamiae lineata*) is the most important nest predator on the rusty lizard. Following the scent left by adult lizards, it may destroy up to 75 per cent of the nests of rusty lizards in some areas of the southwestern United States. The scent of the adult disappears after a few days so that the nest is then safe from predation, but patch-nosed snakes are known to prey on rusty lizards of all life stages (Blair, 1960).

A major share of the predation on snakes is often attributable to other snakes. King snakes (*Lampropeltis*) are notorious for being ophiophagous and are known to eat a variety of venomous and harmless snakes. A king snake will sometimes swallow another snake larger than itself, and, like a number of other kinds, the snakes in *Lampropeltis* are relatively immune to snake venoms. Many, many other snakes are known to prey upon snakes. Among those in the United States are rat snakes (*Elaphe*), garter snakes (*Thamnophis*), rattlesnakes (*Crotalus*), gopher snakes (*Pituophis*), and racers (*Coluber*). Raptorial birds are also important predators on snakes. Marsh hawks, red-shouldered hawks, broad-winged hawks, red-tailed hawks, sparrowhawks, Swainson's hawks, and barn owls are all known to frequently feed on snakes. Crows, roadrunners, jays, and eagles are undoubtedly also major predators on snakes in some areas. Interestingly, Virginia rails and ruffed grouse have been reported to prey upon garter snakes.

A variety of mammals may be nearly as important as birds in preying on snakes. Among mammalian predators are raccoons, skunks, mink, foxes, badgers, opossums, and weasels. McKeever (1958) described an attack on a garter snake by an eastern chipmunk and it seems likely that small species and juveniles of larger snakes are probably preyed upon by a great variety of mammals. Either the mammals attack out of instinct to protect themselves or snakes may be significant kinds of food sought by predators. Because of the common hatred for snakes, man is without question one of the major causes of mortality to snake populations throughout the world. Lizards occasionally eat snakes. Blair (1960) mentions a 70-millimeter rusty lizard which was observed trying to swallow a 220-millimeter green snake (*Opheodrys aestivus*)! Fitch (1963) found the remains of a juvenile blue racer in scats of a slender glass lizard (*Ophisaurus attenuatus*). Large amphibians, such as bullfrogs, may also prey on small snakes from time to time.

Adult crocodilians are largely immune from predation except by man, but their eggs and newly hatched young are heavily preyed upon. Terrestrial predators, such as raccoons and, in the Old World, monitor lizards (*Varanus*), will dig up unprotected crocodilian nests and feed on the eggs and emerging young. Newly hatched crocodilians are also eaten by fish, storks, herons, mongooses, snakes, and larger crocodilians.

Because of their protective shell, adult turtles are also relatively immune from predation except by man. However, a few birds, including gulls, crows, and eagles, are known to carry turtles and tortoises high over rocky ground and then drop them in order to crack open the shells. Turtle eggs and juveniles, like those of crocodilians, are susceptible to predation by a vast array of vertebrates, including birds, mammals, larger reptiles, and fish. A high proportion of young sea turtles is eaten by gulls as they emerge from their nests and attempt to move across open beaches to the water. Once in the ocean, the young turtles are subject to predation by a variety of fishes, birds, and invertebrates.

The geographical isolation of surviving tuataras has resulted in their having few predators other than man and introduced mammals such as cats and dogs. It seems likely that young hatchlings may be preyed upon by larger individuals, and some may succumb to hawks living on the same islands.

Without doubt, man is a most important predator, affecting nearly every species of amphibian and reptile. The impact of man on amphibian and reptilian populations is discussed in Chapter Fifteen.

References

Blair, W. F. 1960. The Rusty Lizard. Austin, Univ. of Texas Press. 185 pp.

Bragg, A. N. 1946. Aggregation with cannibalism in tadpoles of *Scaphiopus bombifrons* with some remarks on the probable evolutionary significance of such phenomena. Herpetologica 3:89–97.

_____ 1962. Predator-prey relationship in two species of spadefoot tadpoles with notes on some other features of their behavior. Wasmann J. Biol. 20: 81–97.

_____ 1964. Further study of predation and cannibalism in spadefoot tadpoles. Herpetologica 20:17–24.

_____ 1965. Gnomes of the Night; The Spadefoot Toads. Philadelphia, Univ. Pennsylvania Press. 127 pp.

Cott, H. B. 1961. Scientific results of an enquiry into the ecology and economic status of the Nile crocodile (*Crocodilis niloticus*) in Uganda and Northern Rhodesia. Trans. Zool. Soc. London 29:211–356.

Doflein, F. 1914. Tierbau und Tierleben, II: Das Tier als Glied des Naturganzen. Leipzig and Berlin.

Farner, D. S. 1947. Notes on the food habits of the salamanders of Crater Lake, Oregon. Copeia 1947: 259–261.

Fitch, H. S. 1963. Natural history of the Racer *Coluber constrictor*. Univ. Kansas Publ. Mus. Nat. Hist. 15(8):351–468.

Gans, C. 1961. A bullfrog and its prey. Nat. Hist. 70(2):26–37.

McKeever, J. L. 1958. Chipmunk and garter snake. Canadian Field Nat. 72:170.

Oliver, J. A. 1955. The Natural History of North American Amphibians and Reptiles. Princeton, N.J., Van Nostrand Reinhold. 359 pp.

Romer, A. S. 1959. The Vertebrate Story. Chicago, Univ. Chicago Press. 437 pp.

Zweifel, R. G. 1949. Comparison of the food habits of *Ensatina eschscholtzii* and *Aneides lugubris*. Copeia 1949:285–287.

CHAPTER 11

REPRODUCTIVE ADAPTATIONS OF AMPHIBIANS

INTRODUCTION TO THE
REPRODUCTIVE BIOLOGY OF
AMPHIBIANS AND REPTILES

It is axiomatic that the perpetuation of any species is dependent upon reproduction by existing individuals and that the number of young produced in stable populations must be sufficient to provide a one-to-one replacement of reproducing individuals from one generation to the next. If the replacement rate of reproducing individuals is less than one to one in succeeding generations, obviously the population will decline in numbers. If the replacement rate is greater than one to one, the population will increase in number. Although a few species of lizards are known to be unisexual, both amphibians and reptiles are typically bisexual organisms; furthermore, they are typically polygynous, since one male generally mates with two to many females. Thus, the maintenance of stable amphibian and reptilian populations is dependent primarily upon a one-to-one replacement of mature females. Males are generally more expendable than females, as the reproductive potential of the population is less dependent upon their number. This fact is importantly related to patterns of reproductive behavior which will be discussed subsequently.

Most natural populations are in balance with their environment and remain relatively stable in size. This reflects the fact that reproductive potentials are under the

control of natural selection, and for any given population (species), the number of young produced will reflect the mortality normally suffered by the prereproductive age classes of that group. Amphibians and reptiles, like other animals, exhibit two general patterns of reproductive effort. One consists of producing vast numbers of young which are, for the most part, left unprotected and which, consequently, suffer a tremendous mortality before attaining adulthood; many, but not all, anuran species fall into this category. The other general pattern, characteristic of some but not all salamanders and reptiles, is the production of relatively few young which are given protection and which, consequently, suffer less mortality than they would otherwise. The "choice," so to speak, is between expending energy in the production of vast numbers of gametes and expending energy in the protection of a smaller number of gametes and resulting zygotes.

Ova can only be fertilized within a very brief period of time after maturation, and, in temperate zones at least, mature gametes are generally produced only during a limited time each year; this period is called the breeding season. Since reproduction by amphibians and reptiles is typically bisexual and since male and female reproductive organs occur in different individuals (they are dioecious animals), mature individuals of the same species and of both sexes must occur together at the proper time if fertilization of ova is to be achieved. Among amphibians fertilization of the ova may occur outside of the female's body (external fertilization), or it may occur within the female's reproductive tract (internal fertilization); fertilization is always internal in reptiles. External fertilization requires the bringing together of members of the opposite sex; this must occur at both the right time and the right place, in a suitable habitat for fertilization. Internal fertilization may occur almost anywhere, but it still requires that members of the opposite sex be brought together at a time when viable sperm are present to be transferred into the female's reproductive tract. Depending upon the sperm-storing capabilities of the species, internal fertilization may occur immediately after or during mating or it may occur at some future time. In the former case, mating must occur when both sexes have mature gametes for release and during the brief period of time when the ova are capable of being fertilized. In the latter case, only the male need have mature gametes (sperm) at the time of mating. Both the timing of the breeding season and the mating behavior that brings members of the opposite sex together at the proper time are important aspects of the general topic of reproduction.

BREEDING SEASONS OF AMPHIBIANS

The environmental factors which influence or control the periodicity of amphibian breeding seasons are poorly known. Rowan (1938) reported that the European common frog (*Rana temporaria*) and the crested newt (*Triturus cristatus*), which normally breed in the spring and summer, can be induced to breed in the winter by artificial light. He also noted that gravid females of the viviparous European salamander (*Salamandra salamandra*) that were exposed to artificially lengthened days gave birth to young whereas others not so exposed retained theirs. Observations such as these indicate that photoperiod, which is of prime importance in triggering gonadal cycles in many mammals, birds, reptiles, and fish, is also an influencing factor affecting am-

phibian reproductive cycles. However, temperature and precipitation are also of major importance in stimulating amphibian breeding activity. The relative importance of photoperiod, temperature, and precipitation to the initiation of amphibian breeding activity undoubtedly varies with the breeding habitats of different species and between tropical and non-tropical populations. Other environmental factors, such as the abundance of the food supply, may also influence reproductive cycles.

Amphibians tend to fall into three general categories in regard to their breeding habitats: One group breeds in permanent bodies of water, the second group in temporary bodies of water, and the third group breeds out of water. Temperature appears to be a major environmental factor influencing, or dictating, the breeding season for non-tropical amphibians reproducing in permanent bodies of water. Generally, these breed in the spring or summer, but the particular temperature required for optimal breeding activity varies considerably from species to species. Wood frogs (*Rana sylvatica*) and northern or high-altitude populations of tiger salamanders (*Ambystoma tigrinum*), for example, breed under extremely cool conditions, often when air temperatures are near 0° C and when ice is just melting from ponds and lakes. The temperature adaptations of such species are reflected in their tendency to breed very early in the spring, often in March and April. Most other species of temperate amphibians in this category require warmer temperatures and tend to breed at later dates in the spring. Because species breeding in permanent bodies of water tend to be species that are aquatic or semiaquatic or have habitats near these same bodies of water, precipitation is not as influential in determining the breeding season for these amphibians as it is for those breeding in temporary waters. However, terrestrial and semiaquatic amphibians always tend to be more active when humidity is high, since this minimizes their water-balance problem, and they are more apt to be actively breeding when there is precipitation. The greatest number of breeding amphibians is generally found on warm moist nights, but fewer individuals are apt to be breeding almost every night or day during the appropriate season. Tropical amphibians breeding in permanent waters may be influenced to do so more by precipitation (or other environmental factors) than by temperature. Multiple breeding during a calendar year occurs in females of some (possibly most) tropical amphibians, and males of many species are reproductively active throughout the year.

The second general group of amphibians, including a diversity of species occupying habitats ranging from deserts to rain forests, breeds in temporary waters produced by heavy precipitation. These amphibians may breed whenever the temperature is high enough and heavy rains fall. Thus, their reproductive season tends to correlate with the rainy season for their particular locality. For instance, both tiger salamanders and western spadefoots (*Scaphiopus hammondi*) breed in late winter or early spring in California and in midsummer in New Mexico and Arizona. Because reproduction, often out of necessity but sometimes out of choice on the part of the species, takes place in temporary water, the larval stages run the risk of having their aquatic environment dry up before they metamorphose. Consequently, through evolutionary adaptation, this group of amphibians tends to have rapid developmental rates, with metamorphosis commonly taking place in two or three weeks as compared to months in the group breeding in permanent water. Such rapid development requires warmer environmental temperatures than the slower growth of those in permanent water; accordingly, there is a general correlation among a higher optimum temperature for breeding activity, a higher temperature for embryonic development, and the habit of breeding in temporary water.

The third group, including most genera of plethodontid salamanders and a large number of anurans in the families Microhylidae, Hylidae, Leptodactylidae, Ranidae, and Bufonidae, plus *Leiopelma* in the family Ascaphidae, do not breed in water but rather lay their eggs on land in moist situations. Reproductive activity in this group, as in temporary-water breeders, tends to correlate with rainy seasons, and precipitation is a major stimulus to breeding activity. In northwestern United States,

some of the plethodontid salamanders belonging to the genus *Batrachoseps* are known to deposit their eggs in late fall and winter, while other western plethodontids generally deposit theirs in the spring and early summer. Differences in reproductive cycles of this nature may reflect differing responses to photoperiod, *Batrachoseps* responding to decreasing daylength and the other plethodontids responding to increasing daylength. On the other hand, Eschscholtz's salamander (*Ensatina eschscholtzi*) may breed at any time during its period of surface activity (fall until spring) but appears to breed most frequently in the spring (Stebbins, 1954). As with other amphibians, the amount of breeding activity at any one time during the breeding season of any member of this group is influenced by temperature, there generally being more activity when environmental temperatures are warm than when they are cool.

In general, oogenesis in non-tropical amphibians is limited to a much shorter period of time than is spermatogenesis. In other words, male amphibians tend to produce viable gametes over a relatively long period of time, whereas females are fertile only during a short period of time. This basic difference arises from the factors stimulating final developmental processes in gamete formation. Activation of the male reproductive system is primarily in response to abiotic environmental factors, such as temperature, humidity, and photoperiod. These factors cause the gonads to become active during a particular time of the year, the reproductive season, but they do not restrict gamete formation or maturation to a short period within the breeding season. Although spermiation, the release of sperm cells, may only occur during a limited period of time, viable sperm are produced in many anurans almost continuously; males can be induced to breed at almost any time that they are normally active except immediately after a breeding season, when there is commonly a latent period.

In an intensive study of spermatogenesis in the European common frog (*Rana temporaria*), Van Oordt (1956) found three general periods of activity in the gonads during any one year: A period of spermiation (1), in which sperm cells were released from tubules in the testes, occurred in this species between February and March. Spermatogenesis (2), the maturation of the sperm, occurred from April to October. This represents practically the entire portion of the year when these frogs are active. A rest period (3) in which gonads were inactive extended from October through January. Using environmental control chambers, Van Oordt conducted a series of experiments to determine the effect of photoperiod on spermatogenesis and found that for this species photoperiod was unimportant. Those individuals kept in total darkness had cycles coinciding with those under natural light conditions. Another series of experiments was made to determine the effects of temperature. In these, the controls were kept at the temperature (in regard to both range and variability) of the natural environment, while other individuals were kept in temperature-controlled units at higher and lower temperatures. Van Oordt found that spermatogenesis could be hastened by subjecting the animals to higher temperatures and that it was retarded by lower temperatures. That these results merely reflected the direct effect of temperature upon the rate of chemical reactions was indicated in Van Oordt's studies, for he found that removal of the anterior lobe of the pituitary had no apparent effect upon spermatogenesis which was already occurring. However, the anterior pituitary controls the activity of the gonads and, therefore, controls the periodicity of spermatogenesis, so Van Oordt concluded that temperature regulated the spermatogenic cycle through its influence on the anterior pituitary. He also found that there was an internal rhythm which controlled the sensitivity of the gonads to the secretions of the anterior lobe of the pituitary, but this rhythm was unexplained. Other studies have indicated that temperature is important in timing the spermatogenic cycle in Grecian frogs (*Rana graeca*), clawed frogs (*Xenopus laevis*), leptodactylid frogs (*Leptodactylus*), and newts (*Triturus*). On the other hand, photoperiod appears to be the most important factor controlling the spermatogenic cycle in spring peepers (*Hyla crucifer*) and the Argentine toad, *Bufo arenarum*. Some species, perhaps many, are affected by a combination of temperature and photoperiod.

The limited amount of evidence available indicates that the initiation and early stages of oogenesis are also controlled by such environmental factors as photoperiod, temperature, and humidity. However, the final ripening of the eggs in the female is dependent upon the stimulus provided by breeding males, and this stimulation is an integral part of the breeding behavior and courtship of amphibians. The dependence of the final stages of oogenesis upon such stimuli is importantly related to the fact that ova can only be fertilized within a brief period of time following their maturation, for it prevents ova from becoming ripe in the absence of males and increases the probability that fertilization will occur, minimizing the wastage of gametes from both sexes.

MATING BEHAVIOR AND COURTSHIP

Some species of Amphibia, particularly those which are permanently aquatic and those which lay their eggs on land, normally breed in the same general area inhabited during the rest of the year. More typically, however, an amphibian population is dispersed over a wide area during the non-breeding seasons, with groups or individuals often some distance from where they breed, and the individuals congregate in large numbers at breeding sites only during the appropriate period. Thus, amphibians must typically make a definite movement from a non-breeding area to a breeding site, and, again, at a time when both males and females have viable and mature gametes. Whether this movement is required or not, species characteristically have discriminatory courtship rituals and mating behavior which involves species and sex recognition. Thus, the general scheme of mating behavior is one of bringing individuals belonging to the same species but of the opposite sex together, promoting final stages of oogenesis, and accomplishing fertilization and deposition of eggs.

SALAMANDERS

Much remains to be learned concerning the courtship of salamanders, but a few general statements can be made. Nothing is known of the courtship of sirens (family Sirenidae); they probably have external fertilization. Hynobiidae and Cryptobranchidae have external fertilization, and this takes place in the water during mating. No courtship is known for *Hynobius* or for *Cryptobranchus,* and males show no apparent interest in females until the egg sacs begin to protrude from their cloacas. In *Hynobius retardatus* this is subsequent to when the female has attached the ends of the egg sacs to some object and is struggling to pull the rest of them out of her oviducts. At this time, several males will approach and push her away from the sacs, and release their sperm as they rub their cloacas over the eggs. In *Cryptobranchus,* the male will trail a female with protruding egg sacs, generally to a position under some large object such as a flat rock, and will fertilize the eggs as they are extruded. In the mountain stream-dwelling *Ranodon sibiricus* (family Hynobiidae), the male deposits a very sticky and elastic spermatophore and the female attaches her eggs to this (Bannikov, 1958). None of these salamanders is known to have any courtship behavior, but the males generally seem to choose the sites where the eggs are laid and may exhibit some form of behavior which entices the female to that site or, in the case of *Ranodon sibiricus,* to where he has deposited his spermatophore. Thorn (1963) has suggested that in *Hynobius nebulosus* male cloacal secretions may serve to attract the female, since males seem to remain near their chosen site periodically undulating their tails.

The courtship of amphiumas (family Amphiumidae) is unique in that females court males. Several females may vie for the attentions of a single male, rubbing their snouts along his body. Upon becoming sexually aroused, the male and female embrace one another and a spermatophore is transferred directly from the male into the female (Baker, 1937; Baker *et al.,* 1947). All of this, including courtship and transfer of spermatophore, takes place in water.

The courtship and mating of terrestrial salamanders other than plethodontids also takes place in water. The males precede the females to the breeding waters and sometimes congregate in large numbers in limited areas. The factors causing salamanders to move to a particular breeding site vary from species to species. They may move downhill, Boy Scout fashion, until they encounter water, or they may move instinctively in a particular direction until they reach the breeding site. In the latter situation, orientation by one means or another is presumably involved, and the ability to orient is indicated by the fact that individuals of a variety of different kinds of salamanders have been observed to return year after year to the same pond or part of a stream at breeding time. It has sometimes been suggested that the salamanders merely make random movements until they encounter a suitable breeding site, but there is no real evidence of this; such a habit would certainly have a low selective value. Having arrived at the breeding site the males swim about, patrolling the shoreline and awaiting the arrival of females.

Although Plethodontidae generally mate on land, their courtship and those of Ambystomatidae, Salamandridae, and Proteidae basically include five stages (Salthe, 1967). During the preliminary stage, A, the male becomes aware of the presence of a potential mate, approaches the other individual, and often makes overtures by nudging or rubbing with his snout. It is during this stage that the male presumably determines the sex of the other individual. Sexually aroused males will generally court and clasp any individual that is the size of a gravid female, but this will be of short duration if the individual proves to be a male. Sex recognition in salamanders involves visual and olfactory discrimination and behavior. Twitty (1955) demonstrated that in redbellied newts (*Taricha rivularis*), males use olfaction to locate females, swimming in the direction from which the odor is coming (generally upstream). This is discussed in more detail in Chapter Thirteen.

If the salamander in question is a female, the courtship will advance to stage B. During this stage the male either clasps the female or blocks her path and continues various rubbing movements, sometimes lashing his tail at the same time. The hedonic glands of the male are rubbed on the head of the female, and their secretions may serve to sexually arouse the female. The lashing of the male's tail creates water currents which seem to have a stimulatory effect on abdominal and cloacal glands of the female. A pair of salamanders may remain in amplexus (clasped) for a period of time ranging up to several days during which the male will continuously nose and rub the female and lash his tail, all of this ceremony being to arouse the female sexually so as to cause maturation of her eggs and, in some cases, to persuade her to follow him to where he will place spermatophores.

Stage C, not present in the courtship of all groups, occurs when the male moves away from the female and she follows him with her attention fixed on some part of his anatomy, usually his cloaca or the base of his tail.

The spermatophore, a packet of sperm on a gelatinous base, is deposited on the substratum by the male in stage D.

During stage E, the male moves ahead about one body length while being followed by or physically pulling the female until her cloaca comes into contact with the sperm cap of the spermatophore. If the female is unresponsive after the male has moved away from the spermatophore, he will return to stage B, rubbing his head on hers and lashing his tail. This sequence may be repeated over and over until the female is sufficiently aroused to follow the male and pick up the spermatophore in her cloaca.

Variations in the above courtship pattern occur in different taxonomic groups. In *Ambystoma*, males appear to deposit more than one spermatophore, repeating the courtship several times, and the females, in turn, pick up at least parts of several sperm caps.

Three different courtship variations occur in the family Salamandridae. In *Euproctus*, the male clasps the female with his tail and the spermatophores are deposited on her body, near or in the cloacal lips. In *Mertensiella, Pleurodeles, Salamandra,* and *Tylotriton*, the male captures the female from below by locking his forelimbs around hers in stage B and then, in a stage with a combination of characteristics of stages B and C, carries the female about on his back. The male deposits the spermatophore while the female is being gripped, and she is maneuvered to it by the male. In *Cynops, Triturus, Taricha,* and *Notophthalmus,*

the courtship pattern is basically the same as in *Ambystoma*. In the latter two genera the male clasps the female from above, but in the former two the male blocks the female's path by moving in front of her and no clasping occurs. During stage C, male *Taricha* dismount and merely deposit spermatophores, but in the other three genera the male leads the female forward. The sperm cap sticks to the female's cloacal lips as she passes over it in *Cynops*, whereas in the other three genera the female seems to deliberately press her cloaca against the spermatophore. Prechtl (1951) reports that female *Triturus alpestris* stuff the sperm caps into their cloacas with their hind limbs.

The Proteidae genus *Proteus*, even though blind and cave-dwelling, has a courtship pattern which is very similar to that of *Triturus*. Courtship is conducted in water, and the territorial males court any females entering their territories. The courtship of *Necturus* is not well known. Salthe (1967) states that the spermatophore has no base and suggests that it may be passed directly from the male into the female's cloaca. However, Bishop (1947), in describing the breeding of *Necturus maculosus maculosus*, indicated that males deposit spermatophores consisting of a gelatinous basal part and a white summit of aggregated sperm.

The courtship of Plethodontidae, which generally occurs on land, follows rather a consistent pattern. In stage A, the male rubs his chin along the female's back or, in *Eurycea*, rubs his snout on her belly. In at least some *Plethodon* species, the male occasionally lifts his chin off the female's back in such a manner that his head snaps back; in *Desmognathus,* he intermittently presses his chin down, arches his back, and scrapes her flanks with his chin as he snaps his body straight (Salthe, 1967). In *Pseudoeurycea belli,* the behavior of the male is as in *Desmognathus* except that he holds his head high and jabs his chin downward (Salthe and Salthe, 1964). Such movements appear to test the female's readiness to mate, since an inadequately stimulated female *Pseudoeurycea* will flee after such strokes (Salthe and Salthe, 1964). In *Plethodon glutinosus, Aneides aeneus,* and possibly other plethodontids, the male will periodically grip the female's back loosely with his jaws. This may explain the significance of the enlarged premaxillary teeth which are often characteristic of breeding males. During stages B and C, the male rubs his hedonic glands on the female and attempts to block her path, eventually passing beneath her chin until it rests on the base of his tail, where additional hedonic glands are located. Stage C is of long duration in plethodontids, and the male leads the female about with her chin pressed against the dorsal surface of the base of his tail. A lateral rocking of the male's pelvic region, accompanied by a lateral wagging of the female's head in the opposite direction, precedes spermatophore deposition. After spermatophore deposition, the male leads the female forward and the sperm cap sticks to her cloacal lips as she passes over it.

Except in Cryptobranchidae and Hynobiidae, fertilization of the salamander eggs occurs as they pass out through the female's cloaca. This may be a considerable period of time after mating. When the species congregates in a breeding pond, the females arrive prior to the final ripening of the eggs. Ripening of the eggs occurs as a result of the physiological and psychological responses of the female to the stimulants of the male's clasping, hedonic glands, cloacal secretions, and so forth; she will then deposit her eggs and leave. Consequently, at any one time a breeding congregation will consist almost entirely of males. This latter point is biologically important because it is during the breeding season that the animals, in an exposed congregation, are most liable to mortality from predation. Females, which are the more critical of the sexes in terms of the reproductive potential of the population, spend the lesser amount of time under the hazardous conditions. In addition, as mentioned, this pattern of courtship and mating assures that ripe eggs will be fertilized because oogenesis will not be completed in the absence of males. In the terrestrial-breeding plethodontids, there are no congregations of individuals and oviposition may be delayed for months after mating has occurred. Oviposition habits vary from species to species. Clutches laid in terrestrial situations may be deposited in small excavations or nests, as with dusky salamanders (*Desmognathus fuscus*), or in decaying vegetation or logs, as with red-backed salamanders (*Plethodon cinereus*). Aquatic eggs may be deposited in nests, as in *Cryptobranchus*, or attached singly or in clusters to aquatic vegetation and other objects in the water, as with rough-skinned newts (*Taricha granulosa*).

ANURANS

In contrast to the salamanders, which are essentially voiceless, sound production plays a very important role in the breeding behavior of most anurans. Just as in the salamanders, anuran populations may migrate to an aquatic site, at which males typically arrive before the females, or breeding may occur in a terrestrial habitat. When at the breeding site, the males emit mating calls either as individuals or, in some species, as choruses. The sound produced is species-specific in that characteristics including frequency of the dominant harmonic (Hertz), note repetition rate, pulse rate, and length are generally uniform within a species and variable between species (Fig. 11–1). As a general rule, no two species breeding at the same locality have similar mating calls. The sound produced by the males is discriminated by the females and is the principal means by which species recognition is accomplished. Upon hearing the mating call of her species, a reproductively active female will respond by moving in the direction of the noise and, accordingly, toward a male or chorus of males of her own kind. A limited number of studies has also indicated that this sound is the stimulus which excites the female and causes her to become physiologically and psychologically ready for breeding. Through these changes in her physiology and nervous system, the final ripening of the eggs takes place. Consequently, when a female anuran arrives at a breeding site, she may already be in a ripened condition. The males, which are either sitting around the edges of the breeding water, resting in nearby vegetation, actually swimming about in the water while calling, or waiting at the terrestrial breeding site, mechanically respond to the presence of females, and clasping takes place. Maintenance of a clasping reflex in the male is dependent upon proper size and shape

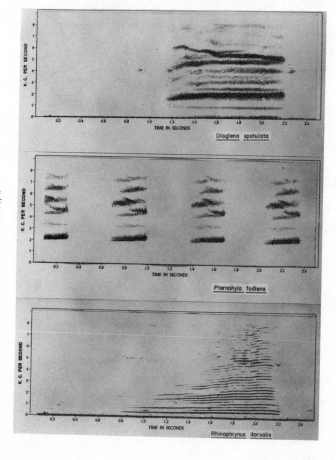

Figure 11–1 Audiospectrograms of the mating calls of three Mexican species of anurans. (From Porter, K. R.: Mating calls and noteworthy collections of some Mexican amphibians. Herpetologica 18:165–171, 1962.)

in the individual being clasped. Males will not clasp females which have already deposited their eggs. Occasionally one male will clasp another male. If this happens, the male being clasped will emit a "release call," a vibratory sound often accompanied by physical vibrations of the body, and this signals the clasping male that the wrong sex has been clasped.

Depending upon the condition of the female, the eggs may be deposited almost immediately after clasping takes place or the amplexus may continue for several hours, occasionally days. While the female is being clasped, the male alternately squeezes her and swims about. This activity presumably hastens the final ripening of the eggs if oogenesis has not yet been completed. In the majority of anurans, fertilization occurs externally as the eggs and sperm are being shed simultaneously. The act of ovulation appears to trigger the release of sperm by the male. Anuran egg masses laid in water

are characteristically shaped and positioned in adaptation to prevailing oxygen tensions. Those anurans which breed in cold waters have more dissolved oxygen available for embryonic respiration than those which breed in warm waters. Similarly, the embryos of those breeding in running waters have more oxygen available than those bred in standing water; the former also have a current to contend with. Because species select a particular type of breeding site, each will characteristically breed in warm or cold water, standing or running water, permanent or temporary water. Egg masses of species breeding in cold water (generally deep permanent water, since deep water tends to be cooler than shallow water) are typically globular in shape and are attached to submerged vegetation or other objects, frequently at some depth below the surface (Fig. 11–2). Because of the temperature of the water, there is usually sufficient oxygen to allow enough to reach even the

Figure 11–2 Three generalized kinds of anuran egg masses. *Top,* a flattened egg mass, characteristic of bullfrogs (*Rana catesbeiana*), deposited at the surface of the water. *Center,* a globular egg mass, such as formed by northern chorus frogs (*Pseudacris triseriata*), attached to vegetation beneath the surface of the water. *Bottom,* eggs, such as those of true toads (*Bufo*), deposited in elongated strings.

centermost embryo in the mass. In contrast, egg masses which are deposited in warm oxygen-poor water characteristically have a large surface-to-volume ratio. In some instances, as with toads (*Bufo*), the eggs are laid in rosary-like strings and are frequently entangled in and around submerged vegetation. Other species breeding in warm water, including bullfrogs (*Rana catesbeiana*) and green frogs (*Rana clamitans*), have flattened egg masses which form a thin film at the surface of the water. In both cases, the large surface area increases the probability that each embryo will obtain sufficient oxygen for normal development. Zweifel (1968) has noted that behavioral

differences determine the form of the egg mass; he poses the interesting question as to whether or not this behavioral pattern is induced by the temperature of the water at the time of oviposition. Egg masses which are subject to heavy current or damage from flash-flooding, such as those of red-spotted toads (*Bufo punctatus*) which are laid in streams, may be deposited as single eggs, in short strands, or in loose scattered masses.

Some anurans exhibit very specialized patterns of breeding behavior which either allow them to reproduce where they could not otherwise or, in other instances, reduce the mortality suffered by embryonic stages. Many members of the family

Leptodactylidae and a few Old World tree frogs in the family Rhacophoridae lay their eggs in masses of foam on the leaves of plants or on the ground uphill from water. The female first emits some fluid which she vigorously kicks into a froth and then emits both eggs and fluid simultaneously, the eggs being fertilized as they are emitted. After oviposition, the female kicks the remaining fluid into a froth until the mass appears as a whipped meringue; within are the fertilized eggs (Fig. 11–3). Eventually, the outer surface of the mass dries and forms a water-resistant membrane. The larvae, upon hatching, live within the liquefied interior of the mass and either metamorphose there or are eventually washed into nearby pools of water by rains.

Some members of the family Microhylidae, some Bufonidae, Dendrobatinae, and many Leptodactylidae deposit their eggs on land in moist situations such as burrows and have a very rapid larval development, with metamorphosis occurring within the egg capsule (Fig. 11–4). The rapid development characteristic of some of these requires a plentiful oxygen supply

which would not always be present in stagnant water. Consequently, their habit of laying eggs on land may have been selected because of the greater amount of oxygen available in air as opposed to what would be available in many breeding ponds. Such a habit could also reflect one of many evolutionary steps towards terrestrialism.

The tree frogs (family Hylidae) are a greatly diversified group. Some are completely arboreal and never descend out of the trees; some are entirely terrestrial and spend their entire life on the surface of the ground. Consequently, the breeding habits of this family are also very diversified. Some hylids build a basin of mud either near the edge of a pool or in the basin of the pool itself. These basins, constructed of balls of mud, catch water and form individual breeding ponds (Fig. 11–5). In some species, the male does all of the nest-building and attracts the female to the basin by calling from its vicinity after the walls are constructed and it is full of water. In general, the tadpoles which develop in such limited volumes of water have tremendous pinnate (feather-

Figure 11–3 Foam nest of the tropical American giant frog (*Leptodactylus pentadactylus*), approximately one-half natural size. (From Mertens, R.: *The World of Amphibians and Reptiles.* George C. Harrap and Co., 1960.)

Figure 11–4 Terrestrial nest and eggs of the African bush-frog (*Leptopelis karissimbensis*). (From Mertens, R.: *The World of Amphibians and Reptiles*. George C. Harrap and Co., 1960.)

shaped) gills which adhere to the surface film on the water. As mentioned in the preceding chapter, another group of tropical arboreal tree frogs lays eggs in small basins of water already existing in nature, such as those which collect between leaves of bromeliads. All *Phyllomedusa* hylids, many *Hyla* (Duellman, 1970), some members of the Old World family Rhacophoridae, and all Centrolenidae have the habit of laying eggs on leaves above water. Pyburn (1970) has written an interesting account of this breeding behavior in *Phyllomedusa callidryas* and *Phyllomedusa dacnicolor*. Males call from vegetation and attract gravid females. In both species, after entering amplexus the female carries the male from his perch down to a pool of water below, takes water into her bladder, and then climbs back up into the vegetation where she lays a clutch of eggs upon a leaf, releasing her bladder water over the eggs as they emerge. She then returns to the pool and fills her bladder again before laying another clutch, this procedure being followed for each clutch. Through experimentation, Pyburn found that the bladder water is necessary for swelling of the jelly and normal development of the embryos; embryos from females deprived of water die. Upon hatching, tadpoles make their way into

water through their own movements or are washed off the leaf by rain. Pyburn suggests that this pattern of leaf-breeding evolved from the habit in pond-breeders of attaching their eggs to emergent vegetation at or near the surface of the water.

The reproductive behavior of Surinam toads (*Pipa pipa*) has been described in detail by Rabb and Rabb (1960, 1963) and Rabb and Snedigar (1960), who observed and photographed breeding of the species at the Chicago Zoological Park. The mating call consists of click sounds, with four clicks per second, and lasts 10 to 20 seconds. The excitation and attraction of males to a gravid female may depend upon an odor secreted by her, and clasping is stimulated by movement or contact of another individual of the proper girth. Clasped pairs remain in amplexus for 24 to 30 hours, and during this time there is a noticeable swelling of the female's dorsal skin and vent lips. Just before the first egg is emitted, the paired animals readjust their position until the female's vent is pressed against the anterior abdomen of the male and his chin is pressed firmly against the swelling skin of the female's back. The amplectant pair then begins making a series of vertical turnovers in the water without breaking the water's surface (Fig. 11–6). Eggs are laid at this

Figure 11–5   Mud nests of a large tree frog, probably related to *Hyla faber*. Notice the dark egg mass in each and how the nest has been made out of balls of mud. (From Mertens, R.: *The World of Amphibians and Reptiles*. George C. Harrap and Co., 1960; photograph by de la Rüe.)

Figure 11–6   The egg-laying turnover maneuver of *Pipa pipa*. *A*, beginning of the sideways ascent: the female has lifted her left side and is about to shove away from the tank bottom with her right hind foot. *B*, the momentary upside-down position in which the eggs are laid. *C*, the headfirst descent to the tilted resting position. The ascending movement involves rotation about the longitudinal axis, and the descent is a half-roll about the transverse axis. Drawn from 16 mm. motion picture film. (From Rabb and Rabb: On the mating and egg-laying behavior of the Surinam toad, *Pipa pipa*.)

time. During the ascent there is a rotation about the longitudinal axis, and in the descent they rotate transversely. During these turnovers, each of which is initiated by the female, there is a sideways upward movement, a momentary pause in a horizontally upside-down position, and a head-first return to the tilted resting position. Three to five eggs are laid at a time, usually when the female is horizontally upside down at the top of the turnover. They are caught by the male's belly and prevented from rolling off by temporary anterior and posterior transverse folds of skin formed through arching of his body. When the toads right themselves at the bottom of the turnover, the eggs drop off the male's belly and onto the female's back, where they stick and are pressed firmly in place by the clasping male. Fertilization seems to occur during the descent phase of the turnover. After completion of egg laying, the female begins a quivering motion and, several minutes later, the two animals separate. The female remains quiet for several hours during which her skin swells and encloses each egg in a cystlike pocket.

The only New World anuran known definitely to have internal fertilization is the tailed frog (*Ascaphus truei*) of northwestern United States. This species inhabits very rapidly flowing mountain streams where external fertilization would be difficult to achieve since sperm would be swept away by the current. The "tail" of the male, an appendage about five to six millimeters in length (Fig. 11–7), is a tubular extension of the cloaca which is inserted into the female's cloaca during amplexus. African toads in the genus *Nectophrynoides* are the only viviparous anurans known, and these too have internal fertilization. Interestingly, the larvae of *Nectophrynoides vivipara* have very long slender tails which are highly vascularized and appear to function as respiratory organs. The tails presumably carry oxygen from the uterine wall to the tadpole, since the uterus may be too crowded with tadpoles (over 100 have been recorded in a single uterus) to allow the bodies to maintain contact with the uterine wall.

CAECILIANS

Very little is known about the reproductive behavior of the Apoda. Because an intromittent organ is present on all males it is assumed that all members of the order are characterized by internal fertilization. This intromittent organ is formed by extrusion of the cloacal region into a semirigid tubular structure as a result of cavernous tissue becoming filled with blood. The completely aquatic species belonging to the family Typhlonectidae (*Typhlonectes*) are ovoviviparous, with the entire embryonic development accomplished within the uterus. There is indirect evidence, based on the difference in size between eggs and newborn young, that the developing young supplement the nutrient provided by the egg with food rasped from the uterine walls (Parker, 1956). More primitive semiaquatic species are oviparous and deposit their eggs in mud or debris near streams. Upon

Figure 11–7 Dorsal view of a male tailed frog (*Ascaphus truei*), showing the tail-like appendage which is used as an intromittent organ during amplexus. (From Blair, W. F., A. P. Blair, P. Brodkorb, F. R. Cagle, and G. A. Moore: *Vertebrates of the United States.* McGraw-Hill Book Co., 1968; photograph by Isabelle Hunt Conant.)

hatching, the young of these forms make their way to water and spend a variable length of time as free-swimming larvae. Many, if not all, terrestrial caecilians are ovoviviparous, with metamorphosis occurring before birth of the young.

PARTHENOGENESIS

Two species of salamanders, *Ambystoma platineum* and *Ambystoma tremblayi*, consist almost entirely of females. Both of these species are triploid (3n=42) and arose through hybridization of the diploid bisexual species *Ambystoma laterale* and *Ambystoma jeffersonianum*. *Ambystoma platineum* seems to only occur together with *A. jeffersonianum*, and *A. tremblayi* apparently occurs only with *A. laterale* (Uzzell, 1964). Cytogenetic studies indicate that *Ambystoma platineum* has two complete genomes from *A. jeffersonianum* and one from *A. laterale*, while *Ambystoma tremblayi* seems to have two sets of chromosomes from *A. laterale* and one from *A. jeffersonianum* (Uzzell and Goldblatt, 1967). MacGregor and Uzzell (1964) examined the lampbrush chromosomes in the oocytes of these triploids and found that they contained 42 bivalents (equivalent to hexaploidy for these salamanders); this finding indicates that there is a premeiotic suppression of cell division. Electropherograms of plasma proteins indicate an absence of genetic recombination (Uzzell and Goldblatt, 1967) which suggests that it is the last premeiotic mitosis that omits cytokinesis and produces the hexaploid chromosomal level. This is followed by normal meiosis and restoration of the triploid chromosomal level in the ova. Development of the ova generally seems to occur by pseudogamy (gynogenesis), requiring stimulation through penetration of the egg membrane by a sperm without fusion of the sperm nucleus with that of the ovum. Thus, the female *Ambystoma platineum* exists as a sexual parasite of *A. jeffersonianum*; *A. tremblayi* is a sexual parasite of *A. laterale*. Populations of both *A. platineum* and *A. tremblayi* exist in central Indiana without their normal bisexual hosts, and these must either reproduce parthenogenetically or be sexual parasites on other bisexual species of *Ambystoma* occurring with them (Uzzell, 1970). There are no other known instances of parthenogenesis among living amphibians.

PARENTAL CARE AND NUMBERS OF EGGS PRODUCED

The number of eggs deposited by amphibians varies greatly from species to species. In general, there is a correlation between the number of eggs laid and the mortality suffered by the species prior to sexual maturity, since in all stable populations each sexually mature adult is replaced by one and only one surviving member of the next generation. Some species of anurans are known to lay up to 30,000 eggs per female per reproductive season (toads in the genus *Bufo*, for example), whereas other species lay as few as a single egg at a time per female (the genus *Sminthillus*). This great disparity in number of eggs is explained by differences among amphibians in sizes of eggs, patterns of development, sizes of females, and reproductive behaviors that result in differences in mortality. Both the numbers of eggs and their sizes may vary geographically, often clinally, within one species, and clinal variation in these is also found in interspecific comparisons (Salthe, 1969).

In general, there is a higher mortality among embryonic stages developing in open water than there is among similar stages developing on land, where they are hidden from many predators. The previously mentioned *Sminthillus*, a tiny Cuban frog and perhaps the smallest anuran in the world, lays its egg in some moist spot on land, and metamorphosis occurs before hatching. Species of *Bufo* characteristically breed in open water and abandon their eggs after oviposition. In addition to the reduction in predation which results, amphibians depositing eggs on land waste fewer gametes during the fertilization process. When eggs and sperm are shed into large masses of water a significant wastage of both is likely because there will not always be contact between egg and sperm. When external fertilization occurs on land the sperm are released directly onto the eggs as they are emitted, and there is less chance of wastage. This conservation of gametes is carried a step further in amphibians utilizing internal fertilization, in which confinement of the gametes in the female reproductive tract virtually assures that eggs will be contacted by sperm. Thus, as a general rule, amphibians depositing eggs on land will produce smaller numbers of them than species depositing eggs in water. Species utilizing internal fertilization will generally produce fewer eggs than those using external fertilization, particularly if the eggs are fertilized externally in water.

Parental care is another factor which reduces mortality of juvenile stages. In fact, much of the success of the higher vertebrates has been the result of evolution of parental care instincts. The most primitive kind of parental care is the brooding instinct, the tendency for one or more parents to remain with the eggs for various periods of time. This instinct, found in various invertebrates and fish, appears to have evolved independently several times in different phylogenetic groups of amphibians. In the family Hynobiidae, the male salamanders remain with the eggs for varying periods of time to guard them and are so possessive of them that they drive even the females away. Members of the family Cryptobranchidae, closely related to the hynobiids, extend the guarding instinct until the eggs hatch. The primitive nature of the brooding instinct of cryptobranchids is revealed by the fact that both sexes devour the eggs but, since the guarding male can only eat a small proportion of the eggs and, in turn, chases the female away, the cannibalistic habit has not prevented the species from being successful.

The brooding instinct is clearly established in the lungless salamanders (family Plethodontidae) and is associated with the habit of depositing eggs on land. The advantages of brooding on land, however, do not lie solely in reducing predation-incurred mortality. The damp body of the parent assures that the eggs will have sufficient moisture, and in addition, her cutaneous secretions appear to inhibit the growth of mold. The additional moisture made available to developing embryos by brooding permits some species, such as the arboreal salamander *Aneides lugubris*, to lay eggs in comparatively dry situations. The brooding instinct has advanced in at least some of the plethodontids to where there is a strong bond between parent and eggs. Almost all females will return to the egg mass after being frightened away, and some will even move their eggs to another location after they have been disturbed. It is not known how most females recognize their own egg masses, but female dusky salamanders (*Desmognathus fuscus*) have been induced to brood the eggs of another female by arranging the eggs in the same manner as her own had been. Thus, at least in this case, the female seems to recognize the pattern of her nest rather than the eggs themselves.

Other salamanders exhibit various levels of brooding instincts. The congo eel (*Amphiuma*, family Amphiumidae), although it is primarily an aquatic animal, lays its eggs under objects on land and has been observed to return to them occasionally. Mudpuppies (*Necturus*) appear to brood their eggs but, since they use their nests as retreats throughout the year and have been found to occupy them after the young have left, this may merely be a reflection of the female's reluctance to leave a favorite hiding place. Most mole salamanders (*Ambystoma*) deposit their eggs in water and then ignore them, but the female marbled salamander (*Ambystoma opacum*) deposits her eggs on land (which subsequently floods) and then curls about them as do terrestrial plethodontids. However, marbled salamanders will not return to their eggs after they have been disturbed.

The males of various species of frogs, including wood frogs (*Rana sylvatica*, family Ranidae), tend to linger in the vicinity of egg masses and appear to be guarding them. This habit may not be a true brooding instinct, however, for it may merely reflect the tendency of males to remain near their calling stations. On the other hand, such species as the Australian foam nest builder *Adelotus brevis* (family Leptodactylidae) exhibit what seems to be real attraction of the male parent toward the eggs. Members of the leptodactylid genus *Eleutherodactylus* typically lay their eggs in ground litter or bromeliads and in crevices in trees and then desert them. However, *Eleutherodactylus caryophyllaceus*, characteristic of cloud forest habitats, places its eggs on exposed leaf surfaces above ground and then broods them until the clutch has hatched (Myers, 1969). Brooding females make no attempt to defend their eggs, and apparently their brooding habit, even in a cloud forest, is to prevent desiccation of the eggs. Myers observed that when water is trickled on the leaf, the brooding female will raise her body to expose the eggs. He also found that untended experimental eggs desiccated and molded, whereas those brooded under the same conditions did not until seven days after being deserted by the female.

Further evolution of the brooding instinct leads to the parental care of young. Among salamanders, only the terrestrial lungless salamanders belonging to *Aneides*, *Hemidactylium*, and possibly *Plethodon* remain with the young after they hatch. Even in these cases, probably little or no protection is given to the young; nevertheless, the water balance problem of the small juveniles, which is more severe than that of the adults because of their relatively large surface area

proportional to mass, may be partially alleviated by moisture from the parent's body surface.

Parental care of young has become very advanced in certain anurans and, in some species, has even led to morphological modifications of the female's body. The midwife toad, *Alytes obstetricans* (family Discoglossidae), has evolved a unique pattern of parental care. During the breeding season, several males will surround a ripe female while she is on land and attempt to clasp her. The successful male, after clasping her, will use his hind legs to stroke her cloacal region and induce her to extend her hind legs so as to form a receptacle for the eggs. As the 20 to 60 large eggs are deposited, the male emits sperm and they are fertilized. After this, the male moves forward on the female and pushes his hind legs through the egg mass until the two strings of eggs are wound about the posterior part of his body. He then releases the female and proceeds with his normal activities, carrying the fertilized eggs about with him (Fig. 11–8). Embryonic water needs are satisfied by dew, and the male may even enter water occasionally. Hatching occurs about one month after fertilization of the eggs. When this time draws near, the male

Figure 11–8   Male (left) and female midwife toads (*Alytes obstetricans*), showing how the male carries the egg strings on his hind limbs. Approximately natural size. (From Mertens, R.: *The World of Amphibians and Reptiles*. George C. Harrap and Co., 1960; photograph by Rosenberg.)

instinctively goes to a pool and submerges the mass; the larvae hatch and swim away; and the male's "baby-sitting" is over. In the neotropical *Phyllobates, Colostethus,* and *Dendrobates* (ranid subfamily Dendrobatinae), the male, or sometimes the female, is also a baby sitter, but in these frogs the parental care extends past hatching. The eggs are laid on land or on vegetation, and the adult remains with them. Hatching occurs after several weeks, and then the tadpoles attach themselves to the back of the adult and remain there until they are well developed. Upon reaching an advanced stage in their development, the larvae take the first opportunity afforded by the adult entering water, drop off, and become independent. Baby-sitting by the adult male is carried to an even greater level of specialization in *Sooglossus* (ranid subfamily Sooglossinae) and in Darwin's frog (*Rhinoderma darwini,* leptodactylid subfamily Rhinodermatinae). Egg deposition and fertilization is on land, as in many leptodactylids, but in this case several males gather around and guard the pile of 20 to 30 eggs deposited by each female for 10 to 20 days, until the larvae are near hatching and are moving about inside the eggs. The movement of the larvae elicits the "snapping reflex" of the males, and each picks up several eggs with his tongue and slides them through the opening in the floor of his mouth into the vocal sac. In these frogs the vocal sac is an elongate structure that extends from the groin up over the flanks and under the chin; the larvae hatch within it and are thus carried about inside the male during the entire tadpole stage;

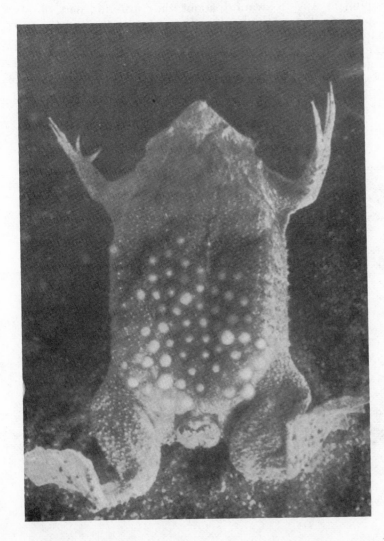

Figure 11–9 Dorsal view of a female Surinam toad (*Pipa pipa*) with eggs embedded in the skin. (From Cochran, D.: *Living Amphibians of the World.* Doubleday and Co., 1966; photograph by Raymond J. Cummins.)

Figure 11–10 Female *Flectonotus goeldii* carrying her eggs exposed on her back. (From Mertens, R.: *The World of Amphibians and Reptiles*. George C. Harrap and Co., 1960; photograph by Oeser.)

the male may or may not even be their father. Upon metamorphosing, the young frogs (about half an inch in length) find their way back to the male's mouth and emerge. Thus, the young pass through the entire dangerous larval period protected by the male in much the same manner as young marsupial mammals who are protected while being carried about in a pouch.

In several species of anurans the female is the baby sitter, a behavioral pattern which has evolved independently in three different families, Pipidae, Rhacophoridae, and Hylidae. In the Surinam toads (*Pipa pipa*), the entire larval life of the young is spent within the mother's skin (Fig. 11–9) so that when they are "born" they appear as metamorphosed young, wriggling out of the pockets after the surface covering pops open.

Hylambates brevirostris (family Rhacophoridae) is the only amphibian known to brood its egg in its mouth. The female takes the fertilized eggs into her mouth and keeps them there until they hatch as metamorphosed frogs, and she goes without eating during the entire period.

In seven genera of hylids, the females protect their young by carrying the egg masses on their backs. In one of these, *Fritziana*, the female has narrow folds of skin on each side of the back which keep the eggs from slipping off sideways but which leave them exposed. *Cryptobatrachus*, *Stefania*, and the unrelated *Hemiphractus* all have individual bowl-shaped depressions for the eggs, but these are not completely enclosed. In *Flectonotus*, the folds of skin on the back meet on the midline (Fig. 11–10). Finally, in the advanced *Gastrotheca* and *Amphignathodon*, there is a brood sac present which completely encloses the eggs except for a slitlike opening at the posterior end. The slit may be relatively large in some species, as in *Gastrotheca ovifera* (Fig. 11–11), or quite small, as in *Gastrotheca marsupiata*. The number of eggs deposited per female varies among these frogs from about four to 50. The young, depending upon the species, may leave the pouch as tadpoles and be free-swimming before metamorphosing or they may undergo metamorphosis within the pouch and emerge as young frogs.

In all of the instances cited above, parental care by either the female or male greatly reduces mortality of young. This accounts for the very low number of eggs

Figure 11–11 Female marsupial frog (*Gastrotheca ovifera*) with eggs in the brood pouch on her back. (From Mertens, R.: *The World of Amphibians and Reptiles*. George C. Harrap and Co., 1960; photograph by Senckenberg Museum.)

produced (from one to 50) compared to those anurans depositing their eggs in open water and deserting them. Again, the reproductive game is played either by putting considerable energy into large numbers of gametes and not expending much energy, if any, on parental care or by putting considerable energy into protecting the young and reducing the amount of energy spent in producing gametes. The effectiveness of parental care is reflected in the need for toads to deposit thousands of eggs per female per reproductive season in order to get the same result, a one-to-one replacement of adults, as is achieved with anywhere from one to a few hundred eggs in amphibians that care for their young.

DEVELOPMENT AND METAMORPHOSIS

The larval period of amphibians lasts from a few days in some species to over three years in others. The differences in duration of the larval stage are associated with breeding habits and reflect adaptation to the embryonic environment. Typically, amphibians breeding in temporary waters have a short larval period because natural selection has favored rapid development and early metamorphosis of larvae before total evaporation of the breeding water occurs. Obviously, no amphibian breeding in temporary water can be permanently aquatic, and in spadefoot toads (*Scaphiopus holbrooki hurteri*), for example, the metamorphic change into a terrestrial toad may occur within 12 or 13 days of when the eggs are fertilized.

Amphibians breeding in permanent water have not faced the same sort of selection because their larvae live in a relatively stable environment. It is even possible for such amphibians to be permanently aquatic, either in the larval form (as neotenic individuals) or as metamorphosed adults. Consequently, amphibians breeding in permanent water tend to spend longer periods of time as free-swimming larvae. They may complete metamorphosis in the water and then remain in it as adults (as in the case of the anuran family Pipidae, for example); they may complete metamorphosis on land and be semi-aquatic or terrestrial as adults (as with the majority of salamanders and anurans); or they may be non-metamorphosing (neotenic).

Those amphibians which breed on land exhibit a trend toward elimination of the larval period and development to the adult form within the egg. The major problem associated with advanced development within the egg is that of nutrition and energy for the developing individual. Consequently, natural selection has favored both large-yolked eggs and the elimination of typical aquatic larval structures such as horny mouth parts and muscular tails; the former adaptation supplies the required energy and nutrition while the latter characteristic conserves both. Among anurans, the trend toward elimination of larval features is carried to the extreme in frogs of the genus *Platymantis* (family Ranidae), which have essentially direct development: the operculum is the only larval structure retained. The embryonic stages of *Eleutherodactylus* (family Leptodactylidae) also generally lack many tadpole characteristics and hatch as fully formed miniature frogs.

Amphibian development, from the fertilized egg through metamorphosis and the attainment of sexual maturity, is a continuous process that is described by embryologists in terms of developmental stages. The purpose of staging amphibian development is to divide the continuous process into progressive steps, the stages, which will allow embryologists making comparative studies to relate what occurs in one species (or population or individual) to what occurs in another. Despite the coordination that is required and the complexity of the changes involved, metamorphosis involves a span of time which is very short as compared with the entire developmental period required to achieve sexual maturity. In fact, what distinguishes metamorphosis from other aspects of development is the dramatic suddenness with which changes occur.

Amphibian metamorphosis is typically associated with the transition from an aquatic larval stage to a terrestrial adult stage, and its general function is to prepare the individual for terrestrial life. It should be noted, however, that the development of some terrestrial adaptations, including limbs and lungs, is gradual and begins early in larval life long before metamorphosis *per se* is initiated. Metamorphosis is generally incomplete when the adult is aquatic, but there are exceptional cases, such as the clawed frog (*Xenopus laevis*), in which metamorphosis is from an aquatic tadpole to an equally aquatic frog. Balinsky (1970) has grouped the organizational changes which occur in metamorphosis into three categories: (1) the reduction or complete disappearance in the adult of organs or structures necessary during larval life but redundant in the adults; (2) the development and functioning of new adult organs during and after metamorphosis; (3) the modification of existing larval structures so that they remain functional in the adult. These changes involve both developmental and degenerative processes of such magnitude that survival of the transforming individual is dependent upon each being coordinated with the others in an orderly sequence. They involve morphology, physiology, and behavior.

DEVELOPMENTAL STAGES

Staging tables have been published for a variety of amphibian species, but few encompass the entire development through metamorphosis. Most staging schemes for anurans are similar for comparable periods of development. Pollister and Moore (1937) defined and illustrated 23 stages in the normal development of wood frogs (*Rana sylvatica*) which begin with the egg at fertilization (stage 1) and end as the opercular fold begins to develop and active spontaneous swimming begins (stage 23). Shumway (1940) defined 25 stages, from fertilization of the egg through complete development of the operculum, of normal embryological development for *Rana pipiens*. Taylor and Kollros (1946) defined 25 additional stages for this species which begin with commencement of independent feeding, following immediately after Shumway's stage 25, through complete resorption of the operculum and tail. Weisz (1945), using slightly different criteria than Shumway, defined 23 stages in the normal development of clawed frogs (*Xenopus laevis*). Weisz's stages begin with the unfertilized egg (stage 1) and end with the third-form tadpole stage in which hind limb buds are evident and the lateral contour of the mouth is wedge-shaped (stage 23). The complete development of this species from fertilization through the end of metamorphosis is given by Nieuwkoop and Faber (1956). Following Shumway's staging, Eakin (1947) defined and illustrated the normal development of the Pacific tree frog (*Hyla regilla*) with 24 stages: stages 15 through 24 differ slightly, but the first 14 are the same as Shumway's and the series ends as the operculum is complete. The development of *Bufo regularis*, the Egyptian toad, is described by Sedar and Michael (1961). Conte and Sirlin (1952) recognized 25 stages in the development of the Argentine toad (*Bufo arenarum*), from fertilization of the egg through complete development of the operculum. Limbaugh and Volpe (1957), using the embryonic staging of Shumway (1940) and the larval staging of Taylor and Kollros as guides, have delineated 46 stages in the normal development of the Gulf Coast toad (*Bufo valliceps*), from egg fertilization through completion of metamorphosis; they have been kind enough to allow reproduction of their illustrations and definition of their stages here. These illustrate the sequence of morphological changes taking place during the development and metamorphosis of a typical anuran. If he desires to do so, the reader will find a more simplified and generalized staging of anuran development in Gosner (1960).

The external morphology of the Limbaugh and Volpe stages from fertilization of the egg through metamorphosis is shown in Figures 11–12 through 11–16; stages in the differentiation and resorption of the larval mouth parts are illustrated in Figure 11–17. The age in hours represents the period of time required for 50 per cent or more of the individuals to reach a particular stage at a temperature of 25° C. The dimension given for each stage is an average value of the total lengths of the individuals that exhibited the characteristics of the stage at the hour indicated.

STAGE 1. The egg at fertilization rotates so that the animal hemisphere is uppermost. The animal hemisphere is brown-black; the vegetal hemisphere is cream-white. The eggs are deposited in a single or double row within a jelly tube. The jelly tube consists of two gelatinous coats, the inner envelope being closely applied to the outer. Measurements of 61 fertilized eggs are as follows: diameter of vitellus, 1.23 mm ± 0.05 (mean and standard deviation); diameter of outer envelope, 2.78 mm ± 0.16; diameter of inner envelope, 2.48 mm ± 0.19. No partition separates the individual eggs. The space between the eggs averages 0.32 mm ± 0.11.

STAGE 2. The second polar body is released, as revealed by a round clear area at the animal pole.

STAGE 3. The first cleavage forms two blastomeres.

STAGE 4. The second cleavage forms four blastomeres.

STAGE 5. The third cleavage forms eight blastomeres.

STAGE 6. The fourth cleavage forms 16 blastomeres.

STAGE 7. The cleavage furrows are irregular after the fourth cleavage. "Early cleavage" is designated as 24 to 64 blastomeres.

STAGE 8. "Mid-cleavage" is determined by the relative size of the blastomeres and position of the pigment border.

STAGE 9. "Late cleavage" is determined by the relative size of the blastomeres and position of the pigment border.

STAGE 10. Involution occurs at the dorsal lip of the blastopore.

STAGE 11. The dorsal lip of the blastopore expands into a semicircle, and involution occurs along the semicircular surface.

STAGE 12. A complete blastopore forms, resulting in a circular yolk plug.

STAGE 13. The embryo flattens dorsally, and the neural plate forms.

STAGE 14. The neural folds form as lateral ridges of the neural groove.

STAGE 15. The embryo rotates slowly inside the egg jelly. The neural folds grow together.

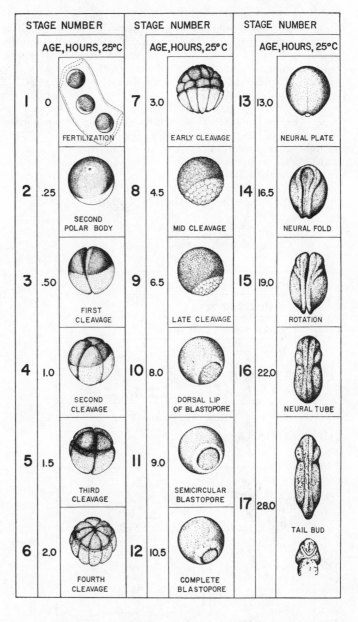

Figure 11–12 Embryonic stages 1 to 17 of the Gulf Coast toad (*Bufo valliceps*). (From Limbaugh and Volpe: Early development of the Gulf Coast Toad, *Bufo valliceps* Wiegmann.)

STAGE 16. The neural folds close to form a neural tube. Gill plates are distinct.

STAGE 17. The tail bud develops. Ventral U-shaped suckers are conspicuous beneath the slight stomodeal depression. Hatching occurs. (An early hatching is characteristic of toad embryos.)

STAGE 18. Muscular movement, constituting a simple flexure in response to stimulation, begins. Pronephric ridges, visceral arches, and olfactory pits are recognizable.

STAGE 19. The heart begins to beat. External gill buds are conspicuous.

STAGE 20. Gill circulation begins, as re-vealed by the movement of corpuscles in the gill filaments.

STAGE 21. The cornea is transparent; the underlying lens is visible as a light spot. The mouth opens, and the suckers begin to disappear.

STAGE 22. Blood corpuscles circulate in the tail fin. The tail epidermis becomes transparent. Melanophores (black chromatophores) appear in the head region and on the dorsal tail musculature.

STAGE 23. The opercular fold covering the gills forms on each side. The number of melanophores increases.

STAGE 24. The operculum closes on the right side. Melanophores extend along the dorsal margin of the tail musculature. Xanthopores (yellow chromatophores) appear at irregular intervals on the most dorsal portion of the tail musculature (represented as clear areas in Figure 11–13).

STAGE 25. The operculum closes on the left side. The melanophores are confined to the dorsal half of the tail musculature. The series of light areas (containing xanthophores) on the most dorsal portion of the tail musculature are found throughout later stages.

STAGE 26. The limb bud appears, marking the transition from the embryonic to the larval period. The length of the limb bud is less than one-half of its diameter. The melanophores on the tail musculature progress ventrally but are still confined to the dorsal half of the musculature.

STAGE 27. The limb bud is equal to or greater than one-half of its diameter. Melanophores appear in the dorsal tail fin (membranous portion).

STAGE 28. The limb bud length is equal to or greater than its diameter. A few melanophores appear on the ventral half of the tail musculature and in the ventral tail fin.

STAGE 29. The limb bud length is equal to or greater than one and one-half times its di-

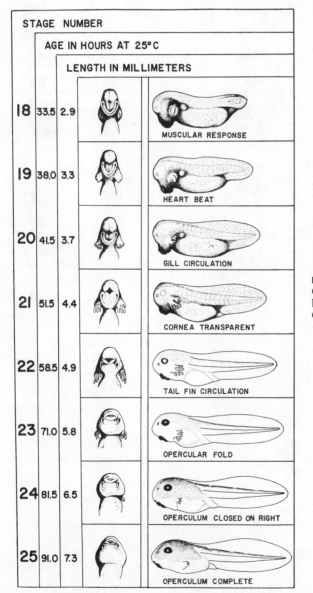

STAGE NUMBER

AGE IN HOURS AT 25°C

LENGTH IN MILLIMETERS

18	33.5	2.9		MUSCULAR RESPONSE
19	38.0	3.3		HEART BEAT
20	41.5	3.7		GILL CIRCULATION
21	51.5	4.4		CORNEA TRANSPARENT
22	58.5	4.9		TAIL FIN CIRCULATION
23	71.0	5.8		OPERCULAR FOLD
24	81.5	6.5		OPERCULUM CLOSED ON RIGHT
25	91.0	7.3		OPERCULUM COMPLETE

Figure 11–13 Embryonic stages 18 to 25 of the Gulf Coast toad (*Bufo valliceps*). (From Limbaugh and Volpe: Early development of the Gulf Coast Toad, *Bufo valliceps* Wiegmann.)

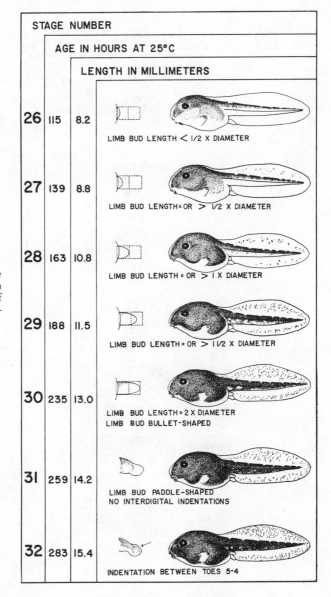

Figure 11–14 Larval stages 26 to 32 of the Gulf Coast toad (*Bufo valliceps*). (From Limbaugh and Volpe: Early development of the Gulf Coast Toad, *Bufo valliceps* Wiegmann.)

ameter. Melanophores are more numerous on the ventral half of the tail musculature and in the tail fins.

STAGE 30. The limb bud is "bullet-shaped;" its length is equal to two times its diameter. The ventral half of the tail musculature is darkened by melanophores except for several areas or patches which contain xanthophores (represented as clear areas in Figure 11–14).

STAGE 31. The distal end of the limb bud is paddle-shaped; no interdigital indentations are present on the paddle margin. The number of melanophores increases in the ventral tail fin.

STAGE 32. The margin of the foot paddle is indented between the fourth and fifth toes. The pigment pattern of the tail is characteristic of the remaining stages of development. Melanophores cover the tail musculature except for a few light areas (containing xanthophores) in the most dorsal portion and in the ventral half. The dorsal and ventral fins are heavily mottled by reticulate melanophores.

STAGE 33. The margin of the foot paddle is indented between the third and fourth and the fourth and fifth toes.

STAGE 34. The margin of the foot paddle is indented between the second and third, the third and fourth, and the fourth and fifth toes.

STAGE 35. The margin of the foot paddle is slightly indented between the first and second toes.

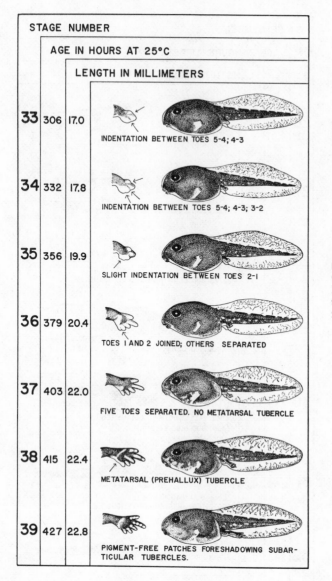

STAGE NUMBER

AGE IN HOURS AT 25°C

LENGTH IN MILLIMETERS

33 | 306 | 17.0
INDENTATION BETWEEN TOES 5-4; 4-3

34 | 332 | 17.8
INDENTATION BETWEEN TOES 5-4; 4-3; 3-2

35 | 356 | 19.9
SLIGHT INDENTATION BETWEEN TOES 2-1

36 | 379 | 20.4
TOES I AND 2 JOINED; OTHERS SEPARATED

37 | 403 | 22.0
FIVE TOES SEPARATED. NO METATARSAL TUBERCLE

38 | 415 | 22.4
METATARSAL (PREHALLUX) TUBERCLE

39 | 427 | 22.8
PIGMENT-FREE PATCHES FORESHADOWING SUBAR-TICULAR TUBERCLES.

Figure 11-15 Larval stages 33 to 39 of the Gulf Coast toad (*Bufo valliceps*). (From Limbaugh and Volpe: Early development of the Gulf Coast Toad, *Bufo valliceps* Wiegmann.)

STAGE 36. The first and second toes are joined; the others are separated.

STAGE 37. The five toes are separated.

STAGE 38. The metatarsal (prehallux) tubercle appears.

STAGE 39. Pigment-free patches appear on the inner surface of the toes, which foreshadow the differentiation of the subarticular tubercles.

STAGE 40. The subarticular tubercles appear. There are two on the first toe, two on the second, three on the third, four on the fourth, and three on the fifth.

STAGE 41. The skin in the area where the forelimb will protrude becomes thin and transparent. The transparent cover is referred to as the "skin window" by Taylor and Kollros (1946). The cloacal tail piece, i.e., the exten-

sion through the ventral tail fin containing the cloacal opening, becomes resorbed. The cloacal tail piece disappears either in this stage or in the following stage.

STAGE 42. One or both forelimbs protrude. The angle of the mouth, as viewed from the side, is anterior to the nostril.

STAGE 43. The angle of the mouth, as viewed from the side, is between the nostril and the midpoint of the eye. The tail begins to regress. Slightly raised warts appear on the back and limbs. A light median dorsal stripe is bordered in the greater part of its length by a narrow black band from which branches extend laterally.

STAGE 44. The angle of the mouth, as viewed from the side, is between the midpoint and the posterior margin of the eye. The tail is considerably reduced.

STAGE 45. The angle of the mouth, as viewed from the side, is at the posterior margin of the eye. A slight stub of the tail remains. Most of the lateral black bands on the back disappear. The bands on the limbs begin to anastomose.

STAGE 46. Metamorphosis is completed. The tail is resorbed. The transformed toad has certain conspicuous features of the adult, but lacks others. It exhibits the wide median stripe, the broad lateral stripes, and the irregular pattern of bands on the limbs but lacks the cranial crests and paratoid glands.

(Limbaugh and Volpe note that individual variations do occur in the distribution of chromatophores on the tail but the variations are not of sufficient magnitude to obscure the basic pigment pattern. The pattern of melanophores at any particular stage may appear a single stage earlier or later.)

The mouth parts begin to differentiate at stage 23. Cornified frameworks for the

Figure 11–16   Larval stages 40 to 46 of the Gulf Coast toad, (*Bufo valliceps*). (From Limbaugh and Volpe: Early development of the Gulf Coast Toad, *Bufo valliceps* Wiegmann.)

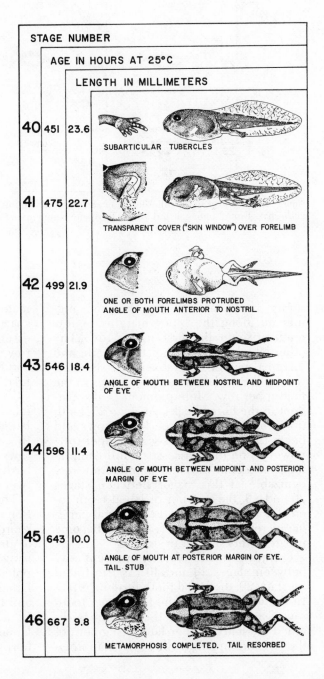

STAGE NUMBER

AGE IN HOURS AT 25°C

LENGTH IN MILLIMETERS

40 451 23.6
SUBARTICULAR TUBERCLES

41 475 22.7
TRANSPARENT COVER ("SKIN WINDOW") OVER FORELIMB

42 499 21.9
ONE OR BOTH FORELIMBS PROTRUDED
ANGLE OF MOUTH ANTERIOR TO NOSTRIL

43 546 18.4
ANGLE OF MOUTH BETWEEN NOSTRIL AND MIDPOINT OF EYE

44 596 11.4
ANGLE OF MOUTH BETWEEN MIDPOINT AND POSTERIOR MARGIN OF EYE

45 643 10.0
ANGLE OF MOUTH AT POSTERIOR MARGIN OF EYE.
TAIL STUB

46 667 9.8
METAMORPHOSIS COMPLETED.   TAIL RESORBED

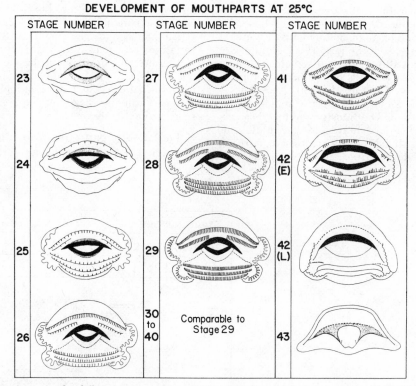

Figure 11–17   Stages in the differentiation and resorption of the larval mouth parts of the Gulf Coast toad (*Bufo valliceps*). (From Limbaugh and Volpe: Early development of the Gulf Coast Toad, *Bufo valliceps* Wiegmann.)

horny beaks appear and become pigmented along the inner margins. A few teeth arise in the first upper row. The other labial ridges do not bear teeth, and marginal papillae are absent. At stage 24, the number of labial teeth increases in the first upper row. More pigment is also evident in the horny beaks. At stage 25, teeth differentiate in the three lower rows and begin to appear in the second upper row; however, no row possesses a full complement of teeth. Marginal papillae are recognizable at this stage. Between stages 26 and 29, the number of labial teeth increases in each row, and the lateral papillae increase in number as well as become progressively smaller in size.

Complete mouth parts are present only between stages 29 and 40. As illustrated for stage 29, the labial tooth formula is 2/3. In the upper labium, the first upper tooth row is continuous and the second is divided medially. Either lateral segment of the second upper tooth row is approxi-

mately twice the length of the median space. The three lower tooth rows are essentially equal in length; the third lower row is slightly shorter than the second, and the second is slightly shorter than the first. Labial papillae are confined to the sides and extend to slightly below the edge of the third lower tooth row. Each lateral fringe bearing the papillae is folded inward between the upper and lower tooth rows. Minor modifications of the above basic pattern may be found in larvae between stages 29 and 40.

The larval mouth parts undergo resorption at stage 41. The third lower tooth row shortens considerably, and teeth begin to drop out from all rows. The papillary fringes become reduced in size. In a tadpole early in stage 42, a large number of labial teeth is lost, the mouth widens, and the lateral papillae become resorbed. In a tadpole advanced in stage 42, all labial teeth are lost and one or both of the horny beaks drop out, the lower generally before the upper. At stage 43, the mouth possesses structures characteristic of the adult.

In the preceding scheme and others like it, stages 1 to 25 compose the embryonic period and stages 26 to 46 represent the larval period. Independent feeding commences sometime between stages 25 and 26. Anuran metamorphosis is frequently regarded as consisting of two phases: prometamorphosis and the metamorphic climax. In the above scheme, prometamorphosis begins with stage 36 and is marked by rapid growth of the hind limbs. Stage 42 is the first of the metamorphic climax and is marked by the appearance of the forelimbs. Degenerative processes occurring during metamorphosis of anurans include resorption of the tail and gills, closure of the gill clefts, disappearance of the peribranchial cavities, shedding of the horny beaks, teeth, and lining of the jaws, reduction of the cloacal tube, and degeneration of some blood vessels (including the fifth pair of aortic arches). Development processes occurring during anuran metamorphosis include the differentiation and growth of limbs, middle ear, tympanic membrane, tympanic cartilage, eyelids, and tongue. The larval skin and intestine are retained in the adult but undergo extensive modification during metamorphosis. The epidermis becomes thicker and is covered by a cornified layer. As noted in Chapter Two, there is a gross reduction in length of the gut and the foregut undergoes radical change.

The embryonic development of urodeles, through hatching, is generally staged in accordance with Harrison's table for the normal development of the spotted salamander (*Ambystoma maculatum*). Harrison's table, made accessible privately to many workers in the field, has never been published, but his stages have been redrawn and included in Rugh (1962); they are reproduced in Figures 11–18 through 11–23. The remainder of urodele development, through metamorphosis, does not involve the drastic changes encountered with anurans and has not been described in the same detail. Wilder (1924) subdivided the time from hatching to transformation of two-lined salamanders (*Eurycea bislineata*) into four stages: a postembryonic stage, the typical larval stage, a premetamorphic stage, and a

metamorphic stage. The postembryonic stage includes the brief period of time after hatching during which the individual subsists on yolk material and its intestine is not fully formed. The ability to ingest food marks the beginning of the typical larval stage, which may last over a year depending upon temperature and the availability of food. This stage is characterized by a fully formed intestine, but neither the nasolacrimal groove nor the os thyreoideum is yet formed. The premetamorphic stage is identified by an open nasolacrimal groove and the existence of the os thyreoideum; there are no vesicular glands in the skin. The metamorphic stage is reached when there are acinous glands in the skin, initially appearing as tiny vesicles, which become larger and more conspicuous. During this stage there are also absorption of larval structures and development of the nasolabial groove and eyelids. The proportionate length of the tail increases gradually throughout this entire stage. Dent and Kirby-Smith (1963) divided the metamorphic stage of the Tennessee cave salamander (*Gyrinophilus palleucus*) into three continuous phases (Fig. 11–24):

During phase 1 (Fig. 11–24, between A and B) the shape of the snout is changed as a result of resorption of lateral soft tissues in the cheek region, constriction of head muscles approximately halfway between the eyes and the gills, shriveling of the gill fimbriae, and a slight reduction in the size of the labial folds. In addition, there is some reduction in the size of the tail fin, the branchial cartilages begin to be resorbed, and the slit beneath the gular fold is fused shut.

During phase 2 (Fig. 11–24, between B and C), the phase of gill resorption, the gills shrink rapidly and become mere stubs within a period of about three weeks; in addition, the labial folds and the tail fin are resorbed and adult structures begin to develop. Parasphenoid teeth and the maxillary bones with their teeth are formed, soft tissues of the tongue begin to thicken and are undercut laterally. Beginning to develop around the nares are the breathing valves which prevent the entrance of water when the adult is submerged. Eyelids begin to form and cover about half of the eye. The skin starts to develop into the adult condition, with

(*Text continued on page 366*)

1

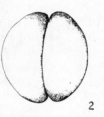

2

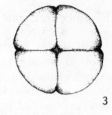

3

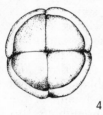

4

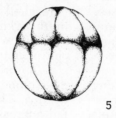

5

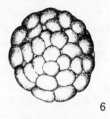

6

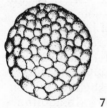

7

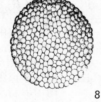

8

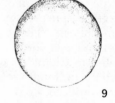

9

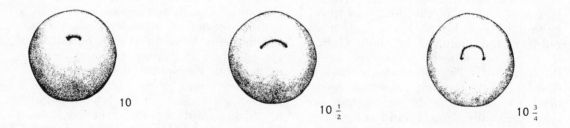

10          $10\frac{1}{2}$ $10\frac{3}{4}$

Figure 11–18 Embryonic stages 1 to 10 of the spotted salamander (*Ambystoma maculatum*), as defined by Harrison. (From Rugh: *Experimental Embryology Techniques and Procedures;* drawn by Naomi Leavitt.)

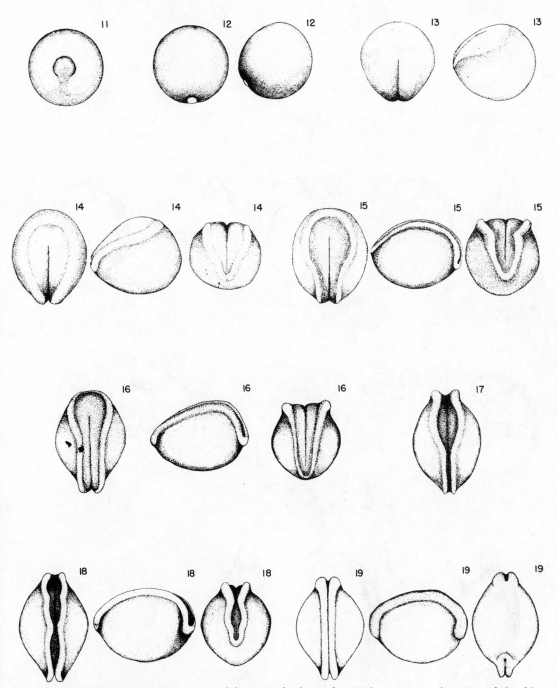

Figure 11–19 Embryonic stages 11 to 19 of the spotted salamander (*Ambystoma maculatum*), as defined by Harrison. (From Rugh: *Experimental Embryology Techniques and Procedures;* drawn by Naomi Leavitt.)

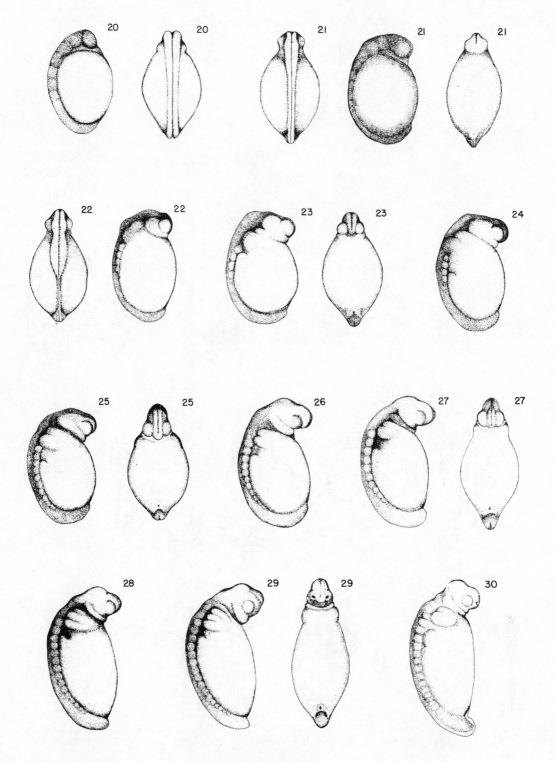

Figure 11–20   Embryonic stages 20 to 30 of the spotted salamander (*Ambystoma maculatum*), as defined by Harrison. (From Rugh: *Experimental Embryology Techniques and Procedures;* drawn by Naomi Leavitt.)

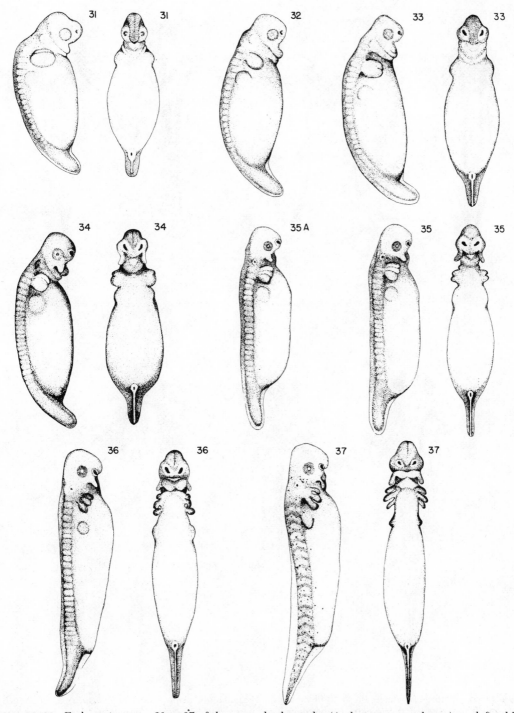

Figure 11–21 Embryonic stages 31 to 37 of the spotted salamander (*Ambystoma maculatum*), as defined by Harrison. (From Rugh: *Experimental Embryology Techniques and Procedures;* drawn by Naomi Leavitt.)

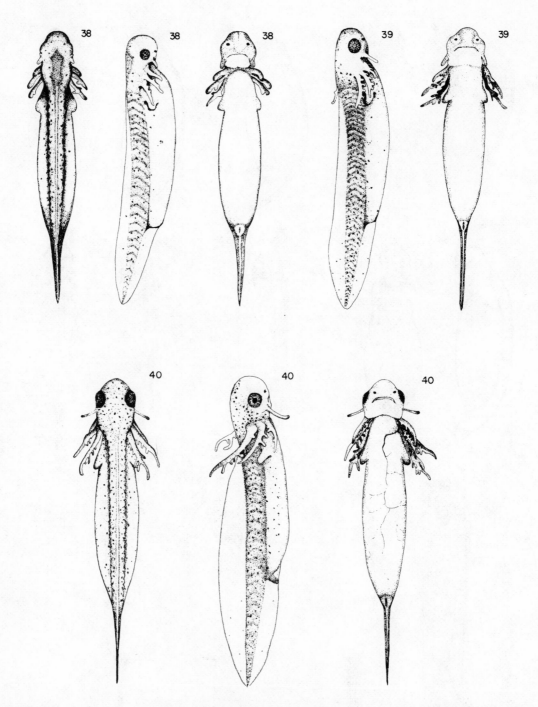

Figure 11–22 Embryonic stages 38 to 40 of the spotted salamander (*Ambystoma maculatum*), as defined by Harrison. (From Rugh: *Experimental Embryology Techniques and Procedures;* drawn by Naomi Leavitt.)

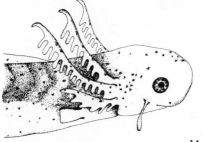
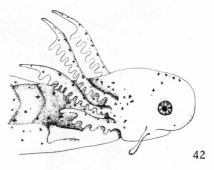

41

42

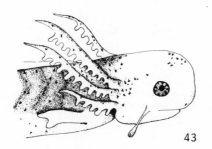
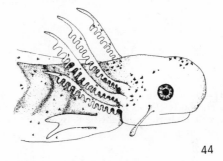

43

44

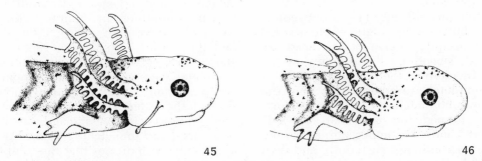

45

46

Figure 11–23   Embryonic stages 41 to 46 of the spotted salamander (*Ambystoma maculatum*), as defined by Harrison. (From Rugh: *Experimental Embryology Techniques and Procedures;* drawn by Naomi Leavitt.)

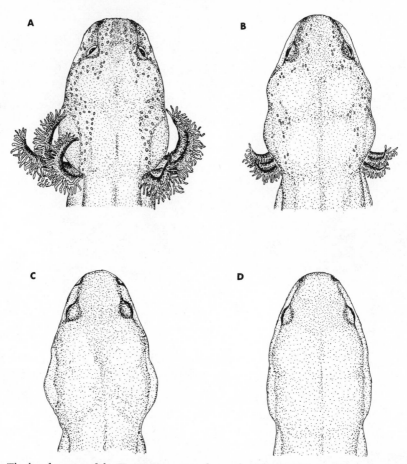

Figure 11–24   The head region of the Tennessee cave salamander (*Gyrinophilus palleucus*) at successive stages of metamorphosis. (From *Metamorphosis: A Problem in Developmental Biology*, edited by William Etkin and Lawrence A. Gilbert. Copyright 1968. By permission of Appleton-Century-Crofts, Educational Division, Meredith Corporation.)

mottled coloration, mucous glands, and the capillary network used during cutaneous respiration.

During phase 3 (Fig. 11–24, between *C* and *D*) bulges at the angles of the jaw and in the occipital region of the head are reduced, smoothing the contours of the head. In addition, the branchial cartilages are completely resorbed, scars from the gills are obliterated, the eyelids are completely formed, the nasolabial groove is evident, and the undercutting of the tongue is continued posteriorly and then anteriorly.

The inhibitory effect of crowding on larval growth and development is well known. Uncrowded larvae grow faster, reach a larger size, and metamorphose sooner than crowded larvae. Furthermore,

crowded larvae grow at different rates, and some will remain stunted even when sufficient food is supplied. Studies of growth inhibition in anuran tadpoles (Richards, 1958, 1962; Rose, 1960; Rose and Rose, 1961, 1965; Akin, 1966) have shown that substances which inhibit or completely stop growth and development can pass from one individual to another through the water. The inhibitory product tends to be species-specific and can affect both larval and prehatching stages. Available evidence indicates that the inhibitor is a volatile substance that is elaborated by the posterior half of the gut and that it has an effect which decreases with time; the half-life of the inhibitor is about 20 days. Akin (1966) and Richards (1958, 1962) found that the inhibitor produced by leopard frog (*Rana pipiens*) tadpoles is associated with algal cells which fill the guts of crowded, stunted tadpoles but are absent or rare in uncrowded tadpoles. Richards (1962) found that the algal cells that had divided *in vitro* were less inhibitory than freshly collected ones, indicating that the inhibitor becomes diluted or is released as the algal cells divide. The transmission of the inhibitor from tadpole to tadpole presumably occurs as they feed on the bottom and ingest the fecal material or algal cells from another tadpole. However, the fact that prehatching in addition to larval stages can be affected means that the inhibitor need not be ingested to retard development. The significance of crowding in nature is unknown, but fecal material frequently accumulates on the bottom of small ponds, so any inhibitory product released by the posterior gut is certain to be present in the water. Extensive crowding would not be necessary for inhibition to occur since Rose and Rose (1965) found that one large growing tadpole could inhibit the growth of six smaller tadpoles in an 80-liter aquarium. The degree to which growth inhibitors promote neoteny in nature, if at all, is not known.

Starvation is known to have an effect on metamorphosis. D'Angelo *et al.* (1941) found that until the early stages of hind limb development, metamorphosis is retarded and then stopped completely by inanition, so that tadpoles can be kept for extended periods of time without any metamorphic change occurring. However, once tadpoles have passed the critical stage in metamorphosis associated with early hind limb development, the rate of metamorphic change is accelerated by inanition, possibly because starvation causes an increase in the degenerative aspects of metamorphosis which provide energy for the transforming individual. These workers found they could induce metamorphosis in inhibited tadpoles by either immersing them in a thyroxin solution or by injecting them with pituitary material; the conclusion drawn was that starved individuals failed to metamorphose because of a decrease in the production and release of thyrotropic hormone from the anterior pituitary. It is possible that starvation is a cause of neoteny in nature, particularly at high elevations where ponds may be relatively sterile and unproductive, but this has never been demonstrated.

HORMONAL REGULATION OF METAMORPHOSIS

Studies of the hormonal regulation of metamorphosis have been conducted almost entirely on a few species of anurans, particularly frogs in the genus *Rana*, but the general mechanism is probably similar in all amphibians. The importance of thyroid hormones was first demonstrated in 1912 by the German biologist Friedrich Gudernatsch, when he discovered that tadpoles would metamorphose if fed extract from the thyroid gland. However, the precise hormones involved and the mechanism by which their levels are controlled during metamorphosis have only recently begun to be understood. The thyroid hormones are iodine-containing derivatives of tyrosine. Shellabarger and Brown (1959) identified mono-iodotyrosine, diiodotyrosine, L-thyroxine, and traces of 3,5-3'-triiodo-L-thyronine in the thyroid gland of clawed frog (*Xenopus laevis*) tadpoles. The development of the synthesis of these substances has been followed in the thyroid glands of tadpoles by Flickinger (1964), and numerous studies have shown that thyroxine is the principal hormone. Implants of thyroxine-cholesterol pellets have demonstrated that thyroxine acts directly, rather than in some indirect manner, on metamorphosing tissue and causes independently unique morphological and biochemical changes at each site which are species-specific and independent of the type of tissue involved. Thus, for example, thyroxine causes a shortening of the tail in anuran tadpoles but not in salamander larvae; when thyroxine is implanted into the striated muscle of a developing anuran limb it causes a marked growth of the limb, but when it is implanted into the striated muscle of an anuran tadpole tail fin it causes resorption of the fin. Transplant experiments have demonstrated that the response of the metamorphosing tissue to thyroxine is an inherent quality of the tissue and is not due to induction by any surrounding tissue: for example, hind limb buds will continue to grow after being transplanted to the tadpole's tail.

As noted, the metamorphic changes which occur must do so in a coordinated and orderly sequence. Related to this is evidence indicating that metamorphosing structures differ from one another with respect to the times at which they first become sensitive to thyroxine. They also differ from one another, and from one stage of development to another, in their relative sensitivity to thyroxine and their rate of response to it. Different structures have different thresholds, and in general, those that metamorphose early are the most sensitive to thyroxine; for example, the threshold for perforation of the operculum in frog tadpoles is lower than that for resorption of the gills (Kollros, 1961). Different rates of response to thyroxine are reflected in the fact that early metamorphic changes are more gradual than are later ones: for example, the growth response of legs, which are very sensitive to thyroxine and begin to develop very early, is very slow compared to the resorption response of the tail, which is one of the more insensitive structures and is changed later in metamorphosis.

Since the later metamorphic changes have higher thresholds in regard to thyroxine sensitivity than early changes, there must be an increase in both thyroid activity and circulating thyroxine concentration as metamorphosis progresses. This implication has been verified by morphological, histological, and chemical investigations. Morphological and histological changes have been described by Etkin (1935, 1964). Prior to the initiation of metamorphosis the thyroid gland is small and histologically appears inactive. During prometamorphosis, it grows rapidly, becoming much larger in proportion to body size, and its epithelium increases in height. These changes, indicating that the gland has been activated and that its rate of activity is increasing, continue until the beginning of the metamorphic climax phase. By this time, the gland is large, epithelial height is maximal, and there is extensive colloid vacuolization. Following the beginning of the metamorphic climax there is a rapid reduction in the height of the epithelium and the glandular follicles become distended with dense acidophilic colloid, indicating a return of the gland to minimal activity.

Radioiodine (I^{131}) has been used in a number of studies to measure the rate of iodine uptake and release by the thyroid, this being a direct measure of its glandular activity. These chemical studies have shown that the premetamorphic gland is functional at a very low level, but they otherwise confirm the histological interpretations. Kaye (1961), for example, found the thyroid uptake of radioiodine was 40 times greater during the metamorphic climax of leopard frog tadpoles (*Rana pipiens*) than during their early larval stages.

As mentioned in Chapter Two, the activity of the thyroid gland is controlled by a pituitary hormone called thyrotropin or thyroid-stimulating hormone (TSH). Thus, explaining the metamorphic changes in thyroid activity becomes a matter of explaining changes in TSH level. Studies on both anuran and salamander larvae indicate that when the part of the hypothalamus connecting the pituitary and the brain is severed or removed the animal will not metamorphose; this suggests that the brain plays a controlling role in metamorphosis by acting on the pituitary. Available evidence at present indicates that this regulatory role of the brain is carried out by way of neurosecretory cells that translate nerve impulses into chemical messages. Fibers carrying neurosecretions have been found which terminate in the eminentia medialis, an eminence in the floor of the fourth ventricle of the brain. Portal veins receive blood from a capillary network in the eminentia medialis and carry it directly to the pituitary. In mammals it is known that the pituitary is activated by a neurosecretion called thyrotropin-releasing factor (TRF) that is produced by the hypothalamus; presumably this chemical messenger is also present in amphibians. Thus, it appears that the amphibian brain controls the pituitary directly by having nerve impulses converted into hypothalamic TRF which is carried by the portal veins into the capillary bed of the anterior lobe of the pituitary, causing the latter to release TSH. The question of thyroid control now has become one of regulation of hypothalamic TRF.

Etkin (1963, 1965a) studied the larval development of the eminentia medialis and the portal veins in *Rana pipiens* tadpoles. He found that these structures are poorly differentiated in premetamorphic tadpoles and that the anterior lobe of the pituitary is broadly attached to the thin floor of the hypothalamic lobes; a diffuse network of capillaries lies between the two structures, conveying blood from the hypothalamic arteries into the capillary system of the anterior pituitary. During prometamorphosis the hypothalamic neurosecretory apparatus differentiates in the floor, becoming shorter and thicker, and the anterior lobe of the pituitary separates from the neural floor except at its anterior tip, where it remains attached by a few large venous channels draining the primary capillary bed; the eminentia medialis is becoming differentiated by this stage, and the above-mentioned venous channels form the portal veins leading to the pituitary. By the time the individual enters the metamorphic climax phase the entire eminentia medialis system is well advanced but not completely differentiated. This developmental sequence indicates that the eminentia medialis system can only transport a minimum of TRF in premetamorphic tadpoles because it is poorly organized. During prometamorphosis, the progressively greater development of the system would allow greater and greater amounts of TRF to be transported to the pituitary, causing it to gradually release more and more TSH, which would in turn cause the thyroid to release more and more thyroxine. Thus, it appears that the normal metamorphosis of a tadpole is dependent upon the development of the eminentia medialis at the proper time; this involves stimulation by thyroxine! Experiments with larvae (Etkin 1963, 1965b) have revealed that the eminentia medialis does not differentiate in thyroidectomized larvae unless they are treated with graded concentrations of thyroxine and have reached a certain stage of maturity. It appears that the hypothalamus becomes sensitive to thyroxine at the beginning of prometamorphosis and then a positive feedback system develops wherein thyroxine stimulates the development of the eminentia medialis, which, by transporting increasing quantities of TRF, stimulates the pituitary to produce and release increasingly greater quantities of thyroxine. Since the premetamorphic activity level of the thyroid is very low,

the cycle would begin at a low rate but accelerate very rapidly through prometamorphosis and the metamorphic climax.

Finally, there is evidence that prolactin, produced by the anterior lobe of the pituitary, plays a role in the regulation of metamorphosis. Prolactin, known to stimulate growth in tadpoles, is inhibited by TRF (or some other hypothalamic prolactin-inhibiting factor). During premetamorphosis, the production and release of TRF is minimal; there is neither a significant stimulation of TSH production nor a significant inhibition of prolactin. Consequently, the tadpole grows rapidly and does not metamorphose. Since the production and release of TRF increases during prometamorphosis and early metamorphic climax, there will be progressively increasing inhibition of prolactin at the same time that there is greater and greater stimulation of TSH production. Consequently, the tadpole's growth slows and eventually stops while simultaneously metamorphic changes proceed at faster and faster rates.

To summarize, each metamorphosing structure has a unique time at which it becomes sensitive to thyroxine, has a unique sensitivity to this hormone, reacts in a unique way, and reacts at a particular rate. These characteristics of the metamorphosing structures are presumably genetically controlled and determine what metamorphic changes will occur and in what sequence. The entire metamorphic sequence of changes is triggered at a critical stage in the tadpole's development when, because of its heredity, the hypothalamus becomes sensitive to the low level of thyroxine circulating during premetamorphosis (Fig. 11-25). The result is differentiation of the eminentia medialis and release of TRF, which stimulates the anterior lobe of the pituitary to secrete TSH, which causes the thyroid to increase its production of thyroxine. Because of the feedback mechanism, the production and release of thyroxine proceeds at a higher and higher rate. The metamorphosing structures that are highly sensitive to thyroxine, such as the limbs, begin changing first. Those, such as the tail, which are relatively insensitive to thyroxin change last. Initial changes tend to be slow; late changes occur rapidly. The inhibition of prolactin by TRF results in

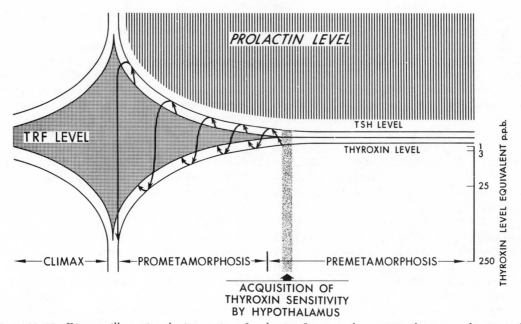

Figure 11–25   Diagram illustrating the interaction of endocrine factors in determining the time and pattern of anuran metamorphosis. In the early premetamorphic period the thyroxin level is very low and remains so until just before prometamorphosis begins. At this time, the hypothalamic TRF mechanism becomes sensitive to positive thyroxin feedback, thereby initiating prometamorphosis. The increase in TRF provoked by the action of the initial thyroxin level upon the hypothalamus stimulates increased TSH release, which acts back to raise the thyroxin level. This leads to a spiraling action which raises the thyroxin level and thereby induces prometamorphosis with its characteristic sequence of changes. The positive feedback cycle leads to maximal activation of the pituitary-thyroid axis, thereby bringing on metamorphic climax. During early premetamorphosis, prolactin is produced at a high rate (vertical lines). With the activation of the hypothalamus, the production of prolactin drops under the inhibitory influence of hypothalamic activity. As the level of TSH rises during prometamorphosis, that of prolactin decreases. The growth rate of the animal therefore falls, and the metamorphosis-restraining activity of prolactin diminishes. Thus, the premetamorphic period in which growth is active and metamorphosis is inhibited is characterized by the predominance of prolactin over TSH. The reverse holds during metamorphosis. The time of shift in hormone balance is determined by the initiation of positive thyroid feedback to the hypothalamus. This varies greatly between species. The pattern of change during metamorphosis is regulated by the pattern of the feedback build-up and is much the same in most anurans. (From *Metamorphosis: A Problem in Developmental Biology,* edited by William Etkin and Lawrence A. Gilbert. Copyright 1968. By permission of Appleton-Century-Crofts, Educational Division, Meredith Corporation.)

the retardation and finally halting of growth during metamorphosis. The absence of growth combined with the resorption of tissues during metamorphosis results in the transformed individual being smaller than the larval form.

NEOTENY

Neoteny is the situation wherein larval characteristics are retained for an extended period of time. Paedogenesis is reproduction by larval individuals. If an entire permanent population is neotenic then it must also be paedogenic; that is, there must be reproduction by the neo-

tenic larvae. No anurans or apodans are known to be paedogenic, but naturally occurring, non-metamorphosing anuran larvae have been reported (Jurand, 1955; Boschwitz, 1957; Saxén, 1957). A number of salamander species exhibit different degrees of neoteny: the families Sirenidae, Cryptobranchidae, Amphiumidae, and Proteidae contain only permanently larval forms that undergo a partial metamorphosis in nature and cannot be caused to complete metamorphosis; *Siredon pisciformis* and *Ambystoma mexicanum* (family Ambystomatidae) and several species of plethodontid salamanders (*Haideotriton wallacei, Eurycea tyner-*

ensis, E. troglodytes, E. tridentifera, E. rathbuni, E. pterophila, E. latitans, E. neotenes, E. nana, and *Gyrinophilus pal-leucus*) are always neotenic in nature but may be caused to undergo metamorphosis. Some species in the families Ambystoma-tidae (*Dicamptodon ensatus, Ambystoma tigrinum, A. talpoideum, A. gracile, Rhy-acosiredon altamirani, R. leorae, R. rivularis,* and *R. zempoalaensis*), Sala-mandridae (*Notophthalmus perstriatus, N. viridiscens, Triturus alpestris, T. cristatus, T. helveticus,* and *T. taeniatus*), Pletho-dontidae (*Eurycea multiplicata*), and Hy-nobiidae (*Hynobius lichenatus*) typically metamorphose in nature, but occasional neotenic individuals or populations occur.

Numerous studies have shown that the thyroid gland controls or causes meta-morphosis through the production of thyroxine. Thus, neoteny could theo-retically result from either hypofunction of the thyroid or loss of sensitivity of met-amorphosing tissues to thyroxine. The fact that neotenes of many of the above spe-cies have been caused to metamorphose by exposure to various thyroxine concen-trations indicates that, at least frequently, the tissues concerned remain sensitive to this hormone and the neotenic condi-tion must be the result of a subminimal production of thyroxine. An inadequate production of thyroxine could conceivably result from hypofunctioning of the thyroid because of endogenous reasons, hypo-functioning of the thyroid because of hypofunction of the pituitary and an in-sufficient production of TSH (thyroid stimulating hormone), or because of an insufficient uptake of iodine (see Chapter Two). The Tennessee cave salamander (*Gyrinophilus palleucus*) has been caused to metamorphose consistently when treated with thyroxine but has also given a variable metamorphic response follow-ing either the injection of mammalian TSH or the implantation of pituitary tis-sue from the leopard frog, *Rana pipiens* (Dent and Kirby-Smith, 1963). Similarly, the Mexican axolotl (*Ambystoma mexi-canum*) has been caused to metamor-phose by pituitary grafts from the tiger salamander, *Ambystoma tigrinum* (Blount, 1950).

Thus, in both the Tennessee cave sala-mander and the Mexican axolotl, neo-teny appears to result from hypofunction

of the pituitary; this may be the reason for neoteny in other species which can be caused to metamorphose. The fact that there are some species that are consist-ently neotenic, some that are occasionally neotenic, and some that are never neo-tenic implies that the potential for neo-teny is genetically controlled. Hypofunc-tioning of the thyroid, pituitary, or both must be part of the animal's heredity in those salamanders which are consistently neotenic but capable of being induced to metamorphose. In the case of the Si-renidae, Cryptobranchidae, Amphiumi-dae, and Proteidae, the fact that meta-morphosis cannot be induced implies that the tissues concerned are not respon-sive to thyroxine and that this lack of sensitivity is genetically determined. Those species which typically meta-morphose but sometimes do not (e.g., the tiger salamander) obviously have the genetic potential for metamorphosing under certain conditions but will not do so when these conditions are not fulfilled. This implies that functioning of the thy-roid or pituitary or both is barely adequate, under normal environmental circum-stances, for inducing metamorphosis and that, when environmental conditions be-come marginal, hypofunctioning will re-sult in a failure to metamorphose. It is also possible that the genetic potential for metamorphosing varies geographically within the range of the species because, under some but not all environmental conditions, natural selection has favored neotenes over metamorphosed individ-uals; this possibility will be discussed subsequently. Species which are never neotenic must be characterized by greater than marginal functioning of the pituitary and/or thyroid under all environmental conditions encountered within their dis-tributional ranges; possibly this is be-cause these species have limited ranges of tolerance and exist under essentially the same environmental conditions throughout their individual ranges, or it may be that selection against neotenes has produced ecotypic variations in thyroid and/or pituitary function which compen-sate for variable environmental conditions.

Within species which commonly meta-morphose neoteny is most prevalent in populations inhabiting high-altitude ponds and lakes. For example, Snyder

(1956) made a comparative study of low- and high-altitude populations of the Northwestern salamander (*Ambystoma gracile*) and found that at sea level many larvae metamorphose at one year of age and that nearly all of the remainder transform at two years of age (the age at which this species becomes sexually mature) but a small number remain as neotenes. In contrast, at high elevations (4300 to 5500 feet) nearly all sexually mature individuals are neotenic and metamorphosis rarely occurs. Temperature decreases with altitude and has long been known to influence the metamorphosis of amphibians. Huxley (1929) found tadpole tissues are not responsive to thyroid hormones when at temperatures lower than 10° C. In a more recent and specific study, Frieden (1968) showed that there is a marked reduction in the response of tail tissue of 3,5-3'-triiodo-L-thyronine at temperatures below 20° C until a temperature of 5° C is reached. At 5° this worker could observe no effect of this hormone even when the tissues were exposed to it for 100 days. Other workers have suggested that cold may inhibit the release of thyroid hormones and, in this way, would also inhibit metamorphosis. Thus, the failure of larvae to metamorphose at high altitudes may be due entirely to temperature effects on the release of hormones and hormonal responses.

There is also evidence of an age threshold at which metamorphic tissues become less sensitive to thyroid hormones (Lipchina, 1929). At high altitudes growing seasons for larvae are of short duration, frequently less than three months, whereas at lower altitudes active feeding and growth may take place throughout the year. Consequently, Snyder (1956) concluded that at high altitudes larvae reach metamorphic size at a relatively late age and they may fail to metamorphose because either their tissues are less responsive to thyroid hormones or too little hormone is being produced. The breeding season for *Ambystoma gracile* extends from January to July and there may be several months' difference in age among individuals produced the same year. Thus, Snyder decided that neotenic individuals in sea-level populations may represent those that pass metamorphic age and size thresholds during the winter and are resistant to the thyroid hormones.

A particular amount of development must take place before metamorphosis will occur. Consequently, even when neoteny does not result, a short growing season may have the effect of adding a year to the duration of the larval stage. Thus, two-lined salamanders (*Eurycea bislineata rivicola*) may metamorphose either in their second or third year; green frogs (*Rana clamitans*) may metamorphose with or without wintering as larvae. When a species is widely distributed geographically and lives under different temperature conditions in various parts of its range, the duration of the larval stage may vary geographically. For example, the bullfrog (*Rana catesbeiana*) may metamorphose the same summer the eggs are laid in the southern part of its range but may spend three winters as a tadpole in the northern part of its range; smooth newts (*Triturus vulgaris*) generally metamorphose at the end of their first summer in southern Europe but spend a winter as larvae in northern Russia.

Environmental factors other than temperature may also cause neoteny. The requirement of iodine for thyroid hormones suggests that neoteny may sometimes reflect an iodine-deficient larval environment. Since montane ponds and lakes generally contain very small quantities of dissolved substances, the probability for an iodine-deficient larval environment would increase with altitude and, thus, could also account for differences in neotenic tendencies between lowland and high-altitude populations. Although no one has been able to demonstrate conclusively that metamorphosis in any natural population is limited by the availability of iodine, metamorphosis has been induced in hypophysectomized and thyroidectomized tiger salamanders ("Colorado axolotls"), *Ambystoma tigrinum*, by subcutaneous implantations of powdered iodine crystals (Ingram, 1928); Blair (1961) reported that one specimen out of two of *Gyrinophilus palleucus*, a neotenic species in nature, metamorphosed after three months' immersion in a solution of sodium iodide.

Clearly, neoteny results from the interaction of genetic and environmental factors. Salamanders have a genetic potential for metamorphosing which is species-specific. Depending upon how great this potential is, the species (population) may never be neotenic, may occasionally be neotenic, or may consistently be neotenic in nature. The genetically determined potential for metamorphosing involves the regulation of thyroid production and adaptation to the larval environmental factors, such as temperature, which affect thyroid production. In most amphibians, natural selection appears to favor metamorphosis over neoteny, for the latter never occurs. In a few salamander species, natural selection apparently favors neoteny over metamorphosis, for metamorphosis never occurs. Finally, a variety of salamander species are sometimes neotenic and sometimes metamorphic, indicating that natural selection may favor neoteny under some environmental conditions and metamorphosis under others. Wherever neoteny is prevalent, the implication is that the larval stage is better adapted to its environment than the transformed adult is to its environment. The propensity for neoteny in salamanders occupying subterranean or high altitude habitats certainly seems to reflect a more favorable environment to the salamander within water than outside of it (Dent, 1968). On the other hand, many populations of *Ambystoma* that normally do not metamorphose will do so if the pond in which they live begins to dry up, apparently in response to crowding and increased salt content of the water. Environmental limitations, such as the availability of iodine, may cause neoteny regardless of the genetic potential for metamorphosing.

GROWTH AND ATTAINMENT OF SEXUAL MATURITY

Very little study has been made of the growth of amphibians, but it appears that all species have a period of rapid growth preceding the attainment of sexual maturity, followed by a period during which growth tends to be intermittent, rather than constant, and the rate of growth gradually slows. The growth of many amphibians appears to continue throughout the life of the individual. Such growth is said to be indeterminate because it results in individuals tending to get bigger as they get older. Indeterminate growth is not constant, however, and its rate is influenced by environmental factors such as temperature and nutrition. It may be intermittent, with periods of noticeable growth separated by periods of very little or no growth. For example, very little, if any, growth takes place during hibernation or aestivation. Thus, populations characterized by such annual periods of inactivity will also consist of individuals exhibiting annual patterns of intermittent growth. In addition to such regular patterns of interrupted growth, some individuals may exhibit irregular interruptions of growth.

Growth rates are influenced by many factors. As noted by Oliver (1955), the total growth potential of an individual is affected by the maximum size of the species to which it belongs and, in general, there appears to be a relationship between maximum size and the size at hatching or birth, the size at sexual maturity, and rate of growth and longevity. Small species tend to have relatively larger offspring, reach sexual maturity at a size close to the maximum for the species, grow at a more rapid rate, and have a shorter life span than larger species. These tendencies are complicated by a number of cause-effect relationships, many of which are poorly understood.

Sexual maturity is reached when the individual's gonads become mature and begin producing gametes. In both amphibians and reptiles, this stage in life is generally associated with completion of a certain amount of growth rather than a particular age. Despite the variations which occur in growth rates, typical ages at which sexual maturity is reached have been determined for populations of a variety of amphibians. This age will frequently differ from population to population, cohort to cohort, and between males and females. The following table indicates what appear to be minimum sizes and ages for sexually mature individuals of various species of amphibians.

SIZE AND AGE AT WHICH SEXUAL MATURITY IS ATTAINED IN VARIOUS AMPHIBIANS

	Snout-Vent Length at Maturity (mm)	Age at Maturity (yrs)
Urodela		
Hynobiidae:		
Hynobius keyserlingi	100–125	3–4
Salamandridae:		
Salamandra salamandra	125–150	4
Salamandra atra	100–125	2–4
Pleurodeles waltl	200	2–3
Taricha granulosa		3
Triturus alpestris	90–100	2–3
Triturus cristatus	125	3
Triturus vulgaris	65–75	3–4
Plethodontidae:		
Batrachoseps attenuatus		2–4
Desmognathus aeneus	18–19	2
Desmognathus fuscus males		3
Desmognathus fuscus females		4
Ensatina eschscholtzi	30–35	4
Leurognathus marmoratus males	50	2
Leurognathus marmoratus females	55–59	3
Plethodon cinereus	32–34	2
Plethodon glutinosus females (north)	59	3–4
Plethodon glutinosus females (south)	45	2
Plethodon glutinosus males (north)	53	3–4
Plethodon glutinosus males (south)	40	2–3
Proteidae:		
Necturus maculosus	205	5
Ambystomatidae:		
Ambystoma jeffersonianum	115	2
Ambystoma macrodactylum croceum	50	2
Ambystoma macrodactylum sigillatum	53–55	3
Ambystoma maculatum		2
Ambystoma opacum		$3\frac{1}{2}$–4
Ambystoma tigrinum		1
Amphiumidae:		
Amphiuma tridactylum females	350	4–5

SIZE AND AGE AT WHICH SEXUAL MATURITY IS ATTAINED IN VARIOUS AMPHIBIANS (Cont.)

	Snout-Vent Length at Maturity (mm)	Age at Maturity (yrs)
Urodela		
Sirenidae:		
Siren intermedia		2
Cryptobranchidae:		
Cryptobranchus alleganiensis		5–6
Anura		
Pipidae:		
Xenopus laevis		1–2
Ranidae:		
Rana catesbeiana males (New York)	85	3–4
Rana catesbeiana females (New York)	90	3–4
Rana clamitans males	60	2–3
Rana clamitans females	65	2–3
Rana pipiens males	52	1–2
Rana pipiens females	54	1–2
Rana temporaria		4–5
Microhylidae:		
Gastrophryne carolinensis		2
Bufonidae:		
Bufo americanus		1–2
Bufo boreas halophilus		2
Bufo canorus	42–47	2–3
Bufo cognatus		2–4
Bufo hemiophrys	38–45	2–3
Bufo valliceps	60–65	10 months
Hylidae:		
Acris gryllus	20	1–2
Hyla regilla		1–2
Hyla versicolor		1–2
Leptodactylidae:		
Eleutherodactylus planisrostris ricordi		1
Syrrhophus marnocki	18–22	1

References

Akin, G. C. 1966. Self-inhibition of growth in *Rana pipiens* tadpoles. Physiol. Zool. 39:341–356.

Baker, C. L., L. C. Baker, and M. F. Caldwell. 1947. Observation of copulation in *Amphiuma tridactylum*. J. Tennessee Acad. Sci. 22:87–88.

Baker, L. C. 1937. Mating habits and life history of *Amphiuma tridactylum* Cuvier and effects of pituitary injections. J. Tennessee Acad. Sci. 12: 206–218.

Balinsky, B. I. 1970. An Introduction to Embryology. Philadelphia, W. B. Saunders. 725 pp.

Bannikov, A. G. 1958. Die Biologie des Froschzahnmolches *Ranodon sibiricus* Kessler. Zool. Jahr. (Syst.) 86:245–249.

Bishop, S. C. 1947. Handbook of Salamanders. The Salamanders of the United States, of Canada, and of Lower California. Ithaca, N.Y., Comstock. 555 pp.

Blair, A. P. 1961. Metamorphosis of *Pseudotriton palleucus* with iodine. Copeia 1961:499.

Blount, B. F. 1950. The effects of heteroplastic hypophyseal grafts upon the axolotl, *Ambystoma mexicannum*. J. Exp. Zool. 113:717–739.

Boschwitz, D. 1957. Thyroidless tadpoles of *Pelobates syriacus* Boettger H. Copeia 1957:310–311.

Conte, E. D., and J. L. Sirlin. 1952. Pattern series of the first embryonary stages in *Bufo arenarum*. Anat. Rec. 112:125–135.

D'Angelo, S. A., A. S. Gordon, and H. A. Charipper. 1941. The role of the thyroid and pituitary glands in the anomalous effect of inanition on amphibian metamorphosis. J. Exp. Zool. 87:259–277.

Dent, J. N. 1968. Survey of amphibian metamorphosis. *In:* W. Etkin and L. I. Gilbert (eds.), Metamorphosis. New York, Appleton-Century-Crofts. Pp. 271–311.

———, and J. S. Kirby-Smith. 1963. Metamorphic physiology and morphology of the cave salamander *Gyrinophilus palleucus*. Copeia 1963:119–130.

Duellman, W. E. 1970. The Hylid Frogs of Middle America. Lawrence, University of Kansas Museum of Natural History. Monograph No. 1. 753 pp. (2 vols.; 324 figs., 72 pls.)

Eakin, R. M. 1947. Stages in the normal development of *Hyla regilla*. Univ. California Publ. Zool. 51: 245–257.

Etkin, W. E. 1935. The mechanisms of anuran metamorphosis. I. Thyroxine concentrations and the metamorphic pattern. J. Exp. Zool. 71:317–340.

——— 1963. The metamorphosis activating system of the frog. Science 139:810–814.

——— 1964. Metamorphosis. *In:* J. A. Moore (ed.), Physiology of the Amphibia. New York, Academic Press. Pp. 427–468.

——— 1965a. The phenomena of amphibian metamorphosis. IV. The development of the median eminence. J. Morphol. 116:371–378.

——— 1965b. Thyroid feedback to the hypothalamic neurosecretory system in frog larvae. Neuroendocrinology 1:45–64.

Flickinger, R. A. 1964. Sequential appearance of monoiodotyrosine, diiodotyrosine, and thyroxine in the developing frog embryo. Gen. Comp. Endocrinol. 4:285–289.

Frieden, E. 1968. Biochemistry of amphibian metamorphosis. *In:* W. Etkin and L. I. Gilbert (eds.), Metamorphosis; A Problem in Developmental Biology. New York, Appleton-Century-Crofts. Pp. 349–398.

Gosner, K. L. 1960. A simplified table for staging anuran embryos and larvae with notes on identification. Herpetologica 16:183–190.

Huxley, J. S. 1929. Thyroid and temperature in cold blooded vertebrates. Nature 123:712.

Ingram, W. R. 1928. Metamorphosis of the Colorado axolotl by injection of inorganic iodine. Proc. Soc. Exp. Biol. Med. 26:191.

Jurand, A. 1955. Neoteny in *Xenopus laevis* Daud. Folia Biol. 3:315–330.

Kaye, N. W. 1961. Interrelationships of the thyroid and pituitary in embryonic and premetamorphic stages of the frog, *Rana pipiens*. Gen. Comp. Endocrinol. 1:1–19.

Kollros, J. J. 1961. Mechanisms of amphibian metamorphosis: Hormones. Am. Zool. 1:107–114.

Limbaugh, B. A., and E. P. Volpe. 1957. Early development of the Gulf coast toad, *Bufo valliceps* Wiegmann. American Mus. Nov. 1842:1–32.

Lipchina, L. P. 1929. The dependence of the process of metamorphosis in axolotls on factors of age, coloring and sex. Zh. Eksp. Biol. Med. 13:73–78.

Macgregor, H. C., and T. Uzzell. 1964. Gynogenesis in salamanders related to *Ambystoma jeffersonianum*. Science 143:1043–1045.

Myers, C. W. 1969. The ecological geography of cloud forest in Panama. American Mus. Nov. 2396: 1–52.

Nieuwkoop, P. D., and J. Faber (eds.). 1956. Normal Table of *Xenopus laevis* (Daudin); A systematical and chronological survey of the development from the fertilised egg till the end of metamorphosis. Amsterdam, North-Holland Publ. 252 pp.

Oliver, J. A. 1955. The Natural History of North American Amphibians and Reptiles. Princeton, N.J., Van Nostrand Reinhold. 359 pp.

Parker, H. W. 1956. Viviparous caecilians and amphibian phylogeny. Nature 178:250–252.

Pollister, A. W., and J. A. Moore. 1937. Tables for the normal development of *Rana sylvatica*. Anat. Rec. 68:489–496.

Prechtl, H. F. R. 1951. Zur Paarungsbiologie einiger Molcharten. Z. Tierpsychol. 8:337–348.

Pyburn, W. F. 1970. Breeding behavior of the Leaffrogs *Phyllomedusa callidryas* and *Phyllomedusa dacnicolor* in Mexico. Copeia 1970:209–218.

Rabb, G. B., and M. S. Rabb. 1960. On the mating and egg-laying behavior of the Surinam Toad, *Pipa pipa*. Copeia 1960:271–276.

——— 1963. Additional observations on breeding behavior of the Surinam Toad, *Pipa pipa*. Copeia 1963:636–642.

———, and R. Snedigar. 1960. Observations on breeding and development of the Surinam toad, *Pipa pipa*. Copeia 1960:40–44.

Richards, C. M. 1958. The inhibition of growth in crowded *Rana pipiens* tadpoles. Physiol. Zool. 31:138–151.

——— 1962. The control of tadpole growth by alga-like cells. Physiol. Zool. 35:285–296.

Rose, S. M. 1960. A feedback mechanism of growth control in tadpoles. Ecology 41:188–199.

———, and F. C. Rose. 1961. Growth controlling exudates of tadpoles. Symp. Soc. Exp. Biol. 15: 207–218.

——— 1965. The control of growth and reproduction in freshwater organisms by specific products. Mitt. Int. Verein. Limnol. 13:21–35.

Rowan, W. 1938. Light and seasonal reproduction in animals. Biol. Rev. Cambridge Philos. Soc. *13*: 374–402.

Rugh, R. 1962. Experimental Embryology Techniques and Procedures, 3rd ed. Minneapolis, Burgess. 501 pp.

Salthe, S. N. 1967. Courtship patterns and the phylogeny of the urodeles. Copeia *1967*:100–117.

_____ 1969. Reproductive modes and the number and sizes of ova in urodeles. Am. Midland Nat. *81*:467–490.

_____, and B. M. Salthe. 1964. Induced courtship in the salamander *Pseudoeurycea belli*. Copeia *1964*: 574–576.

Saxén, L. 1957. Schilddrusenaplasie als Ursache von partieller Neotenie. Acta Endocrinol. *24*:271–281.

Sedar, S. N., and M. I. Michael. 1961. Normal table of the Egyptian toad, *Bufo regularis* Reuss, with an addendum on the standardisation of the stages considered in previous publications. Československ. Morfol. *9*:333–351.

Shellabarger, C. J., and J. Brown. 1959. The biosynthesis of thyroxine and 3:5:3 triiodothyronine in larval and adult toads. J. Endocrinol. *18*:98–101.

Shumway, W. 1940. Stages in the normal development of *Rana pipiens*. I. External form. Anat. Rec. *78*:139–144.

Snyder, R. C. 1956. Comparative features of the life histories of *Ambystoma gracile* (Baird) from populations at low and high altitudes. Copeia *1956*: 41–50.

Stebbins, R. C. 1954. Amphibians and Reptiles of Western North America. New York, McGraw-Hill. 536 pp.

Taylor, A. C., and J. J. Kollros. 1946. Stages in the normal development of *Rana pipiens* larvae. Anat. Rec. *94*:7–23.

Thorn, R. 1963. Contribution à l'étude d'une salamander Japonaise, l'*Hynobius nebulosus* (Schlegel). Comportement et reproduction en captivité. Arch. Inst. Gr-Duc. Luxembourg N.S. *29*: 201–215.

Twitty, V. C. 1955. Field experiments on the biology and genetic relationships of the California species of *Triturus*. J. Exp. Zool. *129*:129–147.

Uzzell, T. 1964. Relations of the diploid and triploid species of the *Ambystoma jeffersonianum* complex (Amphibia, Caudata). Copeia *1964*:257–300.

_____ 1970. Meiotic mechanisms of naturally occurring unisexual vertebrates. Am. Naturalist *104*: 433–445.

_____, and S. M. Goldblatt. 1967. Serum proteins of salamanders of the *Ambystoma jeffersonianum* complex, and the origin of the triploid species of this group. Evolution *21*:345–354.

Van Oordt, P. G. W. J. 1956. Regulation of the spermatogenic cycle in the common frog (*Rana temporaria*). Arnhem, Holland, G. W. van der Wiel. 116 pp.

Weisz, P. B. 1945. The normal stages in the development of the South African clawed toad, *Xenopus laevis*. Anat. Rec. *93*:161–169.

Wilder, I. W. 1924. The relation of growth to metamorphosis in *Eurycea bislineata* (Green). J. Exp. Zool. *40*:1–112.

Zweifel, R. G. 1968. Reproductive biology of anurans of the arid southwest, with emphasis on adaptation of embryos to temperature. Bull. American Mus. Nat. Hist. *140*:1–64.

CHAPTER 12

REPRODUCTIVE ADAPTATIONS OF REPTILES

The breeding season of reptiles is primarily affected by temperature conditions and photoperiod and is influenced very little by seasonal fluctuations in precipitation. Outside of the tropics, reptiles tend to become reproductively active under conditions of increasing length of photoperiod, and breeding generally takes place over a relatively long period of time compared to the length of most amphibian breeding seasons. In the north-temperate areas, reptilian breeding seasons usually extend from April through August. Wade Fox (1958) studied the sexual cycle in the Carolina anole (*Anolis carolinensis*) and found that under normal conditions spermatogenesis occurred through the winter until August. In August, a drastic change in activity was observed; the testes collapsed and regressed, and all cell division in the testicular tissue ceased. Interstitial cells atrophied and the reproductive system went through a period of complete inactivity which lasted until October, when the cycle started over again; initiation of the new cycle was indicated by an increase in the size and activity of the testes. Spermatogenesis began later, under conditions of increasing photoperiod, but carried on into the summer when photoperiods were decreasing in length. By keeping the animals under artificial conditions with controlled photoperiod, Fox found that he could stimulate the gonads to begin spermatogenesis during any season of the year by increasing the day-

length to which the animals were exposed. He also found that the refractory period which normally occurred in August could be eliminated by not reducing daylengths. Consequently, both the initiation of the spermatogenesis and the initiation of the refractory period appear to be controlled by photoperiod. However, the effect of changes in daylength was not the same from one season to another, and Fox found that there was a progressive response from fall to winter to spring. The gradual reduction of daylength to six to nine hours per 24 hours during the fall and winter produced no response, but the same lengths of photoperiod in April stopped spermatogenesis. Fox also found a maximum length of photoperiod which would cause a response: there was an increasing rate of spermatogenesis when he exposed the animals to photoperiods of 14, 16, and 18 hours, but beyond 18 hours no further increase was obtained. In a subsequent study of the same species, Licht (1966) found that the photoperiodic response is very temperature-sensitive and that the body temperature of the lizard must reach the preferred level (about 32° C) during at least part of the daily light period in order for long daylengths to be effective in stimulating testicular recrudescence. Interestingly, high night temperatures and cool days (20° C) may even retard testicular growth more than do continuously low temperatures. Licht interpreted his results as suggesting that at least some of the "high" centers, such as those involving photoreception, neurosecretion, and gonadotropin synthesis and release, must attain a particular temperature level during at least part of the light period in order for the animal to respond effectively to the long daylength stimulus.

Photoperiod, acting alone or in conjunction with temperature, has been shown to also influence reproductive cycles in other reptilian species. Burger (1937) studied sexual periodicity in male pond sliders (*Pseudemys scripta elegans*) and found that, starting in November, exposing these turtles to artificially increased daylengths inhibited the spermatogenic cycle already in progress and induced the initiation of a new cycle. Galgono (1951) exposed the European lizard *Lacerta sicula* to various combinations of temperature and photoperiod and concluded that temperature might control spermatogenesis while photoperiod controls the glandular functioning of the testes. Bartholomew (1950, 1953) studied the reproductive cycles in the yucca night lizard (*Xantusia vigilis*), a nocturnal species, and found that increased daylength was more important than increased temperature in accelerating the growth of both testes and ovaries but that females were less responsive to photoperiodic changes than males. The photoperiodic response of females was insufficient to take the ovaries past the first stages of reproductive development and was independent of temperature; the rate of photoperiodic response of males varied directly with temperature. Baker (1947) found that the reproductive activity of two species of skinks (*Emoia cyanura* and *Emoia werneri*) living on the island of Espiritu Santo in the New Hebrides islands, where seasonal changes in temperature and rainfall are very slight and the difference between the longest and shortest days of the year is only 1 hour and 48 minutes, correlated with seasonal changes in photoperiod. Although both species breed throughout the year, they exhibit distinct peaks of oviposition and enlargement of testes during the period from September to December, the period of longest days, while the season of minimal oviposition and reduction of testes is from March to June, the period of shortest days. Tuataras deposit their eggs in the spring (October to December).

In view of the fact that the length of time a lizard is exposed to daylight in nature is governed by its temperature-regulating behavior, and therefore is temperature-dependent, it is puzzling that the reproductive cycle of these reptiles should be triggered by photoperiod. It would seem particularly perplexing that photoperiod should trigger the reproductive cycle of a nocturnal lizard like *Xantusia vigilis*. However, Miller (1951) observed that, although the greatest activity of this species in spring and summer is at night, during the winter months it feeds during the day. Importantly, this species is not a burrower and is undoubtedly exposed frequently to at least indirect light while it rests or when it moves about among the leaves and bark of the Joshua trees (*Yucca brevifolia*) which it inhabits. The importance of photoperiod seemingly varies considerably from species to species even outside of the tropics: when the Texas horned lizard (*Phrynosoma cornutum*) was exposed to continuous light and a temperature of 35° C for two months during the winter, it showed only slightly greater testicular development than normally occurred during hibernation in the same season (Mellish, 1936).

Although the skinks studied by Baker (1947) have a cyclic pattern of reproductive activity which seems to be responsive to very small changes in photoperiod, this is not true for all tropical lizards. Inger and Greenberg (1966) studied the annual reproductive patterns of four species of lizards (*Cyrtodactylus malayanus*, *Cyrtodactylus pubisulcus*, *Draco melanopogon*, and *Draco quinquefasciatus*) from the rain-forest habitat of Borneo and found no seasonal pattern of egg production and that males produce sperm throughout the year. These Bornean rain-forest lizards live in an area where the annual variation in daylength is only approximately 10 minutes and where climatic conditions are nearly constant throughout the year. Church (1962) studied three species of nocturnal house geckos (*Cosymbotus platyurus*, *Hemidactylus frenatus*, and *Peropus mutilatus*) at Bandung, Java, where the longest day of the year is only 45 minutes longer than the shortest. Although this region has a mild dry season, Church found that males contained sperm throughout the year and that the monthly frequency of gravid females varied but exhibited no regular cyclic pattern. The reproductive activity of tropical lizards living in regions where there are major seasonal fluctuations in precipitation may be affected by the severity of the dry season. Robertson *et al.* (1965) report that the African blue-bellied lizard (*Agama cyanogaster*) of southwestern Tanzania, where there is a severe dry season for six months of the year and an annual photoperiod variation of 55 minutes, deposits eggs only during the first half of the rainy season. The testes of the males undergo cyclic changes, enlarging during the end of the dry season and shrinking after the first half of the wet season. In the same locality the smooth skink (*Mabuya striata*), an ovoviviparous species, gives birth to the majority of its young during the dry season, the fewest being born in the middle of the wet season; however, the testes of the males only vary slightly in size throughout the year. Marshall and Hook (1960) studied the breeding biology of the common agama (*Agama agama lionotus*) in Kenya at a latitude of 0°01′ north, where there is very little variation in daylength but distinct rainy and drought seasons. They found that males contained sperm throughout the year but females contained oviducal eggs only during the period from June to September; this four-month period follows the season of heaviest precipitation. Chapman and Chapman (1964) studied the same species at a latitude of 5°36′ north near Accra, Ghana, where there are two rainy seasons per year, and found that there are two peaks of breeding activity in that region, one following each rainy season. By following periods of very heavy rainfull, the breeding season of the common agama coincides with seasonal abundance of food: this might be the environmental factor to which the reproductive cycle is adapted. The common agama is an omnivore which seasonally feeds almost as much on plants as on animals. However, following heavy rainfall there is a great increase in the reproduction of insects, so protein-rich beetles, caterpillars, and other insects are seasonally very abundant. Cott (1961) observed that in many localities in East Africa the Nile crocodile (*Crocodylus niloticus*) lays its eggs during the dry season and that the young hatch after the onset of rains as lakes and rivers are becoming flooded; this may also represent an adaptation to seasonally abundant food supplies.

The physiological mechanism by which environmental changes stimulate the gonads and initiate the breeding cycle is not clear. The pituitary gland is definitely an intermediary between the environment and the gonads; it seems probable that, as in mammals and birds, the hypothalamus is also involved. Bartholomew (1959) has suggested that the skull openings of lizards and snakes, like those of birds, may allow light to directly reach the hypothalamus or pituitary through the orbits or the otic regions. The relatively heavy and unfenestrated skull of turtles probably prevents light from reaching the hypothalamus-pituitary region; their eyes may be the light receptors involved in gonadal stimulation. The work of Stebbins and Eakin (1958) indicates that in lizards the parietal eye is involved in monitoring the duration of exposure to light, and therefore it seems likely that it may play an important role in the triggering of reproductive cycles. A suddenly improved nutrition and fat deposition may provide the stimulus triggering the reproductive cycle of reptiles such as the common agama.

COURTSHIP AND MATING BEHAVIOR OF REPTILES

The courtship behavior of most species of reptiles is poorly studied. Each undoubtedly possesses a characteristic pattern, since this functions importantly in species and sex recognition and in psychologically and physiologically preparing the participants for actual mating. In contrast to most amphibians, male reptiles tend to establish and defend territories during the breeding season, and there may be much rivalry between those occupying adjacent areas. Sound, which plays such an important role in the courtship behavior of anuran amphibians, is only involved in the mating of crocodilians and (possibly) some turtles. Sex recognition in most reptiles seems to be essentially visual, olfactory, and behavioral.

CROCODILIANS

Except for observations of the American alligator (*Alligator mississippiensis*) and the Nile crocodile (*Crocodylus niloticus*), little is known about the mating behavior and courtship of crocodilians. Crocodiles tend to be territorial, defending their basking sites and a surrounding area during both the breeding and non-breeding seasons, but males become particularly pugnacious during the breeding season and often attack one another viciously. There are several records of unprovoked Nile crocodiles attacking small boats; these attacks have been interpreted as acts of territorial defense. Bull crocodiles and alligators are very vocal during the breeding season, especially at night, and emit deep roars. Generally the sound produced by one individual will stimulate other males to roar, and entire swamplands may echo with the bellowing noise. Because an alligator will move aggressively toward any sound resembling that of another male, the roar has sometimes been interpreted as a challenge to rivals in the territorial defensive behavior of males. However, Cott (1961) found that Nile crocodiles make two kinds of sounds during the breeding season: an abrupt bark or cough and a deep rumbling roar. The bark or cough is often emitted while the animal is basking and may be a warning to other males. Cott interpreted the roar as a mating call functioning to attract females. Female crocodiles are generally not as vocal as males, but they do produce a low growling roar which some have interpreted as a mating call. Thus, it appears that sexually aroused males patrol their territories, driving off other males, and emit roars which function as mating calls to attract females. Once she is near the male's territory, the roar of the female may function to attract the male to her and to distinguish her from immature individuals or sexually inactive females.

The mating of crocodilians apparently always takes place in the water but has seldom been witnessed. Burrage (1965) was able to observe the mating of a pair of captive alligators and noted that the bull caressed the back of the female's head with his own head for several minutes. This seemed to psychologically prepare her for mating, since she then submerged her body and allowed the male to mount. He then depressed his tail and protruded his penis so that it entered her cloaca; copulation lasted for 15 minutes.

CHELONIANS

Turtles exhibit characteristic patterns of courtship behavior that are different for terrestrial as opposed to aquatic fresh-water and marine species. Although all turtles are oviparous and lay their eggs on land, fresh-water and marine species mate in the water and appear to rely almost entirely upon visual means for species and sex recognition. Red-eared pond sliders (*Pseudemys scripta elegans*) and painted turtles (*Chrysemys picta*) have in common an unusual courtship ritual: first, the male swims to a position in front of the female, where he turns and faces her (Fig. 12–1, top). He then maintains his position directly in front of her by swimming backward or forward, depending upon her movements, and periodically extends his forelimbs and uses his long claws to stroke her cheeks and chin with a vibratory motion. When the female is sufficiently excited, she sinks to the bottom and the male follows, swimming on to her back and gripping the rims of her carapace with his long claws. Copulation may last an hour or more. The Florida cooter (*Pseudemys floridana*) has a similar courtship pattern except that, rather than swimming backward in front of the female, the male faces the same direction as the female and swims just above her with his head pressed down close to hers (Fig. 12–1, bottom). He periodically attempts to turn his tail down and under the posterior rim of her carapace and uses his clawed front feet to titillate her cheeks as he swims above and behind her. When the female allows the male to position his tail under her carapace, he grasps her carapace with his claws and the two sink slowly to the bottom, where copulation occurs.

In contrast to other turtles, which usually mate in the spring or early summer and well before oviposition occurs, the mating of marine turtles takes place immediately after the female completes egg-laying. Male marine turtles congregate just offshore on beaches where the eggs are laid and intercept the females as they are heading back out to sea.

Figure 12–1 Courtship position in two species of slider turtles, illustrating species specific differences by which males "titillate" females. *Upper,* red-eared turtle (*Pseudemys scripta elegans*). *Lower,* Florida terrapin (*Pseudemys floridana floridana*). (From Oliver: *The Natural History of North American Amphibians and Reptiles.*)

Hawksbill (*Eretmochelys imbricata*) males have even been observed coming out of the water and pursuing females on the beach. The courtship of marine turtles has apparently not been recorded, but males have a single enlarged and curved claw on each flipper which must be used in grasping the female's carapace.

Terrestrial and semiaquatic turtles mate on land and utilize their vision and possibly olfaction in species and sex dis-crimination. In the spring, males seem to become restless and wander over larger areas than covered normally. They may be drawn to females by odors, or they may merely locate them visually. Auffenberg (1966) has noted that, as illustrated in Figure 12–2, sexually aroused male gopher tortoises (*Gopherus polyphemus*) begin walking in circles approximately 20 feet in diameter. Upon approaching any other tortoise in his vicinity, a male will stop

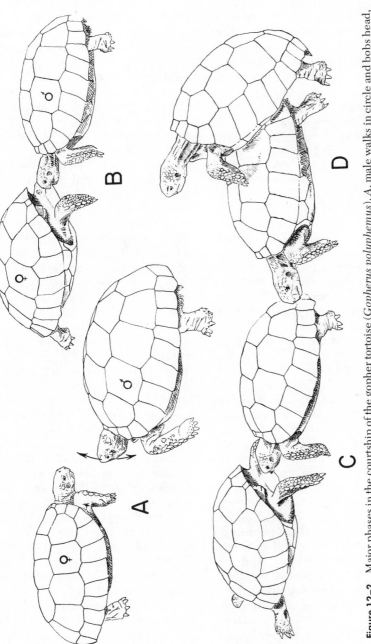
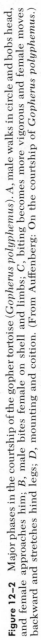

Figure 12–2 Major phases in the courtship of the gopher tortoise (*Gopherus polyphemus*). *A*, male walks in circle and bobs head, and female approaches him; *B*, male bites female on shell and limbs; *C*, biting becomes more vigorous and female moves backward and stretches hind legs; *D*, mounting and coition. (From Auffenberg: On the courtship of *Gopherus polyphemus*.)

and rapidly bob his head. During periods of head-bobbing, the male s swollen subdentary glands are everted; Auffenberg has suggested that the bobbing may function to waft scent through the air as well as to give an obvious visual signal. When the male is observed by an adult female, she approaches him at a quick walk and he in turn starts toward her, stopping occasionally to rapidly bob his head. The female remains still and does not bob her head. The male then bites the female on the legs, the anterior edge of the carapace, the head, and especially the gular projection. The female moves backward in a semicircle, stretching her hind legs. The male moves forward and resumes biting the female; this sequence is repeated several times, with the male attempting to mount at various times. The radius of the semicircle that the female traces as she backs away from the male becomes so small that she begins to pivot 180° around a central point. Eventually the male is successful in mounting the female, and coition occurs. In some species, such as the box tortoises (*Terrapene*), the male must assume an almost vertical position with the posterior margin of his shell resting on the ground and his hind feet gripping the female's shell in between the carapace and the plastron. The male of *Gopherus agassizii*, after positioning himself on top of the female, begins a series of movements in which the tail is thrust forward through his anal notch toward the cloaca of the female and the hind legs are drawn upward so that the posterior edge of his carapace strikes the ground. The vertical bobbing up and down accompanying his motions may be repeated several times. If the female tortoise withdraws into her shell while she is being courted, the male may lunge at her and butt her with his shell or crawl onto her carapace and thump her shell with his forefeet. Copulating male tortoises are frequently quite vocal, emitting noises ranging from grunts to cries and whistles. Copulating male gopher tortoises (*Gopherus agassizii*) are reported to stamp their hind feet on the ground.

LIZARDS

Lizards depend primarily upon visual clues in species and sex recognition. Typically, male lizards establish territories during the breeding season from which other males are excluded. The territory may consist of a single rail on a fence, the trunk of a tree, a particular rock, or a loosely defined area of ground surface. In any case, the territory will include a perch from which the male can make himself prominent. The establishment of territories by males is biologically important, for it tends to spread out the reproducing population rather than concentrate it in a small area, as happens with many amphibians. The size and nature of the lizard's territory is correlated with the nesting requirements of the species and with the degree of competition. In general, the greater the number of individuals competing for territories, the more limited will be the individual's territory. However, the will of an individual to defend an area increases as the territorial boundaries approach his nest and perch, although peripheral areas may be given up rather passively, increased aggressiveness in the defender insures that his territory will not shrink smaller than some critical size.

Sexual dimorphism is common among lizards, particularly those which are diurnally active, and in many species the males possess bright patches of coloration in the gular region, along the sides of the body, or elsewhere. In addition, many have conspicuous head crests, sail-like structures along the back, or large folds of skin in the gular region which, when erected, give the individual a particular silhouette that serves as a ready means of identification. Many species of lizards supplement their color patches and silhouettes with specific patterns of movement. With movement a lizard is more apt to call attention to his presence than if he remains stationary; and each species tends to have a particular rhythm of bobbing which is in itself a means of identification. All of the male lizard's visual clues serve both to warn male conspecifics that the territory is occupied and to identify himself to female conspecifics. The visual identification of a male by another male allows the owner of a territory to challenge intruders at a distance, and this conserves energy which would otherwise be wasted in fighting. Males will generally not attempt to overthrow the owner of a territory, and relatively few fights seem to occur once territories are established.

The courtship behavior patterns of several kinds of lizards have been observed. Physical contact, with or without vocal or chemical stimuli, plays an important role in the courtship of nocturnal lizards such as geckos (family Gekkonidae). Oliver (1955) has described the courtship of the banded gecko (*Coleonyx variegatus*) as follows: The male approaches the female with his tail waving and his head and body close to the ground. He investigates her by poking her with his nose or licking her and then uses his jaws to grasp her by the tail, leg, flank, shoulder, or neck. Holding her with his jaws, he jerkily pushes her forward in a strutting walk. If she does not resist his advances, he will release her and move jerkily forward toward her neck, frequently nipping her on the flank or shoulder as he does so. Once he has grasped her by the neck, copulation will usually follow, with the male retaining his grip on her neck until mating is completed.

The courtship of diurnally active lizards primarily involves visual stimulations. Although the patterns of some species typically include chasing and biting, frequently there is a minimum of physical contact between the male and female until mating actually commences. The courtship consists of two sets of interdependent reflexes, one exhibited by the male and the other by the female, through which the female encourages the male and he stimulates her. Rather than submitting to force, as with geckos, the female is generally persuaded to pair with the male by his displays and attentive behavior. A receptive female may even be the more aggressive member of a pair, approaching the male if he fails to move in her direction. The interrelationship of male and female behavior is well illustrated by the courtship of anoles (*Anolis carolinensis* and *Anolis sagrei*), which has been described by Evans (1938). Both males and females of these species have a bright orange-red or pink fan-shaped flap of skin, called a dewlap, which can be voluntarily extended; the dewlap plays an important role in both courtship and territorial defense. Evans found evidence that both females and males defend territories during the breeding season. If an intruder

enters the territory of a male, the latter will display his dewlap. If the intruder is another male, he will immediately retreat or, occasionally, display his own dewlap and attempt to bluff the owner. If the intruder is a female, she will react to the dewlap display by retreating or nodding her head; her head-movement causes the territorial defender to shift from a fighting to a courtship behavioral pattern. Upon receiving a head-nod from an intruding female, the male will repeatedly extend his dewlap and make a series of spasmodic "push-ups" while approaching the female, often turning sideways to her as he makes his displays. The female responds by nodding her head in reply during the momentary intervals that the male is resting. The male's subsequent pursuit of the female for a short distance is punctuated by intervals of dewlap display and "push-ups" on stiffened forelegs. This causes the female to retreat, but she stops a short distance in front of him, turns, and nods her head at him as he displays his dewlap. The male then approaches the female from her side and grasps her firmly by the neck with his jaws; the female responds by making a slow retreat and often arches her back, seemingly to facilitate his retaining the grip. The male then pushes his tail beneath hers, holds her body with one front foot and one hind leg, and inserts one hemipenis into her cloaca. The female often nods her head during copulation but appears to make no other movements; copulation lasts about 20 minutes. The territorial defense by a female is similar to that by a male. If an intruder enters the territory of a female, she will extend her dewlap. If the intruder is a male, he will respond by displaying his own dewlap and she immediately shifts from defensive to courtship behavior by nodding her head. Upon receiving this cue the intruding male shifts to his courtship behavior, and the interaction proceeds as just outlined.

Courtship patterns similar to that described for anoles have been observed for a variety of other lizards in the family Iguanidae and for those in Lacertidae, Agamidae, and Chamaeleontidae, with combinations of shape, color, and movement being used to communicate species and sex identities. Variations in the basic pattern occur through adaptations to the morphology and ecological niche of the species. Thus, with some species the courtship display takes the form of raising the body so as to expose the belly, in others the trunk is tilted to accomplish the same thing, and in still others the trunk is compressed, thereby exposing ventrolateral colorations. Some species curl their tails in display, while in others the tail is not involved. Male flying lizards (*Draco*) identify themselves by exhibiting their brightly colored wings and dewlap while running among branches or gliding through the air (Bellairs, 1970). Male rainbow lizards (*Agama agama*) challenge intruders and identify themselves to females by changing their head color from a drab brown to a bright orange, their trunk to a deep blue, and their tail to a bright pattern of blue and orange bands within a few minutes, all while performing a series of "push-ups" (Harris, 1964). Carpenter (1967) reports similar color changes for *Agama planiceps, Agama savignyi,* and *Agama agilis.* Head-bobbing by female lacertids (*Lacerta meliselensis, Lacerta agilis,* and *Lacerta viridis*) elicits courtship behavior from the male just as it does in anoles (Kitzler, 1941). Many iguanid males court females by approaching them with a series of fast short bobs of the head and anterior body, and each species appears to have a characteristic pattern of bobbing that varies from others' in duration of time and magnitude of each extension of the front legs. Hunsaker (1962) was able to demonstrate that female spiny lizards (*Sceloporus*) use this pattern of bobbing in choosing a partner and that males react aggressively to the bobbing of other males. The courtship of many lizards ends when the male grasps the female's neck in his jaws (the male grasps a nuchal horn in *Phrynosoma*, as observed by Lynn, 1963) and brings the base of his tail up close to hers so that the cloacas are adjacent; a hemipenis is then inserted.

SNAKES

Sex and species recognition in snakes appears to primarily involve vision and olfaction. Sexually active males are attracted by the movements of other individuals, and the odor and behavior of a receptive female indicate her sex. Chemical perception is important in the sexual stimulation of males and is also a principal means by which males track and locate females. Noble (1937) demonstrated that blockage of the nostrils or inactivation of Jacobson's organ by cutting off of the tips of the tongue generally stops all male courtship behavior. He also showed that a male snake will readily follow a scent trail left across a plate of glass by rubbing it with skin from the back of a female. Other studies have indicated that female cloacal gland secretions may be involved in the tracking and courtship of some species of snakes. Many snakes inhabiting temperate and subpolar regions hibernate in communal dens, breed as they emerge in the spring, and consequently have little difficulty in finding mates. However, most active snakes are solitary, and the problem of finding a mate is potentially more serious in species which are either non-gregarious hibernators or don't hibernate at all.

The basic pattern of courtship behavior is similar in all species of snakes which have been studied and is based on tactile stimulation of the female and olfactory stimulation of the male. Upon locating a female, the male snake will commonly wave his tail slowly back and forth. Following her about, he will approach her from the rear and run his chin and neck over her body while flicking his tongue rapidly in and out, presumably to perceive her odors. Typically, the female reacts to the initial advances of the male by moving away. The male attempts to remain at her side and, as a result, the two snakes move about with their bodies parallel to one another while the male continues to rub his chin along her back. Gradually the male moves his chin anteriorly along the female's back until he reaches her neck; by this time, the anterior part of his body is lying on top of hers. The male now attempts to wind the posterior part of his body around the female's. Frequently, this is accomplished by intertwining his tail about hers and using it to lift the posterior part of her body. This brings the cloacal regions of the two snakes into contact with one another. After reaching this position, the males of many colubrid species (garter snakes in the genus *Thamnophis* and brown snakes in the genus *Storeria*, for example) press their chin on the female's neck and exhibit caudocephalic waves, series of muscle contractions passing from posterior parts of the body to the head. The undulations may continue for 10 minutes or more and result in tactile stimulation of many parts of the female's body. The rubbing of the male's chin on the female's neck and back presumably serves to stimulate his chin-scale organs: if these are artificially covered the courtship behavior of the male will not develop. In some colubrid species, including bull snakes (*Pituophis melanoleucus*) and king snakes (*Lampro-*

peltus getulus), there is relatively little entwining of the bodies and sometimes no caudocephalic undulation, but the male usually bites the female on the back and neck and this may serve to stimulate her. In rattlesnakes (*Crotalus*, family Viperidae), the comparable stimulation of the female occurs as the male thrusts and jerks rather than undulates his body. Paired water moccasins (*Agkistrodon piscivorus*, family Viperidae) entwine the posterior parts of their bodies and then push and rub their heads together. Male boas and pythons (family Boidae) rub their rudimentary hind limbs against

Figure 12-3 *A*, male Bengal monitors (*Varanus bengalensis*) in ritualistic combat; *B*, combat posture of male rattlesnakes (*Crotalus*); *C*, lyre-shaped combat posture of male Aesculapian snakes (*Elaphe longissima*), which possibly provided the model for the caduceus symbol of medicine; *D*, typical combat posture of many colubrid male snakes; *E*, courtship position of king cobra (*Ophiophagus hannah*). (From Bellairs: *The Life of Reptiles; A*, after photograph by R. Y. Deraniyagala; *B, C*, and *D*, after Bogert and Roth; *E*, after photograph by J. A. Oliver.)

the sides of the female to stimulate her; this scratching makes a rasping noise which may be heard several feet away. Regardless of its mode, all of this stimulation of the female serves to persuade her to allow the male to insert his hemipenis. Copulation of snakes may last for an hour or more.

There have been a number of descriptions of "courtship dances" of pit vipers and other snakes. The participants elevate and entwine their foreparts, sometimes erecting one-third or more of the body, and then repeatedly attempt to throw one another to the ground until one becomes exhausted and glides away. In some species, such as water moccasins, there may be such an aspect in the courtship behavior, but in many cases which have been investigated it has been found that both participants were males! Consequently, these "courtship dances" are now interpreted as fighting rather than courting by the participants. Insufficient study has been made of snake behavior to indicate whether or not any are territorial, but it seems reasonable to assume that at least some combats may be over territories. Males of the European snake *Elaphe longissima* assume a lyre-shaped posture while dueling and probably formed the model for the emblem of Aesculapius, the Greek god of healing, which has become the badge of the medical profession (Fig. 12–3). It is perhaps more than coincidental that male monitor lizards (*Varanus*), generally thought to be closely related to snakes, also engage in shoving matches in which rivals grip one another with their front legs and attempt to push the other over.

TUATARAS

The mating of tuataras has not been noted, and nothing is known of their courtship behavior. Males lack a copulatory organ; consequently, insemination must occur through the simple apposition of male and female cloacas.

SPERM STORAGE

It is known that viable sperm can be stored in the female reproductive tracts of a variety of turtles, lizards, and snakes and that successful fertilization can be accomplished by such sperm after remarkably long periods of storage ("delayed fertilization"). Isolated captive individuals have provided records of successful fertilization of eggs by sperm stored for as long as four years in turtles (diamondback terrapin, *Malaclemys terrapin,* and box turtle, *Terrapene carolina*), seven months in lizards (dwarf chameleon, *Microsaura pumila pumila*), and six years in snakes (cat-eyed snake, *Leptodeira annulata polysticta*); neither crocodilians nor the tuatara is known to possess a sperm storage ability. The most common situation is probably for sperm to be stored for several months, such as over winter, and one must consider the possibility of parthenogenesis in instances where fertilization appears to have occurred after several years of isolation of the female (Bellairs, 1970). Nevertheless, sperm storage can effectively lengthen the period of time during which eggs can be fertilized and, perhaps more importantly, allows the female to produce more than one clutch of fertilized eggs after a single insemination. Because of the latter advantage, sperm storage would seem to be an adaptation allowing these reptiles to take advantage of opportunistic meetings of members of the opposite sex and eliminating the necessity for mating congregations and closely coordinated spermatogenetic and oogenetic cycles. Because it allows a female to contribute to the fecundity of her population without mating every breeding season, sperm storage is a feature which would tend to maintain a high level of reproduction in a population regardless of the sex ratio or degree to which individuals are scattered. Thus, it may be a significant factor in the survival of populations living under marginal environmental conditions. The following table on the opposite page gives the records of sperm survival within the female reproductive tract for several different reptiles.

Fox (1956, 1963) studied the morphology of the seminal receptacles of garter snakes (*Thamnophis sirtalis* and *Thamnophis elegans terrestris*) and of the green anole (*Anolis carolinensis*) and the seasonal distribution of sperm in their oviducts. The oviduct consists in lizards and snakes of three major regions: posteriorly, there is a short vagina; the middle segment

DURATION OF SPERM VIABILITY IN FEMALE REPRODUCTIVE TRACT

Species	Maximum Known Period of Sperm Viability[1]
Turtles	
Diamondback terrapin (*Malaclemys terrapin*)	4 years
Common box turtle (*Terrapene carolina*)	4 years
Lizards	
Dwarf chameleon (*Microsaura pumila pumila*)	7 months
Green anole (*Anolis carolinensis*)	6 months
Side-blotched lizard (*Uta stansburiana*)	81 days
Snakes	
Viperids	
European asp (*Vipera aspis*)	over winter
African night adder (*Causus rhombeatus*)	5 months
Prairie rattlesnake (*Crotalus viridis viridis*)	over winter
Colubrids	
Lined snake (*Tropidoclonion lineatum*)	over winter
European smooth snake (*Coronella austriaca*)	over winter
Common grass snake (*Natrix natrix*)	over winter
Natrix subminiata	5 months
Natrix vittata	18 months
Brown snake (*Storeria dekayi*)	4 months
Indigo snake (*Drymarchon corais couperi*)	54 months
Cat-eyed snake (*Leptodeira annulata polysticta*)	6 years
Boiga multimaculata	1 year

[1]Based on evidence of viable sperm in the female reproductive tract or on the production of fertilized eggs by isolated females.

is a long uterus; and the anterior portion consists of the infundibulum (the fimbriated opening of the oviduct) and a funnel-shaped transitional portion called the tube, which connects the infundibulum to the thicker uterus. In the garter snakes, there is a histologically distinct region between the uterus and the infundibulum which contains modified alveolar glands. These glands are simple branched or primitive compound glands connected to the lumen of the oviduct by ciliated ducts and serve as the seminal receptacles. The receptacles occur in clumps, each of which is bound together by a sheath of connective tissue. The clumps are most abundant in the corners of the folds in the oviduct. The cavities of the individual receptacles range from 10 to 20 microns when empty but bulge to a diameter of 20 to 50 microns when filled with sperm. Their location at the anterior end of the oviduct provides a reservoir for sperm that is positioned so as to enable fertilization of the eggs almost immediately after they enter the oviduct. The receptacles may be physiologically important to the nutrition and chemical arrest of sperm activity. The

former point is supported by histological evidence that the cytoplasm of the glandular walls diminishes in the presence of the sperm, especially where it is in contact with the sperm heads. Furthermore, the receptacles provide a refuge for the sperm and prevent them from being swept down the oviduct as the eggs pass posteriorly. As proposed by Fox (1956), it appears that fertilization would occur as a result of the passing eggs stretching the oviduct wall and thus exerting pressure on the receptacles and forcing sperm out into the lumen of the oviduct at the time when they are most needed for fertilization. Each egg would be expected to force out a certain proportion of the stored sperm, and the supply would gradually be depleted, explaining the incomplete fertility within clutches of eggs that has been reported for a variety of snakes. Interestingly, the common garter snake (*Thamnophis sirtalis*) frequently mates in the fall; when this happens, the sperm spend a winter in the posterior two-thirds of the oviduct and do not enter the seminal receptacles until the following February or March. Ludwig and Rahn (1943) report that sperm deposited in fall copulations of the prairie rattlesnake (*Crotalus viridis viridis*) also are retained in the posterior end of the oviduct through January. In contrast is the coast garter snake (*Thamnophis elegans terrestris*), exclusively a spring-breeder: its sperm appear to move directly into the receptacles following copulation, since they have been found in the receptacles in April.

Fox (1963) found that the seminal receptacles of the green anole differ from those of snakes but are similar to those described for chameleons (Saint Girons, 1962). The mucosa of the anterior two-thirds of the vaginal portion of the lizard's oviduct is arranged in longitudinal folds. Small tubules extend deep into the thick lamina propria from the bottom of the grooves between these folds and run parallel to the folds, forming the sperm receptacles. Since fertilization generally occurs in the anterior portion of the oviduct, the sperm receptacles of these lizards would seem to be less advantageously located than those of snakes, which are located near the infundibulum. Fox suggests that possibly at ovulation or when an egg enters the uterus of the lizard, the vaginal tube is stimulated and sperm are forced from the tubules into the lumen of the vagina in time to reach the eggs before the shell membranes are deposited. At any rate, with time since the last insemination there is a depletion of sperm in these receptacles. Fox found the receptacles of females in nature to be bulging with sperm throughout the summer and concluded that spring eggs may be fertilized by sperm from copulations of the previous summer.

Cuellar (1966) made a gross and histological examination of the oviducts of 11 species of iguanid lizards and two geckos. He found that the seminal receptacles of the iguanids occur principally in the anterior segment of the vagina, as described by Fox for the green anole, but that the receptacles of the geckos (*Phyllodactylus homolepidurus* and *Coleonyx variegatus*) appear to be confined to the tube region between the uterus and the infundibulum. The latter are thus more comparable in position to those of snakes. It is noteworthy that seminal receptacles are absent from many, if not all, species of whiptails and racerunners in the genus

Cnemidophorus, a genus containing parthenogenetic unisexual species as well as bisexual species (Cuellar, 1968). There is evidence that bisexual species of the genus *Cnemidophorus* have skewed sex ratios favoring males, and it is possible that an excess of males might function to insure a maximum fecundity in these species just as do sperm receptacles in other reptiles (Cuellar, 1968).

PARTHENOGENESIS

There is indisputable evidence that a polymorphic group of Caucasian rock lizards related to *Lacerta saxicola*, including *L. armeniaca, L. dahli,* and *L. rostombekovi* (family Lacertidae), have populations consisting entirely of females which reproduce by parthenogenesis. That is, reproduction occurs by means of unfertilized eggs (Darevsky, 1966). Cytogenetic studies have shown that the parthenogenetic females of these rock lizards have the diploid number of chromosomes (38), like those which reproduce bisexually, and that, at least in *Lacerta armenica*, two meiotic divisions occur in oogenesis. Any of several mechanisms could result in the production of diploidy in parthenogenetic individuals. Darevsky (1966) suggested that it may be the result of suppression of the second meiotic division. Uzzell (1970) notes that suppression of the second meiotic division would lead rapidly to homozygosity in the population and that this would be disadvantageous to the survival of the lizards because it would result in the expression of recessive lethal alleles and reduced genetic variability for evolution. Uzzell therefore feels that the diploid condition is more likely the result of a premeiotic mitosis without cytokinesis followed by normal meiosis. The latter mechanism would preserve heterozygosity in the population. In either case, the ova mature into diploid egg cells which develop parthenogenetically into lizards that have the diploid number of chromosomes. Normally, the individuals produced in this manner are females, but in *Lacerta dahli* and *L. rostombekovi* male embryos form occasionally. These males generally develop abnormally and die before hatching, but Darevsky (1966: 133) reports records of rare males and speculates that these "are capable of productive mating with females". Other than an occasional male or hermaphrodite, the sexually mature individuals are female. Parthenogenetic females sometimes hybridize with males of bisexual species, and the progeny of such cross-matings are sterile females with rudimentary ovaries and oviducts. These sterile female hybrids have been found to be triploids, with 57 chromosomes; triploidy is the result of fertilization of a diploid egg (having 38 chromosomes) by a normal (haploid) sperm (having 19 chromosomes).

Several forms of whiptails (*Cnemidophorus*, family Teiidae) are also parthenogenetic. Minton (1958) and Tinkle (1959) first noted the absence of records for male specimens of the whiptail *Cnemidophorus tesselatus*. Maslin (1962) later reported that western populations of *Cnemidophorus uniparens* (formerly confused with *C. inornatus*), the subspecies *C. cozumela cozumela* and *C. c. rodecki* (both formerly *C. dippei cozumela*), and all subspecies of the species *C. velox, C. exsanguis* (formerly *C. sacki exsanguis*), *C. tesselatus*, and *C. neomexicanus* (formerly *C. perplexus*) seemed to be predominantly or entirely female, suggesting the possibility of thelytoky (the production of females by parthenogenetic females). He recently reared hatchlings of *C. neomexicanus, C. tesselatus*, and *C. uniparens* to maturity and, in the absence of males, obtained fertile eggs from them (Maslin, 1971). This is conclusive proof of parthenogenesis in these three species. Considerable interest has been generated in uniparental whiptails, and additional taxa have been discovered. Lowe and Wright (1964) applied the names *Cnemidophorus sonorae* and *C. flagellicaudus* to two apparently all-female taxa. Wright (1967) named still another uniparental species, *Cnemidophorus opatae*; and Fritts (1969) named another subspecies, *Cnemidophorus cozumela maslini*, at the same time giving *C. rodecki* specific status. Thus, at least 10 currently recognized species of *Cnemidophorus* are known or strongly suspected to be parthenogenetic. Another teiid genus, *Gymnophthalmus*, appears to include parthenogenetic forms (Thomas, 1965). *Hemidactylus garnotii* (family Gekkonidae), known only from females

and probably triploid (Kluge and Eckardt, 1969), appears to be parthenogenetic. W. P. Hall (unpublished) has found *Leiolepis belliana* (family Agamidae) to be triploid, indicating it is a parthenoform. Finally, Panamanian populations of *Lepidophyma flavimaculatum* (family Xantusidae) appear to be all-female (Telford and Campbell, 1970). This diversity of taxa indicates that parthenogenesis may be a more widespread form of reproduction in lizards, and possibly in reptiles in general, than has been realized; additional research will undoubtedly reveal many more uniparental species.

With very few exceptions, unisexual populations of vertebrates seem to arise through hybridization between distinct bisexual species. Morphological, geographical, karyotypic, and electrophoretic studies of parthenogenetic whiptails have both revealed a hybrid origin for many parthenogens and identified the parental species which produced them (Pennock, 1965; Lowe and Wright, 1966a, 1966b; Neaves, 1969; Neaves and Gerald, 1968, 1969). *Cnemidophorus tesselatus* is an all-female species composed of at least six defined morphotypes designated by the letters A through F (Zweifel, 1965). Morphotypes C through F of these are diploid and resulted from hybridization of the bisexual *Cnemidophorus septemvittatus* and *C. tigris*. Morphotypes A and B are triploids which resulted from the fertilization of the eggs of diploid *C. tesselatus* by sperm from *C. sexlineatus*. *Cnemidophorus uniparens* is a parthenogenetic triploid species which arose through hybridization of the bisexual *Cnemidophorus inornatus* and *C. velox* followed by backcrossing with *C. inornatus* (Neaves, 1969). *Cnemidophorus neomexicanus* is a diploid from the hybridization of *C. inornatus* and *C. tigris*. *Cnemidophorus exsanguis* appears to have triploid heterozygosity (the combination of three different haploid sets of chromosomes) and this may have arisen through hybridization of *C. gularis, C. sexlineatus*, and *C. inornatus* or through hybridization of other species which are at present allopatric to the south (Neaves, 1969).

The ability to reproduce parthenogenetically has played an important role in the speciation of *Cnemidophorus*, for without this ability it is unlikely that either allodiploid or allotriploid populations could have been perpetuated. Cuellar (1971) has shown that in *Cnemidophorus uniparens* the mechanism of chromosomal restitution is as proposed by Uzzell (1970) for the rock lizards: there is a premeiotic doubling of the chromosomes followed by normal meiosis.

In general, the competitive success of the parthenogenetic species lies in their selective advantage over parental species in "hybrid habitats." For example, the bisexual *C. inornatus* is a grassland species, but in many areas of southwestern United States the grassland habitats are being replaced by desert-grassland habitats because of the influence of man and his livestock. The parthenogenetic *C. uniparens* is better adapted to the desert-grassland hybrid habitat than is *C. inornatus* and seems to be replacing it in many areas. The reader will find an interesting discussion of the ecological distribution of parthenogenetic *Cnemidophorus* in Wright and Lowe (1968).

PARENTAL CARE AND NUMBERS OF EGGS PRODUCED

Fertilization is internal in all reptiles and some species are viviparous. However, most reptilian species are oviparous, and their eggs are adapted for development on land; in fact, they will not develop if immersed in water. As mentioned in Chapter Two, these eggs have an outer protective shell and three internal membranes, the amnion, chorion, and allantois, which retard the outward diffusion of water but still allow embryonic respiration to occur. In snakes, some turtles, and some lizards, the protective shell is leathery and pliable; in crocodilians, other turtles, and other lizards, it is calcareous and rigid. Examples of the variation in size and shape among reptilian eggs are illustrated in Figure 12–4.

Because terrestrial eggs may be dispersed and more easily hidden, they generally suffer less predation than eggs in an aquatic environment. Correlated with this, reptiles generally produce significantly fewer eggs per female than do amphibians. However, there is still consider-able variation in numbers of eggs produced by different reptilian species. Differences in this regard tend to correlate with the duration of time that the developing embryo is retained within the female's reproductive tract and with the amount of protection provided by parental care after deposition of developing eggs or birth of hatched young. Within species, as with amphibians, the number of eggs produced by a given female varies proportionately with age and size. Although most reptiles breed once a year and deposit all the eggs in a single clutch, local populations of some species may breed several times a year and two or more clutches of eggs may be deposited at different times throughout an extended breeding season. Frequently, the populations inhabiting warm regions will regularly deposit several clutches of eggs per year, whereas other populations belonging to the same species but inhabiting cooler regions will regularly produce only a single clutch per year. In extreme situations, females belonging to populations inhabiting cold regions where annual periods of activity are short may only breed every other year. Thus, both the number of eggs per clutch and the number of clutches per year per female vary from species to species and from population to population within some species.

CROCODILIANS

All crocodilians are oviparous, and females construct nests for their eggs. American and Nile crocodiles (*Crocodylus acutus* and *C. niloticus*) make their nests by digging a pit in sandy or pebbly soil. These nests are generally located near trees or other sources of shade and may be 200 yards or more from the water's edge. After digging the pit to a depth of about two feet, the female deposits her eggs in the bottom and then covers them over. In regions where they are not molested by man, large numbers of crocodiles may dig their nests in the same general area, but this is not common today. The American alligator (*Alligator mississippiensis*), estuarine crocodile (*Crocodylus porosus*), and some of the caimans construct much more elaborate nests out of vegetation mixed with mud. The Amer-

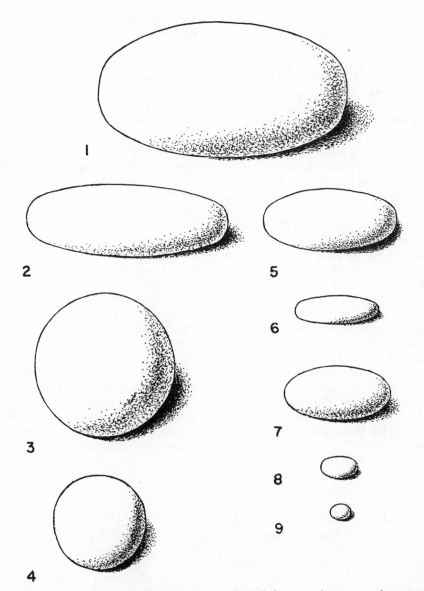

Figure 12–4 Representative eggs of North American reptiles, all drawn to the same scale. *1*, American alligator (*Alligator mississippiensis*); *2*, pilot black snake (*Elaphe obsoleta obsoleta*); *3*, Atlantic loggerhead turtle (*Caretta caretta caretta*); *4*, snapping turtle (*Chelydra serpentina serpentina*); *5*, eastern milk snake (*Lampropeltis triangulum triangulum*); *6*, eastern ring-necked snake (*Diadophis punctatus edwardsi*); *7*, eastern painted turtle (*Chrysemys picta picta*); *8*, Texas horned lizard (*Phrynosoma cornutum*); *9*, reef gecko (*Sphaerodactylus notatus*). (From Oliver: *Natural History of North American Amphibians and Reptiles*.)

ican alligator begins construction of her nest by using her jaws, body, and tail to clear and trample the vegetation in an area about eight feet in diameter. She then piles all of the pieces of loose vegetation into a compact mound which may reach over six feet in diameter and three feet in height. Climbing on top of the mound, she revolves about and uses her hind feet to scoop out a hollow in the center of the mound. Then, using her jaws, she collects and carries mud and more plant debris to the mound and uses this mixture to fill the hollow she just made in it. In this mud-vegetation plug a second hollow is excavated, and the eggs are deposited in it. After oviposition, the female uses material from the edge of the nest, additional mud, and aquatic vegetation, all carried in her mouth, to cover the eggs. Finally, she shapes and packs the nest by crawling back and forth over it until it is smooth and conical-shaped. This entire procedure may extend over a period of two to three days. In addition to providing some protection against predation, the alligator's nest automatically incubates the eggs, since fermentation of the nest material generates heat and warms the eggs. The differences between the shaded sand-pit nest of Nile crocodiles and the fermenting nests of American alligators are undoubtedly correlated with the temperature requirements of the embryonic stages of development and differences in habitat temperatures.

Although the American crocodile is reported to leave her nest unprotected, females are known to return to their nests once or twice during the incubation period; this indicates some sort of parental attachment to the developing young. Female Nile crocodiles and alligators remain in the vicinity of their nests for the duration of the incubation period and may even sprawl right on top of them. There is disagreement as to whether this constitutes active or passive guarding of the nest, but the mere presence of such a large reptile certainly must discourage predators from bothering the nest. Cott (1961) believes that female Nile crocodiles only passively guard their nests, but his observations indicate that this is effective, since monitor lizards and other predators quickly rob the nest if the female leaves it for even a short period of time. The adult female Nile crocodile must also be in the vicinity of the nest when the young are hatching in order to free them from the nest. The foot or more of sand covering the eggs is baked by the sun and may become so crusted that it is impossible for the hatchlings to force their way through it. As they hatch, the young Nile crocodiles make a croaking sound, and this signals the mother to wriggle about on top of the nest and help push the sand to one side. After releasing the young from the nest, the adult female may continue to protect them for varying periods of time. Female Nile crocodiles have been reported to escort their young to the water and lead them around in it like a duck (Bellairs, 1970). American alligators exhibit a similar pattern of behavior. During the incubation period, the female alligator stays near the nest, and when the young hatch, their high-pitched noises stimulate her to help remove the nest material covering them. After the young emerge from their nest, the female may even scoop out a shallow pool for them and protect them from predators for a year or more (Goin and Goin, 1962). Despite the protection provided young crocodilians, many succumb to predation; and the number of eggs per clutch, although very low compared to that of many amphibians, is relatively high for reptiles. American alligators deposit between 15 and 88 eggs per clutch (Oliver, 1955); clutches of Nile crocodiles contain between 25 and 95 eggs, the average being about 60 (Cott, 1961).

TURTLES

Like crocodilians, all turtles are oviparous, and females always bury their eggs in the ground or in piles of plant debris; this may be the only time that highly aquatic marine turtles come to shore as adults. Some turtle species, particularly marine turtles and fresh-water aquatic turtles, congregate to lay their eggs. Communal egg-laying apparently results from a particular area being especially favorable for egg deposition and development, for there is no evidence of cooperation between individuals in the preparation of nests. Thus, certain beaches will be used as nesting areas by large numbers of marine turtles; large numbers of female pond turtles or sliders will deposit their eggs on a particular sandbar or bank; several musk turtles may hide their eggs in the same muskrat mound. The preparation of nests is similar in all turtles that dig pits. The female deliberately selects a site, generally where there is moist sand or dirt, and uses her four limbs to dig a body pit somewhat larger than her own dimensions (Fig. 12–5). This pit serves the function of concealing the female while she completes preparation of the nest and deposits her eggs; this is particularly important for the large marine turtle laying eggs on open expanses of beach where there is no cover. Lying in the body pit, the female next uses her hind limbs to dig a small flask-shaped egg chamber in the floor of the pit. The neck of the egg chamber will be as long as the female's hind limb and, thus, the depth of the chamber will vary from species to species and from individ-

ual to individual. After completing the excavation, the female will lower her cloaca over the opening and rapidly deposit her eggs, singly or two at a time, until the entire clutch is laid. Often a clutch of eggs will completely fill the egg chamber (Fig. 12–6). The female then carefully fills in the egg chamber with her hind limbs, often stopping once in a while to moisten the nest with her bladder water. Bellairs (1970) reports that the European terrapin (*Emys orbicularis*) will even interrupt her egg-laying in hot weather in order to return to water to drink and replenish her supply of bladder water. The habit of using bladder water to moisten the nest may play a significant role in preventing the desiccation of embryos. After the egg chamber is carefully covered, the female will use all four limbs to fill in the body pit and obscure the location of the nest; the pains taken here also have an obvious survival value. Bellairs

Figure 12–5   Green turtle (*Chelonia mydas*) digging body pit. (From Bellairs: *The Life of Reptiles.*)

Figure 12-6 Green Turtle (*Chelonia mydas*) depositing eggs. The egg chamber is nearly full and contained 110 eggs when the photograph was taken. (From Bellairs: *The Life of Reptiles.*)

indicates that the marine green turtle (*Chelonia mydas*) may complete this entire process in less than two hours. The digging of the nest may take about an hour, the laying of eggs (about 100 per clutch in this species) about 20 minutes, and the filling in and concealing of the nest about half an hour.

The only parental care exhibited by turtles is that of carefully selecting the nesting site, preparing the nest, and carefully covering and concealing the location of the nest. Once a female has concealed her nest, she apparently loses all interest in her young, for the nest is left unprotected and no parental care is given to hatchlings. The number of eggs per clutch varies among turtles from one egg for the African tortoise *Malacochersus tornieri* to about 200 for the marine green turtle. Marine turtles are particularly dependent upon the production of large numbers of eggs for their survival, and all have large

clutches. In addition to the 100 to 200 eggs per clutch recorded for green turtles, Atlantic loggerhead turtles (*Caretta caretta*) are known to deposit 70 to 130 eggs per clutch, and leatherback turtles (*Dermochelys coriacea*), 90 to 130 per clutch. Even though individual females appear to only nest every two or three years, the numbers of eggs per clutch laid by these species are high for reptiles, particularly when one considers that green turtles are known to lay at least seven times in a year, that loggerhead turtles are thought to lay two to three clutches in a year, and that Pacific leatherback turtles may produce three or four clutches in a season (Carr, 1952). These high egg productions are obviously an adaptation that balances the tremendous mortality suffered by pre-adult stages and places a premium on the survival of sexually mature adults. As noted by Bellairs (1970), the survival of green turtles is in much greater danger in the West Indies, where man harvests adults for food, than in Malaya, where Muslims object to eating turtle meat and use only the eggs for food. Although the fresh-water snapping turtle of North America (*Chelydra serpentina*) is known to lay clutches of up to 80 eggs, most other fresh-water turtles and land tortoises produce clutches of fewer than 20 eggs. Some species, at least locally, lay more than one clutch a year. For example, the common box turtle (*Terrapene carolina*) generally produces a single clutch of eggs per year in the northern part of its range (New York) but may produce as many as four clutches per year in the southern part of its range (Florida). Snapping turtles (*Chelydra serpentina*), stinkpot turtles (*Sternothaerus odoratus*), and diamondback terrapins (*Malaclemmys terrapin*) are known to deposit at least two clutches per year.

LIZARDS

Reproduction by lizards varies from what is truly oviparity, through all degrees of ovoviviparity, to special cases of viviparity. In other words, some female lizards deposit their eggs within a short period of time after they are fertilized, others retain them in the reproductive tract for varying degrees of development through hatching, and a few, including some skinks (family Scincidae) and night lizards (family Xantusidae), supplement the nourishment in the yolk of the egg with nourishment provided from the maternal blood stream through a placental type of circulatory connection between the embryo and the oviduct. Because of the variability in duration of time that developing eggs are retained and the variability in embryonic dependence upon a maternal energy supply, the terms oviparous, ovoviviparous, and viviparous are not particularly meaningful when applied to lizards. However, eggs that are retained through completion of prehatching development lack an outer shell; in such instances, hatching may take place in the oviduct or shortly after the eggs are deposited. Females of many lizard species normally lay their eggs within a few days or weeks after mating; this length of time varies with whether or not there is sperm storage and how long the eggs are retained after fertilization. Studies such as that of Blair (1960) on the rusty lizard (*Sceloporus olivaceus*) indicate that it is common for a female to deposit her eggs at a site outside of her normal home range. Blair found, in fact, that this occurred more often than not. The reasons for making an extended trip to lay eggs are not clear. In some instances, such a move is probably necessary to find a suitable laying site, but in many cases this is not true. The habit may be of selective value in removing the female from the attention of potential predators that know her habits and habitat at a time when she would be particularly vulnerable, when her attention is focused on nest-making and egg-laying. The move may also be of value in insuring that the young hatch away from the mother's home range and, thus, will not compete with her. However, it does not seem logical that competition for food would be a factor here, since the food habits of young lizards tend to be different from those of adults and there often is virtually no overlap in demands on the food supply. Furthermore, young lizards soon disperse from the nesting site. Another possible explanation for the selection of a distant site is that it contributes to the dispersal of individuals and, consequently, of hereditary characters through the population; nevertheless, the movements of the hatchling lizards will undoubtedly be the most important factor in this regard.

The nesting site chosen by the female depends, of course, upon her species. Some deposit their eggs under rocks or debris, some in burrows, and some in pits they have excavated. Species or populations inhabiting areas subject to sudden and heavy rainstorms, such as southwestern United States, usually nest in an upland situation and on a slope where the eggs will not be flooded or washed away. Some lizards seem to select sites which are exposed to sunlight throughout the day, but sometimes individuals belonging to the same species will have nests in shaded areas. Consequently, it appears that exposure to the sun's radiation is not always a significant factor determining nest location. Lizards which dig nests may or may not show preferences for particular types of soil. However, Blair (1960) did observe that lizards attempting to dig in hard clay would sometimes start two or three nests, give up, and move to another site because of the difficulty of digging. The shape of the hole and its depth vary from one species of lizard to another. The rusty lizard digs a hole which is 50 to 75 millimeters in diameter at the surface, tapers to a smaller diameter at the bottom, is about 100 millimeters in depth, and descends from the surface at an angle of about 45° (Blair, 1960). The lizard digs this hole using its legs much as a dog would, using the forefeet one at a time to throw dirt back under the lizard and then the hind feet to push it farther back. After completing her digging, the female backs into the hole with her tail curled over her back and deposits her eggs, moving forward as the eggs accumulate behind her. After she has deposited the entire clutch, the female rapidly throws dirt into the nest until the hole is filled and then carefully smoothes the surface and scatters debris over it so that the location of the nest is visually indiscernible.

The two-and-a-half-foot Bengal monitor (*Varanus bengalensis*) makes a deep goblet-shaped egg chamber resembling that of turtles. Deraniyagala (1958) has published an extensive report on the nesting behavior of a female of this species that he observed excavating a hole in a flower bed at Akurassa. Digging in much the same manner as a rusty lizard, the monitor took two hours to dig a pit about two feet deep. She then reversed her position so that her hind limbs were within the pit and her head and forelimbs were at the surface. At times she placed one cheek upon the ground and at times closed both eyes as she deposited her eggs. After two and one-half hours, she emerged, rested briefly, and then crept back head first and began to cover the eggs, using her snout and front legs to scrape sand down from the sides of the pit. Ten minutes later, she emerged again and dug at the sides of the entrance, occasionally stopping to creep into the pit and push earth down over the eggs with her snout. She next walked in circles around the pit, breaking in the rim of the entrance and tumbling this earth into the pit and tamping it down with her snout. After 10 minutes of this, the eggs were no longer visible, but the hole was still deep enough to hide the head, neck, and shoulders of the lizard when she crawled in. Continu-

ing to dig with outward horizontal sweeps of her forefeet, she tamped earth into the crevices between the eggs by thrusting her snout with a ramming action. Forty-five minutes later, when the hole was nearly filled, the lizard began to excavate five almost horizontal pits above the buried eggs but only three to four inches deep. The sand surrounding the eggs was thus forced down and tightened between them; in addition, these "false nests" also would tend to mislead predators into exploring superficially above the eggs and burrowing to one side of them. After one "false nest" was completed, another would be dug diametrically opposite to it. After digging several of these, the female rested for a while in the depression formed by the incompletely filled nest and then left, not returning until the next day. She then spent two additional hours filling in the nest depression and digging several more "false nests." The total time spent on all this was 53 hours, 35 minutes! Another variation in nest-making is provided by the flap-necked chamaeleon (*Chamaeleo dilepis*), which digs an oblique tunnel with her hind legs to a depth great enough to conceal her body; she then lays her eggs, covers them with tamped-down earth, and may finally scatter dry grass and twigs over the filled-in nest (Bellairs, 1970).

In contrast to the deliberate nest construction described above, many lizards merely deposit their eggs in sheltered places or in decaying vegetation. The Nile monitor (*Varanus niloticus*) and the lace monitor (*Varanus varius*) deposit their eggs in large moundlike termite nests. These earthen nests have very hard walls except during the rainy season, which coincides with the egg-laying season of these monitors. The monitor tears a hole in the rain-softened wall of the termite nest, lays her eggs, and departs. Then the frenzied termites repair the nest and, in doing so, enclose the eggs. The termites seem to ignore the eggs, and their nest makes a natural incubator; fresh air continuously circulates through the channels and chambers of the nest, maintaining a fairly high and uniform temperature, and as the repaired nest wall hardens in the sun it protects the lizard eggs from at least some predators.

Very few oviparous lizards exhibit any

degree of maternal care; having laid their eggs, they typically show no further interest in them. North American skinks (*Eumeces,* family Scincidae) and glass lizards (*Ophisaurus,* family Anguidae) are well-known exceptions to the general rule, brooding their eggs by curling around them; the skinks will even defend the nest against small rodents or other animals that may approach it. Broad-headed skinks (*Eumeces laticeps*) and five-lined skinks (*Eumeces fasciatus*) exhibit more than a passive maternal instinct. Noble and Mason (1933) observed that females of these species lick and turn their eggs upon returning to the nest after a short absence. Occasionally, a female may use her mouth to pick up an egg and move it to a new position in the clutch. If her eggs have been scattered during her absence, she will gather them up and may even appropriate eggs laid by other individuals of her own or the other species. When Noble and Mason substituted eggs of different kinds of lizards, it was found that females would refuse to brood those of glass lizards and fence lizards (*Sceloporus*). Skink eggs which had been varnished were also refused, as were simulated eggs made of wax. Other experiments revealed that removal of the tongue tips prevented a female from finding her eggs, although a blindfolded female could find and brood her eggs after they had been moved to a different site from where she had left them. Skinks generally seem to utilize their Jacobson's organ and olfactory discrimination to identify mates, food, and rivals, and the experiments of Noble and Mason indicate that they also use the same mechanism in egg recognition. These workers believed that, in addition to protecting the nest, the brooding habit incubates the eggs, since a female's basking in the sun will raise her body temperature and the heat thereby acquired may be used to warm the eggs; the significance of brooding to the temperature regulation of eggs has been questioned by other workers. Vinegar's (1968) study indicates that the brooding of glass lizards has no thermoregulatory effect unless it results from moving the eggs nearer to or farther from the surface of the ground.

An advanced degree of maternal care is exhibited by the Great Plains skink (*Eumeces obsoletus*). In addition to the maternal care of eggs, there is maternal assistance at hatching, and maternal care and grooming of young after they hatch (Evans, 1959). While the eggs are in the nest, the female regularly licks them and turns them over and into different positions. When the young begin to hatch, the female repeatedly touches them, pressing and rubbing them with her head, body, and feet. Each time that the mother touches a young hatchling, it is stimulated to move spasmodically; the end result is to cause the young to wriggle free of the egg shell. For at least 10 days after her young have hatched, the female continues to remain with them. She will literally allow them to gulp food unmolested out from under her snout even though she may be about to snap at the same food item. Periodically, the mother will locate each of her offspring and cleanse it by licking its cloacal region.

Viviparous lizards generally seem to exhibit little maternal care, but females of the South African skink *Mabuya trivittata* are known to use their jaws to help free their young by tearing the embryonic membranes surrounding them at birth. Cowles (1944) observed an almost mammal-like maternal behavior pattern during parturition in the viviparous yucca lizard (*Xantusia vigilis*). When the fetus with its enveloping membranes first protrudes, the mother grasps the fetal envelope with her jaws and rips it open. This seems to activate the young lizard, and its struggles release the tail and hind legs. If the offspring stops struggling, the female will usually nip it on the exposed flanks and legs, causing it to thrash violently and extricate itself from the parental cloaca. The fetal membranes remain in the female's cloaca, and she removes these by grasping them in her mouth and gradually drawing them out. She then swallows them together with their contained liquids and licks up any droplets of fluid that may have fallen on the ground. Cowles noted that this entire procedure requires only about two minutes and that the actual expulsion of the fetus is very rapid, requiring only about one minute.

The number of eggs per clutch varies from species to species. It also varies with the size and age of the female. Blair (1960) found an average of about 11 eggs per clutch for yearling female rusty lizards,

18 per clutch for two-year-old females, and 24 per clutch for three-year-old females. Tinkle (1961) found that the mean reproductive potential, based on the number of enlarged ovarian follicles, increased from 3.3 to 4.9 when body size increased from between 35 and 39 to between 50 and 54 millimeters in desert side-blotched lizards (*Uta stansburiana stejnegeri*). His data also indicate that the size of the clutch declines through the reproductive season and that the last clutch of the season for this species is one egg smaller than the first. Population-to-population variations in clutch size may occur through natural selection and reflect differences in inheritance, as Kramer (1946) demonstrated in *Lacerta sicula*. Finally, smaller species generally have smaller clutches than larger species; viviparous species generally have fewer embryos than similar-sized and closely related oviparous species. The largest quoted clutch size for lizards is the maximum figure of 60 recorded by Fitzsimons (1943) for the Nile monitor (*Varanus niloticus*). In contrast, a variety of lizards deposit as few as one egg per clutch. In at least some species, including the green anole (*Anolis caro-*

linensis), the minimum clutch size of one egg is balanced by the facts that breeding occurs over an extended period of time and up to nine clutches may be laid per season. Bellairs (1970) notes the possibility that there is an inverse relationship between clutch or litter size and the frequency of breeding. Finally, it should be noted that communal egg-laying, in which two or more females deposit their eggs in the same nest, may lead to erroneous conclusions regarding clutch size. For example, Duellman and Schwartz (1958) dissected gravid females of the small reef gecko (*Sphaerodactylus notatus*) and concluded that only one egg matures at a time. However, they noted that a single egg is rarely found in nature, and they discovered one termite-ridden log at Key West, Florida, which housed a composite nest containing 280 eggs. These observations indicate that reef geckos often lay eggs communally, perhaps as a result of a limited number of favorable nesting sites.

The following table presents a summary of egg production data compiled by Oliver (1955), Tinkle (1961), Bellairs (1970), and Fitch (1970) for various species of lizards.

EGG PRODUCTION DATA FOR A VARIETY OF LIZARD TAXA*

	Number of Eggs or Young Per Clutch (Average or Range)	Maximum No. of Clutches Per Season
Gekkonidae (geckos)		
Diplodactylinae	2 eggs or young	
Eublepharinae	2 eggs	
Gekkoninae	1–2 eggs	
Sphaerodactylinae	1–2 eggs	
Iguanidae (iguanids)		
Amblyrhynchus cristatus	2–3 eggs	1
Anolis (165 species)	1–2 eggs	6–9
Basiliscus basiliscus	18 eggs	
Basiliscus vittatus	3–14 eggs	
Callisaurus draconoides	2–7 eggs	5 ?
Corytophanes cristatus	5–8 eggs	
Corytophanes hernandesii	3–4 eggs	
Corytophanes percarinatus	7–8 young	
Crotaphytus collaris	2–21 eggs	2
Crotaphytus wislizenii (in south)	1–9 eggs	2
Crotaphytus silus	2–5 eggs	2
Ctenosaura acanthura	17–28 eggs	
Ctenosaura pectinata	49 eggs	
Ctenosaura similis	20–30 eggs	

EGG PRODUCTION DATA FOR A VARIETY OF LIZARD TAXA* (Continued)

	Number of Eggs or Young Per Clutch (Average or Range)	Maximum No. of Clutches Per Season
Dipsosaurus dorsalis	3–8 eggs	
Holbrookia lacerata	4–12 eggs	2
Holbrookia maculata (Kansas)	2–8 eggs (ave. 4.95)	
Holbrookia maculata (Mexico, Arizona)	4–10 eggs (ave. 7.00)	
Iguana iguana	24–45 eggs	
Laemanctus longipes	3–5 eggs	
Laemanctus serratus	3–5 eggs	
Liolaemus magellanicus	2 young	
Liolaemus fuscus	2–3 eggs	
Liolaemus lineomaculatus	2–3 young	
Liolaemus monticola	3–4 eggs	
Liolaemus gravenhorstii	5–6 young	
Liolaemus tenuis	6 eggs	
Liolaemus multiformes	5–6 young	1
Ophryoessoides ornatus	2 eggs	
Phrynosoma asio	7–21 eggs	
Phrynosoma coronatum	6–21 eggs	
Phrynosoma cornutum	14–37 eggs	1+ ?
Phrynosoma douglassi	5–31 young	
Phrynosoma m'calli	6–7 eggs	
Phrynosoma modestum	9 eggs	
Phrynosoma orbiculare	12–13 young	
Phrynosoma platyrhinos	6–10 eggs	
Phrynosoma solare	17–25 eggs	
Phymaturus palluma	4 young	
Plica plica	2 eggs	
Polychrus marmoratus	7–8 eggs	
Sauromalus obesus	7–10 eggs	biennial cycle ?
Sceloporus chrysostictus	1–4 eggs	
Sceloporus formosus	3–9 young	
Sceloporus graciosus	2–5 eggs	1
Sceloporus grammicus	4–12 young	
Sceloporus malachiticus	2–9 young	
Sceloporus pyrocephalus	3–4 eggs	2 ?
Sceloporus scalaris	9–12 eggs	
Sceloporus carinatus	7–10 eggs	
Sceloporus clarki	4–22 eggs	2+ ?
Sceloporus horridus	8–15 eggs	2+ ?
Sceloporus magister	7–19 eggs	2+ ?
Sceloporus melanorhinus	5–20 eggs	
Sceloporus olivaceus	11–24 eggs	4
Sceloporus orcutti	11 eggs	
Sceloporus cyanogenys	6–18 young	
Sceloporus jarrovi	5–13 young	2
Sceloporus poinsetti	7–11 young	
Sceloporus torquatus	6 young	
Sceloporus occidentalis	3–14 eggs	
Sceloporus undulatus	3–13 eggs	
Sceloporus variabilis	1–5 eggs	several
Tropidurus albemarlensis	1–6 eggs	
Uma inornata	2–4 eggs	
Uma notata	1–5 eggs	2–3
Uranascodon superciliosa	6–11 eggs	2
Urosaurus graciosus	4 eggs	

EGG PRODUCTION DATA FOR A VARIETY OF LIZARD TAXA* (Continued)

	Number of Eggs or Young Per Clutch (Average or Range)	Maximum No. of Clutches Per Season
Urosaurus ornatus	3–13 eggs	2 ?
Urostrophus torquatus	5–6 eggs	
Uta mearnsi	2–6 eggs	
Uta stansburiana	1–6 eggs	
Agamidae (agamids)		
Agama agama	3–12 eggs	several
Agama bibronii	9–12 eggs	
Agama caucasica	6–14 eggs	
Agama hispida	4–20 eggs	
Agama sanguinolenta	8–10 eggs	3
Agama tuberculata	7–9 eggs	2+
Amphibolurus maculatus	4–5 eggs	
Amphibolurus pictus	8 eggs	
Amphibolurus barbatus	8–24 eggs	
Calotes cristatellus	1–2 eggs	
Calotes mystaceus	7 eggs	
Calotes rouxi	4–9 eggs	
Calotes versicolor	1–23 eggs	
Draco dussumieri	4 eggs	
Draco melanopogon	1–2 eggs	
Draco obscurus	1–4 eggs	
Draco quinquefasciatus	1–4 eggs	
Goniocephalus grandis	1–5 eggs	
Goniocephalus liogaster	1–4 eggs	
Moloch horridus	6–8 eggs	
Otocryptis wiegmanni	3–4 eggs	
Phrynocephalus luteoguttatus	2–4 eggs	
Sitana ponticeriana	6–14 eggs	
Uromastix acanthinurus	8–14 eggs	
Chamaeleontidae (chamaeleons)		
Brookesia (all species)	3–6 eggs	
Chamaeleo basiliscus	23 eggs	
Chamaeleo bitaeniatus	9–22 young	
Chamaeleo dilepis	23–50 eggs	
Chamaeleo hohneli	8–11 young	
Chamaeleo senegalensis	7–60 eggs	
Microsauria ventralis	13–20 young	
Rhampholeon marshalli	12–18 eggs	
Xantusidae (xantusids)		
Klauberina riversiana	4–9 young	
Xantusia vigilis	1–3 young	
Scincidae (skinks)		
Ablepharus wahlbergii	4–6 eggs	
Carlia fusca	2 eggs	
Egernia whitii	1–5 young	
Emoia atrocostata	2 eggs	
Eumeces dicei	3 young	
Eumeces callicephalus	6 eggs	
Eumeces fasciatus	4–15 eggs	
Eumeces inexpectatus	11 eggs	
Eumeces laticeps	13–16 eggs	
Eumeces multivirgatus	3–5 eggs	
Eumeces obsoletus	7–17 eggs	
Eumeces septentrionalis	5–13 eggs	
Eumeces skiltonianus	2–6 eggs	

EGG PRODUCTION DATA FOR A VARIETY OF LIZARD TAXA* (Continued)

	Number of Eggs or Young Per Clutch (Average or Range)	Maximum No. of Clutches Per Season
Hemiergis peronii	2–4 young	
Lampropholis guichenoti	2–3 eggs	
Leiolopisma aeneum	2–3 young	
Leiolopisma zelandica	2–8 young	
Lipinia noctua	1–2 young	
Lygosoma quadrupes	3 eggs	
Mabuya bayonii	4–7 young	
Mabuya brachypoda	2–6 young	
Mabuya dissimilis	6–7 eggs	
Mabuya macularia	3 eggs	
Mabuya maculilabris	5–13 eggs	
Mabuya megalura	4–13 young	
Mabuya multifasciata	2–8 young	2 ?
Mabuya quinquetaeniata (north)	6–18 young	
Mabuya quinquetaeniata (south)	10 eggs	
Mabuya rudis	2–4 eggs	
Mabuya striata	2–11 young	
Mabuya varia	1–12 young	
Neoseps reynoldsi	2 eggs	
Scincella assata	1–4 eggs	
Scincella cherriei	2–3 eggs	
Scincella laterale	1–7 eggs	5 ?
Tiliqua scincoides	5–18 young	
Tropidophorus microlepis	7–9 young	
Cordylidae (cordylids)		
Cordylus cordylus	1–3 young	
Gerrhosaurus flavigularis	4–5 eggs	
Teiidae (teiids)		
Ameiva festiva	1–4 eggs	
Ameiva undulata	2–7 eggs	
Cnemidophorus species	1–7 eggs	2+
Echinosaura horrida	1–2 eggs	
Kentropyx calcaratus	1–5 eggs	
Leposoma	1–2 eggs	
Neusticurus	1–2 eggs	
Prionodactylus	1–2 eggs	
Anadia	1–2 eggs	
Proctoporus	1–2 eggs	
Tupinambis teguixin	6–32 eggs	
Lacertidae (lacertids)		
Acanthodactylus pardalis	2–4 eggs	
Cabrita leschenaulti	6 eggs	
Eremias pleskei	2 eggs	2
Lacerta agilis	6–13 eggs	2
Lacerta jacksoni	3–5 eggs	
Lacerta muralis	3–8 eggs	
Lacerta viridis (France)	6–19 eggs	1
Lacerta viridis (Italy)	6–19 eggs	2
Lacerta vivipara	5–8 young	
Latastia longicaudata	3 eggs	
Takydromus tachydromoides	1–9 eggs	6
Varanidae (monitor lizards)		
Varanus bengalensis	19–24 eggs	
Varanus komodoensis	15 eggs	
Varanus niloticus	40–60 eggs	

EGG PRODUCTION DATA FOR A VARIETY OF LIZARD TAXA* (Continued)

	Number of Eggs or Young Per Clutch (Average or Range)	Maximum No. of Clutches Per Season
Anguidae (lateral fold lizards)		
Abronia graminea	4 young	
Anguis fragilis	4–22 young	
Diploglossus costatus	3–9 young	
Diploglossus curtissi	2–4 young	
Diploglossus bilobatus	6 eggs	
Gerrhonotus coeruleus	2–15 young	
Gerrhonotus liocephalus	5–22 eggs	
Gerrhonotus monticolus	3–10 young	
Gerrhonotus multicarinatus	1–20 eggs	
Ophisaurus attenuatus	6–17 eggs	1
Anniellidae (shovel-snouted limbless lizards)		
Anniella pulchra	1–4 young	
Xenosauridae (xenosaurids)		
Xenosaurus grandis	1–5 young	
Helodermatidae (beaded lizards)		
Heloderma suspectum	3–7 eggs	biennial cycle ?
Heloderma horridum	4–15 eggs	
Amphisbaenidae (ringed lizards)		
Tomuropeltis pistillum	4 eggs	
Trogonophis wiegmanni	5 young	
Amphisbaena dubia	3 eggs	
Rhineura floridana	2 eggs	

*Numbers of eggs are for oviparous species; numbers of young are for species in which hatching occurs in the oviduct or shortly after deposition of the eggs (ovoviviparous and viviparous species). In cases where no number of eggs or young or of clutches is supplied, this information is as yet unknown.

Examination of the data in the preceding table reveals the close relation between clutch size and body size of females. At the family level, for example, the family Varanidae includes the largest living lizards and is characterized by large clutch sizes; the family Gekkonidae includes relatively small lizards and is characterized by small clutches. Within families, the same relationship holds true. For example, the microteiids (*Echinosaura, Leposoma, Neusticurus, Prionodactylus, Anadia,* and *Proctoporus*) all have 1–2 egg clutches while the giants of the family, *Tupinambis*, have clutches of up to 32 eggs. Blair's (1960) data on *Sceloporus olivaceus* exemplify the relationship between body size (age) and clutch size within a single species. The data also tend to reflect larger clutches in oviparous lizards than broods in ovoviviparous or viviparous species of the same general kind and size. However, one of the more prolific lizards is the short-horned lizard (*Phrynosoma douglassi*), which is viviparous but averages about 15 young per litter. Viviparous lizards, which commonly occur in the cool habitats of high latitudes and high altitudes, probably produce broods less frequently than oviparous lizards of the same general kind because of the longer utilization of the female's reproductive tract by each brood.

SNAKES

As with lizards, snakes vary interspecifically in their mode of reproduction. Many snakes are oviparous and deposit their leathery-shelled eggs in moist situations (rotting logs, stumps, and banks, or humus) where, frequently, they will be exposed to the warming effects of sunshine or decomposing organic material. Other snakes are ovoviviparous or viviparous and give birth to young which are enclosed only in thin membranes. As with lizards, there is often a correlation between viviparity in snakes and the occupation of cool high-altitude or high-latitude habitats.

Relatively little is known about the reproductive behavior of snakes; only a few oviparous forms are known to go to the trouble of making a nest for their eggs. A pair of Indian cobras (*Naja naja,* family Elapidae) was observed in a zoo to cooperate in the construction of an egg chamber (Smith, 1937). The male and female both burrowed into a pile of dirt until their noses met in the middle and then they formed a cavity which was large enough to hold them both. The female deposited her eggs in this chamber and then guarded them throughout most of the incubation period. Whenever she would leave, generally to feed or drink, the male would keep guard over the eggs. The king cobra (*Ophiophagus hannah*) is the only snake known to construct a nest from vegetation. Oliver (1956) observed the reproductive behavior of this species in the New York Zoological Park. The female constructed a nest out of sand and litter (bamboo stalks and dried magnolia leaves). Using a coil of her body for a drag, she first piled up the material, making several trips to drag in more loose vegetation. Next, she made an egg chamber in the middle of the pile by coiling tightly and revolving her body. After depositing her eggs in the nest, she covered them with leaves and remained coiled on top to guard them. In this species, the male did not assist in either the construction of the nest or the guarding of the eggs. In fact, the female would drive him off if he approached too closely while she was making the nest.

It is uncertain how many snakes brood their eggs. Individuals belonging to a variety of species have been found coiled around their eggs, but in general it has not been determined whether these individuals were brooding eggs or were merely females that had just deposited their eggs and not yet left. Oliver (1955) mentions a case where both members of a pair of pilot blacksnakes (*Elaphe obsoleta*) were observed to sun themselves and then incubate a clutch of eggs deposited in a sawdust pile by carrying heat absorbed by their bodies to the buried eggs. In this case, the male was found actually encircling the eggs. Pythons are also known to brood their eggs. After depositing her eggs, the female python uses her body and tail to draw them into a pile and then coils around them until her body, surrounding them on all sides, forms an egg chamber. This chamber is covered on top by her head, and the result is that the eggs are within an enclosed space and protected from drafts and so forth. Female pythons have been observed to brood their eggs for as long as six weeks and to only rarely leave them in order to drink. As mentioned in Chapter Eight, the brooding female Indian python (*Python molurus bivittatus*) has a limited ability to raise her body temperature metabolically and, thus, to warm the eggs to a temperature above the ambient. Burrowing worm snakes (*Leptotyphlops*) are also reported to brood their eggs (Bellairs, 1970).

Once the eggs of oviparous species have hatched, there appears to be little, if any, parental care or protection for the young. Similarly, no ovoviviparous or viviparous species of snake is known to exhibit more than a transitory attachment to the young. One of the most popular snake myths is that a mother will swallow her young to protect them and later disgorge them when the danger is passed. Generally, those species to which this is attributed are ovoviviparous or viviparous, so perpetuation of the folk myth may be the result of observations of pregnant females killed at a time when their young are about to be born. The shock of the female's sudden death or chance coincidence may result in the young crawling out of her cloaca shortly after her death. The myth may also be perpetuated through observations of young captive snakes seeking shelter under the mother's body when frightened. Such behavior may merely reflect a snake's basic instinct to seek shelter under something or it may actually represent some degree of parental protective behavior. At any rate, there is no basis in fact for female snakes protecting their young by swallowing them!

Snakes tend to deposit larger numbers of eggs per clutch and to have larger broods than lizards, but the number varies considerably from species to species. Although small clutches of two to eight eggs are the general rule for the tiny wormlike snakes (families Typhlopidae and Leptotyphlopidae), large species do not always have larger clutches than small species. Viviparous and ovoviviparous species do not always have smaller broods than oviparous forms. The following table presents observed clutch sizes for various species of snakes and, where known, the maximum number of clutches (or broods) produced per female per year. The majority of these data were compiled by Fitch (1970).

EGG PRODUCTION DATA FOR A VARIETY OF SNAKE SPECIES*

	Number of Eggs or Young Per Clutch (Average or Range)	Maximum No. of Clutches Per Season
Boidae (boids)		
Boa constrictor	20–64 young	
Charina bottae	1–8 young	1
Corallus enydris	10–30 young	
Epicrates striatus	12 young	
Eryx conicus	1–11 young	
Eryx jaculus	6–12 eggs	
Eunectes murinus	4–82 young	
Lichanura roseofusca	5–10 young	
Python curtus	10–16 eggs	
Python molurus	15–54 eggs	
Python reticulatus	16–103 eggs	
Python sebae	20–100 eggs	
Typhlopidae (blind snakes)		
Typhlops avakubae	6 eggs	
Typhlops bibroni	5–8 eggs	
Typhlops braminus	2–7 eggs	
Typhlops diardi	5–14 young	
Typhlops punctatus	10–19 eggs	
Typhlops schlegeli	12–60 eggs	
Leptotyphlopidae (thread snakes)		
Leptotyphlops dulcis	2–7 eggs	
Leptotyphlops humilis	2–6 eggs	
Acrochordidae (wart snakes)		
Acrochordus javanicus	25–30 young	
Acrochordus granulatus	6–8 young	
Uropeltidae (rough-tails)		
Uropeltis ocellatus	3–5 young	
Colubridae (colubrids)		
Arizona elegans	3–23 eggs	
Aspidura brachyorrhus	2–5 eggs	
Coluber constrictor	1–28 eggs	1
Carphophis amoenus	1–12 eggs	
Coniophanes imperialis	4–5 eggs	
Dendrelaphis punctulatus	5–12 eggs	
Diadophis punctatus	1–10 eggs	
Drymarchon corais	3–11 eggs	
Elaphe carinata	12 eggs	
Elaphe climacophora	4–24 eggs	
Elaphe conspicillata	1–7 eggs	
Elaphe dione	4–11 eggs	
Elaphe guttata	3–21 eggs	
Elaphe obsoleta	5–44 eggs	
Elaphe quadrivirgata	4–15 eggs	

EGG PRODUCTION DATA FOR A VARIETY OF SNAKE SPECIES* (Continued)

	Number of Eggs or Young Per Clutch (Average or Range)	Maximum No. of Clutches Per Season
Elaphe rufodorsata	4–21 eggs	
Farancia abacura	4–104 eggs	
Farancia erytrogramma	20–52 eggs	
Gongylosoma baliodeira	1–3 eggs	
Lampropeltis getulus	5–17 eggs	
Lampropeltis triangulum	5–16 eggs	
Leimadophis reginae	1–8 eggs	
Lycodon aulicus	3–11 eggs	
Lycophidion capense	1–8 eggs	
Masticophis flagellum	4–24 eggs	
Masticophis lateralis	6–9 eggs	
Masticophis taeniatus	3–12 eggs	
Mehelya capensis	5–8 eggs	
Ninia maculata	5 eggs	
Opheodrys aestivus	3–12 eggs	
Opheodrys vernalis	2–8 eggs	
Philothamnus hoplogaster	3–8 eggs	
Philothamnus irregularis	4–8 eggs	
Pituophis melanoleucus	3–24 eggs	1
Pseudaspis cana	30–50 young	
Ptyas korros	1–9 eggs	2+
Ptyas mucosus	9–14 eggs	
Rhinocheilus lecontei	4–9 eggs	
Salvadora grahamiae	6–10 eggs	
Salvadora hexalepis	5–10 eggs	1
Sonora semiannulata	4–6 eggs	
Spalerosophis diadema	3–16 eggs	
Boiga multomaculata	4–7 eggs	
Chrysopelea ornata	6–12 eggs	
Crotaphopeltis hotamboeia	3–10 eggs	
Dispholidus typus	10–14 eggs	
Hypsiglena ochrorhyncha	4–6 eggs	
Hypsiglena torquata	3–12 eggs	
Leptodeira annulata	2–7 eggs	
Oxyrhopus petolus	5–10 eggs	
Psammodynastes pulverulentus	6–10 eggs	
Psammophis angolensis	3–5 eggs	
Psammophis sibilans	10–15 eggs	
Psammophis subtaeniatus	9–20 eggs	
Psammophylax rhombeatus	30 eggs	
Tantilla coronata	2–3 eggs	
Tantilla gracilis	1–4 eggs	
Trimorphodon biscutatus	20 eggs	
Amphiesma craspedogaster	1–7 eggs	
Amphiesma vibakari	4–10 eggs	
Natriciteres olivacea	2–8 eggs	
Natrix erythrogaster	5–27 young	
Natrix fasciata	22–57 young	
Natrix natrix	8–53 eggs	
Natrix rhombifera	14–62 young	
Natrix sipedon	8–99 young	
Regina alleni	4–34 young	
Regina grahami	9–39 young	
Rhabdophis tigrina	2–26 eggs	
Storeria dekayi	3–27 young	1

EGG PRODUCTION DATA FOR A VARIETY OF SNAKE SPECIES* (Continued)

	Number of Eggs or Young Per Clutch (Average or Range)	Maximum No. of Clutches Per Season
Storeria occipitomaculata	1–13 young	
Thamnophis butleri	4–16 young	
Thamnophis crytopsis	25 young	
Thamnophis elegans	6–16 young	
Thamnophis marcianus	6–18 young	
Thamnophis ordinoides	3–15 young	
Thamnophis proximus	6–17 young	
Thamnophis radix	5–92 young	
Thamnophis sauritus	3–20 young	
Thamnophis sirtalis	3–85 young	1–2
Virginia striatula	2–8 young	
Xenochrophis piscator	20–85 eggs	
Xenochrophis vittata	5–8 eggs	
Heterodon platyrhinos	4–61 eggs	1
Enhydris enhydris	10–18 young	
Homalopsis buccata	2–20 young	
Sibynophis chinensis	2–4 eggs	
Dasypeltis scabra	8–14 eggs	
Elapidae (cobras and coral snakes)		
Austrelops superba	18–21 young	
Callophis maccellandi	4–14 eggs	
Dendroaspis	12–14 eggs	
Elaps	6 eggs	
Micrurus fulvius	2–4 eggs	
Naja naja	8–45 eggs	
Naja melanoleuca	15–26 eggs	
Naja haje	8–20 eggs	
Ophiophagus hannah	41–51 eggs	
Hydrophidae (sea snakes)	2–14 young	
Viperidae (vipers and pit vipers)		
Agkistrodon acutus	20–26 young	
Agkistrodon contortrix	2–17 young	2–10
Agkistrodon piscivorus	1–15 young	1
Bothrops atrox	8–71 young	
Crotalus atrox	3–19 young	1
Crotalus horridus	4–25 young	
Crotalus ruber	3–20 young	
Crotalus viridis	3–21 young	0.5–1
Lachesis muta	2–12 young	
Sistrurus catenatus	2–19 young	
Sistrurus miliarius	2–32 young	
Trimeresurus flavoviridis	3–17 eggs	
Bitis arietans	23–56 young	
Causus rhombeatus	11–26 eggs	
Vipera aspis	4–12 young	biennial cycle
Vipera berus	6–20 young	
Vipera russelli	1–63 young	

*Numbers of eggs are for oviparous species; numbers of young are for species in which hatching occurs in the oviduct or shortly after deposition of the eggs (ovoviviparous and viviparous species). In cases where no number of eggs or young or of clutches is supplied, this information is as yet unknown.

Ecotypic variation in clutch size is known to be present in some species of snakes. For example, *Diadophis punctatus* is reported to have an average clutch of 3.5 eggs in Michigan, 4.2 in Kansas, and 5.2 in Florida (Fitch, 1970). Clutches of *Crotalus viridis* vary in a pattern which is not latitudinal, averaging 11.4 on the northern Great Plains and declining to only 2.6 on Los Coronados (islands off the northwestern tip of Mexico); intermediate populations of the Great Basin, southern Rocky Mountains, and interior California average between 7 and 8 (Fitch, 1970). This northeast-to-southwest decline in numbers is paralleled in *Agkistrodon contortrix*, with broods of 6.2 in northeastern states, 5.3 in Kansas, and 3.0 in western Texas. An east-to-west decline in average clutch size occurs in *Coluber constrictor*, the averages being 16.8 in northeastern United States, 11.8 in the central states, and 5.8 on the West Coast (Fitch, 1970).

In contrast to at least some tropical lizards which reproduce throughout the year, snakes generally seem to have well-defined breeding seasons, and probably few produce more than a single clutch per year. Neill (1962) found that even tropical British Honduras snakes, representing 11 genera and four families, appeared to have well-defined breeding cycles and concluded that reproduction was curtailed, as in non-tropical areas, by minimal temperatures.

TUATARA

The female tuatara (*Sphenodon punctatus*) deposits her eggs in a shallow depression which she makes about four or five inches deep in loose soil. She then covers the clutch, consisting of 9 to 14 eggs, with loose debris. No guarding of the nest or parental care of young occurs in this species.

EMBRYONIC DEVELOPMENT AND STAGES

The period of embryonic development in reptiles is usually described in terms of gestation period and incubation period. Gestation period is the duration of em-

bryonic development within the female's reproductive tract. The incubation period is the time between deposition of the eggs outside of the female's body and the rupturing of the egg membranes at the initiation of hatching. Field observations and laboratory hatching of eggs have shown that the duration of the incubation period is highly variable, even for eggs of the same species. For example, Blair (1960) found that the incubation of rusty lizard (*Sceloporus olivaceus*) eggs varied from 43 days to 83 days. Various factors contribute to intraspecific variability in incubation time. Embryonic development begins with the fertilization of the egg, and the time that fertilization occurs is not necessarily correlated with when the eggs are deposited. Thus, periods of embryonic development do not coincide with incubation periods. One female might retain her eggs until they have reached a later stage of development than another female would, with the result that the former's eggs will hatch after a shorter incubation period. Nest-to-nest variability in temperature would also result in varying rates of embryonic development and differing embryonic development periods and incubation periods. Despite nest-to-nest differences, the majority of the eggs in a single clutch will usually hatch within a short period of time, reflecting the fact that they were fertilized at nearly the same time and also the fact that embryonic development rates are characteristic for a given species under a given set of temperature conditions The acceleration of development by heat can cause significant differences in incubation or gestation periods. For example, Blanchard and Blanchard (1941) found that in southern Michigan the gestation period for the common garter snake (*Thamnophis sirtalis sirtalis*) might vary from 87 days during an exceptionally hot summer to 116 days during an unusually cold summer. An increase of 1° C in the average temperature during gestation shortens the gestation period by four and one-half days. The embryonic development (development up to hatching or birth) for most reptiles requires 10 to 12 weeks under normal temperature conditions, but the incubation period for the tuatara is 13 to 14 months (Sharell, 1966) and the gestation period for the viviparous mesquite lizard (*Sceloporus grammicus*) is listed by Oliver (1955) as five to six months. When eggs are fertilized late in the summer or when unusually cool weather prevails, particularly at the higher latitudes of a species, the young may winter in the egg and undergo what is sometimes termed embryonic hibernation. Embryonic hibernation seems to be the general rule for the tuatara, explaining the extremely long incubation period, and also occurs at least occasionally in the North American snapping turtle (*Chelydra serpentina*).

DEVELOPMENTAL STAGES

Illustrated tables of the normal development have been prepared for a variety of reptilian species. In order to illustrate both the similarities and differences in the embryonic growth of two morphologically very different kinds of reptiles, stages in the normal development of the snapping turtle and the European common lizard are presented on the following pages. Comparable stages in the development of a crocodilian (*Alligator mississippiensis*) and of a snake (*Thamnophis sirtalis sirtalis*) may be found in Reese (1915) and Zehr (1962), respectively.

Stages in the Embryonic Development of *Chelydra serpentina*

The following stages in the development of the common snapping turtle (*Chelydra serpentina*) are as defined and illustrated by Yntema (1968); he has kindly granted permission to reproduce them here. These stages are based on timed intervals at a constant temperature of 20° C and, because growth rates and size are not consistent, are defined in terms of surface anatomical characteristics occurring at typical time intervals. The series of 26 stages forms three groups: the early or presomite period (stages 0 to 3), the somite period (stages 4 to 10), and the limb period (stages 11 to 26). The somite period overlaps the limb period, but staging by the limb becomes feasible by stage 10 or stage 11 and the counting of somites becomes more difficult. The larger stages are based on conditions in the forelimb and the final stages on those in the first digit of the forelimb.

STAGE 0 (the egg at time of laying). The embryonic disc lies at an end of the ovoid pellucid area. The blastopore is a relatively large dorsal transverse slit in the posterior portion of the disc. The chordamesodermal canal runs as a tunnel which opens on the ventral surface. Extending anteriorly from the canal, a head process is indicated ventrally.

A sheet of cells forms a sector on the deep side of the embryonic disc. Along its radii, irregular extensions run laterally and anteriorly. Clusters of cells on the ventral surface are seen beyond the margin of the disc. The cells of the primitive plate posterior to the blastopore overlap the vitelline area; this is more extensive in the ventral view. The randomly placed spheres on the ventral surface are yolk granules.

In the extraembryonic area, the pellucid area is set eccentrically in relation to the embryonic disc. This is underlaid by some mesoderm and is surrounded by the vitelline or opaque area.

Variation at time of laying is illustrated by three ventral views of embryonic discs (Fig. 12–7, *O, O', O''*). Each of these embryos is characteristic of other siblings observed from the same clutch. There is variation as to configurations of the cell groups radiating from the chordamesodermal canal. That labeled *O''* is the oldest of the three embryos, to judge by the migration of mesoderm into the pellucid area. In addition, the ventral opening of chordamesodermal canal is constricted as in stage 1.

STAGE 1 (Fig. 12–7; one day old at 20° C). In the embryonic disc, the blastopore is narrower and slightly arched. The lateral borders of the ventral opening of the chordamesodermal canal make a more acute angle than in stage 0.

The mesoderm has become more conspicuous beyond the border of the embryonic disc in the pellucid area. Its borders make a ring incomplete anteriorly but conspicuous posteriorly so as to present a sicklelike appearance. In the ventral view a triangular condensation of mesoderm extends forward from the primitive plate and is bisected by the head process.

STAGE 2 (Fig. 12–8; two days old at 20° C). The borders of the embryonic disc have become obscured by mesoderm. The mesoderm forms an eccentric ring about the embryonic disc. The posterior half of the ring is more conspicuous and is confluent with the primitive plate. Posteriorly, cells spread out dorsally over the vitelline area, but ventrally the junction between the vitelline area and the primitive plate is concise.

STAGE 3 (Fig. 12–8; three days old at 20° C). The embryonic disc is elongated and the blastopore is smaller. The neural groove and head fold are shallow dorsal depressions. The slight

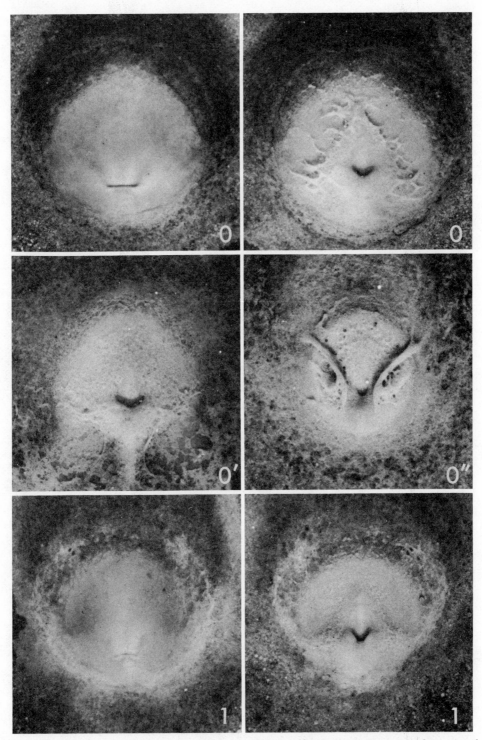

Figure 12–7 Stage *O*, dorsal and ventral views of embryo at time of laying (× 20). Stage *O'* and *O''*, ventral views of embryos at time of laying (× 20). Stage *1*, dorsal and ventral views of 1 day embryo (× 20).

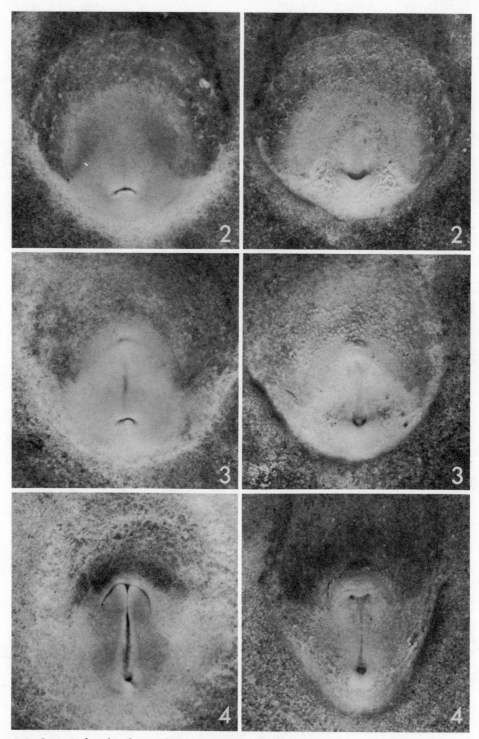

Figure 12–8   Stage *2*, dorsal and ventral views of 2 day embryo (× 20). Stage *3*, dorsal and ventral views of 3 day embryo (× 20). Stage *4*, dorsal and ventral views of 4 day embryos (× 20).

chordamesodermal groove extends anteriorly from the canal on the ventral surface. The mesoderm is diffuse anterior to the embryonic disc; posteriorly and laterally it forms a sickle confluent with the primitive plate.

STAGE 4 (Fig. 12–8; four days old at 20° C). Three pairs of somites are present but cannot be seen in reflected light. The neural folds have risen along the length of the embryo. They surround the blastopore posteriorly and are approximated at what are probably the midbrain and anterior hindbrain levels. The chordamesodermal canal is now to be called the neurenteric canal. The ventral surface contains a well-formed head process. The chordamesodermal groove extends forward from the neurenteric canal and flares out anteriorly. There are several extraembryonic structures apparent: The chorioamniotic fold covers the anterior ends of the neural folds. The extraembryonic mesoderm is diffuse dorsally but appears to form a concise boundary with the vitelline area ventrally. Anterior to the embryo, the mesoderm is sparser in a relatively clear area.

STAGE 4+ (Fig. 12–9; between four and five days old at 20° C): Four pairs of somites are present. The neural folds are slightly fused in the posterior head region and over the blastopore. In the ventral surface, a well-defined notochord is bordered laterally by somitic mesoderm whose segmentation is not clearly illustrated. The anterior intestinal porta has formed in conjunction with flexure of the head rudiment. Among the extraembryonic structures, the outer margin of the area vasculosa is marked by an ovoid ring of mesoderm, and the chorioamniotic membrane covers forebrain and midbrain rudiments.

STAGE 5 (Fig. 12–9; five days old at 20° C): Five pairs of somites are present. The neural folds are open anteriorly and are closely placed but unfused in much of the spinal region. Development of extraembryonic structures is seen: The chorioamniotic fold covers the anterior hindbrain and lies at the level of the anterior intestinal porta. The former pellucid area is vague in the dorsal view but in the ventral view of this specimen it is outlined as an irregular ovoid about the embryo. Lateral to the embryo, the inner margin of the yolk-bearing cells of the opaque area is obvious. The pellucid area is much smaller than at the time of laying. Condensations of mesoderm indicating blood islands can be seen anterior to the embryo in the dorsal view.

STAGE 5+ (Fig. 12–9; six days old at 20° C): Six pairs of somites are present. The neural folds are raised but open anteriorly. The mandibular pharyngeal arch is indicated. The cranial flexure is approximately 180°.

STAGE 6 (Fig. 12–9; seven days old at 20° C): Eight pairs of somites are present. (In order to show embryonic structures, some of the ex-

traembryonic membranes have been removed at this and later stages. In this case, the membranes anterior to the anterior intestinal porta were removed, including the amnion. The amnion was cut along the sides of the embryo and removed with the chorion over most of the embryo. The posterior portion of the chorioamnion was left intact; its free posterior margin does not cover the posterior end of the neural tube.) Cranial structures are apparent: a small anterior neuropore persists; optic vesicles are indicated; the mandibular arch is delimited posteriorly. In the extraembryonic area, blood islands occur in the opaque region, and their extent marks the vascular area. Body folds have developed: lateral body folds extend from the anterior fold around the anterior half of the embryo. The vitelline area has closed in along the sides of the embryo. The primitive plate is prominent ventrally.

STAGE 7 (Fig. 12–10; nine days old at 20° C): Ten pairs of somites are present. Among the cranial structures, the optic vesicles and mandibular arch are more clearly seen, the otic rudiment forms a depression seen best in the dorsal view; the neural folds are still incompletely fused anteriorly; the hyoid arch is interposed between the mandibular arch and the heart, which is S-shaped. Changes in the extraembryonic structures have taken place, with the posterior edge of the chorioamnion lying a little beyond the margin of the vascular area and the formation of the posterior amniotic tube begun; blood islands are conspicuous, and vessels of the vitelline plexus can be seen lateral to the embryo. The lateral body folds bound a longitudinal chordamesodermal recess in which somites and notochord are seen. The primitive plate and internal opening of the neurenteric canal are conspicuous.

STAGE 8 (Fig. 12–10; 12 days old at 20° C): Fourteen pairs of somites are present. Cranial structures continue to progress: the neural folds are completely fused anteriorly; the otic rudiment is cup-shaped and its opening is not constricted; ventral to it, the second pharyngeal groove separates the hyoid from the third pharyngeal arch; the mandibular arch is prominent laterally and ventrally. The lateral body folds have moved medially; the chordamesodermal recess is reduced; the posterior body fold has formed and has raised the primitive plate off the level of the extraembryonic membranes; a tail process results and the chordamesodermal recess is limited posteriorly.

STAGE 9 (Fig. 12–10; 16 days old at 20° C): Nineteen pairs of somites are present. In the cranial area, the roof of the fourth ventricle has become thin; a lens pit is present; the opening of the otic cup is constricted; the first pharyngeal slit is open; the third pharyngeal arch is delimited posteriorly; the buccal membrane has been resorbed. The circulatory sys-

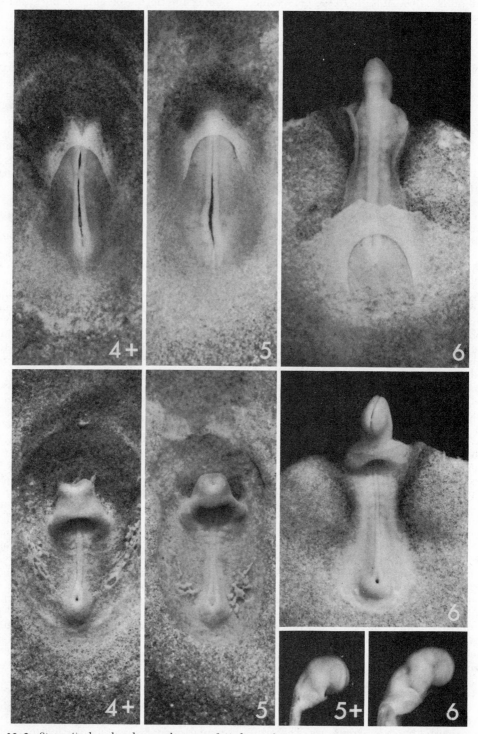

Figure 12–9   Stage 4⁺, dorsal and ventral views of 4⁺ day embryo (×20). Stage 5, dorsal and ventral views of 5 day embryo (×20). Stage 5⁺, lateral view of head process of 6 day embryo (×20). Stage 6, dorsal, ventral, and lateral views of 7 day embryo (×20).

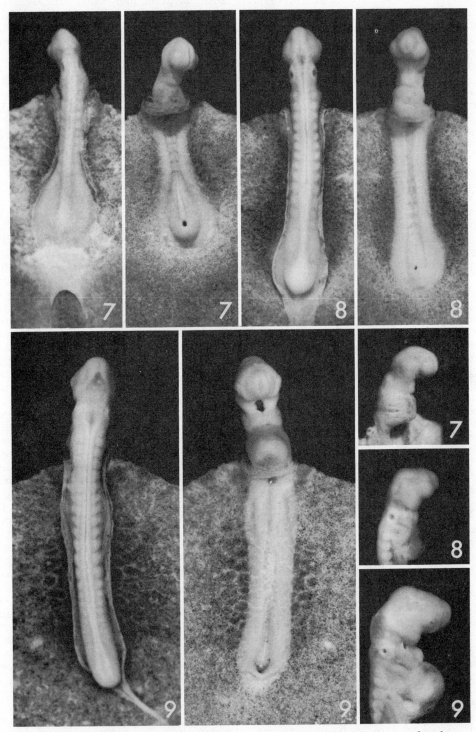

Figure 12–10   Stage 7, dorsal, ventral, and lateral views of 9 day embryo (× 20). Stage 8, dorsal, ventral, and lateral views of 12 day embryo (× 20). Stage 9, dorsal, ventral, and lateral views of 16 day embryo (× 20).

tem has developed: the heart is prominent; circulation started two or three days previously; the vitelline plexus is conspicuous, with vessels anastomosing in the ventral midline between the intestinal portae. The rudiment of the forelimb in the lateral body fold is indicated by a dilation of the amniotic cavity. Among the body folds, the posterior intestinal porta has formed and is overhung by the tail process; the anterior porta is reduced in size; between the portae a fusion in the midline has obliterated the intervening chordamesodermal recess and carried the medial part of vitelline plexus to a midline anastomosis.

STAGE 10 (Fig. 12–11; 20 days old at 20° C): Twenty-four pairs of somites are present. Cranial and cervical structures progress: the lens pit has disappeared; the olfactory pit is present; the otic vesicle is closed; its endolymphatic sac extends dorsally; the first two pharyngeal slits are open; the fourth pharyngeal arch is delimited posteriorly; the mandibular process extends caudally to the first slit.

In the circulatory system, the right and left anterior vitelline veins are well formed (not illustrated at this stage). The limbs are developing, and dilations of the amniotic cavity occur at limb levels where the limb buds are present in the lateral body folds. The forelimb rudiment is more extensive anteroposteriorly than the hind limb rudiment, which is three somites wide at its base. The tail process has become elongate, and its mesoderm is segmented only at its base.

STAGE 11 (Fig. 12–11 and 12–15; 25 days old at 20° C): Thirty-one pairs of somites are present. Cranial and cervical structures show the first pharyngeal slit still open dorsally, the second slit covered by the hyoid arch, the fifth arch delimited posteriorly, and the maxillary process extending toward the eye. The developing circulatory system has a right anterior vitelline vein that is relatively small, and just anterior to the embryo, much of the vein's flow is shunted to the large left vein. The tail process has increased in length, and its mesoderm is segmented for more than half of its length. The hind limb rudiment spans the distance of five somites at its base; the forelimb rudiment and the crest of Wolff anterior to it stretch over nine somites. The cervical flexure has increased prior to the turning of the embryo onto its left side.

STAGE 12 (Figs. 12–11 and 12–15; 30 days old at 20° C): The embryo lies on its left side; this is associated with formation or increase in cervical, dorsal, and sacral flexures. Of the cranial and cervical structures, the pharyngeal slits have disappeared, the maxillary process extends as far ventrally as the mandibular, and the hyoid arch limits the cervical sinus anteriorly; the retina has been pigmented for

three days. In the circulatory system, the anterior vitelline vein crosses under the head posterior to the otic vesicle. The distance between the intestinal portae is now about one-third the length of the embryo. The allantois protrudes between the hind limb rudiments and is approximately the same size as those rudiments. The forelimb bud is about three somites wide and slightly shorter than it is wide, the crest of Wolff anterior to the limb is no longer obvious, and the apical ridge is starting to form; the axis of the limb runs ventroposteriorly.

STAGE 13 (Figs. 12–12 and 12–15; five weeks old at 20° C): Due to increasing flexures, the crown-rump length has decreased. The cranial and cervical structures have changed: the maxillary process extends beyond the mandibular and limits a well-marked nasolacrimal groove posteriorly; this leads to a deep olfactory pit; the operculum extends over the anterior part of the cervical sinus. The anterior vitelline vein crosses under the head between the eye and ear. The opening of the yolk sac is constricted to about four times the diameter of the anterior vitelline vein. The allantois is enlarged so that its diameter is about one-third the crown-rump length of the embryo. The forelimb bud is slightly longer than it is wide; it points more caudally than ventrally; the apical ridge is at its maximum.

STAGE 14 (Figs. 12–12 and 12–15; six weeks old at 20° C): The maxillary and lateral nasal processes are fused; the mandibular process is inconspicuous; the cervical sinus is reduced. The diameter of the allantois is more than the crown-rump length of the embryo. The anterior vitelline vein runs ventrally from the umbilical region; the allantoic vessels lie posterior to the vein and form an acute angle with it. The opening of the yolk sac is about twice the diameter of the anterior vitelline vein. The axis of the forelimb runs caudally: it is an early paddle stage with the digital plate vaguely indicated; the apical ridge is present. A groove on the lateral trunk indicates a demarcation of the carapace.

STAGE 15 (Figs. 12–13 and 12–15; seven weeks old at 20° C): The mandibular process extends to the posterior eye level. The cervical sinus is closed; the frontal process is evident. The extraembryonic portions of the allantois and circulatory system have been dissected away and are no longer used as criteria for staging. With the constriction of the yolk sac from the intestine, part of the latter is herniated through the body wall, forming an intestinal loop. The digital plate of the forelimb is well formed, with no digital grooves present; random pigment cells occur at the base of the limb. The carapace is clearly limited laterally but indistinctly so in front and behind; the central and lateral laminae are vaguely out-

(Text continued on page 420.)

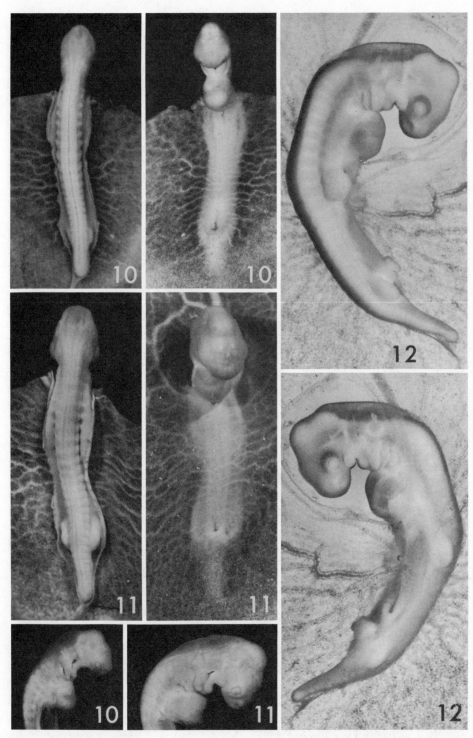

Figure 12-11 Stage *10*, dorsal, ventral, and lateral views of 20 day embryo (× 10). Stage *11*, dorsal, ventral, and lateral views of 25 day embryo (× 10). Stage *12*, right and left views of 30 day embryo (× 10).

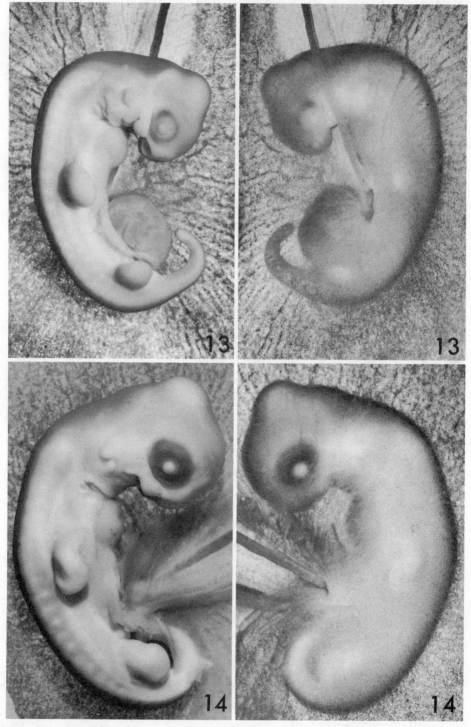

Figure 12–12   Stage *13*, right and left views of 5 week embryo (× 9). Stage *14*, right and left views of 6 week embryo (× 9).

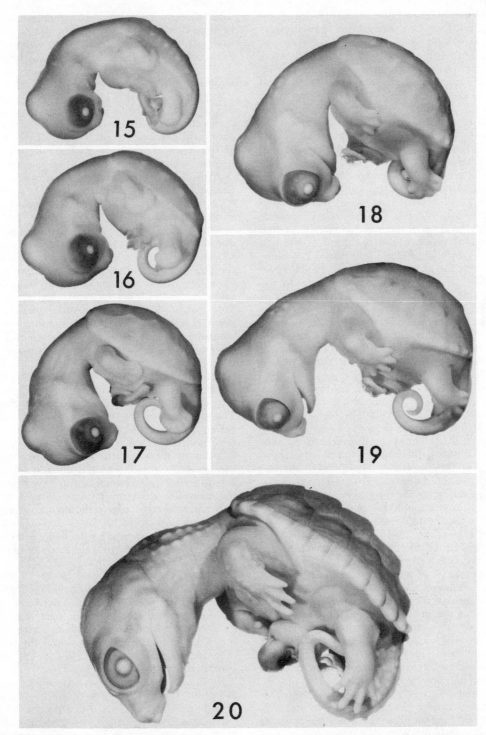

Figure 12–13   Stage *15*, left view of 7 week embryo (× 4.5). Stage *16*, left view of 8 week embryo (× 4.5). Stage *17*, left view of 9 week embryo (× 4.5). Stage *18*, left view of 10 week embryo (× 4.5). Stage *19*, left view of 11 week embryo (× 4.5). Stage *20*, left view of 12 week embryo (× 4.5).

lined; pigmented areas have formed on some of the central laminae along both sides of the midline (not illustrated).

STAGE 16 (Figs. 12–13 and 12–15; eight weeks old at 20° C): Among the cranial structures, the lower jaw ends just behind the level of the lens, and scleral papillae are indicated. The gut remains herniated, retaining the intestinal loop, through stage 22. The digital plate of the forelimb is relatively large and has a smooth periphery and slight indications of digital ridges; pigmentation on the limb and its base is sparse. The carapace is clearly limited around its periphery; the central and lateral laminae are still vaguely outlined; pigmented areas have formed on some of the lateral laminae (not illustrated).

STAGE 17 (Figs. 12–13 and 12–15; nine weeks old at 20° C): The lower jaw extends beyond the level of the lens but does not reach the frontal process; the rudiment of the egg tooth or caruncula is a slight dorsal protuberance near the anterior end of the frontal process; the scleral papillae are distinct. On the forelimb the periphery of the digital plate is slightly serrated, the five digits are indicated by five ridges and four intervening grooves, and pigment cells are scattered over the plate. The central and lateral laminae of the carapace are distinctly demarcated, but the marginal laminae are not.

STAGE 18 (Figs. 12–13 and 12–15; 10 weeks old at 20° C): The cranial and cervical structures have further developed: the lower jaw ends at the frontonasal groove, which is conspicuous; the lower eyelid is indicated; cutaneous papillae are indicated on the dorsum of the neck. On the forelimb, the digital plate has clearly indicated digits which protrude to make deep serrations in the periphery; pigmentation is slight. The marginal laminae of the carapace are more clearly delimited.

STAGE 19 (Figs. 12–13 and 12–16; 11 weeks old at 20° C): Of the cranial and cervical structures, the free end of the lower jaw is pointed, the frontonasal groove persists, and the lower lid still remains short of the scleral papillae; rows of cutaneous papillae are indicated on the dorsum of the neck. The forelimb has continued to develop: the central digits project beyond the webs a distance slightly greater than their thickness at the web; pigmentation is sparse and not evident at a low magnification. On the carapace there is a medial row of pigment spots on the central laminae lateral to the midline and a lateral row on five or six of the lateral laminae near their medial margins; the marginal laminae are delimited.

STAGE 20 (Figs. 12–13 and 12–16; 12 weeks old at 20° C): The frontonasal groove has disappeared; the lower lid reaches the level of the lens; the scleral papillae are no longer conspicuous; cutaneous papillae form rows of projections on the dorsum of the neck. The central digits of the forelimb project beyond the webs a distance about twice as great as their thickness at the webs; pigmentation is evident down to the bases of the digits; cutaneous papillae are formed on the dorsum of the antibrachium. The central and lateral laminae of the carapace are gray with pigment; on the marginal laminae the pigmentation is light.

STAGE 21 (Figs. 12–14 and 12–16; 13 weeks old at 20° C): The lower eyelid crosses the lower margin of the lens. On the forelimb the claws are delimited from the rest of the digits at the level of the webs and are lightly pigmented; skin folds are present on the preaxial dorsum. The central and lateral laminae of the carapace are slightly corrugated; pigment spots are found in the grooves between the marginal laminae.

STAGE 22 (Figs. 12–14 and 12–16; 14 weeks old at 20° C): The lower eyelid covers most of the pupil. On the forelimb, pigmentation of the claws is still light but obvious; the preaxial claws are darker than the postaxial; cutaneous folds are on the preaxial digits and along the postaxial border of the forearm as well as the preaxial dorsum. The pigmentation of the carapace has increased but it does not yet approach being black; the corrugation of the laminae has increased.

STAGE 23 (Figs. 12–14 and 12–16; 15 weeks old at 20° C): The lower eyelid is separated from the upper by a slit in the fixed specimen. The profile view of the claw shows a homogeneous structure; pigmentation of the claws is heavy; cutaneous folds are present over the dorsum of the forearm. The changes in the carapace are not differential enough for staging; the central and lateral laminae are rougher; the posterior marginal laminae form a markedly serrated edge; pigmentation has increased to produce a dark but not black shell. The loop of gut which has been herniated since stage 15 has been drawn back into the body of the turtle; the yolk sac remains herniated.

STAGE 24 (Figs. 12–14 and 12–16; 17 weeks old at 20° C): On the carapace the central laminae form a dorsal keel; increased pigmentation results in a black carapace. The cutaneous folds on the forelimb are more prominent; the claw is blunt and extends the length of its rudiment; it is enclosed in a sheath which is relatively translucent.

STAGE 25 (Figs. 12–14 and 12–16; 19 weeks old at 20° C): The dorsal keel on the carapace is reduced. The claw is differentiated within its rudiment to form a pointed structure which is shorter than the rudiment. The umbilical hernia is present and contains a part of the yolk sac (not illustrated).

STAGE 26 (Fig. 12–14; 20 weeks old at 20° C): The carapace flattens when released from the confines of the egg shell. With use the sheaths

(Text continued on page 424.)

Figure 12–14 Stage *21*, left view of 13 week embryo (× 2). Stage *22*, left view of 14 week embryo (× 2). Stage *23*, left view of 15 week embryo (× 2). Stage *24*, left view of 17 week embryo (× 2). Stage *25*, left view of 19 week embryo (× 2). Stage *26*, left view of hatchling immediately after leaving shell (× 2).

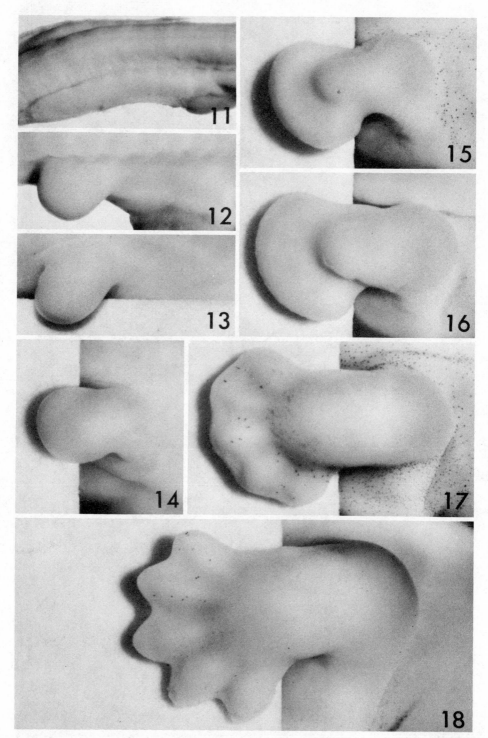

Figure 12–15   Stage *11*, view of right forelimb region of 25 day embryo (× 20). Stage *12*, view of right forelimb bud of 30 day embryo (× 20). Stage *13*, view of right forelimb bud of 5 week embryo (× 20). Stage *14*, view of right forelimb bud of 6 week embryo (×20). Stage *15*, view of right forelimb of 7 week embryo (× 20). Stage *16*, view of right forelimb of 8 week embryo (× 20). Stage *17*, view of right forelimb of 9 week embryo (× 20). Stage *18*, view of right forelimb of 10 week embryo (× 20).

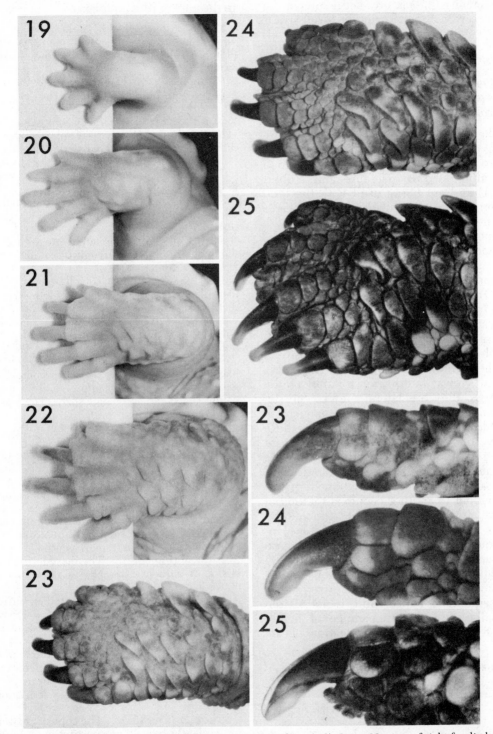

Figure 12–16   Stage *19*, view of right forelimb of 11 week embryo (× 8). Stage *20*, view of right forelimb of 12 week embryo (× 8). Stage *21*, view of right forelimb of 13 week embryo (× 8). Stage *22*, view of right forelimb of 14 week embryo (× 8). Stage *23*, views of dorsal antibrachium and medial side of digit I of right forelimb of 15 week embryo (× 8 and × 16). Stage *24*, views of dorsal antibrachium and medial side of digit I of right forelimb of 17 week embryo (× 8 and × 16). Stage *25*, views of dorsal antibrachium and medial side of digit I of right forelimb of 19 week embryo (× 8 and × 16).

about the claws on the forelimb are worn off. The umbilical hernia may not be present: its position is then represented by a soft area in the plastron; in other instances, a relatively small yolk sac may persist as illustrated.

Stages in the Embryonic Development of *Lacerta Vivipara*

The following stages in the development of the viviparous European common lizard (*Lacerta vivipara*) are as defined and illustrated by Defaure and Hubert (1961) and are being used with the kind permission of these workers. These stages are not based on time, which varies with temperature, but on morphological development. The following table indicates the approximate dates of certain sets of reproductive stages, as given by Defaure and Hubert, but these are subject to variation from year to year, from place to place, and according to environmental temperatures. Each stage is figured by a photograph of an embryo in its natural position. In the initial stages, the photographs are dorsal views; later, after rotation of the embryo, they are lateral views. The descriptions sometimes include changes not visible in the photographs and, in these cases, the notations "v.v" and "v.h" are given to indicate, respectively, "visible ventrally" and "visible histologically." The authors note that variation from one individual to another occurs at the end of neurulation. For example, the neural folds close earlier or later in some individuals. The dimensions of the embryo up until neurulation are also subject to variation. The number of pairs of somites, commonly used to characterize stages subsequent to neurulation, is not visible dorsally until the appearance of nearly the tenth pair and after rotation of the

CHRONOLOGICAL DATA ON DEVELOPMENT TIME

	April	May	June	July	August
Mating	?–20-----14				
Ovulation		4---29			
Stages 1 to 4		11-------5			
Stages 5 to 33		23------------------5			
Stages 34 to 40			20---------31. . . ?		
Parturition				15–31. . . ?	

embryo; thus, the somites do not appear in the earlier illustrations.

STAGE 1 (Fig. 12–17): Beginning of segmentation: one central blastomere is completely isolated: six large furrows radiate around it.

STAGE 2 (Fig. 12–17): The center of the plate is paved with little blastomeres. The periphery is hollowed out by about 20 large furrows.

STAGE 3 (Fig. 12–17): The embryonic disc exhibits three categories of blastomeres of different sizes. Some large peripheral blastomeres encircle a zone in which the division of blastomeres is more rapid from one "pole" to the other.

STAGE 4 (Fig. 12–17): End of segmentation: the embryonic disc is entirely paved with very little blastomeres.

During all the segmentation (stages 1 to 4) the diameter of the embryonic disc ranges from 2.5 to 3 millimeters.

STAGE 5 (Fig. 12–17): The embryonic bud is distinguished. The blastopore is formed: it is rectilineal and marks the posterior region of the bud. The blastoporal plaque, of considerable size, forms an excrescence posterior to the bud.

STAGE 6 (Fig. 12–18): Reduction of the blastoporal plaque. The extraembryonic zone is divided into two concentric zones: a transparent space and an opaque space. The blastoporal canal is outlined (v.h.).

STAGE 7 (Fig. 12–18): The blastoporal plaque is in contact with the bud. The dorsal lip of the blastopore is prominent and curved. The blastoporal canal is clearly formed and it is closed ventrally (v.h.).

STAGE 8 (Fig. 12–18): The cephalic region is individualized. The bud presents a mesodermal crescent from which extremities are directed toward the front of the embryo. The blastoporal canal is open ventrally (v.h.).

STAGE 9 (Fig. 12–18): The mesoderm thickens, as is evident externally by the formation of two flaps on the embryo.

STAGE 10 (Fig. 12–18): A median furrow marks the first outline of the neural groove. The mesodermic flaps reach the base of the cephalic region.

STAGE 11 (Fig. 12–18): The anterior extremity of the cephalic region begins to bend ventrally. The blastopore takes the form of an upside-down U. The notochord is individualized (v.v.).

STAGE 12 (Fig. 12–18): The cephalic extremity is clearly bent, thus constituting a ventral swelling (v.v.) and a limiting anterior fold (the first individualization of the cephalic zone of the embryo is clearly denoted). The neural swellings form on each side of the outline of the neural groove.

STAGE 13 (Fig. 12–18): The limiting anterior

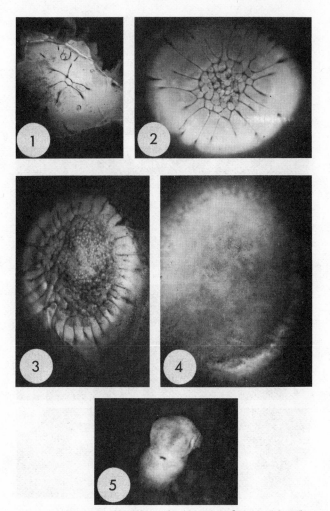

Figure 12–17 Stages *1* through *4* (× 20) and stage 5 (× 15).

fold becomes falciform. The neural groove deepens. The neural swellings are formed.

STAGE 14 (Fig. 12–18): The limiting anterior fold deepens. Formation of the anterior amniotic fold; lateral prolongations almost reach the surface of the blastopore. The neural swellings are of equal width all along the neural groove.

STAGE 15 (Fig. 12–18): The limiting anterior fold lengthens toward the back and from each side of the cephalic region by the lateral limiting folds.

With the appearance of lateral swellings in connection with the anterior swelling an ectodermic pocket is formed; the anterior falciform swelling regains an ectodermic pocket, the first outline of the pharynx (v.v.).

The neural swellings differentiate themselves anteroposteriorly: they widen in the anterior region which constitutes the first outline of the brain, remaining more contracted in the posterior or trunk region.

The neural groove does not adjoin the blastopore and the latter is again distinct.

From stages 5 to 15, the length of the embryo varies between 1 and 2 millimeters.

STAGE 16 (Fig. 12–19): Two pairs of somites (v.v.). The limiting folds are always located in the cephalic region. The difference in size between the neural cephalic swellings and the neural swellings of the trunk region is accentuated. The trunk region is elongated. The neural groove has adjoined the blastopore; thus, the blasticanal becomes the neurenteric canal (v.h.).

STAGE 17 (Fig. 12–19, *17a* and *17b*): Four pairs of somites (v.v.). The amniotic hood begins to cover the cephalic region. The neural swellings reunite posteriorly. The anterior

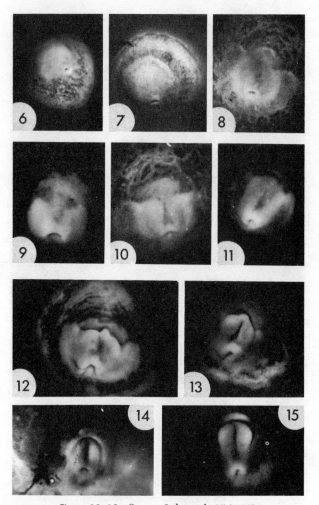

Figure 12–18   Stages 6 through 15 (× 12).

end of the head is individualized. During this stage, the embryo begins to pedunculate. Islets of Wolff appear in the posterior extraembryonic zone.

STAGE 18 (Fig. 12–19): Four pairs of somites (v.v.). The amnion completely covers the cephalic region.

STAGE 19 (Fig. 12–19, 19a and 19b): Six pairs of somites (v.v.) (19b). The amnion covers the anterior half of the embryo. Two large seroamniotic cavities extend from each side of the embryo. The islets of Wolff are very apparent anteriorly.

STAGE 20 (Fig. 12–19): Seven pairs of somites (v.v.). The neural tube is formed in the anterior region.

From Stage 16 to 21, the length of the embryo varies from 1.5 to 2.2 millimeters.

STAGE 21 (Fig. 12–20): Somites: 10 pairs. The amnion covers the embryo almost entirely. A large posterior amniotic swelling still

exists. Twisting: the head of the embryo lies on its left side. The trunk region remains flat on the vitelline membrane. The primary optic vesicles are outlined. The trunk region is largely open ventrally. Total length of embryo: 3 millimeters ± 0.1.

STAGE 22 (Fig. 12–20): Somites: 10 pairs. Twisting occurs mainly in the trunk region. The caudal region is individualized and a tail bud is present. The total length is 3.2 millimeters ± 0.1.

STAGE 23 (Fig. 12–20): Somites: 15 pairs. The posterior amniotic swelling remains. The telencephalon, diencephalon, mesencephalon, and rhombencephalon of the brain are differentiated. Neural groove: the swellings are joined together in the trunk and caudal regions. The first cardiac outline is formed. The primary optic vesicles are differentiated. The total length is 3.5 millimeters ± 0.1.

STAGE 24 (Fig. 12–20): Somites: 15 pairs.

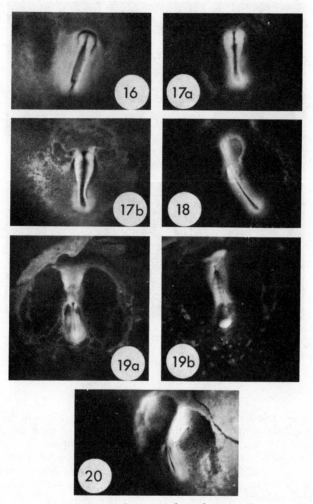
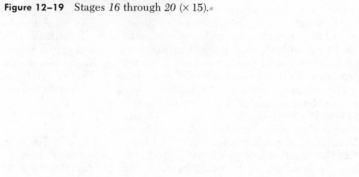

Figure 12–19   Stages *16* through *20* (× 15).

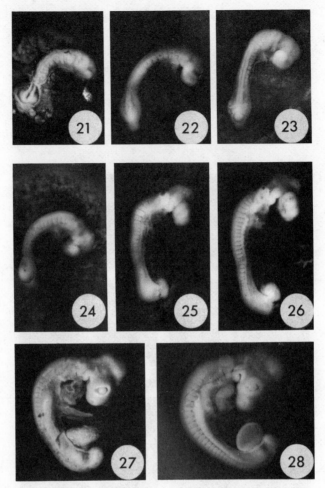

Figure 12–20   Stages *21* through *28* (× 15).

The twisting reaches the caudal region. The neural groove is closed all along its length. The lateral walls of the trunk region draw together ventrally. The total length is 3.6 millimeters ± 0.1.

STAGE 25 (Fig. 12–20): Somites: 18 pairs. The amnion completely covers the embryo, which is now lying entirely on its left side. The secondary optic vesicles are outlined. The auditory vesicles are differentiated with an auditory outline. The individualization of the mandibular and hyoid arches brings on the appearance of the hyomandibular branchial slit. The allantois forms a little posterior vesicle. The total length is 4 millimeters ± 0.1.

STAGE 26 (Fig. 12–20): Somites: 25 pairs. The secondary optic vesicles and the lens are differentiated. Branchial slits 1 and 2 begin to open. The trunk region is almost closed ventrally. The total length is 5 millimeters ± 0.2.

STAGE 27 (Fig. 12–20): Somites: 33 pairs.

The optic capsule is in the form of a horseshoe. The choroid fissure is open. Branchial slits 1 and 2 are open. The trunk region is closed ventrally. Olfactory outlines appear: they are differentiated, but do not connect with the outside. The trigeminal nerve is visible by transparence. Some lateral protrubances mark the first outline of anterior and posterior limbs. The total length is now 7 millimeters ± 0.2.

STAGE 28 (Fig. 12–20): Somites: 35 pairs. The third branchial slit begins to open. The limbs are much as in the preceding stage. The ventricle and auricle of the heart are differentiated. The total length is 7.2 millimeters ± 0.1.

STAGE 29 (Fig. 12–21): Somites: 40 pairs. Branchial slits 1, 2, and 3 are open. The outlines of limbs have the form of stumps. The anterior stumps are 0.25 millimeter long. Total length is 8.8 millimeters ± 0.2.

STAGE 30 (Fig. 12–21): Somites: 50 pairs.

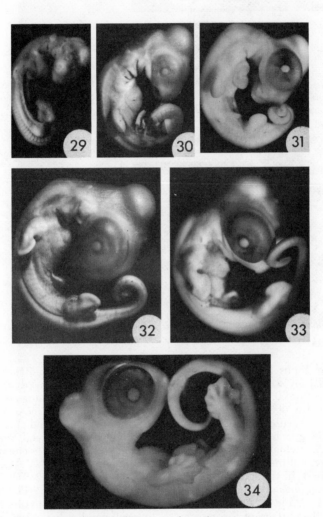

Figure 12–21   Stages *29* through *34* (× 13).

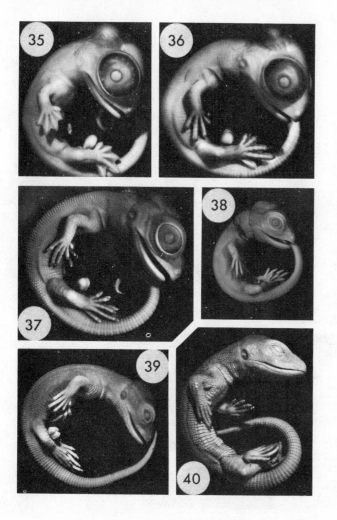

Figure 12–22 Stages 35 through 37 (× 8) and stages 38 through 40 (× 4).

The stumps of the limbs (Fig. 12–23) are fringed longitudinally by a little comb. The anterior stumps measure 0.5 millimeters in length. In the eye, the lips of the choroid fissure are joined. The optic capsule is spherical and its edges large. The eye begins to be pigmented. The five branchial slits are open. Among the auditory outlines, the endolymphatic canal is defined. In the parietal region the outlines of the epiphysis and the parietal eye are joined and visible in transparency. The allantois joins the surface of the mesencephalon. The total length is 10 millimeters ±0.1.

STAGE 31 (Fig. 12–21): The stumps of the limbs (Fig. 12–23) are flattened at the end. This flattened zone or "paddle" makes up the first outline of the autopodium. The comb on the stumps in the preceding stage makes up the edge of the paddle. The anterior limbs are 1 millimeter long. The eye is pigmented and prominent. Branchial slits 4 and 5 are closed again. The appearance of the first outline of the penis is noted. The total length is 15.9 millimeters ±0.5.

STAGE 32 (Fig. 12–21): The paddle on each limb (Fig. 12–23) is distinctly differentiated into the zeugopodium and the stylopodium. The anterior limbs measure 0.4 millimeter in length. Two branchial slits are again open. An outline of the parietal region has distinctly begun to appear. The total length is now 17 millimeters ±0.2.

STAGE 33 (Fig. 12–21): The outline of the limb paddles (Fig. 12–23) is meandering. Three fingers begin to differentiate themselves in each paddle; the anterior limbs measure 1.6 millimeters in length. A circular eyelid begins to develop. On the mouth, the lower jaw begins to develop. The penes are well-developed and alike in the two sexes. Total length has reached 19.2 millimeters ±0.2.

STAGE 34 (Fig. 12–21): Five digits are differentiated in each limb paddle (Fig. 12–23). The anterior limbs measure 2.2 millimeters. The eye presents some sclerotic papillae. All the branchial slits are closed. Among the olfactory outlines, the nares are formed. The total length is 20 millimeters ±0.1.

STAGE 35 (Fig. 12–22): The digits on the

limb paddles (Fig. 12–23) are individualized and joined by a palm. The anterior limbs are 2.7 millimeters long. The internal edge of the eyelid is serrated. The pineal eye presents a pigmented circle. The tympanum of the ear is formed. The digits are covered with pigment spots. A concentration of these spots is present at the end of the digits. Some pigment spots cover the head. The lower jaw is completely formed. The total length is 22 millimeters.

STAGE 36 (Fig. 12–22): The palms (Fig. 12–23) are formed. Each progressively becomes withdrawn from its outer toward its inner digit, more rapidly on the anterior limb than on the posterior limb. The digits are slightly annulated. Some nails are outlined at the extremities of the fingers. The anterior limbs

measure 3.2 millimeters. An important concentration of pigment occurs on the margins of the nails and on each segment of the digits. Some scales begin to form on the body. The head shows no outline of scales. The dimorphism of the penis in males and females becomes obvious. The total length is now 23 millimeters ±0.6.

STAGE 37 (Fig. 12–22): The digits of the anterior and posterior limbs (Fig. 12–24) are completely free from the palm; the nails are differentiated. Some scales are outlined on the feet. The anterior limbs measure 3.6 millimeters. The ocular eyeballs are less prominent. The eyelid is developed; its internal edge is oval. The head is covered with very crowded pigment spots. The body scales are

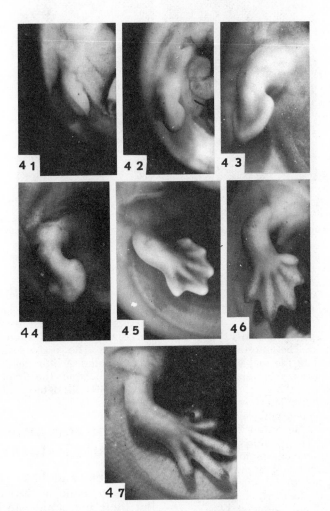

Figure 12–23   No. *41* shows development of the limb at stage 30 (× 20). *42*, development of the limb at stage 31 (× 20). *43*, development of the limb at stage 32 (× 20). *44*, development of the limb at stage 33 (× 20). *45*, development of the limb at stage 34 (× 20). *46*, development of the limb at stage 35 (× 20). *47*, development of the limb at stage 36 (× 20).

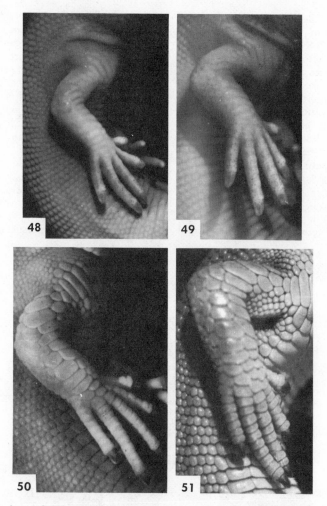

Figure 12–24   No. *48* shows development of the limb at stage 37 (× 20). *49*, development of the limb at stage 38 (× 20). *50*, development of the limb at stage 39 (× 20). *51*, development of the limb at stage 40 (× 20).

very apparent. The total length is 25 millimeters ±0.6.

STAGE 38 (Fig. 12–22): The digits (Fig. 12–24) are very annulated. The anterior limbs are 4 millimeters long. The internal edge of the eyelid becomes almost crystalline. The total length is 26 millimeters ±0.2.

STAGE 39 (Fig. 12–22): The feet and digits (Fig. 12–24) are covered with incompletely differentiated scales. The anterior limbs are 5.3 millimeters long. The eyelid is scaly. The body scales begin to become pigmented. The large scales are differentiated on the snout and the lower jaw. The rest of the head shows no scales. The collar is distinctly formed. The total length is now 35 millimeters ± 1.0.

STAGE 40 (Fig. 12–22): The scales on the limbs (Fig. 12–24) are completely differentiated and pigmented. The anterior limbs are

6.7 millimeters long. The embryo has acquired the pigmentation that it will show at birth. Large cephalic scales are differentiated on the head. The body scales are completely differentiated. A little deciduous tooth slightly projects beyond the upper jaw. The total length is 40 millimeters ± 1.0.

GROWTH AND ATTAINMENT
OF SEXUAL MATURITY

As with amphibians, very little study has been made of the growth of reptiles but they, too, appear to have a period of rapid growth preceding the attainment of sexual maturity, followed by a period during which growth tends to be intermittent, rather than constant, and occurs at a gradually declining rate. Reptiles also appear to be characterized by indeterminate growth so that individuals tend to get bigger as they get older.

The rate at which reptiles grow may be influenced by a number of environmental factors. For example, Legler (1960) was able to correlate the growth of ornate box turtles (*Terrapene ornata ornata*) with temperature, rainfall, and the abundance of grasshoppers. He also noted that individuals that grew rapidly in the season of hatching tended to also grow more rapidly than others during the subsequent two or three years. The amount of growth achieved by hatchlings is dependent upon the duration of favorable conditions between hatching and onset of their first winter. During favorable summers, hatching occurs early enough in the summer to allow the animals to feed and grow for eight or more weeks before hibernation. Unfavorable summer weather retards embryonic development and delays hatching so that the young either emerge late in the summer and have little time to feed or they remain over winter in the nest. In either case, little growth is possible until the following summer. The effect of hatchling growth on future growth is so great that individuals growing rapidly the season of hatching may reach sexual maturity three or four years earlier than the average.

Intermittent growth has been observed in a variety of reptiles. Very little, if any, growth of reptiles takes place while they are hibernating or aestivating. Thus, many reptilian populations exhibit annual patterns of intermittent growth. Irregular interruptions of growth have also been observed in reptiles. For example, Oliver (1955) cites an example of an eastern box turtle (*Terrapene carolina carolina*) in New York which grew rapidly to a measurement of four and three-quarters inches, had no measurable growth during the next three years, and then grew to a length of five and three-eighths inches in six years. The complexity of growth patterns is exemplified by the perplexing fact that Dodge and Wunder (1963) used carapace length as a measure of growth and discovered that young red-eared terrapins (*Pseudemys scripta elegans*) which were placed in centrifuges for five weeks and subjected to a five-times-normal gravitational force grew about twice as fast as control specimens.

As is true of amphibians, reptiles seem to reach sexual maturity after completing a certain amount of growth rather than reaching a particular age. However, what are typical ages at which sexual maturity is reached have been determined for various reptiles. These ages and/or sizes at which sexual maturity is attained are indicated in the following table.

SIZE AND AGE AT WHICH SEXUAL MATURITY IS ATTAINED IN VARIOUS REPTILES

	Snout-Vent Length at Maturity (mm)	Age at Maturity (yrs)
Chelonia		
Chelydridae:		
Sternothaerus odoratus males	65–73 (carapace)	3
Sternothaerus odoratus females	82–96 (carapace)	3
Testudinidae:		
Gopherus agassizi	230–250 (carapace)	15–20
Malaclemys terrapin		5–6
Pseudemys scripta elegans males	80–100 (carapace)	2–5
Pseudemys scripta elegans females	150–195 (carapace)	3–8
Terrapene ornata	100–110 (plastron)	4–6
Chrysemys picta males	70–90 (plastron)	1–7
Chrysemys picta females	110–130 (plastron)	
Cheloniidae:		
Eretmochelys imbricata		3
Trionychidae:		
Trionyx japonicus		6

SIZE AND AGE AT WHICH SEXUAL MATURITY IS ATTAINED IN VARIOUS REPTILES (Continued)

	Snout-Vent Length at Maturity (mm)	Age at Maturity (yrs)
Crocodilia		
Crocodylidae:		
Alligator mississippiensis	1.8 meters	6–10
Crocodylus niloticus males	2.9–3.3 meters	10–20
Crocodylus niloticus females	2.19 meters	10–20
Rhynchocephalia		
Sphenodontidae:		
Sphenodon punctatus		20
Squamata (Sauria)		
Iguanidae:		
Corytophanes percarinatus		1–2
Anolis carolinensis	145	1 or less
Crotaphytus collaris females	90–95	1–3
Liolaemus multiformes		1–2
Phrynosoma solare		1–2
Sceloporus graciosus	51–65	
Sceloporus horridus		1 or less
Sceloporus magister		1–2
Sceloporus olivaceus males	65	1
Sceloporus olivaceus females	80	1
Uma inornata		1–2
Uta stansburiana males	42	1–2
Uta stansburiana females	46	4 months
Agamidae:		
Calotes versicolor		9–12 months
Scincidae:		
Eumeces fasciatus	65–70	1–3
Eumeces septentrionalis	65	2–3
Lygosoma laterale		1
Teiidae:		
Cnemidophorus sexlineatus	60–65	1–2
Xantusidae:		
Xantusia vigilis females		3
Anguidae:		
Anguis fragilis		3–4
Anniellidae:		
Anniella pulchra		3
Squamata (Serpentes)		
Boidae:		
Charina bottae		2–3
Colubridae:		
Masticophis taeniatus		3
Pituophis melanoleucus		2–3
Elaphe obsoleta	900	4
Coluber constrictor		2
Storeria occipitomaculata	220	2
Thamnophis sirtalis (Kansas)	400	1–2
Tropidoclonion lineatum (Oklahoma)	220	21 months
Viperidae:		
Agkistrodon piscivorus (in captivity)		2
Agkistrodon contortrix		3–4
Crotalus viridis oreganus	700	3

References

Auffenberg, W. 1966. On the courtship of *Gopherus polyphemus*. Herpetologica 22:113–117.

Baker, J. R. 1947. The seasons in a tropical rain forest. Parts 6 and 7. J. Linn. Soc. London 41:243–258.

Bartholomew, G. A. 1950. The effects of artificially controlled temperature and day length on gonadal development in a lizard, *Xantusia vigilis*. Anat. Rec. 106:49–59.

———. 1953. The modification by temperature of the photoperiodic control of gonadal development in the lizard, *Xantusia vigilis*. Copeia 1953:45–50.

———. 1959. Photoperiodism in reptiles. *In:* R. B. Withrow (ed.), Photoperiodism and Related Phenomena in Plants and Animals. Washington, D.C., American Assoc. for the Advancement of Science, Publ. No. 55. Pp. 669–676.

Bellairs, A. 1970. The Life of Reptiles. Volume II. The Universe Natural History Series. New York, Universe Books. Pp. 283–590.

Blair, W. F. 1960. The Rusty Lizard. Austin, Univ. Texas Press. 185 pp.

Blanchard, F. N., and F. C. Blanchard. 1941. Factors determining time of birth in the garter snake *Thamnophis sirtalis sirtalis* (Linnaeus). Pap. Michigan Acad. Science. 26:161–176.

Burger, J. W. 1937. Experimental sexual photoperiodicity in the male turtle, *Pseudemys elegans* (Wied.). Am. Naturalist 71:481–487.

Burrage, B. R. 1965. Copulation in a pair of *Alligator mississipiensis*. Brit. J. Herpetol. 3(8):207–208.

Carpenter, C. C. 1967. Aggression and social structure in iguanid lizards. *In:* W. W. Milstead (ed.), Lizard Ecology; A Symposium. Columbia, Univ. Missouri Press. Pp. 87–105.

Carr, A. 1952. Handbook of Turtles—The Turtles of the United States, Canada, and Baja California. Ithaca, New York, Comstock Publ., Cornell Univ. Press. 542 pp.

Chapman, B. M., and R. F. Chapman. 1964. Observations on the biology of the lizard *Agama agama* in Ghana. Proc. Zool. Soc. London 143:121–132.

Church, G. 1962. The reproductive cycles of the Javanese house geckos, *Cosymbotus platyurus*, *Hemidactylus frenatus*, and *Peropus mutilatus*. Copeia 1962:262–269.

Cott, H. B. 1961. Scientific results of an enquiry into the ecology and economic status of the Nile crocodile (*Crocodilus niloticus*) in Uganda and Northern Rhodesia. Trans. Zool. Soc. London 29:211–356.

Cowles, R. B. 1944. Parturition in the yucca lizard. Copeia 1944:98–100.

Cuellar, O. 1966. Oviducal anatomy and sperm storage structures in lizards. J. Morphol. 119:7–20.

———. 1968. Additional evidence for true parthenogenesis in lizards of the genus *Cnemidophorus*. Herpetologica 24:146–150.

———. 1971. Reproduction and the mechanism of meiotic restitution in the parthenogenetic lizard *Cnemidophorus uniparens*. J. Morphol. 133:139–166.

Darevsky, I. S. 1966. Natural parthenogenesis in a polymorphic group of Caucasian rock lizards related to *Lacerta saxicola* Eversmann. J. Ohio Herpetol. Soc. 5:115–152.

Deraniyagala, P. E. P. 1958. Reproduction in the monitor lizard, *Varanus bengalensis* (Daudin). Spolica Zeylanica 28(2):161–166.

Dodge, C. H., and C. C. Wunder. 1963. Growth of juvenile red-eared turtles as influenced by gravitational field intensity. Nature 197:922–923.

Duellman, W. E., and A. Schwartz. 1958. Amphibians and reptiles of southern Florida. Bull. Florida State Mus. Biol. Sci. 3(5):181–324.

Dufaure, J. P., and J. Hubert. 1961. Table de développement du lézard vivipare: *Lacerta* (*Zootoca*) *vivipara* Jacquin. Arch. Anat. Microscop. Morphol. Exp. 50:309–328.

Evans, L. T. 1938. Courtship behavior and sexual selection of *Anolis*. J. Comp. Psychol. 26:475–497.

———. 1959. A motion picture study of maternal behavior of the lizard, *Eumeces obsoletus* Baird and Girard. Copeia 1959:103–110.

Fitch, H. S. 1970. Reproductive cycles of lizards and snakes. Univ. Kansas Mus. Nat. Hist. Misc. Publ., No. 52:1–247.

Fitzsimons, V. F. 1943. The lizards of South Africa. Pretoria, Transvaal Museum.

Fox, W. 1956. Seminal receptacles of snakes. Anat. Rec. 124:519–539.

———. 1958. Sexual cycle of the male lizard, *Anolis carolinensis*. Copeia 1958:22–29.

———. 1963. Special tubules for sperm storage in female lizards. Nature 198:500–501.

Fritts, T. H. 1969. The systematics of the parthenogenetic lizards of the *Cnemidophorus cozumela* complex. Copeia 1969:519–535.

Galgano, M. 1951. Prime recerche intorno all'influenza della luce e della temperature sul ciclo sesuale di *Lacerta sicula campestris* (Betta). Bool. zool. 18:108–115.

Goin, C. J., and O. B. Goin. 1962. Introduction to Herpetology. San Francisco, W. H. Freeman. 341 pp.

Harris, V. A. 1964. The Life of the Rainbow Lizard. London, Hutchinson Tropical Monographs.

Hunsaker, D., II. 1962. Ethological isolating mechanisms in the *Sceloporus torquatus* group of lizards. Evolution 16:62–74.

Inger, R. F., and B. Greenberg. 1966. Annual reproductive patterns of lizards from a Bornean rain forest. Ecology 47:1007–1021.

Kitzler, G. 1941. Die Paarungsbiologie einiger Eidechsen. Z. Tierpsychol. 4:353–402.

Kluge, A. G., and M. J. Eckardt. 1969. *Hemidactylus garnotii* Duméril and Bibron, a triploid all-female species of gekkonid lizard. Copeia 1969:651–664.

Kramer, G. 1946. Veränderungen von Nachkommenziffer und Nachkommen grosse sowie der Alterswerteilurg von Inseleidechsen. Z. Naturforsch. 1:700–710.

Legler, J. M. 1960. Natural history of the Ornate Box Turtle, *Terrapene ornata ornata* Agassiz. Univ. Kansas Publ. Mus. Nat. Hist. 11(10):527–669.

Licht, P. 1966. Reproduction in lizards: influence of temperature on photoperiodism in testicular recrudescence. Science 154:1668–1670.

Lowe, C. H., Jr., and J. W. Wright. 1964. Species of the *Cnemidophorus exsanguis* subgroup of whiptail lizards. J. Arizona Acad. Sci. 3:78–80.

———. 1966a. Evolution of parthenogenetic species of *Cnemidophorus* (Whiptail Lizards) in western North America. J. Arizona Acad. Sci. 4:81–87.

———. 1966b. Chromosomes and karyotypes of

cnemidophorine teiid lizards. Mamm. Chrom. News. 22:199–200.

Ludwig, M., and H. Rahn. 1943. Sperm storage and copulatory adjustment in the prairie rattlesnake. Copeia 1943:15–18.

Lynn, R. T. 1963. Comparative behavior of the horned lizards, genus Phrynosoma, of the United States. (Ph.D. thesis, Univ. Oklahoma.)

Marshall, A. J., and R. Hook. 1960. The breeding biology of equatorial vertebrates: reproduction of the lizard Agama agama lionotus Boulenger at Lat. 0°01' N. Proc. Zool. Soc. London, 134:197–205.

Maslin, T. P. 1962. All-female species of the lizard genus Cnemidophorus, Teiidae. Science 135:212–213.

——— 1966. The sex of hatchlings of five apparently unisexual species of Whiptail Lizards (Cnemidophorus, Teiidae). Am. Midland Nat. 76:369–378.

——— 1971. Conclusive evidence of parthenogenesis in three species of Cnemidophorus (Teiidae). Copeia 1971:156–158.

Mellish, C. H. 1936. The effects of anterior pituitary extract and certain environmental conditions on the genital system of the horned lizard (Phrynosoma cornutum, Harlan). Anat. Rec. 67:23–33.

Miller, M. R. 1951. Some aspects of the life history of the Yucca Night Lizard, Xantusia vigilis. Copeia 1951:114–120.

Minton, S. A., Jr. 1958. Observations on amphibians and reptiles of the Big Bend region of Texas. Southwest. Naturalist. 3:23–54.

Neaves, W. B. 1969. Adenosine deaminase phenotypes among sexual and parthenogenetic lizards in the genus Cnemidophorus (Teiidae). J. Exp. Zool. 171:175–184.

———, and P. S. Gerald. 1968. Lactate dehydrogenase isozymes in parthenogenetic teiid lizards (Cnemidophorus). Science 160:1004–1005.

——— 1969. Gene dosage at the lactate dehydrogenase b locus in triploid and diploid teiid lizards. Science 164:552–559.

Neill, W. T. 1962. The reproductive cycle of snakes in a tropical region, British Honduras. Quart. J. Florida Acad. Sci. 25:234–253.

Noble, G. K. 1937. The sense organs involved in the courtship of Storeria, Thamnophis and other snakes. Bull. American Mus. Nat. Hist. 73:673–725.

———, and E. R. Mason. 1933. Experiments on the brooding habits of the lizards Eumeces and Ophisaurus. American Mus. Nov. No. 619:1–29.

Oliver, J. A. 1955. The Natural History of North American Amphibians and Reptiles. Princeton, N.J., Van Nostrand-Reinhold. 359 pp.

——— 1956. Reproduction in the King Cobra, Ophiophagus hannah Cantor. Zoologica 41:145–152.

Pennock, L. A. 1965. Triploidy in parthenogenetic species of the teiid lizard, genus Cnemidophorus. Science 149:539–540.

Reese, A. M. 1915. The Alligator and Its Allies. New York, G. P. Putnam's Sons.

Robertson, I. A. D., B. M. Chapman, and R. F. Chapman. 1965. Notes on the biology of the lizards Agama cyanogaster and Mabuya striata striata collected in the Rukwa Valley, Southwest Tanganyika. Proc. Zool. Soc. London 145:305–320.

Saint Girons, H. 1962. Presence de receptacles séminaux chez les caméleons. Beaufortia 9:165–172.

Sharell, R. 1966. The Tuatara, Lizards and Frogs of New Zealand. London, Collins. 94 pp.

Smith, M. 1937. Breeding habits of the Indian cobra. J. Siam. Soc. Nat. Hist. Suppl. 11:62–63.

Stebbins, R. C., and R. M. Eakin. 1958. The role of the "third eye" in reptilian behavior. American Mus. Nov. No. 1870:1–40.

Taylor, H. L., J. M. Walker, and P. A. Medica. 1967. Males of three normally parthenogenetic species of teiid lizards (genus Cnemidophorus). Copeia 1967:737–743.

Telford, S. R., and H. W. Campbell. 1970. Ecological observations on an all female population of the lizard Lepidophyma flavimaculatum (Xantusidae) in Panama. Copeia 1970:379–381.

Thomas, R. 1965. The smaller teiid lizards (Gymnophthalmus and Bachia) of the southeastern Caribbean. Proc. Biol. Soc. Washington 78:141–154.

Tinkle, D. W. 1959. Observations on the lizards Cnemidophorus tigris, Cnemidophorus tesselatus, and Crotaphytus wislizeni. Southwest. Naturalist 4:195–200.

——— 1961. Population structure and reproduction in the lizard Uta stansburiana stejnegeri. Am. Midland Nat. 66:206–234.

Uzzell, T. 1970. Meiotic mechanisms of naturally occurring unisexual vertebrates. Am. Naturalist 104:433–445.

Vinegar, A. 1968. Brooding of the eastern glass lizard, Ophisaurus ventralis. Bull. S. California Acad. Sci. 67:65–68.

Woodbury, A. M., and R. Hardy. 1948. Studies of the desert tortoise. Gopherus agassizii. Ecol. Monographs 18:145–200.

Wright, J. W. 1967. A new uniparental whiptail lizard (genus Cnemidophorus) from Sonora, Mexico. J. Arizona Acad. Sci. 4:185–193.

———, and C. H. Lowe. 1968. Weeds, polyploids, parthenogenesis, and the geographical and ecological distribution of all-female species of Cnemidophorus. Copeia 1968:128–138.

Yntema, C. L. 1968. A series of stages in the embryonic development of Chelydra serpentina. J. Morphol. 125:219–252.

Zehr, D. R. 1962. Stages in the normal development of the common garter snake Thamnophis sirtalis sirtalis. Copeia 1962:322–329.

Zweifel, R. G. 1965. Variation in and distribution of the unisexual lizard, Cnemidophorus tesselatus. American Mus. Nov. No. 2235:1–49.

CHAPTER 13

REPRODUCTIVE ISOLATION OF AMPHIBIAN AND REPTILIAN SPECIES

The modern concept of a sexually reproducing species is that of a group of actually or potentially interbreeding populations which are, as populations, effectively isolated reproductively in nature from other such groups so as to maintain their genetic identity. This concept has both a negative and a positive aspect regarding interbreeding and, hence, regarding gene flow: (1) all populations of the same species have at least a potential ability to interbreed with other populations of the same species and, therefore, have at least a potential access to all the genetic variability found within that species; (2) populations of one species are in nature effectively prevented from interbreeding with populations of another species and, therefore, do not have access to the genetic variability found in another species.

The second aspect of the above concept implies that species are genetically isolated from other species by one or a combination of isolating mechanisms which impede the formation of viable and fertile hybrids. Such mechanisms, acting singly or in combinations to separate species in nature, result in one species not being able to contribute genetically to the evolution of another. This is so simply because the formation of a significant number of viable and fertile hybrids would provide an avenue for gene flow between species

through the mechanism of backcrossing. If sufficient numbers of viable and fertile hybrids were produced, each of the gene pools of the hybridizing populations would become diluted with genetic material from the other and the two populations would merge genetically into one.

Various types of isolating mechanisms exist which work in a variety of ways and differ in terms of selective advantage to the species involved. As is probably true for all sexual species, no species of amphibian or reptile is known to be separated from other species by a single isolating mechanism unless that mechanism is geographical isolation. Reproductive isolation of sympatric species generally occurs as the result of a combination of isolating mechanisms which reinforce one another. These mechanisms are subject to natural selection and, in general, have been evolved over a long period of time representing the history of the species concerned. The following table presents a classification of the kinds of reproductive isolating mechanisms encountered among amphibians and reptiles. (See table on the following page.)

The difference between premating mechanisms and postmating mechanisms is a fundamental one since premating mechanisms prevent the wastage of gametes (eggs and sperm); postmating mechanisms do not prevent the wastage

A CLASSIFICATION OF REPRODUCTIVE ISOLATING MECHANISMS

I. *Premating mechanisms* (Interspecific crosses are prevented.)
   A. Potential mates are physically isolated:
      1. Geographical isolation of breeding populations exists.
      2. Ecological isolation of breeding populations exists.
      3. Temporal isolation of breeding populations exists.
   B. Potential mates are not physically isolated:
      1. Sexual isolation exists (potential mates do not attempt to mate).
      2. Mechanical isolation exists (potential mates attempt unsuccessfully to mate but no transfer of sperm occurs).

II. *Postmating mechanisms* (Interspecific crosses are unsuccessful.)
   A. Gamete mortality prevents fertilization of egg.
   B. Zygotic mortality prevails (individual dies after fertilization of egg and before reaching sexual maturity).
   C. F-1 hybrids are viable but sterile.
   D. F-2 hybrids are inviable or sterile.

of gametes. Consequently, premating isolating mechanisms are more efficient in that they do not involve a wastage of reproductive potential. Of the various premating mechanisms, mechanical isolation is the least efficient, since it does involve wastage of energy by the individuals attempting to copulate. Thus, when potential mates belonging to different species are not physically isolated, natural selection will result in the eventual evolution of sexual isolating mechanisms because they conserve both energy and gametes. Premating mechanisms which involve the physical isolation of potential mates need only be operative during the breeding season to be effective. This is a particularly important point to keep in mind when dealing with sympatric species which, although possibly coexisting throughout the year, may be temporally isolated by restricted and non-overlapping breeding seasons. Contemporaneous sympatric species may breed at the same time of the year but be ecologically isolated because they have different kinds of breeding sites. The isolation which results in either of these situations, like that stemming from mechanical isolation

or sexual isolation, has at least a partial genetic basis. If for no other reason, populations which are permanently isolated temporally, ecologically, or geographically are reproductively isolated simply because the extrinsic factors of time and space prevent members of one species from encountering members of the other species. For this reason, Mayr (1970) restricts the term isolating mechanism to mechanisms which are biological properties of individuals preventing the interbreeding of populations that are actually or potentially sympatric.

Postmating isolating mechanisms all operate after the actual mating process and are probably rarely, if ever, established or intensified through the action of natural selection (Mecham, 1961). They have no selective basis since they do not prevent cross-matings. Postmating isolating mechanisms generally fall under the heading of genetic incompatibility. Genetic incompatibility results from chromosomal and genetic differences of such a magnitude that there is either gametic incompatibility (no fertilization of the egg), hybrid inviability (the hybrid dies before reaching sexual maturity), or hybrid sterility (the hybrid survives but is, either in the F-1 or F-2 generation, not able to backcross to either parent). Genetic incompatibility between two related but isolated populations may arise through the gradual accumulation of genetic differences because of differences in natural selection and chance, or there may be an abrupt attainment of isolation through the production of polyploid individuals which are not compatible with the parental population.

Once genetic incompatibility comes into existence it is, for all practical purposes, irreversible. When there is gamete incompatibility, hybrid inviability, or hybrid sterility, natural selection operates against cross-matings and causes the evolution of sexual isolating (premating) mechanisms: the tendencies for individuals to mate with individuals of another species are being selected against, since such matings do not leave fertile offspring in the next generation. Simultaneously, whatever tendencies there are for individuals to mate with conspecifics are being selected for, since the progeny of such intraspecific matings are the viable and fertile offspring of the next generation. Thus, the occurrence of cross-matings between genetically incompatible species will lead to the evolution of sexual isolating mechanisms and reinforcement of reproductive isolation. Such sexual isolating mechanisms will not evolve between species which are prevented from cross-mating except by chance or as the by-product of selection for something else about the organisms. Thus, it is generally expected that two closely-related but geographically isolated species will freely cross-mate when they are first brought together. Sexual isolating mechanisms involve behavioral barriers to cross-matings and center around features which are discriminated in species and sex recognition. As discussed in Chapter Eleven and Twelve, such features include visual, auditory, and chemical stimuli.

POSTMATING ISOLATING MECHANISMS AMONG AMPHIBIANS

Studies of the postmating isolating mechanisms of amphibians have been primarily concentrated on anurans. Naturally occurring interspecific hybrids have been reported within the anuran genera *Bufo* (true toads), *Gastrophryne* (narrow-mouthed toads), *Rana* (true frogs), *Hyla* (tree frogs), and *Scaphiopus* (spadefoot toads); in the literature are numerous references to hybridization in each of these. Much additional information has resulted from artificial crosses of species in the laboratory. No genus of anurans has been completely analyzed to determine the presence or absence of postmating isolating mechanisms between its member species, but such an analysis is nearing completion for the genus *Bufo* (Blair, 1963 and elsewhere). The wealth of data on interspecific genetic compatibility of toads has revealed that postmating reproductive isolation takes a long time to evolve in *Bufo*, for what are obviously distinct species commonly produce viable hybrids when crossed in the laboratory. Porter (1968) studied the relict Wyoming population of Dakota toads (*Bufo hemiophrys baxteri*), a southern disjunct population that reflects south-

ward displacement in the Wisconsin glaciation and subsequent return northward of the main body of the species and one that has been geographically isolated for approximately 10,000 years. He found that a measurable amount of divergence has occurred in both mating call and morphology but that this period of isolation has been insufficient to produce a new species. The results from intrapopulational and interpopulational crosses of toads from Wyoming, the Dakotas, and Manitoba exhibited little difference in fertility, mortality, or rate of development. In contrast to the general situation in toads, the data which are accumulating on *Rana* (Moore, 1941 and elsewhere; Volpe, 1954; Zweifel, 1955; Volpe and Harvey, 1958; Porter, 1961, 1969) indicate that the genetics of true frogs and their embryological requirements are such that hybrid development often fails even in crosses of closely related species. Porter (1969) found that a relict population of wood frogs (*Rana maslini*), isolated for what has probably been the same length of time as the Wyoming population of Dakota toads, exhibits a high degree of incompatibility when crossed with its parental species (*Rana sylvatica*).

Viable hybrids have been obtained in artificial intergeneric crosses within the families Hylidae (tree frogs) and Microhylidae (narrow-mouthed toads). Blair (1941) and Gosner (1956) were able to obtain a small percentage of metamorphosed hybrids in crosses of *Pseudacris triseriata* (northern chorus frog) and *Hyla crucifer* (spring peeper). Littlejohn (1959) crossed female *Gastrophryne olivacea* (=*Microhyla olivacea*) with male *Chiasmocleis panamensis* and found that hybrids of these narrow-mouthed toads survived beyond metamorphosis. Thus, the expression of genetic incompatibility does not always correlate with taxonomic relationships.

Relatively little study has been made of postmating isolating mechanisms within most genera of Urodela. Naturally occurring hybrids of European newts are known from cross-matings of *Triturus helveticus* and *Triturus vulgaris* (Benazzi, 1957) and of *Triturus cristatus* and *Triturus marmoratus* (Lantz, 1947). Hybrids of the latter have been referred to as *Triturus blasii* and *T. trousessarti* in older literature. Most of the experimental data bearing on postmating isolating mechanisms among urodeles are also on newts. Experimental hybridization of *Triturus cristatus* and *Triturus marmoratus* (White, 1946; Lantz, 1947) has shown that F-1 female hybrids are fertile in backcrosses with either parental species but that F-1 males are generally sterile. The sterility of male hybrids is due to the failure of chromosomes to pair in meiosis and to the degeneration of spermatids. Several workers, including Hamburger (1936), have been able to make artificial reciprocal crosses of *Triturus cristatus* (=*Triton cristatus*) and *Triturus helveticus* (=*Triton palmatus*). Oyama and Nakamura (1939) were able to artificially fertilize *Triturus pyrrhogaster* (=*Cynops pyrrhogaster pyrrhogaster*) eggs with *Triturus ensicauda* (=*Cynops pyrrhogaster ensicauda*) sperm. The resulting hybrids developed normally and metamorphosed within five months. Later, Kawamura (1952) reported that four allotriploids reared for several years had poorly developed ovaries and no capacity for ovulation. Fischberg (1948) was able to obtain viable hybrids by artificially crossing *Triturus helveticus* (=*Triton palmatus*) females with *Triturus alpestris* (=*Triton alpestris*) males. The genetics of this is interesting since Fischberg found that the *helveticus* traits were generally dominant over those of *alpestris*. Benazzi (1943) artificially crossed *Triturus cristatus* with *Triturus taeniatus* and obtained preponderantly female hybrids. The general picture from the above data is that postmating isolating mechanisms are not generally well developed within the genus *Triturus*. The homogametic sex (female) is frequently fertile in F-1 hybrids, although the heterogametic sex (male) is commonly sterile. In contrast, Callan and Spurway (1951) made an intraspecific study of *Triturus cristatus* and found the chromosome complement of *T. cristatus karelinii* differs from that of *T. cristatus carnifex* by at least one translocation and from that of *T. cristatus cristatus* by at least two. These translocations give rise to genetic duplications and deficiencies upon segregation at meiosis; the subspecies also differ by inversions. Despite the large zones of intermediacy which occur in nature between these subspecies

and the fact that F-1 hybrids are vigorous, F-2 and backcross hybrids have a high larval and metamorphic mortality. This would, accordingly, seem to represent a situation where the selection against hybrids will lead to the evolution of premating isolating mechanisms and the divergence, rather than covergence, of the three taxonomic entities.

In the New World, the late Victor Twitty and his associates have made comprehensive studies of the genetics and mating behavior of western newts belonging to the genus *Taricha*. There is no evidence that the three species belonging to this genus hybridize in nature, even where two or more coexist in the same region or even where they choose a common breeding site, as frequently happens (Twitty 1939, 1959). However, Twitty (1955) was able to obtain large numbers of F-1 hybrids by using artificial fertilization to cross *Taricha rivularis* with *Taricha torosa* (*T. torosa torosa* and *T. torosa sierrae*). These F-1 hybrids were released as larvae or newly metamorphosed juveniles into sections of Pepperwood Creek, which was being used as an experimental stream. Subsequently, sexually mature F-1 hybrid adults were recaptured and used for tests of their fertility and for genetic analyses of species' characters (Twitty, 1961). Twitty found both the fertility of hybrids and the viability of the second generation offspring were high in most of the backcrosses and F-2 generations. This indicates that the evolutionary separation of the two species and the preservation of their integrity, through ethological and other mechanisms of reproductive isolation, have been achieved without the differentiation of basically incompatible genomes. Despite the effectiveness of the reproductive isolation of these species in nature, Twitty found that his F-1 hybrids would mate voluntarily and effectively with native *T. rivularis* adults; this indicates that the functional isolating mechanisms, effective as they are, are delicate indeed. Twitty concluded that the species differences are each referable to mutations affecting a very limited number of genes. The character differences segregated independently of one another and there was little, if any, indication of reconstitution of parental species in backcrosses or interbreeding of F-1 adults.

Other workers (e.g., Connon, 1947) have produced hybrids artificially by crossing *Taricha torosa* with the third species, *Taricha granulosa*. Thus, as with the Old World newts (*Triturus*), New World newts in the genus *Taricha* have not evolved very efficient postmating isolating mechanisms, the implication being that all species are genetically very similar.

Virtually nothing is known about postmating isolating mechanisms within other genera of urodeles or within the order Apoda.

POSTMATING ISOLATING MECHANISMS AMONG REPTILES

Compared to the Amphibia, reptiles seem to hybridize very infrequently. Clark (1935) reported on the occurrence of a probable hybrid between eastern and western box turtles (*Terrapene carolina* and *Terrapene ornata*) found in 1923 in Indiana, and Mertens (1964) mentions hybridization of the tortoises *Testudo marginata* and *Testudo graeca*. However, turtle hybrids seem to be very rare.

No crocodilian hybrids are known and, of course, there are no similar species with which the tuatara might hybridize.

Snakes have been known to hybridize, both in the wild and in captivity. Naturally occurring hybrids have been found in Europe from cross-mating of diced water snakes (*Natrix tessellata*) with grass snakes (*Natrix natrix*) and of sand vipers (*Vipera ammodytes*) with adders (*Vipera berus*); hybridization of these species has also been recorded in captivity (Mertens, 1960, 1964). Klauber (1956) lists three specimens collected in the wild which, because of the intermediacy of their characters, appear to be hybrids of rattlesnakes; he also discusses three instances of interspecific matings of captive rattlesnakes which produced viable hybrids. The three wild specimens appear to be hybrids resulting from (1) a cross between an eastern diamondback (*Crotalus adamanteus*) and the canebrake rattlesnake (*Crotalus horridus atricaudatus*), (2) a cross between a red diamond rattlesnake (*Crotalus ruber ruber*) and a southern Pacific rattlesnake (*Crotalus viridis helleri*), and (3) an intergeneric cross between an eastern massasauga (*Sistrurus catenatus catenatus*) and the timber rattler (*Crotalus horridus horridus*). A brood of nine hybrids was produced in the San Diego Zoo by the artificial association of a male southern Pacific rattlesnake with a female red diamond rattlesnake: only one of these hybrids took food, and it lived to be an age of over nine years; the other eight were preserved as juveniles. Attempts to backcross the surviving hybrid failed and its fertility could not be determined. However, the hybrid's behavior of courting males and "dancing" with females indicated it was incapable of sex recognition! Another litter of four hybrids was produced at the San Diego Zoo through the association of a male Mojave rattlesnake (*Crotalus scutulatus scutulatus*) with a female Aruba Island rattler (*Crotalus unicolor*); these species are geographically isolated in nature. All four of the hybrids (one female and three males) survived and were kept together and isolated from other snakes. Matings of the F-1 siblings produced five different broods of F-2 hybrids, none of which survived to reach sexual maturity. The third instance of hybridization by captive rattlesnakes occurred at the University of California at Berkeley. Here the artificial association of a male northern Pacific rattlesnake (*Crotalus viridis oreganus*) with a female Mojave rattlesnake resulted in a brood of 12 hybrids; three of these hybrids survived for over four years, but their fertility could not be verified.

The natural hybridization of parthenogenetic lizards with closely related bisexual forms is known to occur in both the Old World and the New World. Study of the Old World parthenogenetic lizards related to *Lacerta saxicola* has shown that parthenogenetic females of *Lacerta armeniaca*, *L. dahli*, *L. rostombekovi*, and *L. unisexualis* sometimes mate with neighboring bisexual forms of *Lacerta saxicola*, producing hybrids that are always sterile triploid (3n=57) females (Darevskii, 1967). The triploidy results from the union of the diploid parthenogenetic ovum and a haploid sperm. In this situation, the absence of males plays the role of an isolating mechanism and maintains the integrity of the parthenogenetic species. Darevskii and Danielyan (1969) found that the mating of parthenogenetic females of *Lacerta armeniaca* and *L. unisexualis* with males of *Lacerta saxicola valentini* usually results in broods containing both diploid and triploid individuals; the triploids represent about 40 per cent of the broods for *L. unisexualis* and 43 per cent for *L. armeniaca*. The mixed broods of diploids and triploids are explained by the fact that when normally non-mating parthenogenetic females are mated there is generally only fertilization of a portion of the diploid eggs that have been ovulated.

In the New World, Wright and Lowe (1967) have shown that there is widespread hybridization of the sympatric parthenogenetic whiptail lizard *Cnemidophorus neomexicanus* with the bisexual *C. inornatus*. These workers also (Lowe and Wright, 1967) have proposed that *Cnemidophorus neomexicanus* is an allodiploid with a complement of 46 chromosomes that originated by hybridization of the sympatric and bisexual *C. tigris* (n=23) and *C. inornatus* (n=23). Thus, the aforementioned hybridization of *C. neomexicanus* with the bisexual *C. inornatus*, which produces triploids originally described as *C. perplexus*, represents backcrossing and the fertilization of a diploid egg by a haploid sperm as described above for *Lacerta*. Lowe and Wright (1967) propose in addition that the triploid parthenogenetic *Cnemidophorus uniparens* (3n= 69) originated by hybridization of two bisexual species, *C. inornatus* and *C. gularis*, to form an intermediate (and now apparently extinct) allodiploid parthenogenetic species; this species subsequently backcrossed with *C. inornatus* to form the existing triploid *C. uniparens*. Thus, *C. uniparens*, like "*C. perplexus*," is viewed as an allotriploid. Karyological and morphological evidence indicates that *C. cozumela* is an allodiploid hybrid between *C. deppei* and *C. angusticeps*, and it is presumed that *C. rodecki* is also the result of hybridization of these two species (Fritts, 1969).

The ability to reproduce parthenogenetically has allowed the survival and perpetuation of hybrid lizard populations. However, postmating isolating mechanisms have prevented gene flow between the hybridizing parental species, and the hybridizing parental species, and their integrity has been maintained. The general lack of hybrids of other reptilian species implies efficient reproductive isolation of most species, but whether the mechanisms involved are in the premating or the postmating category is difficult to ascertain. The fact that artificial association of viperid species has led to the production of hybrids indicates that their premating isolating mechanisms are not too efficient and that they may break down in nature. However, available evidence based on a very limited number of hybrids indicates that postmating isolation of rattlesnakes would operate through infertility of F-1 or F-2 hybrids in the event of such a breakdown of premating mechanisms.

PREMATING ISOLATING MECHANISMS AMONG AMPHIBIANS

GEOGRAPHICAL ISOLATION

As indicated in the introductory remarks, premating isolating mechanisms may be geographical, ecological, temporal, sexual, or mechanical. Most closely related species come into existence through the gradual genetic divergence of geographically isolated sister populations. Hence, geographical isolation is frequently the most important and effective isolating mechanism separating sister species. Obviously, if two species do not occur in the same geographical region they are unable to interbreed or jeopardize the genetic integrity of one another. In view of the slow evolution of postmating isolating mechanisms among amphibians and the lack of selection for premating isolating mechanisms between allopatric (geographically isolated) species, there may be many species of amphibians which are reproductively isolated only by geography. This surely is the case with the sister pair of Mexican toads *Bufo mazatlanensis* and *Bufo gemmifer* (see distribution map in Figure 6–27, page 266), since their mating calls are indistinguishable (Porter, 1964), and Blair (1966) has demonstrated a high degree of genetic compatibility within the *Bufo valliceps* species group. One might question whether such populations should be considered different species.

ECOLOGICAL ISOLATION

Ecological isolation of amphibian species can result from their utilizing different breeding sites or, in the case of anurans, from different species utilizing different calling positions about the same general breeding site. Ecological isolation of breeding sites appears to be a widespread type of isolating mechanism among amphibians. Those species which breed on land, such as lungless salamanders (family Plethodontidae) and many anurans in the families Leptodactylidae and Microhylidae, are effectively isolated from other amphibian species which occur in the same area but which breed in water. Similarly, there are only remote possibilities for cross-matings of species which breed in running water with those which breed in standing water. The Mexican toads *Bufo marmoreus* and *Bufo canaliferus*, sympatric species on the Isthmus of Tehuantepec with very similar mating calls (Porter, 1964, 1966), exemplify the latter situation: they have little opportunity for cross-mating since *B. canaliferus* always breeds in running water and *B. marmoreus* typically breeds in quiet water, both temporary and permanent. The author has observed these two species breeding within approximately 100 yards of one another along the Rio Ostuto in Oaxaca; *Bufo canaliferus* was calling from the river bank near the main channel and *Bufo marmoreus* was utilizing the still backwater of a pool formed by the same river. However, I have never found the two species breeding together. The breeding site of amphibians, whether on land, in running water, or in standing water, is a species-specific characteristic, since there are fundamental embryological adaptations associated with these ecological differences. Consequently, ecological isolation of amphibian breeding sites is probably more effective in preventing interbreeding than such isolation is in other classes of vertebrates.

As noted by Mecham (1961) differences in calling position of male anurans may operate as an isolating mechanism in two related ways: by spatially isolating the breeding congregations of different species and by making males of one species relatively inaccessible to females of another species. If one species consistently calls while floating in the water, another while sitting on the ground near the water's edge, and another from nearby vegetation, as illustrated in Figure 13–1, the spatial isolation of the species may be sufficient to deter cross-mating. Mecham (1965) found that isolation of the closely related tree frogs *Hyla gratiosa* and *H. cinerea*, sympatric in southeastern United States, may to a large degree be the result of different calling sites. Males of *H. gratiosa* typically call while floating in the open water and only rarely call from shore or from vegetation. Males of *H. cinerea*, on the other hand, call from the ground or from vegetation but never from the water. Thus, males of *H. gratiosa* probably have little opportunity to clasp females of *H. cinerea*. The existence of natural fertile hybrids and introgression between the two species indicates that all reproductive isolating mechanisms break down occasionally. Littlejohn (1958) described another example of isolation through calling position of three species of Australian leptodactylids. *Crinia sloanei* usually calls while floating in the water, *C. signifera* calls from the near shore while partly immersed in water and under cover of overhanging grasses, and *C. parinsignifera* calls from clumps of grass out of water. Since nearly all anuran species call from characteristic locations of this type, the spatial isolation of males probably acts as a filter in restricting cross-mating, but it is probably never totally effective because of different species sharing the same calling sites and because of various degrees to which males are inaccessible to females. Duellman (1967) studied the courtship isolating mechanisms separating 10 hylid species breeding in the same Costa Rican pond and found that two or three species shared the same kind of calling site. The degree to which a male of one species will be inaccessible to females of another species depends upon the psychological differences between the species which might, for example, inhibit the female of a shore-calling species from entering the water where she could be clasped by males of a water-calling species. A female belonging to a water-calling species might be inhibited from climbing

Figure 13–1 The typical calling sites of breeding males of eight species of anurans which often breed simultaneously at the same pond in Florida. Shown, clockwise, from bottom center: a southern toad *(Bufo terrestris)* in shallow water near edge of pool; a green tree frog *(Hyla cinerea)* on a limb below overhanging foliage; a squirrel tree frog *(Hyla squirella)* clinging to low vegetation near pond's edge; a pine woods tree frog *(Hyla femoralis)* on a low bush; a leopard frog *(Rana pipiens)* floating on the surface of deeper water; an oak toad *(Bufo quercicus)* under vegetation on shore (usually in a more obscure position); a narrow-mouthed toad *(Gastrophryne carolinensis)* in shallow water near the edge of pool (usually with little more than the head extended above the water); and a southern cricket frog *(Acris gryllus)* on shore near the water's edge. (Drawn by Carol Elsaesser, after Bogert: The influence of sound on the behavior of amphibians and reptiles.)

vegetation where the male of another species might be calling (Mecham, 1961).

TEMPORAL ISOLATION

Differences in breeding season, reflecting differing responses to precipitation, photoperiod, and/or temperature, serve as an isolating mechanism helping to separate some sympatric amphibian species. For example, where both the American toad *(Bufo americanus)* and the closely related Woodhouse's toad *(Bufo woodhousei)* are sympatric, the Woodhouse's toads begin breeding about two weeks after the peak in breeding activity of the American toads. The breeding season of wood frogs *(Rana sylvatica)* begins as soon as ice

melts from breeding ponds and generally much earlier than the breeding of other frogs in the same region. Despite such examples, temporal isolation of species is probably rarely completely effective since similar species in a given region are apt to have similar adaptations to temperature and photoperiod. This, together with a common requirement for availability of water (which is apt to be seasonal for species breeding in temporary water), makes complete separation of breeding seasons difficult to achieve. Consequently, temporal isolation is generally effective in separating peaks of breeding activity rather than totally isolating populations. It, like premating isolating mechanisms in general, is not very efficient without reinforcement from other types of isolating mechanisms.

SEXUAL ISOLATION

Sexual isolating mechanisms are generally the most important of the premating isolating mechanisms affecting amphibians. Among anurans, the principal sexual isolating mechanism involves emission of a species-specific mating call by breeding males and discrimination of this call by essentially voiceless females (Blair, 1958). With a few possible exceptions, such as a territorial function, the only function of the anuran mating call is to attract a mate. Because a sexually excited male will grasp indiscriminately any object of appropriate size, animate or inanimate (frogs have been found clasping apples!), the discrimination which takes place is almost entirely a function of the female. Among the call characteristics which appear to be discriminated by female anurans are the trill rate (notes per second), frequency(ies) of the dominant harmonic (Hertz, constant or changing), the length of the individual calls, and the interval between individual calls. The call functions as a homing device. As females move toward the source of the sound, they are drawn to males of their species. Thus, natural selection favors the call which attracts a female of the same species and operates against, through inviable or infertile hybrids, the call which attracts a female of another species. Natural selection, therefore, promotes uniformity of mating calls throughout the interbreeding population and differentiation of the calls of different species which breed at the same time and place.

The mating call is affected measurably by both the body temperature of the calling animal and the size of the animal's vocal structures. A linear increase generally occurs in both trill rate and frequency of harmonics with an increase in temperature. The general effect of increased size of vocal structures (in practice, usually measured by increased body size) is to decrease the trill rate and lower the frequency of the harmonics. Consequently, some variation in mating calls is found among individuals of the same species but, as a rule, this variation does not approach the variation between mating calls of different species which are sympa-

trically breeding. An example of the magnitude of call variation throughout the range of a widely distributed species may be found in Porter (1964).

Because natural selection favors differentiation of calls of species which breed together, mating calls of closely related species are most differentiated among those which breed at the same time and place. The magnitude of the call difference, to be functional, must only be sufficient so as to allow the female to discriminate it from other similar calls. Littlejohn and Michaud (1959) demonstrated experimentally that sexually stimulated females of Strecker's chorus frog (*Pseudacris streckeri*) could readily discriminate between recorded mating calls of their own species and those of the closely related and sympatric spotted chorus frog (*Pseudacris clarki*); Littlejohn (1961) later demonstrated that female spotted chorus frogs could equally well discriminate between calls of their species and those of the Strecker's chorus frog. Sound spectrographic analyses of the recordings actually used in these discrimination tests allowed quantification of the call differences; these differences are indicated in the following table. It is not known whether this discrimination depends upon a single factor or a combination of the differences, but in either case, it involves small magnitudes of difference. Although the *P. clarki* call is five times longer than that of *P. streckeri*, the difference in length is only 0.16 seconds. The 650 Hertz (cycles per second) difference in frequency of the dominant harmonic is within the range of individual variation for many anuran species. As noted by Bogert (1960), the differences between trilled and untrilled calls are not actually dichotomous but, rather, trill rates which are above or below that aurally detectable by the human ear or visually detectable on spectrograms of calls reproduced at normal speeds. When "untrilled" calls are reproduced at lower-than-normal speeds, they are found to consist of separate pulses. Thus, the difference between trilled and untrilled calls is really one of a faster trill rate in the untrilled call. (See table on the following page.)

Some anuran species have either lost their call or are in the process of doing so. These species are found in regions or habi-

PHYSICAL CHARACTERISTICS OF CALLS OF *PSEUDACRIS CLARKI* AND
PSEUDACRIS STRECKERI USED FOR DISCRIMINATION TESTS*

Place of Recording	Temperature		Duration	Dominant Frequency	Trill Rate
	Air	Water	(seconds)	(Hertz)	(notes per second)
P. clarki Welder Refuge, San Patricio Co., Texas	18° C	21° C	0.20	2850	75
P. streckeri 6 mi SSE of Luling, Caldwell Co., Texas	19° C	21.3° C	0.04	2200	not trilled

*From Littlejohn and Michaud, 1959.

tats where no other similar species occurs and, consequently, few or no opportunities for cross-mating exist. Among anurans of North America which fall into this category are western toads (*Bufo boreas*), Colorado River toads (*Bufo alvarius*), and mountain yellow-legged frogs (*Rana muscosa*).

The reproductive biology of urodeles is varied and relatively poorly known, but glandular secretions are known to play a major role in the mating patterns of this order. Thus, sexual isolating mechanisms of these, at least in some species, probably involves olfactory discrimination. Twitty (1955) demonstrated that in red-bellied newts (*Taricha rivularis*) a chemical secretion released by the female informs the male of her presence in the water. The male responds by swimming in the direction from which the odor is coming (generally upstream) until contact is made with the female. Twitty was able to demonstrate this by attracting males to sponges which he had soaked in water in which females had been stored. Presumably, the secretions discriminated in this manner are exposed to natural selection of the same sort as are mating calls. A secretion which attracts males of the same species will be selected for, through the production of viable and fertile offspring; secretions which attract males of another species will be selected against, through the production of inviable or sterile offspring. Thus, one would expect the chemicals involved in olfactory discrimination to be remarkably uniform throughout an interbreeding population and to differ from those of another similar species breeding at the same place and time. Some urodeles appear to have characteristic mating dances and, in these, visual discrimination may play an important role in sexual isolation of species.

Reproductive isolating mechanisms have not been studied in apodans, but it seems likely that olfactory discrimination may be important in sexually isolating species of this kind since their sense of vision is so poorly developed.

MECHANICAL ISOLATION

Mechanical isolation is important in separating those species lacking sexual isolating mechanisms and as a supplement to the latter in case they fail to prevent cross-mating. Mechanical isolating mechanisms may operate through the presence of interspecific morphological differences which are discriminated by the male when he attempts to clasp a female of another species. Such features as size, skin texture, or body proportions may be sufficient to cause a male anuran to release a female of another species. However, the differences between species generally have to be drastic to function in this manner since, again, sexually aroused male anurans will clasp a great range of objects. The author once observed three male Rocky Mountain wood frogs (*Rana maslini*) simultaneously clasping a single male western toad (*Bufo boreas*); one frog was clasping each hind leg of the toad and the third frog was clasping the toad about its groin. This mixed mating lasted for several minutes while the toad tried feebly to swim around on the surface and under water. There are obviously many differences in skin texture and body proportions between wood frogs and toads, but none of these was sufficient to prevent this bizarre mating from taking place. The quartet disappeared from view when my attention was on a male wood frog, so it is not known how long the frogs continued to clasp the toad. When male anurans are clasped by other males they generally emit a vibratory "release call" which signals the clasping male that the wrong sex has been gripped. Upon receiving the release call signal, the clasping male will normally release his grip immediately and swim off to look for another mate. It was interesting in the wood frog-toad incident to observe that the release call mechanism seemed to be non-communicative; apparently the release calls of toads are not always meaningful to frogs!

Inger and Greenberg (1956) found that there is a sexually dimorphic difference in skin texture of the African toads *Bufo funereus* and *B. regularis* which seems to function in sex recognition but may also be an isolating mechanism between these two species. The dorsal skin of the female *Bufo funereus* is covered with spiny warts, whereas that of the male is smooth. In *Bufo regularis*, the reverse is true; the female's skin is nearly smooth, while that of the male is very spiny. Thus, the skin may function intraspecifically in sex recognition and interspecifically as an isolating mechanism.

The most effective form of mechanical isolation stems from size differences. Because female amphibians are normally considerably larger than males, a minimal size is more effective than a maximal size in isolating species which clasp while mating. The size difference may operate psychologically and cause the male to release the female before gametes are shed, or it may operate in a truly mechanical fashion and make clasping impossible. As examples of the former situation, A. P. Blair (1942) observed that male Woodhouse's toads (*Bufo woodhousei*) were reluctant to clasp females of the smaller American toad (*Bufo americanus*). He also found (1950) that male eastern narrow-mouthed toads (*Gastrophryne carolinensis*) are reluctant

to clasp the smaller females of great plains narrow-mouthed toads (*Gastrophryne olivacea*).

PREMATING ISOLATING MECHANISMS OF REPTILES

GEOGRAPHICAL ISOLATION

Much remains to be learned about reptilian premating isolating mechanisms. There are undoubtedly many instances in which geographical isolation is the only form of premating isolating mechanism preventing species from interbreeding. For example, the hybridization of a Mojave rattlesnake with an Aruba Island rattler in the San Diego Zoo (see p. 441) indicates that these two species might also interbreed in nature if their ranges were not separated. Experimental hybridization may be used to evaluate the importance of geographical isolation to allopatric species, but this has seldom been attempted with reptiles.

ECOLOGICAL ISOLATION

Because their breeding site requirements are less rigid than those of amphibians, ecological isolation of entire breeding populations of otherwise similar and sympatric reptilian species is less probable. This is because, having both internal fertilization and, frequently, an ability to store sperm, reptiles may breed opportunistically and under varied environmental conditions. Ecological isolation of similar species is potentially possible if an aquatic species, such as a marine turtle, selects a particular beach for depositing its eggs and mates in the vicinity of that same beach. The spatial isolation of aquatic species using different beaches would be comparable to that which isolates amphibian species using different waters for breeding. Obviously, merely using a different beach for egg deposition will not reproductively isolate a population unless mating occurs subsequent to movement to that particular place.

The territorial behavior exhibited by many reptiles may isolate relatively similar species in the same manner that differences in calling position of male anurans may operate as an isolating mechanism. A surface-dwelling lizard, for example, may hesitate to climb to the vicinity of the male of another rock-climbing species whose perch is atop a large boulder.

Ecological isolation of dissimilar species undoubtedly does reduce the number of opportunities for cross-mating of these. For example, arboreal species have almost no opportunity for cross-mating with burrowing species; aquatic species have little opportunity for cross-mating with terrestrial species. The important difference between amphibians and reptiles in this regard is that reptiles have the ability to mate in the same environment in which they normally occur; most amphibians with external fertilization generally lack this ability and usually must mate in water regardless of where they occur during the nonbreeding season. Thus, ecologically dissimilar amphibians may breed in the same body of water.

TEMPORAL ISOLATION

The importance of temporal isolation as a barrier to gene flow between sympatric reptilian species is unknown. Many species breed upon emerging from hibernation and, if the species involved have restricted breeding seasons, it is possible that the early emergence of one species could prevent it from cross-mating with another species emerging at a later date. Goin and Goin (1962) suggest, for example, that such an isolating mechanism might separate eastern hognose snakes (*Heterodon platyrhinos*) and southern hognose snakes (*Heterodon simus*). The early emergence of eastern hognose snakes may mean that they copulate earlier in the year than southern hognose snakes, which are hot weather snakes. However, the actual time at which southern hognose snakes breed is unknown. It should also be noted that many hibernating species breed both in the fall and in the spring, and in such cases temporal isolation is less likely than when there is a single brief mating season each year.

The ability of many reptiles to store sperm means that mating need not coincide with ovulation but, rather, can occur at almost any time that males are producing viable sperm. In at least some species, males appear to be fertile throughout their seasonal period of activity; this reduces the probabilities for temporal isolation of sympatric species.

SEXUAL ISOLATION

Few species of reptiles have been studied to determine the behavioral traits that function as sexual isolating mechanisms. Instances where there is sexual dimorphism in turtles indicate that visual stimuli may be discriminated in species and sex recognition. Differences in courtship patterns may also be effective in sexually isolating turtles. For example, the sympatric pond sliders (*Pseudemys scripta*) and coastal plain cooters (*Pseudemys floridana*) have very different courtship patterns; the male pond slider swims backward in front of the female, whereas the male coastal plain cooter swims in the same direction of the female and just above her (Fig. 12–1). Since the courtship functions to make the female receptive to mating, such a difference in the courting patterns may result in sexual isolation because a female will not be stimulated by the actions of a male of a different species. Sound production is a normal part of the mating behavior of male tortoises, such as the Galapagos giant tortoise (*Geochelone elephantopus*), and may represent another sexual isolating mechanism functioning like the mating calls of male anuran amphibians; on the other hand, the sound production may merely be used by the tortoises to challenge rival males.

Sound production is also a conspicuous feature of crocodilian breeding behavior; Cott (1961) believes that one sound, a deep roar, produced by bull Nile crocodiles (*Crocodylus niloticus*) is a mating call to attract females. Thus, this may function to sexually isolate that species from others with which it might cross-mate.

At the present time, the only reptiles with which tuataras are sympatric are a few geckos and skinks. Thus, there must not be any selective pressure for sexual isolating mechanisms and it seems doubtful that this relict species would still retain any such mechanisms possessed by ancestors when similar species existed.

The order Squamata differs from the other orders of reptiles in possessing many more living species. Concomitantly, most lizards and snakes have geographical ranges which overlap those of other similar species, and opportunities for cross-mating are typically greater than for species in the more restricted orders. The more opportunity there is for cross-mating, the greater is the selective value of sexual isolating mechanisms. Thus, it is not surprising that the most highly developed forms of courtship and territorial behaviors in reptiles are found among the Squamata. Among lizards, which are typically diurnal, visual stimuli are primarily discriminated and form the basis for sexual isolation. Male lizards commonly have prominent patches of color which they exhibit from their perches and which are discriminated by females of the same species. Among species of lizards which have similar coloration, sexual isolation may operate through female discrimination of a species-specific pattern of head bobbing performed repeatedly by the male. As discussed in Chapter Twelve, head bobbing is a common feature of lizard behavior; it has been shown by Hunsaker (1962) to be of importance as an isolating mechanism between Mexican species of spiny lizards in the *torquatus* group of the genus *Sceloporus*. After observing that each of the species he studied (*S. cyanogenys, S. poinsetti, S. jarrovi, S. ornatus, S. dugesi, S. torquatus,* and *S. mucronatus*) had a characteristic pattern of head bobbing involving variations in the duration of time and magnitude of each extension of the front legs, Hunsaker mimicked these patterns with plastic model lizards attached by threads to rods which were moved up and down by cams driven by electric motors. He then allowed females in his experimental cage to select between three compartments which were either empty or contained models bobbing in random or species-specific patterns. In by far the majority of trials, the females selected the model which bobbed as their species would. In another series of observations, Hunsaker was able to verify that the female selects the partner with which to share a territory. Thus, the discrimination of the proper head bob attracts the female lizard to a male of her kind, and territorial behavior results in her remaining with him.

Comparatively little is known about sexual isolating mechanisms of snakes because they have not been studied directly. However, certain inferences may be drawn from observations on mating behavior. Noble's (1937) observations of a male snake following the scent trail left by a female indicates that olfactory discrimination by the male of odors left by the female may result in sexual isolation of species. Visual stimuli have been observed to be of some importance in species and sex recognition and may also provide a basis for sexual isolation.

What is not known is how discriminatory snakes are when they are given a choice of two or more odors or visual stimuli, such as might be provided by closely related species, or a single choice of a stimulus which is similar but not identical to that of their own species. It is obvious that the rattlesnakes which hybridized in captivity were not deterred from cross-mating despite any species differences they might have discriminated, but whether they would have cross-mated in nature is uncertain.

MECHANICAL ISOLATION

As with amphibians, mechanical isolation of reptiles may result because of the male's inability, or reluctance, to mate with a female of an improper size. Because female reptiles are generally larger than males of their species, a female's small size is probably more effective as an isolating mechanism than a large size. Although mechanical isolation of this type would seem to be a certainty when there is a great disparity in the size of the male and female, it is not known what minimum size difference must exist to effectively isolate one species from others.

All reptiles have internal fertilization and, in contrast to nearly all amphibians, male reptiles other than the tuatara possess copulatory organs which must be compatible with the female's cloaca if sperm transfer is to occur. This requirement allows for another type of mechanical isolation not applicable to amphibians: incompatibility of genitalia. The hemipenes of Squamata vary considerably in dimensions and spinosity from species to species but are remarkably uniform within species. Since the morphology of this structure must be adapted to the species' requirement for successful intraspecific mating, hemipenes differences from species to species may be sufficient to prevent successful mating and transfer of sperm in interspecific matings. Goin and Goin (1962) suggest that this may be the primary isolating mechanism separating two Asiatic species of lancehead snakes, *Trimeresurus stejnergeri* and *T. albolabris* (family Viperidae). These two species are very similar morphologically, but the hemipenes of *T. stejnegeri* are short, thick, and not deeply forked and bear heavy spines. Those of *T. albolabris* are longer, more slender and more deeply forked and lack spines. Coinciding with these masculine differences, the female cloaca of *T. stejnegeri* is shorter, not deeply lobed, and lined with thicker walls than that of *T. albolabris*. Thus, the cross-mating of a male *T. stejnegeri* with a female *T. albolabris* could result in damage to the female's cloaca serious enough, perhaps, to jeopardize her life. Cross-mating of a male *T. albolabris* with a female *T. stejnegeri* might fail because the short cloaca and lack of spines on the hemipenis might prevent its being held properly during copulation.

References

Benazzi, M. 1943. Sviluppo e sessualità in ibridi di tritoni. Boll. Soc. Ital. Biol. Sperim. *18(5)*:156–157.

———— 1957. Sulla ibridazione fra *Triturus helviticus* e *T. vulgaris*. Boll. Zool., *24(2)*:235–242.

Blair, A. P. 1941. Isolating mechanisms in tree frogs. Proc. Natl. Acad. Sci. U.S.A. *27(1)*:14–17.

———— 1942. Isolating mechanisms in a complex of four species of toads. Biol. Symposia 6:235–249.

———— 1950. Notes on Oklahoma microhylid frogs. Copeia *1950*:152.

Blair, W. F. 1958. Mating call in the speciation of anuran amphibians. Am. Naturalist *92*:27–51.

———— 1963. Evolutionary relationships of North American toads of the genus *Bufo*: a progress report. Evolution *17*:1–16.

———— 1966. Genetic compatibility in the *Bufo valliceps* and closely related groups of toads. Texas J. Science *18*:333–351.

Bogert, C. M. 1960. The influence of sound on the behavior of amphibians and reptiles. *In:* W. W. Lanyon and W. N. Tavolga (eds.). Animal Sounds and Communication. Washington, D. C., A. I. B. S. Publ. No. 7. Pp. 137–320.

Callan, H. G., and H. Spurway. 1951. A study of meiosis in interracial hybrids of the newt *Triturus cristatus*. J. Genetics *50(2)*:235–249.

Clark, H. W. 1935. On the occurrence of a probable hybrid between eastern and western bay turtles, *Terrapene carolina* and *T. ornata* near Lake Maxinkuckee, Indiana. Copeia *1935*:148–150.

Connon, F. E. 1947. A comparative study of the respiration of normal and hybrid *Triturus* embryos and larvae. J. Exp. Zool. *105*:1–24.

Cott, H. B. 1961. Scientific results of an enquiry into the ecology and economic status of the Nile crocodile (*Crocodilus niloticus*) in Uganda and Northern Rhodesia. Trans. Zool. Soc. London *29*:211–356.

Darevskii, I. S. 1967. O taksonokiche skom range partenogeneticheskikh form skal'noi yashcheritsy (*Lacerta saxicola* Eversmann) o svyazi s voprosom o primenenii vidovykh kriteriev k agamnym vidam. Zool. ZH *46(3)*:413–419.

————, and F. D. Danielyan. 1969. Diploidyne i triploidyne osobi v potomstve partenogeneticheskikh samok skal'nykh yashcherits, estestvenno sparivayushchikhsya s samtsami blizkikh biseksual'nykh vidov. Dokl. Akad. Nauk SSR *184(3)*:727–730.

Duellman, W. E. 1967. Courtship isolating mechanisms in Costa Rican hylid frogs. Herpetologica *23*:169–183.

Fischberg, M. 1948. Bestehen inder Ausbildung der Artmerkmale Unterschiede zwischen den diploiden und triploiden Bastarden von *Triton palmatus* ♀ und *Triton alpestris* ♂ ? Rev. Suisse Zool. 55: 304–310.

Fritts, T. H. 1969. The systematics of the parthenogenetic lizards of the *Cnemidophorus cozumela* complex. Copeia *1969*:519–535.

Goin, C. J., and O. B. Goin. 1962. Introduction to Herpetology. San Francisco, W. H. Freeman, 341 pp.

Gosner, K. L. 1956. Experimental hybridization between two North American tree frogs. Herpetologica *12(4)*:285–289.

Hamburger, V. 1936. The larval development of reciprocal species hybrids of *Triton taeniatus* Leyd. (and *Triton palmatus*, Duges) × *Triton cristatus*, Laur. J. Exp. Zool. *73*:319–374.

Hunsaker, D., II. 1962. Ethological isolating mechanisms in the *Sceloporus torquatus* group of lizards. Evolution *16*:62–74.

Inger, R. F., and B. Greenberg. 1956. Morphology and seasonal development of sex characters in two sympatric African toads. J. Morph. *99*:549–574.

Kawamura, T. 1952. Triploid hybrids of *Triturus pyrrhogaster* and *T. ensicauda*. Annot. Zool. Japonenses *25(1/2)*:218–224.

Klauber, L. M. 1956. Rattlesnakes; Their Habits, Life Histories, and Influence on Mankind. Berkeley and Los Angeles, University of California Press. 2 vols.

Lantz, L. A. 1947. Hybrids between *Triturus cristatus* Laur. and *Triturus marmoratus* Latr. Proc. Zool. Soc. London *117(1)*:247–258.

Littlejohn, M. J. 1958. A new frog of the genus *Crinia* Tschudi from southeastern Australia. Proc. Linn. Soc. New South Wales *83*:222–226.

———— 1959. Artificial hybridization within the Pelobatidae and Microhylidae. Texas J. Science *11(1)*:57–59.

———— 1961. Mating call discrimination by females of the Spotted Chorus Frog (*Pseudacris clarki*). Texas J. Science *13(1)*:49–50.

————, and T. C. Michaud. 1959. Mating call discrimination by females of Strecker's Chorus Frog (*Pseudacris streckeri*). Texas J. Science *11(1)*:86–92.

Lowe, C. H., and J. W. Wright. 1967. Evolution of parthenogenetic species of *Cnemidophorus* (whiptail lizards) in western North America. J. Arizona Acad. Sci. *4(2)*:81–87.

Mayr, E. 1970. Populations, Species, and Evolution;

an abridgment of Animal Species and Evolution. Cambridge, Mass., Belknap, Harvard Univ. Press. 453 pp.

Mecham, J. S. 1961. Isolating mechanisms in anuran amphibians. *In:* W. F. Blair (ed.), Vertebrate Speciation. Austin, Univ. Texas Press. Pp. 24–61.

——— 1965. Genetic relationships and reproductive isolation in southeastern frogs of the genera *Pseudacris* and *Hyla*. Am. Midland Naturalist 74:269–308.

Mertens, R. 1960. The World of Amphibians and Reptiles. Translated by H. W. Parker. London, George C. Harrap. 207 pp.

——— 1964. Über Reptilienbastarde, III. Senckenberg. biol. 45:33–49.

Moore, J. A. 1941. Developmental rate of hybrid frogs. J. Exp. Zool. 86:405–422.

Noble, G. K. 1937. The sense organs involved in the courtship of *Storeria*, *Thamnophis* and other snakes. Bull. American Mus. Nat. Hist. 73:673–725.

Oyama, J., and D. Nakamura. 1939. On the hybrid between *Triturus pyrrhogaster* (Boie) ♀ and *T. ensicauda* (Hallowell) ♂. Japanese J. Genetics 15(2):78–79.

Porter, K. R. 1961. Experimental crosses between *Rana aurora aurora* Baird and Girard and *Rana cascadae* Slater. Herpetologica 17:156–165.

——— 1964. Morphological and mating call comparisons in the *Bufo valliceps* Complex. Am. Midland Naturalist 71(1):232–245.

——— 1966. Mating calls of six Mexican and Central American toads (genus *Bufo*). Herpetologica 22(1):60–67.

——— 1968. Evolutionary status of a relict population of *Bufo hemiophrys* Cope. Evolution 22(3):583–594.

——— 1969. Evolutionary status of the Rocky Mountain population of Wood Frogs. Evolution 23(1):163–170.

Twitty, V. C. 1939. Correlated genetic and embryological experiments on *Triturus*. I and II. J. Exp. Zool. 74:239–302.

——— 1955. Field experiments on the biology and genetic relationships of the California species of *Triturus*. J. Exp. Zool. 129:129–147.

——— 1959. Migration and speciation in newts. Science 130:1735–1743.

——— 1961. Second-generation hybrids of the species of *Taricha*. Proc. Natl. Acad. Sci. U. S. A. 47(9):1461–1486.

Volpe, E. P. 1954. Hybrid inviability between *Rana pipiens* from Wisconsin and Mexico. Tulane Studies in Zool. 1:111–123.

———, and S. M. Harvey. 1958. Hybridization and larval development in *Rana palmipes* Spix. Copeia 1958:197–207.

White, M. J. D. 1946. The spermatogenesis of hybrids between *Triturus cristatus* and *T. marmoratus* (Urodela). J. Exp. Zool. 102:179–208.

Wright, J. W., and C. H. Lowe. 1967. Hybridization in nature between parthenogenetic bisexual species of whiptail lizards (genus *Cnemidophorus*). American Mus. Nov. 2286:1–36.

Zweifel, R. G. 1955. Ecology, distribution, and systematics of frogs of the *Rana boylei* group. Univ. California Publ. Zool. 54:207–292.

CHAPTER 14

POPULATION DYNAMICS

Both stabilization and change in the size and composition of a population result from the dynamic equilibrium between two opposing forces. One of these forces is the *biotic potential* of the population, the inherent capacity to multiply and survive, which tends to cause the population to grow. Pitted against the biotic potential is the *environmental resistance,* the restriction imposed upon the numerical increase of the population by all the physical and biological factors of the environment in which the population exists. The biotic potential of any population tends to be influenced by unique characteristics of that population, such as its age composition, sex ratio, and the genetic make-up of the particular individuals comprising it. Consequently, there will be variation in biotic potential both from one population to another within a species and through time within a single population. Because there is greater genetic similarity among individuals of the same species than among individuals of different species, the biotic potentials of populations representing the same species tend to be more similar than those of populations representing different species. The factors contributing to environmental resistance, including nutrient and energy supplies, competition, predation, disease, temperature, moisture, and space, tend to vary from place to place and through time in the same place. However, if a species has narrow ecological tolerances and always occurs under essentially the same environmental conditions, the components of environmental resistance and the relative importance of each are probably similar for all populations; if a

species has wide ecological tolerances so that different populations occur under a variety of environmental conditions, the components of environmental resistance and the importance of each may vary considerably from population to population.

From the above discussion, it should be apparent that the analysis of a population's dynamics requires examination of the numerous factors affecting its biotic potential and identification and evaluation of numerous environmental factors comprising the environmental resistance of the habitat. The study of population dynamics is one of the most difficult aspects of biology and, in herpetology, is an area especially in need of study. The causes of mortality, whether predation or other factors, generally are poorly understood. Furthermore, accurate age-specific data on natality and mortality of amphibian and reptilian populations are almost non-existent. The general problem is one of learning about the interaction of populations with their environments and the movement of individuals within populations. The following is a general outline of the principal questions to be answered in a population dynamics study.

I. What is the geographical and ecological range of the population?
  A. What is the principal habitat of adults? of juveniles? of larval or embryonic stages?
    1. What are the abiotic conditions?
    2. What are the other components of the biotic community?
  B. What are the principal barriers isolating this population from others of its kind?

452

II. What is the age and sex composition of the population?
   A. What is the sex ratio in each mature age group? In juveniles? in larvae?
   B. What is the sex ratio of adults participating in reproduction?
   C. What seasonal changes occur in the proportion of juveniles to adults? in the sex ratio of adults?
   D. What oscillations or cycles occur in the composition of the population over an extended period of time?
III. What is the pattern of distribution of the population?
   A. What is the density?
   B. Are either individuals or mated pairs territorial?
   C. What is the size of the home range? territory?
   D. What dispersal (emigration, immigration, migration) occurs?
      1. How many individuals? Of what age classes? Of what sex?
      2. Under what conditions and when does dispersal occur? How far?
IV. How is the population maintained?
   A. What is the reproductive potential of the population?
      1. At what size or age is sexual maturity attained in females? in males?
      2. What is the duration of reproductive activity in the life of females? of males?
      3. What is the potential number of young or eggs produced per year per female?
         a. How many eggs or young may be produced per brood?
         b. How many broods may be produced per year?
         c. Is there a correlation between the potential number of eggs per brood or potential number of broods per year and the age or size of females?
      4. What is the significance of sex ratio to reproductive potential of the population?
   B. What is the realized reproductive performance of the population?
      1. How many eggs or young are produced per brood per female in each size or age group?
      2. How many broods are produced per year by females in each size or age group?
      3. What fertilization rates occur?
   C. What age-specific survival occurs?
      1. What is the survival of larvae?
      2. What is the survival of juveniles?
      3. What is the age-specific survival of adults?
   D. What is the environmental resistance (potential minus realized productivity of the population)?
      1. What environmental factors influence natality?
      2. Is there a correlation between reproductive performance and stress in previous generations?
      3. What environmental factors produce mortality?
         a. Are these factors density-dependent or density-independent?
         b. What age classes are primarily affected by each factor?
   E. What is the importance of immigration to maintenance of the population?

The main emphasis in most amphibian and reptilian studies to date has been the determination of social relations of individuals (calculation of the size of home ranges and territories) and of the survival of individuals. Both of these kinds of information have been obtained through the use of marking and recapture data. Amphibians are most frequently marked by toe-clipping, turtles by notching marginal carapace plates, lizards by tatooing or toe-clipping, and snakes by clipping caudal scutes. Population studies have been made on a variety of anurans, few salamanders, no caecilians, a handful of turtles, the American alligator and Nile crocodile, an assortment of lizards, and a few snakes. Quantitative studies of the population dynamics of tropical amphibians and reptiles are generally lacking, and apparently no one has attempted to study the population dynamics of the tuatara.

POPULATION DYNAMICS OF ANURANS

Although there is considerable published information on the breeding behavior of anurans and numbers of eggs contained in their masses, few quantitative studies have been made of how size and age affect natality, the structure of populations, sources of mortality and ages affected (larval survival is virtually unknown in nearly every instance), or the magnitude and causes of year-to-year fluctuations in the size and structure of populations. The following discussions are based on studies of various species of north-temperate and south-temperate anurans; much of this information was compiled and summarized by Turner (1962).

NATALITY

Because the number of eggs deposited per mass tends to vary directly with the size of the female, the natality of an anuran population is influenced by the number of age groups participating in reproduction. It is also affected by the duration of the breeding season to the extent that this affects the number of times each female breeds per year. In his study of a population of fire-bellied toads (*Bombina bombina*), Bannikov (1950) concluded that the females composed one sexually mature age group (one to two years old) which bred once a year with a natality rate of 300 eggs per female per year. Studies of other species indicate that such a simple and fixed age structure with a constant natality rate per individual rarely, if ever, occurs. Various techniques, including the monitoring of growth rates of marked individuals and the use of anatomical measurements, have been used to determine age structures of populations and, in general, at least three age classes of sexually mature females seem to exist in anuran populations; these classes make different contributions to the natality of the population. Anderson (1954) found that two-, three-, and four-year-old females were reproductively active in Louisiana populations of narrow-mouthed toads (*Gastrophryne carolinensis*). The number of eggs deposited by individual females varied from 152 to 1089 (mean of 510), and the average number of eggs produced by each age-group varied. Because of the relative number of breeding females in each age class, Anderson found that the three-year-olds accounted for the majority of the nata-

lity of the population even though the larger four-year-olds tended to produce more eggs per individual. In contrast to the stabilized age structure of Bannikov's fire-bellied toads, Martof (1956) found that the age structure of populations of green frogs (*Rana clamitans*) fluctuated considerably. Turner's (1960) data also indicate major fluctuations in the age structure of a population of spotted frogs (*Rana pretiosa*): for example, 12.8 per cent of the 937 individuals he aged in 1954 were two-year-olds, whereas 32.8 per cent of the 531 individuals aged in 1955 were two-year-olds. Fluctuations in the age structure of populations, particularly in regard to sexually mature females, undoubtedly result in great variations in the numbers of eggs produced by anuran populations per year. While the number of eggs produced by females belonging to different size groups may be determinable, this information is insufficient to describe the natality of any population unless one also knows how many females in each size group actually participate in reproduction each season. This is affected by how many females of each size class are present, which seems to fluctuate from year to year, and by the frequency with which individual females breed.

Anuran populations may be divided into three categories in regard to the periodicity of breeding by individual females. One category is composed of populations which have very extended or even continuous breeding seasons during which an individual female may breed two or more times each year. The Texas cliff frog (*Syrrhophus marnocki*) populations studied by Jameson (1955) are in this grouping. Their breeding season extends from February to early December, and Jameson found evidence that females may deposit eggs up to three times in one year. The second category includes populations whose females breed once a year regardless of the duration of the breeding season. This category is exemplified by the Australian spotted marsh frog (*Limnodynastes tasmaniensis*), which is in breeding season from August to January in the Melbourne area (Littlejohn, 1963) although individual females apparently only breed once a year (Harrison, 1922). The third category includes populations with annual breeding seasons but in which individual females do not breed annually. This category, which may include many non-tropical species of frogs and toads, is represented by the Yellowstone Park population of spotted frogs (*Rana pretiosa*) studied by Turner (1960); he concluded that females breed every other year or possibly every third year, although males probably breed every year. Although other factors complicate the issue, it appears that there may often be an inverse relationship between the number of eggs per oviposition and the number of ovipositions per year. At one extreme are females of the tiny Cuban dendrobatin *Sminthillus limbatus*, which lay one egg at a time but probably breed more than once each year; at the other extreme are females of various toad (*Bufo*) populations, which lay thousands of eggs at a time but which probably do not breed every year (Bragg, 1940; Blair, 1943). Whenever it is possible, natural selection seems to favor the development of small numbers of larval and young frogs at different times of the year rather than the synchronous development of a large number (Pearson, 1960).

Finally, it should be noted that some anuran populations inhabiting arid regions may completely by-pass breeding during unfavorable years when a lack of precipitation results in the lack of any suitable breeding site.

The combined effect of the variables influencing natality is such that, at least outside of the tropics, anuran populations are probably subject to wide year-to-year fluctuations in natality. Such fluctuations may be expected to be manifested in fluctuations in other parameters of the populations, including age structure, dispersal, and biomass.

EMBRYONIC AND LARVAL SURVIVAL

With one exception, the proportion of eggs fertilized is not known for any natural population of anurans; the fraction of zygotes which actually hatch and the survival of larvae are only surmised in most instances. In one of the few quantitative studies of anuran larval survival, Herreid and Kinney (1966) estimated that the average fertilization success by wild populations of Alaskan wood frogs (*Rana sylvatica*) was 86.7 per cent, but one pond had only a 32.8 per cent success. Cursory examination of egg masses of a variety of

North American anurans has led the author to believe that fertilization rates are generally very high in nature but this conclusion has never been quantified; fertilization rates by amplectant frogs and toads in the laboratory are commonly above 90 per cent. Low fertilization rates in nature may result from temperature effects, from reduced fertility of males which breed successfully with different females, or from abnormalities in sperm and eggs (Herreid and Kinney, 1966).

From data on 53 wood frog egg masses, Herreid and Kinney estimated an average mortality through gastrulation of four per cent. They observed that abnormalities in development occurred throughout the embryonic period but were unable to determine the magnitude of embryonic mortality after gastrulation. Bragg and Bresler (1951) estimated that about 34 per cent of the Great Plains toad (*Bufo cognatus*) eggs hatched in what appeared to be an uncontaminated pond. When compared to what is often 100 per cent survival of laboratory-reared embryos, 34 per cent seems atypically low, but the rigors of nature may mean that the figure is more realistic than it initially appears. This author has often observed entire strings of boreal toad (*Bufo boreas*) eggs which have failed to hatch but, on the other hand, has observed what appeared to be 100 per cent hatching of egg masses of other species. The causes of embryonic mortality are varied. Herreid and Kinney (1966) noted that there was increased mortality when wood frog egg masses were laid at lower depths in ponds and suggested that reduced oxygen supply, increased pressure, reduced temperature, and increased silt deposition may have been the causes. Extensive predation by diving beetle larvae (*Dytiscus*) was also apparent in all of the wood frog breeding ponds. Ferguson (1954) has suggested that wave action of snow-slide origin is a factor in the destruction of boreal toad eggs. Fungal infestations are generally characteristic of embryos which fail to hatch but, rather than being parasitic, such growths may represent saprovores.

A number of workers have observed great variations in the survival of anuran larvae and noted that premetamorphic mortality is sometimes tremendous. Herreid and Kinney (1966) used neutral red solutions to dye samples of tadpoles red and then used mark and recapture methods to estimate larval survival. They found the greatest mortality of wood frog larvae occurred during the first month of development and that it had declined by the time the hind legs were appearing. The median survival time was about 20 days, and an average of only 3.7 per cent of the original number of zygotes survived to climax metamorphosis in the four ponds studied; all survival rates were less than 10 per cent. Turner (1960) marked a sample of tadpoles by clipping tail fins and used the Lincoln Index to obtain data on larval survival from egg to late larval stage in the Yellowstone population of spotted frogs. He estimated that out of approximately 25,000 eggs deposited in late May and early June in three different ponds, only about 800 larvae survived to late August and early September. The survival in each of the ponds varied, ranging from nil at one pond to about 8.5 per cent at another. In contrast, Bannikov (1950) estimated that the larval survival in his population of fire-bellied toads was about 54 per cent.

A major reason that Turner's spotted frog larvae survived in such small numbers compared to Bannikov's fire-bellied toad larvae is that the latter occurred in relatively stable permanent ponds whereas the former were in ponds subject to desiccation. Thus, the survival of anuran larvae depends very much upon where oviposition occurs. It is also dependent to a great measure upon seasonal climatic conditions, for ponds which may produce large numbers of anurans one year may be dry the next year; the net result is great year-to-year fluctuations in larval survival and input into the first year age class. For example, Martof (1956), using quadrate counts to estimate numbers of green frog (*Rana clamitans*) larvae and their survival, estimated that in a Michigan population from 60,000 to 75,000 larvae died during the evaporation of some ponds in the summer of 1948. In 1949, an estimated 10,700 larvae successfully metamorphosed in the same ponds. Bragg (1965) has noted instances where spadefoot larvae (*Scaphiopus*) have all perished in a particular pond one year, whereas large numbers from the same pond metamorphosed another year. In general, the anurans which habitually breed in temporary water may be expected to have greater year-to-year

fluctuations in larval survival than the anurans which habitually breed in permanent water, where the larval environment is more stable from one year to the next. Similarly, it is to be expected that those anurans which provide parental care for their young will have a higher and more constant survival of larval stages than others. In any population, larval survival is a major factor regulating population size; Savage (1952) believes it to be the primary regulator of British anuran populations. Yet, as noted by Turner (1962), the estimation of survival of larval anurans remains one of the most challenging problems in anuran demography.

POSTMETAMORPHIC SURVIVAL, ATTAINMENT OF SEXUAL MATURITY, AND LONGEVITY

Mark and recapture studies of anurans have provided some information on the survival of transformed individuals and the ages at which sexual maturity is attained. Although the number of marked individuals recaptured has usually been very small and most studies have been of short duration, the general indication from available data is that anuran populations are characterized by a high rate of turnover in practically all age groups with very few individuals, if any, living to senility. Martof (1956) noted that the proportion of marked green frogs which were recaptured at a later date was the lowest (six per cent) for those which were marked when they were newly transformed and that as the size of the frogs increased their rate of recapture became progressively higher, reaching 42 per cent for 65- to 75-millimeter (snout-vent) frogs and about 50 per cent for 75- to 85-millimeter mature frogs; frogs which were larger than 85 millimeters showed a slight decline in proportion recovered. These data imply that metamorphosed frogs follow a survivorship curve in which mortality mainly affects the very small individuals and the mean life expectancy of any individual increases with age; the median life expectancy is very short compared with the mean. Bannikov's population of fire-bellied toads appears to follow a similar survivorship curve. Using the age distribution of 4208 toads to estimate their age-specific mortality, Bannikov concluded that only about two per cent of the metamorphosed individuals survive to an age of one year. About 40 per cent of the one-year-olds survive to an age of two, and almost none of the two-year-olds reach an age of three.

In contrast to the above situations in which the one-year-olds receive the brunt of the annual mortality, Turner (1960) concluded that the annual mortality rate of western spotted frogs remained constant (39±10 per cent) through six year-classes. Except for possibly the second and fifth year-classes, the proportions of each age group which he recaptured were nearly the same, and the data certainly indicate little probability that the one-year-olds suffer a significantly higher mortality than the other age classes. A similar conclusion was reached by Green (1957) in a study of a West Virginia population of mountain chorus frogs (*Pseudacris brachyphona*); the annual mortality in that population was estimated to be a constant 70 per cent for all age groups. In both of these populations, because a constant fraction of the animals alive die at each age, the mean life expectancy at any age remains constant throughout the postmetamorphic life of the animal.

Studies of *Rana blythi, R. erythraea, R. ibanorum,* and *R. macrodon* indicate that there is a relatively stable age structure in tropical frog populations (Alcala, 1955; Brown and Alcala, 1970; Inger and Greenberg, 1963, 1966). They also indicate that tropical species become sexually mature earlier in time following metamorphosis (9 months to a year after metamorphosis) and at an earlier stage of growth than temperate species. Brown and Alcala (1970) estimated the longevity for both sexes of *Rana erythraea* to be about four years, supporting Turner's (1962) suggestion that tropical anurans have a shorter life span and higher turnover rate than temperate forms. For *Rana erythraea,* the relatively stable age structure over a four-year period suggests that survival for all age groups is about the same (Brown and Alcala, 1970).

Although not age-specific, other recapture data have been published which indicate the turnover in anuran populations. Turner (1959) marked 34 red-spotted toads (*Bufo punctatus*) in a Death Valley population which he subsequently estimated at

about 37 individuals; he recaptured 14 of these marked individuals a year later. Using Hayne's (1949) method for estimating population size, he calculated that the sampling efficiency in the second year was about 95 per cent and that 15 of the original group of 34 toads survived one year. Blair (1953) marked a group of 357 newly transformed Gulf Coast toads (*Bufo valliceps*) and recaptured 25 of them a year later; 20 of the recaptured were males. Assuming that the sex ratio in the original group of 357 was 1:1, Blair estimated that a minimum of 11 per cent (20/178) of the marked males survived one year; he noted that more searching over a larger area probably would have resulted in the recovery of additional marked toads. Jameson (1956a) found evidence that there is a very rapid turnover in an Oregon population of Pacific tree frogs (*Hyla regilla*). He marked 1156 newly transformed individuals but was able to recover only 30 of these two to eight weeks later; two were recovered a year later. Out of 373 adults marked and released, 38 were recovered the following year, indicating a minimum survival of about 11 per cent from one year to the next. The same worker (1956b) studied survival in a variety of central Texas anurans. Fifteen of 27 male Great Plains spadefoot toads (*Scaphiopus couchi*) were recovered a year after marking, and one of these was again recaptured two years later; of 17 marked females, nine were recovered a year later but none two years later. Jameson marked eight male and six female spotted chorus frogs (*Pseudacris clarki*) and recovered six of these males and two of the females a year later. One of two Great Plains narrow-mouthed toads (*Gastrophryne olivacea*) marked one year was recovered 12 months later. Of 43 male and 18 female Strecker's chorus frogs (*Pseudacris streckeri*) marked one year, 26 males and 11 females were recaptured a year later; fifteen of the males and one female were again recaptured two years later. All recapture data of this kind must be interpreted as minimum survival; not all individuals necessarily breed each year and dispersion may contribute to apparent mortality. Pearson (1955), in a study of the eastern spadefoot (*Scaphiopus holbrooki*), noted that two major reasons other than mortality may affect recapture data: individuals may not be recaptured because they have moved from the area, or they may have had periods of surface activity which were not observed. For these reasons, he categorized all individuals which he failed to recapture under "disappearance" rather than "mortality."

The recapturing of individuals which were marked when newly metamorphosed has provided data on the attainment of sexual maturity. Blair's (1953) study of the Gulf Coast toads indicates that they reach sexual maturity in the breeding season of the year following the one in which they hatched and at an age as young as 10 months. Jameson's (1956a) data on Pacific tree frogs indicate that this species attains sexual maturity within one year of transformation. Ryan (1953) studied the growth of several ranids under natural conditions and noted that transformation size and date both play a role in determining when a frog will attain maturity. He found that green frogs that are small (26 to 27 millimeters) at transformation in late June may not become sexually mature until the next season during late summer and after the breeding period has ended; they thus may not be sexually active for another year. Similarly, Tester and Breckenridge (1964) found that the Manitoba toad (*Bufo hemiophrys*) reaches sexual maturity at a snout-vent length of 38 to 45 millimeters; this size is reached by most toads during the latter part of their second summer, and consequently they normally do not breed until the start of their third summer. Ryan (1953) found leopard frogs (*Rana pipiens*) may occasionally breed in the year following hatching and most will breed two years after hatching. Some bull frogs (*Rana catesbeiana*) might possibly breed late in the summer a year after they metamorphose, but most probably take two years to reach sexual maturity after transforming. In general, a minimum size must be reached before sexual maturity is attained; therefore, the length of growing seasons and food and temperature conditions will determine when individuals in a particular population mature. In general, anurans seem to grow rapidly during the summer in which they metamorphose, grow very little or not at all during the following winter, and then grow rapidly again until they reach sexual maturity; after maturity is reached, growth slows again.

The maximal longevity of anurans in nature seems to vary specifically and probably also varies from population to population within the same species. A male Great Plains narrow-mouthed toad (*Gastrophryne olivacea*) has been reported to have attained an age of seven or eight years (Fitch, 1956a). Both eastern spadefoot toads and the Yellowstone spotted frogs appear to attain an age of at least nine years (Pearson, 1955; Turner, 1960). Mountain chorus frogs and green frogs both appear to reach an age of five to six years (Green, 1957; Martof, 1956). Florida oak toads (*Bufo quercicus*), Texas cliff frogs, and Oregon Pacific tree frogs all appear to survive only to an age of three years (Hamilton, 1955; Jameson, 1955, 1956). Gulf Coast toads (*Bufo valliceps*) and Woodhouse's toad (*Bufo woodhousei*) are known to live at least four years in southeastern Texas (Thornton, 1960). Turner (1960) concluded that spotted frog females live up to three years longer than males (10 to 12 years); he also noted (Turner, 1962) that older females outnumber older males consistently about 3.5 to 1 in populations all over the southeastern portion of the range of the species. It seems reasonable to assume that in all anuran species males suffer a greater mortality than females during the breeding season, but no sexual difference in the survival of marked individuals has been found.

The composition of any anuran population may be expected to vary in any given place throughout the year. This is well exemplified by Martof's (1956) study of the green frog. He found that during the breeding season (mid-May to mid-August) most male green frogs were in ponds, whereas 80 per cent of the adults taken along streams were female; away from breeding areas he found a ratio of 56 females to 44 males. In August and September, newly transformed frogs composed nearly all of the pond populations. These young frogs quickly moved to locations along the streams, where they formed about 95 per cent of the population shortly after the peak of transformation. Later in the fall, the proportion of juveniles steadily declined because of mortality and dispersion, reaching about 75 per cent of the population in mid-October. In the spring, the ratio of juveniles to adults continued to decline and reached approximately 1:1 in May. The juvenile population became the lowest in July, just before metamorphosis of the next year-class began.

SOCIAL BEHAVIOR

The *home range* of an animal, as the term will be used here, is the total area over which the individual habitually travels while it is engaged in its usual daily activities. A *territory*, on the other hand, is an area which the animal defends against intruders of the same species; it is generally only a portion of the home range. Home ranges are important in that they restrict individuals to definite localities where there is sufficient cover and food. Because an animal becomes intimately familiar with it, the home range contributes immeasurably to avoidance of enemies and survival. Home ranges also limit the available individuals with which an animal may mate unless the entire population migrates to a breeding site and intermingles during the breeding season. Limitations on the movement and number of available mates significantly affect the rate of gene flow through a population. Territories tend to organize populations into relatively well-spaced aggregations with more efficient spacing of breeding and feeding functions than would otherwise exist. By limiting movement, territories also promote monogamy and significantly affect gene flow.

Since there is no evidence that they tend to be nomadic, most amphibians and reptiles appear to have fairly well defined home ranges. Some anurans are known to establish and defend territories, but the extent of territoriality within most families is not clear because it is difficult to determine whether or not an animal is actively defending an area without direct observation of social interactions.

Quantitative studies of the home ranges of *Bufo boreas* (Carpenter, 1954), *Bufo woodhousei fowleri* (Stille, 1952), *Bufo punctatus* (Turner, 1959), *Rana clamitans* (Martof, 1953), *Rana pretiosa* (Carpenter, 1954; Turner, 1960), *Scaphiopus holbrooki* (Pearson, 1955), and *Syrrhophus marnocki* (Jameson, 1955) indicate that anurans tend to remain in a particular area for extended periods of time and have small home ranges, ranging from as little as 100 square feet in the spadefoot toads (*Scaphiopus*) to about 4000 square feet in the cliff frogs (*Syrrhophus*). The size and permanence of an anuran's home range is influenced by a variety of behavioral, populational, and environmental factors. The Death Valley population of spotted toads studied by Turner (1959) has only about five acres of suitable habitat available to it, and the spatial limitations of the habitat undoubtedly restrict the magnitude of individuals' movements. Turner found that the maximum distance traveled by an individual toad was about 1200 feet and the average distance was about 300 feet. This much movement took place within 11 to 20-day intervals and did not increase with any greater elapse of time. The movement which did occur appeared to be random back-and-forth shuttling, with centers of activity around bodies of water.

Jameson (1955) found that the size of the home ranges of cliff frogs varied seasonally, being larger during the fall and spring than during the winter and summer when rainfall was low and environmental conditions adverse. The size of the home ranges also varied with the habitat, being the largest (0.139±0.014 acres) on open slopes and smallest (0.052±0.006 acres) on taluses. The trend was for the size of the home range to vary inversely with the amount of rock surface and cover available. Jameson's data indicate that once cliff frogs establish their residence, they remain in the area throughout their life; many individuals studied by Jameson utilized the same groups of rocks or cracks in a cliff for over two years. However, by artificially altering the density of populations through removal or introduction of individuals, Jameson determined that an individual cliff frog will move its home range when there are local areas of low population density. A prime factor which seems to affect any anuran's choice of a home range is the proximity of an escape area (water, crevice in rock, burrow, and so forth); the movement of home ranges in low density situations may reflect the availability of a more desirable escape area than was previously available.

Clearly, the permanence of an animal's home range is dependent upon continual satisfaction of the individual's ecological demands; if some environmental factor fluctuates seasonally, it may cause seasonal shifts in home ranges. For example, Turner (1960) found that where the ecological demands of the spotted frogs were satisfied in a relatively small area, such as the vicinity of a pool, some frogs established what appeared to be permanent home ranges. Other frogs carried out major periodic spring migrations to upland situations from the stream, returning to it later in the summer. Fitch (1956a) found that Great Plains narrow-mouthed toads (*Gastrophryne olivacea*)) also make shifts from one home range to another; such shifts vary in length and may occur abruptly or by gradual stages. The home ranges first established by subadult green frogs (*Rana clamitans*) don't seem to be as permanent as those of adults and also differ ecologically (Martof, 1953a). The home ranges of subadults include shallow water with a dense growth of vegetation, whereas those of adults (frogs at least 60 to 75 millimeters in length) are in deeper water where aquatic vegetation is not so dense.

Many adult anurans, especially those outside of the tropics which are semiaquatic to terrestrial, travel seasonally from their home ranges to breeding areas or wintering areas. Thus, there are temporary abandonments of home ranges and, at least in many species, seasonal periods of gregarious behavior. The population is particularly vulnerable to predation during the breeding season because there is a massing of individuals outside of the home ranges with which they are so familiar.

There is a great shortage of data bearing on the presence or absence of territoriality among anurans. Most species are generally thought to be non-territorial but, in reality, very few attempts have been made to actually determine this. Limited observations on aggressive behavior indicate that different degrees of territoriality, ranging from primitive to highly developed, have evolved in some aquatic species and species having specialized breeding habits. Using Nice's criteria for territoriality (an animal's demonstration of advertisement, isolation, intolerance, and fixation), Martof (1953b) concluded that adult male green frogs maintain a small proportion of their home range as a primitive type of territory during the breeding season. This incipient territoriality results in a rather even spacing of the individual males and, by reducing strife, results in conservation of the males' energies. Spacing of the males probably also facilitates rapid detection of females as they appear. Emlen (1968) demonstrated conclusively that male bullfrogs (*Rana catesbeiana*) have an advanced degree of territoriality during the breeding season and defend an area surrounding their calling location against intrusion by other males. The defensive behavior observed consists of an initial vocal challenge followed by an advance toward the intruder, another vocal challenge, another advance of a few feet, and so on until the intruder leaves; if the intruder fails to leave, the two frogs eventually jump directly at one another and grasp each other's pectoral regions, each attempting to throw the other on its back. As soon as one frog is forced onto its back, contact is broken and the winner begins calling again. After remaining submerged for several seconds, the loser usually swims away some distance under water before surfacing. By using models, Emlen was able to demonstrate that the inflated high posture assumed by calling males will elicit the aggressive challenges from the owner of a territory whereas a low posture, such as assumed by a female or non-calling male, will not elicit this response. Wiewandt (1969) also used a model frog and playback technique to study bullfrog behavior and found that he could estimate the on-shore dimensions of each male's territory by noting the frog's response to playback as he moved the speaker. The three most responsive frogs he observed defended from 9 to 25 meters of shoreline, but the territories were neither clearly defined nor stationary during the course of a summer. The most dominant male had a relatively small territory, and Wiewandt concluded that perhaps the location of a territory is a more important feature than its size.

Test (1954) described territorial defense by the South American frog *Colostethus trinitatis* (subfamily Dendrobatinae). In this species, territories are defended especially by adult females, but adult males and subadults also may be territorial. When the territorial resident first observes the intruder, it orients to face the intruder. The resident pulsates its bright yellow throat as a challenge, and frequently this is sufficient to cause the intruder to move away. If the intruder remains, the resident may hop in the direction of the intruder and repeat the challenge. If this fails to drive the intruder away, the resident will jump on the back of the intruder and a wrestling match will ensue. The forelimbs of each animal are extended laterally at right angles to the body axis; the hind limbs are extended so that both frogs appear to be standing. While in this position, each frog pushes against the other and attempts to knock the opponent off balance. The conflict ends when one of the frogs leaves the area. Following Test's observations, Sexton (1960) made a detailed study of the physical nature of the territories of these frogs and found that the total area defended by adult females averages about 1.07 square feet and that each territory typically contains elevations which have crevices beneath them and which are alongside or surrounded by mountain streams. The elevation, streamside location, and crevices appear to all fulfill important ecological requirements. Feeding frogs perch on top of elevations from which they can observe insects flying above the water and can leap out at them. From such perches, frogs are also easily observable to potential intruders and can observe any trespassing which may occur. The streamside location results in an abundance of insects, isolation of females from one another, and provision of an escape area. Crevices also appear to function as refuges, and there tends to be an increase in the number of frogs inhabiting an elevation with an increase in the number of crevices.

Duellman (1966) observed aggressive behavior by males of two other species of dendrobatin frogs, *Dendrobates pumilio* and *Colostethus panamansis*, in Panama. The combat behavior of the *Dendrobates* consists of wrestling as described for the female *Colostethus trinitatis*. Because the wrestling followed one male's calling from a distance of about 15 centimeters from another's calling perch and stopped when the former jumped to another plant, it would appear to have represented territorial defense, but this could not be positively determined. Duellman's observations of the larger *Colostethus panamansis* seem to confirm territorial defense by breeding males of this species. Males of this kind call from the tops of small

boulders, and Duellman observed three different instances in which resident frogs drove off intruding males. In one instance, the resident merely hopped toward the intruder and the intruder left. In the other two instances, the resident literally charged into the intruder and knocked him off the rock by hitting him with his head.

The only other anurans which have been observed aggressively defending a territory are males of the Smith frog (*Hyla faber*). As mentioned in Chapter Eleven, this large Brazilian tree frog has an unusual pattern of reproductive behavior which includes construction of individual breeding pools by males. Lutz (1960) observed that fights may break out between breeding males, either inside or outside of a nest. These fights, which consist of wrestling matches during which the males drive their spurlike rudimentary pollices into the opponent, appear to represent defense of a territory that includes the nest and some surrounding area.

Sexton (1962) observed breeding behaviors in nest-building frogs (*Leptodactylus insularum*) which indicates that they may be territorial. Upon two different occasions when males that had been disturbed from their calling sites (foam nests or burrows) returned to them, seemingly chance movement of two into the same nest resulted in their making physical contact beneath the foam. Subsequent to making contact, one frog withdrew while the other remained. Although this would seem to indicate that a male of this species occupying a foam nest or burrow will defend it against intrusion by other males, it was also observed that there may be some shifting in the occupancy of these nests. Thus, whatever territorial instinct is present is primitive, and the males apparently do not feel attached to a particular site but rather defend that which they happen to be occupying at the time. Their behavior may be interpreted as being merely antisocial toward other males rather than as territorial-defensive.

Zweifel (1969) found that males of the New Guinea frog *Platymantis papuensis* (family Ranidae) use vocalizations in what appeared to be territorial defense. When a normal (mating?) call is played back to a calling frog, the latter immediately responds with a "response" call consisting of a mixture of squeaking and quacking noises; on two occasions the calling frog was observed to move toward the loud-speaker in a manner interpreted as a terri-torial response. "Response" calls could be elicited from distances of 10 to 30 feet, suggesting the approximate size of territory.

As noted by Duellman (1966), the terri-torial behavior of dendrobatid frogs and of the Smith frog is associated with special-ized breeding behavior. The former do not migrate outside of their normal home range to breed and, thus, may be more inclined to establish territories than would anurans which seasonally leave their home ranges for breeding or hibernation. The nest-building behavior of male Smith frogs presumably preceded or evolved si-multaneously with their territorial in-stincts; having constructed a nest, the male would ostensibly be reluctant to relin-quish it to another male. Territoriality in green frogs and bullfrogs seemingly evolved, as with the dendrobatins, as a concomitant of breeding taking place within the normal home range. In fact, it could be speculated that many, if not most, fully aquatic anurans exhibit some degree of territorial behavior.

POPULATION DYNAMICS OF SALAMANDERS

Because of their especially secretive habits, analysis of population dynamics for salamanders is a task even more form-idable than it is for anurans; thus, most species remain unstudied. Members of the family Plethodontidae have received the most attention; consequently, these are emphasized in the following discus-sion.

NATALITY

The relationship between female body size and fecundity seems to vary among salamanders. Blanchard (1928) found that there was no correspondence between the size of female red-backed salamanders (*Plethodon cinereus*) in Michigan and the total number of eggs produced; the smaller females he observed deposited as many eggs as the larger females. Tilley (1968) found that there was a positive correlation between body size and clutch size, in-traspecifically and interspecifically, in four species of *Desmognathus* (dusky salaman-ders, *D. fuscus;* black-bellied salaman-ders, *D. quadramaculatus;* mountain sala-manders, *D. ochrophaeus;* and pigmy salamanders, *D. wrighti*) but an insignifi-cant (actually negative) correlation in seal salamanders (*D. monticola*). Harrison (1967) concluded that clutch size is di-rectly related to body size in cherokee sal-amanders (*D. aeneus*). Stebbins (1954) found a positive correlation between number of eggs and size of ensatinas (*En-satina eschscholtzi*) up to a snout-vent length of 69 to 70 millimeters; very large females in the 71- to 76-millimeter size class showed a decline in number of eggs produced which Stebbins interpreted as a possible reduction in reproductive vigor. Anderson (1960) found that the mean clutch size for Brooks Island and mainland populations of California slender salaman-ders (*Batrachoseps attenuatus*) was 8.0 and 8.7, respectively, despite the fact that the island females averaged a larger size. In contrast, he found that the clutch size of both island and mainland arboreal sala-manders (*Aneides lugubris*) was corre-lated with size, smaller females producing smaller clutches. A significant correlation between body size and number of mature eggs has also been reported for the two-lined salamander (*Eurycea bislineata*) by Wood and Duellman (1951).

High density in mammalian populations is known to reduce natality. Anderson (1960) found evidence that high popula-tion density may also result in reduced natality in salamanders by causing repro-ductive inhibition in males and reduction in clutch size of reproducing females. This conclusion is supported by the coinci-dence of high density and reproductive inhibition on Brooks Island and Red Rock Island. The reduction in fecundity of the California slender salamanders is inde-pendent of body size, while that of the ar-boreal salamanders is associated with a smaller size of sexually active females on the islands as compared to those on the mainland.

Tilley (1968) found that female dusky salamanders from the Balsam Mountains of southwestern Virginia lay more eggs than females of the same species from Licking County, Ohio. Furthermore, this is attributable to both the larger size and a different size-fecundity relationship of the Virginia females whereby, for any given size, females of this population lay more eggs than females from Ohio. Tilley proposed that the lower fecundity of the Ohio dusky salamanders may be attributable to the fact that they are free of predation by large aquatic members of the same genus, a major source of mortality to the Balsam Mountains females. This conclusion assumes that natural selection favors greater reproductive efforts per season where females are subjected to heavier mortality and have shorter life expectancies (Williams, 1966). Anderson's and Tilley's findings exemplify the complexity of population dynamics and point out that natural selection can alter natality with or without a change in body size of females.

The number of clutches of eggs produced per female per year is not clear for some species of salamanders. Organ (1961) concluded that five Appalachian species of *Desmognathus* (*D. quadramaculatus, D. monticola, D. fuscus, D. ochrophaeus,* and *D. wrighti*) have biennial reproductive cycles in which, although there is an annual breeding season, individual females only deposit eggs every other year. This conclusion was based on the observation that mature females fell into two distinct groups: (1) those with large white ova in the ovaries and (2) those with small white ova and usually one or two large yellowish spherical bodies in the ovaries and with enlarged convoluted oviducts indicating they were spent. Organ assumed that the spent females found in the spring before egg-laying season must represent females that produced eggs during the preceding year and felt it improbable that the ovaries of such females are capable of forming large-yolked ova in time for egg deposition during the current year: thus, the biennial breeding cycle in the females. Tilley and Tinkle (1968) reinterpreted the reproductive cycle of *Desmognathus ochrophaeus* as consisting of two peaks of egg laying per year, one peak occurring in the spring and early summer and the other approximately six months later in the fall or early winter. This interpretation was based on plotting of mean follicle size against time which indicated two semidistinct groups of females in the spring. One group had follicles of 0.5 to 1.7 millimeters in diameter from

April 17 to May 8, and what appeared to be the same group had follicles of 0.9 to 2.0 millimeters from May 7 to 16 and follicles of 1.7 to 2.9 millimeters from May 22 to 31. This group of females with follicles nearing ovulatory size was absent from samples collected June 7 to 11, and it was presumed that they were laying and brooding eggs at that time. The second group of females contained small follicles in June and ovulated in the fall and early winter. The rate of growth of follicles, furthermore, was such that it seemed possible for a female depositing eggs in the winter to develop and ovulate another clutch by the next fall. This conclusion is supported by the fact that females brooding fully developed embryos or larvae have been found with yolking follicles (Bishop and Chrisp, 1933; Martof and Rose, 1963; Tilley and Tinkle, 1968). The presence of convoluted oviducts in the females examined by Organ also is evidence that they had only recently deposited eggs, since the oviducts tend to straighten rapidly after egg deposition. The evidence accumulated by Tilley and Tinkle indicates that, rather than a biennial cycle, each female mountain salamander is capable of depositing one and possibly two clutches per year. If a female deposited a clutch of eggs in the spring and another in the fall, her ability to lay another clutch the following spring would depend upon how rapidly vitellogenesis proceeded during the colder winter months. If the growth of follicles decreased or stopped during winter, as it does in *Plethodon glutinosus* in Maryland (Highton, 1956), then the female would be unable to ovulate until further development occurred in the spring; consequently, she would be limited to a single clutch that summer. Such a pattern would result in the female producing three clutches over a two-year period, two clutches being produced every other summer. Tilley and Tinkle's findings suggest that other species of *Desmognathus* may also be annual or biannual breeders. In fact, Spight (1967) concluded previously that dusky salamander (*D. fuscus*) females in eastern North Carolina have an annual breeding cycle, and Harrison (1967) found that female Cherokee salamanders (*D. aeneus*) also deposit eggs annually.

Some species of salamanders appear definitely to be biennial breeders. Sayler

(1966) found that it takes a total of two years for eggs measuring 0.1 millimeter in spent red-backed salamander (*Plethodon cinereus*) females to reach a diameter of 3.0 millimeters or more in females ready to deposit eggs. Interestingly, the follicular growth was found to be slow but uniform during both warm and cold months. Thus, mature females of this species must deposit eggs every other year, and, presumably, about half the females in a population deposit eggs each year. Martof (1962) found evidence that female shovel-nosed salamanders (*Leurognathus marmoratus*) spawn every other year; in September females fall into two equal-sized groups consisting of those with eggs smaller than 1 millimeter in diameter and those with eggs larger than 2 millimeters. Similarly, Stebbins (1954) found that a considerable number of adult female ensatinas (*Ensatina eschscholtzi*) of differing sizes lack maturing ova late in the summer. He interpreted this as indicating that they would not deposit eggs the following spring and concluded that all adult females do not lay eggs every summer.

It appears that salamanders have three general patterns of reproductive cycles: biannual breeding, annual breeding, and biennial breeding. These differences are critical to the dynamics of salamander populations since, assuming clutches of the same size, the fecundity of a biannual breeding female is four times that of a biennial breeding female. Because of adaptation to differing mortality rates and because of habitat differences in regard to temperature or other factors which affect follicular development, it might be expected that natality of salamanders will vary intraspecifically as well as interspecifically. Natality differences are particularly probable among populations of widely distributed species, such as the tiger salamander (*Ambystoma tigrinum*). There is a tendency for aquatic salamanders to suffer a higher mortality and to compensate for this with higher egg production. Consequently, one might also expect the natality of neotenic populations to be higher than that of metamorphosing populations of the same species. Again, there is much to be learned about this aspect of population dynamics and regulation.

SURVIVORSHIP, ATTAINMENT OF SEXUAL MATURITY, AND LONGEVITY

No one has succeeded in accurately determining the survivorship of either embryonic stages or of larvae in any natural population of salamanders. To do this, one must know how many eggs have been produced per year by the population, how many hatch, and how many larvae metamorphose or mature. The number of eggs produced by the population, normally used to indicate the number of individuals of age 0 on a survivorship curve, is an unknown quantity for all salamander populations because of the difficulty of determining the number of reproductively active females, inadequate knowledge of their breeding cycles, or both. The consequence of this is that the survivorship of embryonic stages cannot be measured. Some workers, including Organ (1961), have attempted to measure posthatching survival by sample sizes of individuals of different age or size classes. However, as Organ noted, the small size and secretive habits of the larvae (and of transformed one- and two-year-old salamanders in his study) reduces the accuracy of survival estimates.

Assuming adequate samples are available, size-frequency data provide an accurate means of determining the number and relative abundance (survival) of year-classes up to when sexual maturity is reached. Diminished growth after sexual maturity results in the merging together of different year-classes and a "piling-up" of two or more classes in one size group, so a different method is required to separate the year-classes of mature specimens. Humphrey (1922) provided such a method for distinguishing ages of mature salamanders in the genus *Desmognathus* that may be applicable to other groups as well. He found that there is a distinct relationship between the number of lobes on a multiple testis and the age of mature males of dusky salamanders (*D. fuscus*); a male with two full mature lobes per testis is in its third year of sexual activity, a male with three mature lobes per testis is in its fifth year of sexual activity, and a male with four mature lobes per testis is in its seventh year of sexual activity.

Organ (1961) avoided the necessity of determining the stage of development of the lobes by counting all lobes equally so that the one-lobe category included males which were immature or in their first breeding year, the two-lobe category included males in their second and third breeding years, the three-lobe category included males in their fourth and fifth breeding years, and the four-lobe category included males in their sixth and seventh breeding years. He next plotted the number of lobes on a testis against number of individuals in such a way that the number of males with one lobe per testis in each sample was converted to a sample size of 100; the numbers of males with two, three, and four lobes per testis were then proportionally converted to the number surviving on the basis of an initial population of 100 males with one lobe per testis. The partial survivorship curves which resulted varied among the five species Organ studied (see Figure 14–1). Male black-bellied salamanders (*D. qua-*

dramaculatus) and male seal salamanders (*D. monticola*) both have curves indicating a constant rate of survivorship; that is to say, a constant fraction of the animals alive die at each age so that the mean life expectancy at any age remains constant throughout the life of the individual. The curve for seal salamander males is not as steep as that for black-bellied salamander males, indicating that the fraction of seal salamander males which die at each age is smaller than that of black-bellied salamander males. The survivorship curves for male dusky salamanders (*D. fuscus*), male mountain salamanders (*D. ochrophaeus*), and male pygmy salamanders (*D. wrighti*) respectively show a progressive shift from a situation in which a constant fraction of the animals alive die at each age to a situation in which mortality is concentrated on the old animals and in which the mean life expectancy remaining for each individual decreases with age. The progressive increase in survivorship through the early years of sexual maturity is paralleled by

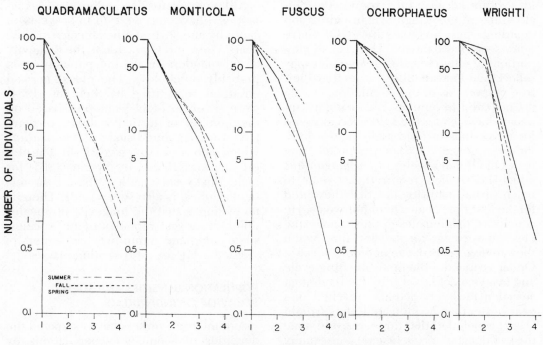

Figure 14–1 Survivorship curves for males belonging to five species of *Desmognathus* from Virginia and based on the number of lobes per testis. The vertical axis is on a logarithmic scale and the points on the survivorship curves were calculated by the method outlined above. (After Organ: Studies of the local distribution, life history, and population dynamics of the salamander genus *Desmognathus* in Virginia. Reproduced by permission of the Duke University Press.)

a progressively greater terrestrialism in *D. quadramaculatus, D. monticola, D. fuscus, D. ochrophaeus,* and *D. wrighti.* Thus it appears that the more terrestrial the population, the higher the survivorship through the early years of sexual maturity. Unfortunately, female salamanders have no known structure comparable to the testicular lobes of males which will differentiate age classes, so their survivorship is not clear.

A number of workers have used mark and recapture methods to study salamander populations, but, while this has provided information on movements, it has generally not contributed to an understanding of survivorship. As noted by Test and Bingham (1948) and Gordon (1952), the number of animals present on the surface and *available* for capture (or recapture) at any one time represents only a portion of the total population of salamanders present in an area. This is manifested in the fact that samples tend to contain disproportionately large numbers of adult males, small proportions of adult females, and at least seasonally small proportions of juveniles. Females tend to be unavailable for capture because they spend considerable time in breeding areas brooding and guarding eggs; juveniles frequently have more secretive habitats. Thus, small proportions of either females or juveniles may reflect their unavailability rather than their true proportions of the population.

Compared to anurans, salamanders appear to have longer life spans and to spend longer periods of time as immature individuals. Organ (1961) concluded that males of all five species of *Desmognathus* which he studied required 3.5 years to reach sexual maturity and that they bred for the first time at an age of four years. He concluded that females of these same species deposit eggs for the first time when they are five years of age or older. Because Organ assumed a biennial breeding cycle and two years of larval life, it is likely that sexual maturity is actually reached one year earlier than he indicated; thus, Spight (1967) concluded that three-year-old male dusky salamanders (*D. fuscus*) are sexually mature. Martof (1962) concluded that male shovel-nosed salamanders (*Leurognathus marmoratus*) require at least two years to reach sexual maturity, while females mature when they are three years of age.

Pope and Pope (1949) reported that slimy salamanders (*Plethodon glutinosus*) mature in Michigan in a little under three years. Highton (1962) has stated that Florida slimy salamanders mature in two years but that females and most males probably do not breed until they are nearly three years old. He concluded that more northern populations (in Maryland, Pennsylvania, and southwestern Virginia) with biennial breeding cycles do not mature until they are four years old and may not breed until they reach an age of five years. Anderson (1967) found evidence, based on size distributions, that long-toed salamanders (*Ambystoma macrodactylum*) attain maturity after one year and breed their second year of life in lowland populations (*A. m. croceum*) but take a year longer and breed at the end of their third year in highland populations (*A. m. sigillatum*). Stebbins (1954) concluded that ensatinas (*Ensatina eschscholtzi*) reach sexual maturity during their fourth year. Hendrickson (1954) found that California slender salamanders (*Batrachoseps attenuatus*) require two to four years to reach sexual maturity.

Although giant salamanders (*Andrias japonicus*) have survived up to 55 years in captivity and several other species over 20 years (Goin and Goin, 1962), the longevity of salamanders in natural populations is probably much less. The oldest marked ensatinas recaptured by Stebbins (1954) were seven to eight years old. Based on the number of testicular lobes present, Organ (1961) found male *Desmognathus* which were six to seven years old. The oldest age class listed by Spight (1967) for male dusky salamanders, based on testicular lobes, is also seven years. There is no evidence that the longevity of females differs from that of males in any species, since adequate sampling has generally yielded a 1:1 sex ratio in adult classes.

POPULATION DENSITIES AND SOCIAL BEHAVIOR OF INDIVIDUALS

A number of workers have estimated the densities of salamander populations by collecting a known surface area. Published densities have been summarized by Spight (1967), who calculated them in terms of animals per square meter in cases where authors had not done so. These values are given in the following table together with the original literature citation. (See table below.) As noted by Spight (1967), a typical density for salamanders living in selected habitats seems to be 0.43 to 1.42 salamanders per square meter.

Movements of marked individuals in the field indicate that many salamanders remain within a very small area during a given season but they frequently move seasonally. The pattern of seasonal movement described by Gordon (1952) for green salamanders (*Aneides aeneus*) is probably typical for many salamanders except that the specific types of habitat will vary from species to species. Gordon noted that during the breeding period (mid-May to early September) there is little fluctuation in the total visible population until hatching of young occurs in September. There is then a dispersal and aggregation period from late September to November during which the young and adult females leave breeding crevices, the young salamanders moving to moss-filled crevices and ledges where they become well concealed. By October the adults begin to congregate in the vicinity of deep interconnecting crevices, where this species is believed to spend the winter; this is referred to as the prehibernation aggregation. In April and early May, there is a posthibernation aggregation and dispersal. The salamanders first congregate in the vicinity of the hibernation crevices and then the adults move to breeding crevices. Other species follow similar patterns of movement between hibernating areas and

breeding areas, particularly when adult life is relatively terrestrial and breeding takes place in water. For example, rough-skinned newts (*Taricha granulosa*) make mass movements to breeding waters in late winter and spring and then disperse to more terrestrial situations in forests and grasslands. The tiger salamander (*Ambystoma tigrinum*) inhabits relatively arid habitats throughout the Great Plains; these also migrate in large numbers back and forth from breeding ponds, often covering distances of several hundred yards or more. Organ (1961) observed that there was a seasonal vertical movement of aquatic and semiaquatic species of *Desmognathus* one fall. This movement was less than 100 feet and involved following the headwaters of the stream down the mountain as water levels dropped during a dry period. A comparable displacement of populations of other species probably occurs elsewhere when there are seasonal fluctuations in water levels.

Mark and recapture data indicate salamanders have very small home ranges. Terrestrial plethodontids have been found repeatedly under the same shelter (Hendrickson, 1954; Stebbins, 1954; Highton, 1956; Cunningham, 1960). Cagle (1948) found that even the aquatic three-toed amphiumas (*Amphiuma tridactylum*) restrict their movements in Audubon Park, Louisiana, to an area where food is available although they are capable of moving about freely within the park. Generally, the larger the salamander the greater the distance traversed in its normal activities

ESTIMATED DENSITIES OF SALAMANDER POPULATIONS*

Species	Salamanders/m²	Author
Aneides aeneus	0.0025–0.01	Gordon, 1952
Batrachoseps attenuatus	1.0–1.5	Anderson, 1960
Batrachoseps attenuatus	0.6	Hendrickson, 1954
Batrachoseps pacificus	0.76	Cunningham, 1960
Desmognathus ochrophaeus	0.017	Gordon et al., 1962
Ensatina eschscholtzi	0.15–0.17	Stebbins, 1954
Plethodon cinereus	0.050	Test and Bingham, 1948
Plethodon cinereus		
in dry litter	0.09	Heatwole, 1962
in wet litter	0.89	Heatwole, 1962
Plethodon jordani	0.022	Gordon et al., 1962
Plethodon glutinosus	0.004	Gordon et al., 1962
Plethodon yonahlossee	0.007	Gordon et al., 1962

*From Spight, 1970.

(Cunningham, 1960), and the home ranges of individuals and species may vary accordingly. Hendrickson (1954) found that 59 per cent of the 133 California slender salamanders (*Batrachoseps attenuatus*) he recaptured at least once never changed shelter, and those which did move tended to stay under a new shelter rather than to move about freely. In Cunningham's (1960) study of the larger Pacific slender salamander (*Batrachoseps pacificus*), 92 per cent of the 375 individuals he recaptured at least once were recorded as having moved to other cover. Sometimes these movements were merely to nearby shelter but many individuals made more than one move exceeding three feet. Both workers concluded that the salamanders seem to have a "cruising radius" from their home shelter and the proximity of other shelter is a deciding factor affecting whether and how often they move. For example, Hendrickson found that the cruising radius of the California slender salamander is about five feet and that when no other shelter is located within that radius an individual will tend to return to spend each day under the same shelter. When shelter objects are located farther than five feet from the home shelter, a salamander will tend to remain under such a distant object once it has arrived there rather than return to the original home shelter. Thus, periodically the salamanders establish new home ranges and in most instances never return to their original home range. Cunningham noted a similar behavior in the Pacific slender salamanders except that their cruising radius is larger; of those individuals captured at least five times, the average maximum distance moved was 19.8 feet, indicating that the home ranges of this species are much larger than those of the California slender salamanders. In all instances, the size and shape of the home range varies with the microenvironmental conditions. Salamanders living under the most favorable conditions of food, moisture, and cover tend to have smaller home ranges than salamanders living under less favorable conditions. The Pacific slender salamanders were also noted to have smaller home ranges when there were numerous earthworm burrows present, these being used by the salamanders for shelters. The size of the home range also tended to be smaller in densely

populated areas than in sparsely populated regions, indicating competition may result in restriction of movement. These influences are consistent with factors affecting the size of home ranges of other vertebrates. As has also been reported for other vertebrates, slender salamanders occasionally make extended (exploratory?) excursions outside of their normal home range, but these are of short duration and the animal returns rapidly to its original site.

Cunningham (1960) found that some changes in home range by Pacific slender salamanders can be attributed to seasonal variations in vegetation. During the growing seasons when plants were abundant, salamanders with home ranges that included a plant seemed to restrict their activities to the vicinity of the plant's shelter. At other times of the year, when the plants were dead or were pruned, the salamanders sometimes moved and established another home range elsewhere. He found that normal gardening activities, such as weeding, planting, cultivation, and irrigation have a profound influence on home range relationships of salamanders and, in fact, most of the changes in home range which he observed seemed directly attributable to man's alteration of the habitat.

Territoriality seems to be developed to different degrees among salamanders. The fact that Cunningham (1960) found numerous individuals of all sizes and presumably of different sexes within a single burrow indicates that the Pacific slender salamander is not territorial. He also observed that the home ranges of different individuals sometimes overlap in a given area and noted that up to 20 individuals were often seen at one time in the beam of a flashlight. Grant (1955) studied the home range and territorial behavior of four-toed salamanders (*Hemidactylium scutatum*) and two-lined salamanders (*Eurycea bislineata*) in terraria which provided habitats of 18 by 9 inches. Individuals of both species would establish home ranges of five to six inches (radius) from their home sites, which were under pieces of bark. When two individuals were placed in the same terrarium, their home ranges would slightly overlap. When more than two individuals were placed in the same terrarium, those that failed to establish home ranges wandered at random through the ranges of the others but would not feed and eventually starved to death even though food was provided throughout the terrarium. Thus it appeared that in these species the home range corresponds roughly to the general feeding area of the animal, just as with Cagle's three-toed amphiuma. Different degrees of territoriality were exhibited by the four-toed and two-lined salamanders. Territorialism seems to only be weakly developed in the four-toed salamanders. When one individual's home site was approached by another, the resident would merely advance toward the intruder and use intimidation to defend its territory. Upon some occasions, an intruder was allowed to climb over the home site of another without retaliation by the resident. The two-lined salamanders, on the other hand, actively defended an area within about two inches of their home site. Typically, this territory was defended by the resident advancing toward the intruder and placing its snout in direct contact with that of the intruder. After physical contact of the snouts the intruder generally would back away, but occasionally the resident would aggressively bite the intruder's snout or tail, sometimes biting off large portions of the other's tail.

Stebbins (1954) could find no evidence of territorial behavior in the ensatinas (*Ensatina eschscholtzi*). Gordon (1952) observed that brooding green salamanders (*Aneides aeneus*) defend their eggs and, to that degree, are territorial during at least the incubation period. Four-toed salamanders (*Hemidactylium scutatum*) are also known to remain with their eggs and may defend their nests. It is not known how widespread nest-defending is among other plethodontids, but it does not seem to be characteristic of other salamander groups.

POPULATION DYNAMICS OF TURTLES

No complete study of the dynamics of a turtle population has been made, and much remains to be learned about even the more familiar species which have received the greatest amount of attention. Although they tend to be diurnal animals and are often relatively easy to catch and mark, turtles are difficult subjects for a population dynamics study because they are very long-lived (some have been known to reach ages of 100 years or more [Flower, 1925]), and generally only the younger members of a population can be accurately aged. Thus far, the movements and dynamics of marine turtles can only be surmised from the recapture of tagged individuals that come ashore. Although reproductive patterns and homing tendencies are becoming apparent, the life of marine turtles at sea remains a mystery for the most part. Very little is known about the turtles of the Southern Hemisphere or of the tropics. Consequently, much of the following section deals with aspects of the population dynamics of northern tortoises and fresh water turtles.

NATALITY

Among different species of turtles, the average number of eggs per clutch tends to increase with the size of the turtle. The giant green turtles (*Chelonia mydas*) deposit an average of 100 eggs and up to 200 eggs per clutch, while many of the smaller land tortoises and fresh-water turtles deposit less than 10 eggs per clutch. However, within a population there tends to be considerable variation in clutch size. Hendrickson's (1958) data indicate how extreme this variation may be: the size of 8147 green turtle clutches deposited on the Sarawak Islands of southeastern Asia varied from three to 184 eggs; the mean was 104.7 eggs. Intrapopulational variation in clutch size may or may not be correlated with the size of the female turtle. In a study of painted turtles (*Chrysemys picta*), Cagle (1954) examined 48 egg-containing females from southern Illinois and found that the number of eggs varied from three to eight (mean 6.3), in 13 egg-containing females from Tennessee the number of eggs varied from three to six (mean 4.3), and in 12 females from northern Michigan the number of eggs varied from two to seven (mean 4.7). In none of these groupings was there a correlation between the number of eggs and the size of the female. Gibbons (1967) and Gibbons and Tinkle (1969) have studied three populations of the same species in southern Michigan and found that the three populations differ significantly in both the sizes of individuals and the sizes of clutches. However, they, like Cagle, were unable to correlate differences in clutch size with differences in body size (plastron length) of females. Instead, both body size and clutch size seem to be affected by dietary differences among the populations. The Kalamazoo River turtles were found to be more carnivorous, to have the largest average body size, and to have a larger average number of eggs per clutch (8.2±0.4 eggs). The Sherriff's Marsh turtles were more herbivorous, averaged the smallest size, and had the smallest average number of eggs per clutch (5.2±0.3 eggs). The Wintergreen Lake turtles were intermediate. Gibbons (1967) was able to find evidence that the large river turtles were the same age or younger than much smaller marsh turtles; therefore, the size difference between the populations does not appear to be the result of an age structure difference. Gibbons and Tinkle suggest that the dietary differences are the primary cause of both the differences in growth rate and differences in reproduction among these three populations.

There is evidence that within some species the clutch size is higher in large turtles than in small turtles. Cagle (1950) found the number of oviducal eggs in 126 female pond sliders (*Pseudemys scripta*) from Illinois and Louisiana varied from 4 to 20 and was correlated with the plastron length of the female (r = 0.74). Einem (1956) reported that mudturtles (*Kinosternon bauri*) from Florida contained 2 to 4 oviducal eggs and noted that their number varied with the size of the females. Tinkle (1961) examined the oviducts of 30 southern (Alabama, Arkansas, Florida, Georgia, Louisiana, Tennessee, and Texas) and 15 northern (Connecticut, Illinois, Indiana, Kansas, Maryland, Michigan, New Jersey, New York, and Pennsylvania) common musk turtles (*Sternothaerus odoratus*) and found that there is a significantly higher clutch size in large than in small turtles in both northern and southern populations. However, Tinkle also found that the average number of oviducal eggs in southern turtles was 2.2 (range, one to four) whereas in northern turtles it was 4.6 (range, two to seven) and, despite the fact that the northern turtles were generally larger than the southern ones, the size difference was not sufficient to explain the north-south difference in clutch size.

When the turtles were arbitrarily divided into two size groups of 61 to 90 millimeters and 91 to 120 millimeters, the average clutch size in the smaller size group was 2.0 for southern turtles and 3.5 in northern turtles, an average of 75 per cent more eggs in the northern clutches. Southern females in the larger size group averaged 3.2 eggs per clutch whereas the average for large northern females was 5.5 eggs, the latter number being 72 per cent greater than the former. In contrast, large southern turtles only produced 60 per cent larger clutches than small southern turtles and large northern turtles only produced 57 per cent larger clutches than small northern turtles. Thus, even when turtles of the same size are compared, the northern turtles are more productive than southern turtles. Cagle's (1950) data on sliders (*Pseudemys scripta*) also indicate a north-south difference in clutch size. The average number of eggs found in 129 nests examined by Cagle in southern Louisiana was 7 (range, two to 19), whereas the numbers in Illinois and Tennessee clutches were higher: the average of 102 Illinois nests was 9.3 (range, four to 18), and the average in 47 Tennessee nests was 10.5 (range, five to 22). Thus, it appears that turtles, like birds and mammals, produce larger broods in the northern parts of their ranges than in more southern areas.

Tinkle next questioned whether natural selection has favored a greater net gain in clutch size in northern turtles by simultaneously favoring both larger individuals in the northern populations and individuals with a higher reproductive potential. He found indications that such a coupled selection does occur, for the northern turtles must generally reach a greater carapace length before attaining maturity than southern turtles. Similarly, Cagle (1954) found that painted turtles (*Chrysemys picta*) mature at a smaller size in Louisiana than in Illinois or Michigan.

Assuming that the larger size and greater clutch size of northern turtles were necessitated by higher mortality in harsher environments, Tinkle next grouped carapace length and clutch size data according to isotherm zones. Graphs of carapace length versus isotherm zone are shown in Figures 14–2 and 14–3. Female turtles between isotherms 65 and 75 are significantly smaller than those from above isotherm 55;

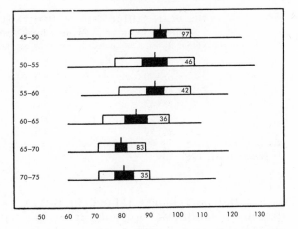

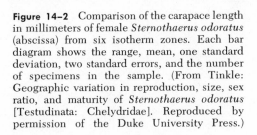

Figure 14-2 Comparison of the carapace length in millimeters of female *Sternothaerus odoratus* (abscissa) from six isotherm zones. Each bar diagram shows the range, mean, one standard deviation, two standard errors, and the number of specimens in the sample. (From Tinkle: Geographic variation in reproduction, size, sex ratio, and maturity of *Sternothaerus odoratus* [Testudinata: Chelydridae]. Reproduced by permission of the Duke University Press.)

males show a steady cline in body sizes from south to north (the apparent reversal above isotherm 50 was not statistically significant), and the mean size of males between isotherms 70 and 75 is significantly different from those of all other isotherm groups. The average clutch size (number of oviducal eggs) and isotherm data are presented in the following table. These data indicate that the average clutch size is increased progressively as one goes from the warmer southern isotherms to the cooler northern isotherms; the average clutch size triples between the isolines of 75 and 45. (See table on the following page.)

The number of clutches produced per reproductive season by an individual female turtle varies considerably among those species which have been studied. Mitsukuri (1895) found that Japanese softshelled turtles (*Trionyx japonicus*) deposit

as many as four clutches each season. Hildebrand (1932) reported that diamondbacked terrapins (*Malaclemys terrapin*) lay one to five clutches per year. Deraniyagala (1939) indicated that the Ceylonese soft terrapin (*Lissemys punctata graosa*) and the starred tortoise (*Testudo elegans*) lay several groups of eggs within a few weeks. Hendrickson (1958) found that the marine green turtles (*Chelonia mydas*) nesting on the Sarawak Islands lay an average of seven times and up to 11 times in a breeding season, successive clutches being produced at about 10-day intervals; Carr and Ogreen (1960) found the same species re-nests up to six times at intervals of 12 to 14 days in the Caribbean. Einem (1956) concluded that the mudturtle (*Kinosternon bauri*) is capable of depositing two clutches per season. Tinkle (1961) found that the egg-laying season of common musk turtles in the south begins

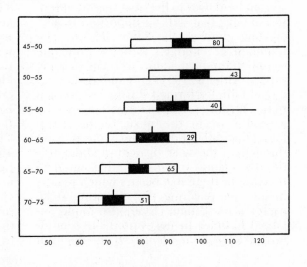

Figure 14-3 Comparison of the carapace length in millimeters of male *Sternothaerus odoratus* (abscissa) from six isotherm zones. Each bar diagram shows the range, mean, one standard deviation, two standard errors, and the number of specimens in the sample. (From Tinkle: Geographic variation in reproduction, size, sex ratio, and maturity of *Sternothaerus odoratus* [Testudinata: Chelydridae]. Reproduced by permission of the Duke University Press.)

THE MEAN CLUTCH SIZE OF TURTLES FROM SEVERAL ISOTHERM ZONES*

Isotherm Zone	Mean Clutch Size	Number of Specimens
70–75	1.8	8
65–70	2.3	22
60–65	2.8	4
55–60	3.0	2
50–55	3.5	2
45–50	5.5	10

*From Tinkle, 1961.

earlier and lasts longer (March to July) than in the north (May to June). If this is generally true for all turtles, one would expect the ability to ovulate more than once per season to be greater in the south than in the north. The deposition of two or more clutches per season by southern turtles would compensate for the smaller number of eggs per clutch in southern populations. Although Tinkle (1961) found that the number of enlarged ovarian follicles in postreproductive southern females was 66 per cent greater than in northern females, he and Risley (1933) both concluded that the common musk turtle produces only a single clutch per season throughout the range of the species. In other species, however, the fecundity of different populations must be significantly varied because of differing numbers of clutches per female per season.

Basically, the potential fecundity of a female depends upon how many ovarian follicles can be ovulated during one season. An accurate determination of this requires a knowledge of how many follicles fail to mature and whether or not enlarged follicles formed during one season are held over to the next season or represent eggs that will be deposited in succeeding clutches the same season. The retention of enlarged follicles from one season to the next seems to occur in some turtles. Altland (1951) concluded that enlarged follicles formed during one season in box turtles (*Terrapene carolina*) may be carried over to the next. Cagle (1944) found that enlarged follicles were present throughout the year in mature slider turtles (*Pseudemys scripta*) and that a gradual increase in their size occurs which results in one set reaching the ovulating stage each season. Einem (1956) also found enlarged follicles in postreproductive mud-turtles (*Kinosternon bauri*). If one does not

know how many clutches may be produced in a season or how many follicles fail to mature for some reason or another, counting the number of enlarged follicles will not provide an accurate measure of the potential reproduction of an individual. However, an estimate of the *maximum* fecundity of a female may be obtained by counting the number of large yolk-filled follicles that could be ovulated during one season and adding to this the number of corpora lutea representing recent ovulations and the number of oviducal eggs present. The relative numbers of each of these will vary with the season, depending upon whether the female is prereproductive, in the season of ovulation, or postreproductive. Tinkle (1961) calculated the average number of enlarged follicles in prereproductive and postreproductive females from the northern and southern populations of common musk turtles and then, using the average clutch size as a measure of actual fecundity, determined the percentage realization of reproductive potential. Assuming that these turtles lay one clutch of eggs per year, the average reproductive potential for all southern turtles Tinkle examined was 4.9 eggs per female and the percentage realization of this potential was 51 as opposed to an average potential of 9.5 for northern turtles with a 45 per cent realization. Like clutch size, Tinkle found the reproductive potential increased with a south-to-north change in isotherm zones; this trend is apparent in the following table on the opposite page. The average reproductive potential for the 45–50 isotherm zone (10.8) is over three times that of the 70–75 isotherm zone, nearly the same relative increase as for clutch size. Above isotherm 50, the realization of potential increases steadily through the warmer isotherm zones; below 50, the trend is reversed and the realization is about the same as that of isoline 70. Although these data are based on rather limited samples, the general indication is that northern populations of common musk turtles are much less efficient reproductively than southern populations.

There is very little information bearing on reproductive cycles in turtles. Cagle (1954) and Gibbons (1968) indicate that painted turtles have an annual cycle, with females depositing one or two clutches each year. Harrisson's (1956) and Hendrickson's (1958) recapture data of marked green turtles (*Chelonia mydas*) indicate this marine species follows a triennial cycle of breeding off the coast of Borneo, tagged females having returned to the same place to breed after a two-year absence. Carr and Ogreen (1960) found the same species also has a strong three-year cycle in the Caribbean but in a minority of cases there is nesting on two-year cycles. In no case did a turtle return to nest during the season following one in which it had nested before. Turtles are generally recognized as being very long-lived; this probably means that their reproductive life is also of very long duration. Consequently, the fecundity of a female that ovulates as infrequently as every three years can still, over an extended period of time, be awesome indeed. Harrisson (1956) reported that one tagged female green turtle had deposited over 1000 eggs in just two seasons (1953 and 1956)! To speculate, it seems that natural selection would be more likely to favor a triennial or biennial cycle in turtles which deposit several clutches per season and have a very long reproductive life than it would in turtles which deposit one or two clutches per season and have a shorter duration of reproductive life; the latter may be more likely to have an annual cycle.

SURVIVORSHIP, ATTAINMENT OF SEXUAL MATURITY, AND LONGEVITY

Although field observations have revealed that turtle eggs and hatchlings are very susceptible to predation, no one has succeeded in quantitatively determining the survival rate of turtles in a natural population. Hendrickson (1958) was able to estimate the survival of green turtle embryonic stages by transplanting clutches to hatchery enclosures where the eggs were reburied. Whenever it was possible to do so, the eggs were transported and placed in the artificial nests in the same position they had occupied when found. Hendrickson found that, of 354 clutches transplanted, at least some turtles emerged from all but eight per cent of the nests; 47.1 per cent of all the eggs transplanted produced young. He also noted that the highest survival occurred in clutches deposited during the peak of the breeding season and that the lowest survival was in clutches transplanted during the monsoon season. The heavy monsoon rains packed the soil very tightly and this appeared to be a factor contributing to the mortality of embryos and hatchlings attempting to leave the nests. The duration of time from oviposition to emergence of hatchlings varied from 48 to 80 days but, significantly, averaged 65 days during the monsoon season compared to 55 days during the rest of the breeding season. The minimal time between hatching and emergence from the nest was found to be four to five days.

REPRODUCTIVE POTENTIAL IN COMMON MUSK TURTLES AND THE PERCENTAGE REALIZATION OF POTENTIAL IN SIX ISOTHERM ZONES*

Isotherm Zone	Mean Reproductive Potential	Number of Specimens	Percentage Realization of Potential
70–75	3.2	6	56%
65–70	4.7	26	49%
60–65	5.9	9	47%
55–60	8.2	13	37%
50–55	9.7	7	36%
45–50	10.8	20	51%

*From Tinkle, 1961.

Despite attempts to follow large numbers of hatchlings, Hendrickson learned nothing of their survival rate. He did note that eggs and hatchlings on the beach are preyed upon by crabs, rats, monitor lizards (*Varanus salvator*), gulls and herons, house cats, wild dogs, tigers, the mangrove snake (*Boiga dendrophila*) and reticulated pythons (*Python reticulatus*). Once they reach the water, the turtles are preyed upon by sharks and large fish. Sharks seem to be the most important predators on adult turtles, and a large shark such as *Carcharodon* can inflict heavy damage on adult turtles. Comparable studies of embryonic and hatchling survival have not been made for other species of turtles.

The analyses of size classes and apparent age classes have provided data indicating when turtles reach sexual maturity. Although Agassiz (1857) concluded that female turtles do not generally begin to lay eggs until they are 10 to 11 years old, more recent studies have indicated that much less time, usually four to six years, is required for maturation. Hendrickson (1958) states that Asian green turtles usually mature in six years but may require as few as four years. Mitsukuri (1905) found that the Japanese soft-shelled (*Trionyx japonicus*) also may reach sexual maturity in six years. Tinkle (1961) found that common musk turtles (*Sternothaerus odoratus*) from different geographical localities reach maturity in three years but that southern turtles mature at a smaller size (mean of 65 millimeters for males and 82 millimeters for females) than northern turtles (mean of 73 millimeters for males and 96 millimeters for females). Other species seem to mature at a particular size, regardless of age. Cagle (1954) concluded that painted turtles (*Chrysemys picta*) females become sexually mature at a plastron length of 120 to 130 millimeters and males reach sexual maturity at plastron lengths of 70 to 90 millimeters. In southern populations males may become sexually mature in one complete growing season, whereas in the north two or possibly three seasons are required for males to reach maturity. Gibbon's (1968) study of Michigan painted turtles also indicated that sexual maturity is correlated with length and not age. He found males matured at a length of 80 millimeters and females reached maturity at a plastron length of 110 to 120 milli-

meters. However, males from one population (Wintergreen Lake) matured in their third to fifth year, whereas males from the other population (Sherriff's Marsh) reached maturity in their sixth or seventh year. Gibbons found that some Sherriff's Marsh females as old as 12 years were immature, whereas some seven-year-old individuals from Wintergreen Lake were mature. Thus, at least in some species, the reproductive potential of a population is affected by environmental conditions which promote or impede growth. Cagle (1946) has noted that great individual variation in growth occurs in turtles and that differences in growth rate of turtles occupying different habitats are conspicuous. He found, for instance, that the growth of turtles in shallow water was more rapid than those in deeper water, probably because shallow water tends to be warmer.

The longevity of adult turtles in natural populations can only be surmised, since no one has found a method to accurately age turtles beyond an age of about ten years. Most populations have yielded approximately 1:1 ratios of males and females in samples of younger age classes, but more females than males are usually present in older or larger groupings. For example, Tinkle (1958) found a ratio of three females per male in his samples of *Sternothaerus depressus*. Hildebrand (1929) reported a ratio of 4.4 females per male in 1300 specimens of diamond-backed terrapins (*Malaclemys*). Risley (1933) found 2.3 females per male in 255 specimens of common musk turtles from Michigan, but Tinkle (1961) reported 52 per cent females and 48 per cent males in 647 specimens of the same species from various latitudes. Thus there are indications, but not proof, that male turtles may not survive as long on the average as females. On the other hand, the greater proportion of females in collections may reflect a sexual difference in probability of being caught or of seasonal activity.

SOCIAL BEHAVIOR AND POPULATION DENSITIES

Much remains to be learned about the social behavior of most turtles and tortoises, but a few species have been studied in considerable detail. Studies of marked painted turtles (*Chrysemys*), snapping turtles (*Chelydra*), musk turtles (*Sternothaerus*), sliders (*Pseudemys*), box turtles (*Terrapene*), desert tortoises (*Gopherus*) and pond turtles (*Clemmys*) have consistently shown that each adult turtle has a home range to which it will attempt to return and within which it will confine its activities for extended periods of time. The size of the home range varies with the species and the extent of suitable habitat. Nichols (1939) marked and recaptured box turtles (*Terrapene carolina*) near his Long Island home; he concluded that the normal diameter of an individual's home range is 250 yards or less but noted that individuals do sometimes wander much farther and may shift their home ranges over a period of years. Stickel (1950) studied the same species in Maryland and concluded that the average maximum home range diameter of adult males was 330 feet while that of adult females was 370 feet; the difference was not statistically significant. She also found that some turtles have two home ranges and travel between them at infrequent intervals. Desert tortoises (*Gopherus agassizi*) in southwestern Utah appear to have home ranges covering about 10 to 100 acres (Woodbury and Hardy, 1948). In contrast, Cagle (1950) reported that it is not unusual to observe 50 to 100 sliders (*Pseudemys scripta*) in 100 square feet of water surface!

Many workers have observed that turtles will sometimes make trips outside of their home ranges. These trips include egg-laying trips by females and movements to hibernation sites or merely seem at times to reflect restless wanderings (Cagle, 1944). Turtles may be forced from their home ranges by habitat changes such as drying-up of water, and in such instances they may move in search of suitable habitat or follow the last remnants of water (Cagle, 1944). Turtles forced from their home ranges by such habitat changes may eventually return to them when conditions allow.

There is no evidence of any territoriality among turtles, and the home ranges of turtles of different ages or sex frequently overlap grossly. Woodbury and Hardy (1948) found that an individual tortoise does not always use the same den or summer hole but may shift from one to another within the same home range. Dens are generally utilized simultaneously by groups of tortoises (these workers found up to 17 in a den at one time), but summer holes are rarely shared with other tortoises. The only tortoises found sharing a summer hole were members of a pair (male and female). Although the solitary use of summer holes would seem to imply territoriality, Woodbury and Hardy could find no evidence to indicate that defended territories existed. As noted by Cagle (1950), the lack of aggression between individual turtles, their ability to go without food for long periods, and their omnivorous feeding habits make it possible for high density populations to exist. The density of turtle populations will, therefore, tend to vary with the restriction of the habitat. Stickel (1950) estimated the density of her Maryland population of box turtles at four to five turtles per acre. Woodbury and Hardy, in contrast, concluded that the density of the Utah population of desert tortoises was about one tortoise for each four acres of land and noted that this is a relatively high concentration compared to the concentrations which might occur elsewhere. The density of aquatic fresh-water turtles is undoubtedly many times either of these figures in some populations.

POPULATION DYNAMICS OF CROCODILIANS

Only scattered observations exist regarding the population dynamics of crocodilians and, unfortunately, because of man's onslaughts few or no stable populations exist anywhere and breeding populations are continually being depleted in most areas.

NATALITY

There is considerable variation in the number of crocodilian eggs per clutch and some evidence that larger females deposit more eggs per clutch than smaller females. Cott (1961) found that the range in clutch size for 775 nests of the Nile crocodile (*Crocodylus niloticus*) was 25 to 95 eggs with an average of 60.4 eggs and believed there was some indication that larger females deposit larger clutches. American alligators (*Alligator mississippiensis*) may lay 15 to 88 eggs per clutch (Oliver, 1955), and McIlhenny's (1934) data indicate a rough correlation between the size of the female and the number of eggs in a clutch for this species. McIlhenny studied a marked population of American alligators which he released as hatchlings on Avery Island, Louisiana, and was able to record three clutches together with the length and weight of the females depositing them: one clutch contained 42 eggs and was deposited by a female 9 feet 1½ inches long that weighed 163½ pounds; the second clutch contained 34 eggs and was deposited by a female 7 feet 3 inches long that weighed 116½ pounds; the third clutch contained 41 eggs and was deposited by a female 7 feet 8 inches long that weighed 129½ pounds.

The breeding cycles of crocodilians have not been studied, but it is generally assumed that mature females deposit a single clutch annually.

SURVIVORSHIP, ATTAINMENT OF SEXUAL MATURITY, AND LONGEVITY

No age-specific survivorship data are available for any crocodilian population. However, Cott (1961) did note heavy mortality of eggs and hatchlings from predators, especially Nile monitors (*Varanus niloticus*) and larger crocodiles, and observed that cannibalism results in an internal regulation of maturing Nile crocodile populations. Parental guarding of the young lasts until they reach the water, and then cannibalism is so important a source of mortality that it appears to have led to the evolution of distinctly different habitats for young and adults. Young Nile crocodiles are usually found in weed-choked shallow waters and, being proficient climbers, will climb six to eight feet into trees to lie on overhanging branches. Adult crocodiles occur in and around deeper, relatively open bodies of water. In addition, Cott and others have noticed the conspicuous absence of two- to five-foot crocodiles during daylight hours, and he suggests that this size grouping probably goes into retreat from the diurnally-active mature individuals because of cannibalism. Other than man, the only animals dangerous to adult crocodiles appear to be hippopotamuses, lions, and African elephants. The differences in habitat and be-

havior correlate with three rather distinct food patterns for hatchling, middle-aged, and mature Nile crocodiles (see page 328).

The survival of male and female Nile crocodiles appears to be about the same, for out of 651 crocodiles that Cott examined there were 324 males and 327 females. McIlhenny (1934) has provided the only accurate data on growth rate and age of maturity for any natural population of crocodilians. Having toe-clipped each of his original population when they were hatchlings, he subsequently recaptured, weighed, and measured different individuals and noted when females first nested. His data indicate that American alligators grow a little more than one foot per year up to and including the fifth year in females and up to at least the ninth year in males; growth continues more slowly in females after the fifth year. He recorded the first nesting of a female 9 years and 10 months after she had hatched, indicating that this is about the age that sexual maturity is reached in this species. Cott (1961) found that sexual maturity in female Nile crocodiles is attained at a minimum length of 2.19 meters, while males mature at a length between 2.9 and 3.3 meters. Thus, either the growth rates of male Nile crocodiles are faster than those of females or the females mature at an earlier age than the males. Cott thought it positive that female Nile crocodiles do not mature before an age of 10 years and thought it probable that they might require 19 years.

The longevity of crocodilians in nature is not known. Captive American alligators (*Alligator mississippiensis*) have survived for 56 years, Chinese alligators (*Alligator sinensis*) for 50 years, and crocodiles of various kinds for 16 to 31 years (Flower, 1925). Cott (1961) believes that the longevity in nature is potentially much longer than for captive individuals and suggests the largest known specimens of Nile crocodiles must be over 100 years old.

Little is known of territoriality in crocodilians, although most seem to defend their basking area and feeding place. Territoriality, however, is probably not as important a factor in regulating the population size of any species as is cannibalism. There apparently are no published estimates of population densities for any crocodilian species.

POPULATION DYNAMICS OF LIZARDS

The greatest volume of information on reptilian population dynamics applies to lizards, but only a very small fraction of the total number of species of lizards has been studied, and generally, even the best studies have been limited to certain aspects of life histories. To emphasize the latter point, Tinkle (1969) chose 12 topics, such as age at first breeding, age-specific fecundity, adult life expectancy, and so forth, and determined the frequency with which each item was reported in a sample of 50 publications. Although Tinkle's sample of publications was biased by the fact he chose all of what he considered to be the best papers, he found the probability for all 12 of the pertinent points being included in a particular study was only 8.2×10^{-5}! The point that is constantly emphasized throughout this chapter is the need for detailed quantitative life history studies.

NATALITY

Clutch size in lizards ranges from one or two eggs to 60 eggs (see table pages 400–404). Much of the following discussion of this variation in clutch sizes is based on Tinkle, Wilbur, and Tilley (1970), and the reader is urged to peruse this reference for details not presented here. Tinkle and his collaborators analyzed literature data on clutch sizes and found that clutch size varies independently of body size in late-maturing and tropical species, but otherwise there is a highly significant correlation between snout-vent length at maturity and mean clutch size and also a significant correlation between clutch size and age at first reproduction. Large lizards tend to have larger clutches than small lizards and lizards which mature early in life tend to have smaller clutches than those maturing late in life. In their analyses of data on 37 species, representing the families Iguanidae, Agamidae, Anguidae, Teiidae, Chamaeleontidae, Xantusidae, and Scincidae, Tinkle and his collaborators indicated that the mean increase in clutch size with increase in body size was approximately one egg per 10 millimeters in snout-vent length. The mean clutch size for early-maturing lizards, defined as those which ma-

ture and reproduce in the next season after hatching, was 4.27 eggs, whereas that for late-maturing lizards, defined as those which mature in a season subsequent to the one after hatching, was 6.48 eggs. Approximately 26 per cent of the variance in clutch size among the 37 species was accounted for by interspecific variation in these two characteristics.

Another correlation noted in the same study is that lizard species which produce a single clutch per season have larger bodies and produce larger clutches than species which produce more than one clutch per season; this difference is not entirely due to size. The increase in clutch size with increase in body size occurs at nearly the same rate in both groupings but at any given size, single clutch lizards produce nearly twice as many eggs per clutch as multiple clutch lizards (means 7.07 and 3.96, respectively). No significant differences in mean clutch sizes or mean body sizes were found between oviparous and viviparous or between tropical and temperate species.

The interrelationships of the characteristics mentioned above are such that the lizard species which mature early are those which are smaller when they mature, produce relatively small clutches, and deposit more than one clutch per season. Species which mature late are relatively large at maturity, produce larger clutches, and deposit a single clutch per season. In addition, Tinkle *et al.* note that the former tend to be tropical and temperate in distribution as opposed to a primarily temperate distribution of the latter. The number of clutches of eggs produced per year by tropical lizards is not generally known. Some species, such as the Australian skink *Leiolopisma rhomboidalis*, appear to breed thoughout the year (Wilhoft, 1963) while other species, such as the common agama (*Agama agama*), have cyclical reproductive patterns (Marshall and Hook, 1960). However, in no instance has it been demonstrated that the same individual female will produce successive clutches thoughout the year.

Both populational and individual differences may contribute to intraspecific variation in clutch sizes. Tinkle (1967) compared Colorado and Texas populations of the side-blotched lizard (*Uta stansburiana*) and found that females in Texas lay clutches averaging four eggs each while in Colorado three eggs per clutch is normal. This difference appears to be associated with the fact that Colorado females mature at a smaller size and do not become as large as the Texas females. Population-to-population variation in clutch size probably occurs commonly, at least in widely distributed species, but it remains unstudied. A number of workers have observed that clutch size varies with the age and growth of females. As examples of this intrapopulational variation, Blair (1960) found an average of about 11 eggs per clutch for yearling female rusty lizards (*Sceloporus olivaceus*), 18 per clutch for two-year-old females, and 24 per clutch for three-year-old females. Tinkle (1967) found that the mean clutch size for side-blotched lizards was 2.7 for 35- to 39-millimeter females, 3.1 for 40- to 44-millimeter females, and 3.5 for 45- to 49-millimeter females. Thus, the potential reproduction of a population depends not only on the number of mature females present but also on the age classes of these females and how long they remain part of the breeding population.

SURVIVORSHIP, ATTAINMENT OF SEXUAL MATURITY, AND LONGEVITY

The causes of mortality of lizards are only poorly known, and few age-specific data are available indicating survivorship curves. Two studies have shown that the mortality of embryonic stages may be very high in lizard nests. Blair (1960) reported that 75 to 78 per cent of the clutches of rusty lizards (*Sceloporus olivaceous*) on a ten-acre study area in Texas were lost to predators, desiccation, or other mortality-inducing factors. Fitch (1956b) observed that only about 50 per cent of the eggs in nests of the collared lizard (*Crotaphytus collaris*) actually hatch. Thus, at least in these species, the mean life expectancy at fertilization must be extremely short.

The survival of hatchling lizards appears to vary widely in different populations and species. Blair (1960) estimated that 6 to 20 per cent of the hatchlings produced in his population of rusty lizards survived to reach sexual maturity. Tinkle (1967), who marked 482 male and 585 female hatchlings of side-blotched lizards (*Uta stansburiana*) in Texas, found very high

mortality during the first few weeks of life. He estimated that about 15 per cent of the females hatched survive to sexual maturity. The survival of males was even lower. The average life expectancy of hatchlings was found to be about 18.5 weeks. Crenshaw (1955) found a similar pattern of mortality in eastern fence lizards (*Sceloporus undulatus*): 50 per cent of a sampling of hatchlings died during their first six weeks and only 32 per cent survived the first two months. On the other hand, Fitch (1956b) reported a much higher survival of collared lizards (*Crotaphytus collaris*), estimating that 40 per cent of the hatchlings reach sexual maturity.

The exact age at which lizards become sexually mature is known for less than 10 species (Tinkle, 1967). This is because, generally, the age at which maturity is attained has been determined from the distribution of size groupings, rather than by following hatchlings through to maturity, and at least upon some occasions, errors have occurred in assigning ages to sizes. A few species seem to mature when they are less than one year old, but most require at least one year. As noted previously, species which are viviparous or have the larger clutches tend to mature at an age greater than one year, and some, such as the yucca night lizard (*Xantusia vigilis*), appear to require three years to reach maturity (Miller, 1951; Zweifel and Lowe, 1966).

Data on the survivorship of adult lizards indicate considerable variation in longevity of individuals among different species. Tinkle (1969) found that 90 or more per cent of the adults in Texas populations of side-blotched lizards live one year or less, but in Colorado nearly one third of the adults of one season will survive to reproduce during another season and many will reproduce for three seasons. Blair (1960) estimated that about 20 per cent of the adult rusty lizards in his Texas population survived to reproduce a second season. Hirth (1963) was only able to recapture 10 (four males, six females) Costa Rican beach lizards (*Ameiva quadrilineata*) out of 115 marked adults after a period of twelve months; none were found two years after marking. Hirth, in the same study, recaptured eight (six males, two females) of 100 adult basilisks (*Basiliscus vittatus*) after an interval of 13 months, and two of the males were still alive two years after marking. At the other extreme, Fitch (1956b) found about one-half of the sexually mature collared lizard adults survive to a second season; he also (1958) reported a 50 per cent reduction each year in adult populations of the six-lined racerunner (*Cnemidophorus sexlineatus*).

SOCIAL BEHAVIOR AND POPULATION DENSITIES

Studies of a variety of lizard species, particularly in the families Agamidae and Iguanidae, indicate that after an individual has dispersed from its parental area it is closely restricted to a home range and will remain in the same area throughout its life. The home range often will center around a very favorable basking site or principal perch, but sometimes the perch will be located near the edge of the range.

Rusty lizards studied by Blair (1960) have a pattern of social behavior which is probably representative of most lizards. Using peripheral sight records for individual lizards to estimate home ranges, Blair found that those of juvenile males are larger (0.027 acre on the average) than those of juvenile females (0.017 acre). Home ranges of adult lizards averaged 0.072 acre for females, about four times the size of those of juvenile females, and 0.169 acre for males, about six times the size of those of juvenile males. Thus, the home ranges of males are considerably larger than those of females, indicating that males tend to wander more than females. Both juveniles and adults center their home ranges around one or more basking sites (large trees, outbuildings, board fences, stumps, logs, and trash piles). In contrast to females and juveniles, which concentrate their activities around a principal station, males move about freely from station to station within their home ranges and show no preference for a particular perch. The type of perch utilized by juveniles is distinctly different from those of adults. Blair found that juveniles tend to use relatively small perches, such as mesquites, fence posts, or tree seedlings, whereas adults showed a strong preference for the larger cedar elm trees. The distribution of these different kinds of basking sites was such that few juvenile

home ranges overlapped those of adults. The distribution of basking sites also meant that, with growth and maturity, the change from juvenile home range to adult home range frequently involved dispersal to a completely different area. At other times, it involved enlargement of the juvenile home range or merely a change in the choice of perches within the same home range. As appears to be typical of lizards in general, once a rusty lizard reaches sexual maturity it tends to remain in the same area for the duration of its life, although the boundaries of the home range may fluctuate from year to year in response to population density and changes in the environment.

Details of the home ranges of other lizard species vary with their habitat requirements and environmental conditions, but the general pattern exhibited by the rusty lizards compares favorably with those of others. Fitch (1954) found that the home ranges of five-lined skinks (*Eumeces fasciatus*) vary considerably in size and shape, depending upon the individual's age and sex and on the presence and distribution of essential features of the habitat. For instance, he found that a rock pile measuring approximately 70×30 feet must have constituted the entire home range for many five-lined skinks living in it, since the surrounding areas were not suitable for the species, and 212 individuals were collected from this pile over a four-season period. Obviously, many were there simultaneously and must have shared essentially the same home range. The home ranges of five-lined skinks in open woods were more elongated and perhaps larger. In general, the scattered distribution of flat rocks and outcrops, stumps, logs, and glades with patches of sunlight possibly induces larger home ranges in this species just as a scattering of suitable perches promoted larger home ranges in Blair's rusty lizard population. Fitch did not estimate the actual size of home ranges but found that the average movement between captures for all juveniles was 33 feet, for adult males it was 49 feet, and for adult females it was 22 feet. The comparable figures for the Great Plains skink (*Eumeces obsoletus*) given by Fitch (1955) were 47.0, 89.8, and 55.0 feet, respectively. Carpenter (1959) found that the mean maximum distance between captures of eastern fence lizards (*Sceloporus undulatus*) was 126 feet for males, 76 feet for females, and 25 feet for juveniles. The latter figures indicate a much larger home range for each category than exhibited by sagebrush lizards (*Sceloporus graciosus*), for which Stabbins (1944) reported mean maximum distances between captures of 82.1 feet and 59.0 feet, respectively, for males and females.

Home range sizes for adult males of other species which have been based on fairly large numbers of recaptures generally range from 0.50 acre, reported by Jorgonson and Tanner (1963) for western whiptails (*Cnemidophorus tigris*), to 0.03 acre, reported by Hirth (1963) for the Costa Rican basilisk (*Basiliscus vittatus*). Female home ranges reported vary from 0.24 acre for western whiptails to 0.03 acre for *Basiliscus vittatus;* juvenile home ranges vary from 0.09 acre for western whiptails to the 0.027 acre value given by Blair (1960) for rusty lizards.

All of the above data on movement and home range size reflect the propensity of adult male lizards to wander over greater areas than either adult females or juveniles. The precise values obtained in any study of home ranges, however, will be subject to variation depending upon how the home range size is estimated. Tinkle (1967) found in his study of the side-blotched lizard, for example, that when he connected the outermost recapture points for 143 males and 171 females for which he had an average of 10 and 9 recaptures, respectively, he obtained an estimated home range size of 0.11 acre for males and 0.03 acre for females. When only the 59 females and 57 males for which he had 11 or more recaptures were considered, his estimated home range sizes were 0.05 acre for females and 0.15 acre for males. Using the probability density function, obtained by finding the geometric center of all recaptures on an individual and measuring from this point to all capture points to obtain a mean recapture radius which defines an assumed circular home range, he estimated the home range of males as 0.49 acre and that for females as 0.15 acre. In general, one would expect the estimated size of the home range to vary with the number of recaptures; Tinkle noted that few individuals captured less than 10 times had large home ranges, but increas-

ing the number of captures beyond 11 did not result in an appreciable increase in the mean size of home range. The use of the mean recapture radius seems to overestimate the size of home ranges but is useful as a means of comparing degree of movement in different populations (Tinkle, 1967; Tinkle and Woodward, 1967). When a sufficient number of observations can be obtained, the use of peripheral sight records, as employed by Blair (1960), probably provides the most accurate measure of an individual's home range.

Aggressive behavior, indicative of territoriality, has been observed in nearly all species of lizards studied; the New Zealand skink *Leiolopisma zelandica* is exceptional in being non-aggressive (Barwick, 1959). Aggressive behavior may be characteristic of males only or may be involved in interactions between individuals of both sexes. As noted in Chapter Twelve, most male lizards appear to be territorial at least during the breeding season. Bustard (1965) observed that female chameleons (*Chameleo hohneli*) will attack other lizards of either sex if they approach too closely; Tinkle (1967) noticed the same behavior in female side-blotched lizards. Although such aggression need not always reflect territoriality, Harris (1964) observed aggression between two male rainbow lizards (*Agama agama*) in Nigeria which certainly was a manifestation of territoriality. The two males had adjacent home ranges which were separated by a path and several times were observed fighting over this path. Gradually the violence of their feud diminished as each stayed more and more on his own side of the path, intimidating the other lizard across the path by bobbing at him. When violations of the territorial boundary did occur, the lizard whose territory had been entered would invariably win the ensuing fight, reflecting the psychological advantage a resident has over an invader in a fight.

The area defended by an individual lizard may be all or only a portion of the home range. Tinkle (1967) found that the side-blotched lizards defend their entire home range but noted that some portions of the area may be defended more vigorously than others. Bustard (1970) indicated that adults of the Australian gecko *Heteronotia binoei*, which find shelter under bark on the ground and at the base of stumps, only defend their home sites; he concluded that the competition for these is primarily responsible for regulating the size of populations. This regulation is comparable to the role of basking sites (stations) in the regulation of Blair's rusty lizard population. In any case, the aggressive nature of lizards is a major factor in determining their population densities and contrasts with the typically non-aggressive behavior of turtles.

POPULATION DYNAMICS OF SNAKES

The population dynamics of snakes are difficult to determine because of their secretive nature, frequent periods of inactivity, and longevity. Considerable information is available on certain populations of a few species, but much less is known about populations of snakes than of lizards.

NATALITY

Snakes tend to produce larger clutches of eggs or broods of young than lizards, but considerable variation in clutch size occurs within and between species. Most frequently clutch or brood sizes range between 6 and 30, but at least some species deposit up to 100 eggs per clutch and other species produce clutches with as few as one or two eggs (see table pages 406–408).

As with other reptiles, larger females tend to produce larger clutches or broods than smaller females of the same species. Fitch (1963b) found an 870-millimeter (snout-vent) female black rat snake (*Elaphe o. obsoleta*) deposited a clutch of six eggs; a 983-millimeter female contained seven eggs ready for ovulation; a 960-millimeter female laid eight eggs; a 1028-millimeter female contained nine large eggs; and a 1170-millimeter female contained 11 eggs. Thus, with an increase of 300 millimeters in snout-vent length there was nearly a doubling of the clutch size. Blanchard (1937a) found the clutch size of the female eastern ringneck snake (*Diadophis punctatus edwardsi*) increases with the size of the female up to a length of about 400 millimeters, but beyond this there was a tendency for fewer eggs per clutch. In another study, Blanchard (1937b) averaged the number of young per brood reported for the viviparous red-bellied snake (*Storeria occipitomaculata*) and found that the number increased with the length of the female up to a length of 235 millimeters, but between the lengths of 235 and 260 millimeters the average scarcely varied; in larger size classes the broods again increased in size. An observation made by Blanchard is probably applicable to snakes in general: the minimal number in a clutch increases more steadily with the length of the female than does the maximal number in a clutch. The decrease in clutch size in very large eastern ringneck snakes is paralleled, apparently, in the garter snake (*Thamnophis sirtalis*), since Fitch (1965) found that the brood size increased from an average of 12 in primiparae (504 to 580 millimeters in length) to about 16 in six-year-olds (795 to 815 millimeters in length) but averaged only 14 in four individuals that ranged from 844 to 950 millimeters in length, perhaps as a reflection of reduced reproductivity in very old individuals. Blanchard and Blanchard (1942) mentioned a captive female of the same species that produced six litters in successive years but then ceased to reproduce, apparently as a result of having become sterile with senility.

The sizes of clutches and broods appear to vary geographically in widely distributed species. Fitch (1960) found evidence that there may be local differences in sizes of broods of copperheads (*Agkistrodon contortrix*), even within one part of the geographic range, but concluded that within a local population the number of young per litter is proportional to the size of the female and that differences in average numbers per litter in different populations may be associated with geographical differences in size of individuals. He (1963) also found geographic trends in the number of eggs per clutch of racers (*Coluber constrictor*) such that the larger and bulkier eastern subspecies (*C. c. constrictor*) produces nearly three times as many eggs per clutch as the small western subspecies (*C. c. mormon*); the centrally located *C. c. flaviventris* was found to be somewhat intermediate in size and numbers of eggs produced. In both racers and copperheads, it appears that geographical differences in clutch size have evolved through selection for body size.

Several workers have observed that a significant proportion of the adult females in any one population of snakes do not produce young or eggs during any one year. In some instances, large numbers of non-breeding females seem to reflect biennial reproductive cycles or, rarely, cycles that may even be longer than two years. Biennial cycles seem to be particularly characteristic of northern populations of viviparous viperids. The adder (*Vipera berus*) is thought to have a biennial reproductive cycle in Finland and northern Sweden (Saint Girons, 1966). Rahn (1942), Klauber (1956), Fitch (1949) and others have studied developing ova in rattlesnakes (*Crotalus viridis*) of the United States and found eggs in two different states of development during the breeding season; these have been interpreted as being indicative of a two-year reproductive cycle. It is assumed that small eggs present in the ovaries in the summer and fall of one year begin to grow rapidly during the summer of the next year and reach a large size (still in the ovaries) when the snake enters hibernation that fall. Ovulation occurs when the snake emerges from hibernation and, at that time, fertilization takes place through stored sperm from mating during the previous summer or fall. Thus, the biennial cycle of northern areas is a consequence of the long hibernation season during which little or no development of ova occurs. This long period of inactivity results in about a year and a half of calendar time being required for the maturation of ova prior to ovulation, and young are produced every second year because the summer season is too short to allow complete development of egg and young in one year. Longer warm seasons in southern areas would presumably allow the comparable development to occur within a calendar year; Klauber (1956) suggests that there are probably annual reproductive cycles in *Crotalus viridis* in Texas and other portions of its southern range. The European adder is also thought to have an annual cycle in the southern part of its range. Klauber suggests, furthermore, that there are probably territories in which both annual and biennial cycles

occur in rattlesnake populations because smaller females might reproduce at two-year intervals while the larger ones might be able to bring a brood to term within a year. Regardless of the female cycle, all evidence indicates that the reproductive cycle of the male rattlesnake is annual, spermatogenesis not being as time- or energy-consuming as oogenesis.

Although the incidence of non-breeding females may be rather high at times, all colubrid snakes for which there are data appear to normally have annual female reproductive cycles. Blanchard (1937b) found that 90 per cent of the adult females of the red-bellied snake (*Storeria occipitomaculata*) collected in Michigan were pregnant during the embryo-bearing season; such a high percentage would not be expected if a biennial cycle prevailed. Blanchard and Blanchard (1940), through breeding experiments, showed definitely that in the vicinity of Ann Arbor, Michigan, the common garter snake (*Thamnophis sirtalis sirtalis*) is capable of an annual reproductive cycle when kept in outdoor pits. Carpenter (1952) found a female of both the common garter snake and the ribbon snake (*Thamnophis sauritus sauritus*) which contained embryos for two succeeding years but also noted that the proportions of adult females which were not gravid were 35, 35, and 33 per cent, respectively, for common garter snakes, ribbon snakes, and Butler's garter snake (*Thamnophis butleri*). He suggested that most of the adult females which were not gravid had not been inseminated. Fitch (1965) found the incidence of non-breeding females in Kansas populations of the common garter snake was correlated with the age (size) of the snakes as follows:

of 62 two-year-olds (504 to 580 millimeters in length), 42 per cent were gravid;

of 57 three-year-olds (585 to 660 millimeters in length), 58 per cent were gravid;

of 40 four-year-olds (668 to 738 millimeters in length), 93.5 per cent were gravid;

of 12 five-year-olds (740 to 786 millimeters in length), 83 per cent were gravid;

of 6 six-year-olds (795 to 815 millimeters in length), 100 per cent were gravid;

of 4 seven years old or more (844 to 950 millimeters in length), 100 per cent were gravid.

Fitch concluded that in garter snakes and probably in most other common snakes of the United States, the reproductive cycle is normally annual and the non-breeders are those females which have not reached sexual maturity or are mature snakes which have not been inseminated, are undernourished, or are pathologically prevented from producing eggs.

From the little information available, it is clear that the fecundity of a snake population is dependent upon the number of mature females present which actually reproduce in a given year and the average size of their clutches or litters. Some information on clutch sizes is available for most species of snakes but the factors affecting the proportion of mature females which will actually reproduce have not been studied. The data of Woodbury et al. (1951) indicate great fluctuations in the percentage of gravid females in a Utah hibernation den over a period of 10 years and imply that the natality of snake populations may be subject to considerable variation from year to year. Fitch's (1965) data imply that the largest snakes, which tend to have the largest clutches, are also most likely to reproduce regularly, and consequently, that longevity may be extremely important in increasing the natality of snake populations.

SURVIVORSHIP, ATTAINMENT OF SEXUAL MATURITY, AND LONGEVITY

It has been frequently observed that some of the eggs of viviparous snakes which are ovulated are not fertilized, some embryos develop abnormally, some abortions occur, and even some of the full-term young are born dead so that the average litter produced is probably smaller than the average number of enlarged eggs counted in females (Fitch, 1965). Similarly, predation and other causes of mortality take a toll of eggs deposited by oviparous snakes. However, the severity of embryonic mortality in natural populations is not known.

The few data available on proportions of different size classes indicate that there is heavy mortality during the first year in many snake populations. Fitch (1949) found, for example, that in a sample of 702 individuals from a central California population of northern Pacific rattlesnakes (Crotalus viridis oreganus), 22.5 per cent

were first year young but only 9.4 per cent were second year young; 13.4 per cent were third and possibly fourth year young, and 54.7 per cent were subadults and adults. His study of marked individuals of this same subspecies also indicated an increase in life expectancy with larger size. Of 679 snakes marked between 1938 and 1941, about 29 per cent were old adults, but among those recaptured in 1946–1947, the ones already old adults when marked constituted 43 per cent of the sample. Young in their first year constituted 39 per cent of all those marked originally but only 18 per cent of those recaptured in 1946–1947. In the same study, Fitch (1949) found that approximately one third of the spring population of gopher snakes (Pituophis melanoleucus) consisted of small young only a few months old but that the mortality of this age group was very high prior to their first hibernation. Of those that did survive through their first hibernation, Fitch estimated that 78 per cent were eliminated during the subsequent year.

Probably the most complete analysis of the age structure of a snake population was made by Woodbury et al. (1951) in a ten-year study of a Utah hibernation den. These workers recorded the following total numbers of Great Basin rattlesnakes (Crotalus viridis lutosus) of different ages over the ten-year study; 930 new snakes (less than one year old), 289 one-year-olds, 183 two-year-olds, 152 three-year-olds, 110 four-year-olds, 81 five-year-olds, 86 six-year-olds, 41 seven-year-olds, 39 eight-year-olds, 32 nine-year-olds, and 18 ten-year-olds. Based on the percentages given for known survivors, the survivorship curve for this population would be as in Figure 14-4. The data for this population, like those of Fitch, indicate that the life expectancy of individuals goes up with age.

The high mortality of young snakes seems to reflect the importance of predation as a mortality factor. Several workers have noted the susceptibility of young snakes to predation by coyotes, birds of prey, and other carnivores but have noted that mature adults, particularly of large or venomous species, tend to be relatively immune from attack. Humans are important predators on mature snakes. The result of high juvenile mortality and increased life expectancy with age is that the

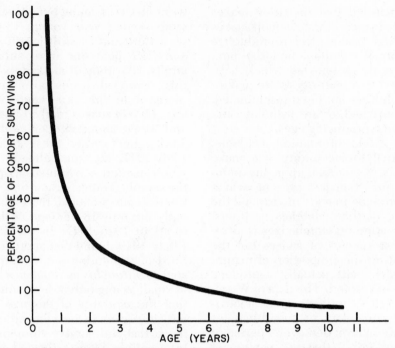

Figure 14–4 Survivorship curve for Great Basin rattlesnakes (*Crotalus viridis lutosus*), based on recapture data of Woodbury *et al.* (1951) from a hibernation den in Tooele County, Utah. The total number of individuals captured which were less than one year old (930) was converted into a sample size of 1000 (100 per cent) and the numbers of recaptured snakes which were between 1 and 2 years old, 2 and 3 years old, etc., through those at 10 years of age were proportionately converted to the number surviving on the basis of an initial cohort of 1000 individuals.

adult population of snakes represents the accumulation of many years' reproduction. The natural longevity of snakes appears to commonly exceed 10 years, and the turnover in populations is very slow.

It would appear that the relative survivorship of males and females varies among different snake populations and species. Many investigators have found a higher percentage of males than females in collections, but this may reflect either a higher survival of males or a tendency for greater activity and larger individual home ranges of males which makes them more likely to be found. The sex ratios of hibernators recorded by Woodbury *et al.* (1951) indicate that males live longer than females. Among the Great Basin rattlesnakes recorded 59.5 per cent were males, and of the striped whipsnakes (*Masticophis t. taeniatus*) recorded 62 per cent were males. In contrast, Fitch (1965) concluded that the survival of female common garter snakes is higher than that of males. He found the sex ratio in samples varies

according to the season. Of 519 snakes trapped in the fall, over 15 years, 301 were males and 218 females; in contrast, of 408 snakes trapped in the spring, 154 were males and 254 females. Fitch believes the high proportion of males in the autumn samples may reflect sexual activity, since females caught in traps may serve as bait to attract one or more males. A large proportion of the snakes he caught in the spring were second-year adolescents which were often small enough to escape from his mesh wire traps. Thus, he suggests the higher proportion of females captured in the spring may reflect their larger size and reduced ability to escape compared to males. However, when all the second-year individuals were eliminated from consideration, the remaining 255 adults consisted of 99 males and 156 females, indicating a sex ratio of 1:1.58. Because males are more apt to be caught than females, Fitch suggests that males may be even less numerous than the ratio of actual captures indicates.

In his study of the racer (*Coluber constrictor*), Fitch (1963a) obtained a ratio of 51.6 per cent females in a total sample of 734 snakes. However, when he assigned each individual to an arbitrary age on the basis of its size, he found that the proportion of females increased in higher age groups. Of the supposed two-, three-, and four-year-olds combined, females constituted 51.2 per cent; however, they totaled 55.6 per cent of those more than four years old and 61.3 per cent of those more than five years old. He suggested that the greater activity of males may result in a higher mortality. In his study of the copperhead (*Agkistrodon contortrix*), Fitch (1960) also noted a higher mortality of males, especially in their first year of life, but was unable to explain why the young males should be subject to heavier mortality than the young females. Interestingly, he notes that the sex ratio both in his sample of 126 young wild copperheads and in young born in captivity is approximately 2.8 to 1, males over females. In view of the significant number of adult female snakes which apparently do not reproduce because they have not been inseminated, the sex ratio of snake populations may be very important in influencing the realized natality, especially in populations where females outnumber males.

There is much variation among different species of snakes in the time required to reach sexual maturity. As with other reptiles, this seems to involve a minimum amount of growth; in general, the larger the species, the longer it takes to reach sexual maturity. Among snakes he studied in Kansas, Fitch (1960, 1963a, 1963b, 1965) found that common garter snakes grew relatively rapidly as compared with other snakes and reached maturity during their second year. He also found that two-year-old racers may breed, but concluded that copperhead females breed for the first time in their third year and black rat snakes (*Elaphe obsoleta*) attain sexual maturity in the fourth year. The range of two to four years seems to encompass the time required for other species to mature. Northern female rattlesnakes (*Crotalus viridis*), for example, probably first breed when they are four (rarely five), while southern species (*Crotalus atrox* and *Crotalus ruber*) produce their first young at an age of three (Klauber, 1956).

SOCIAL BEHAVIOR AND POPULATION DENSITIES

Literature statements regarding the movements and home ranges of snakes vary from Blanchard and Finster's (1933) conclusion that wanderings of snakes may or may not be extensive and are not predictable, to the conclusion of Stickel and Cope (1947) that snakes have definable home ranges. In part, these conflicting statements reflect the fact that in some studies, like the former, marked snakes have not been released where originally captured but, rather, were introduced into areas where the individual snakes had not been before. In studies, like the latter, where marked snakes have been released at the original point of capture, recapture data indicate that the majority of individuals have rather definite home ranges. However, in most studies there appear one or more records of snakes traveling long distances. Stickel and Cope (1947) found a racer (*Coluber c. constrictor*) 1.1 miles from its first locality after an elapsed time of two years. Blanchard and Finster (1933) mention a common garter snake (*Thamnophis sirtalis*) that traveled 1.5 miles in 41 days. Imler (1945) recorded a two-year-old female bull snake (*Pituophis melanoleucus*) that moved 1.5 miles between June and September. In general, it appears that longer movements over a period of years represent situations in which the snake is shifting its range or hibernaculum or both. Long movements within a few months may represent movement to or from hibernacula or the occasional, perhaps exploratory, sallies outside of home ranges such as have been noted with other vertebrates. As with lizards, male snakes generally have been found to travel more extensively than females and to have larger home ranges. The home ranges of juveniles are smaller than those of adults.

The size of the home range and its permanency vary from species to species and, presumably, with differences in habitat (cover, food availability, basking sites, and so forth). In general, larger species have larger home ranges than smaller species. In a study made in Kentucky, Barbour, Harvey, and Hardin (1969) used Co[60] wire tags and a survey meter to locate eastern worm snakes (*Carphophis amoenus amoenus*) and were able to locate each of 10 individuals a minimim of 46 times. Applying the minimum area method of connecting the outermost locations, they found the home range sizes varied from 23 square meters for a small (110-millimeter) juvenile female to a maximum of 726 square meters for an adult (225-millimeter) male; the average home range size for the 10 individuals was 253 square meters. The fact that individuals were found at or near the same place after a year indicates these worm snakes occupy the same home ranges in successive years.

Although no individual was recaptured a sufficient number of times to accurately determine the extent of its home range, Fitch (1960) assumed the average distance between any two capture points to be approximately equivalent to half the diameter of the home range of copperheads in Kansas and concluded that the average home range for males had a diameter of 1162 feet and an area of 24.4 acres while that for females had a diameter of 690 feet and an area of 8.5 acres. Using the same method, he found (1963a) that the home ranges for male racers in the same general area had an average diameter of 1190 feet and an area of 26.3 acres while those for female racers had an average diameter of 1148 feet and an area of 23.8 acres. His (1963b) data on black rat snakes on the same grounds yielded male home ranges with an average diameter of 1252 feet and an area of 29 acres and female home ranges with an average diameter of 1152 feet and an area of 24 acres. Male common garter snakes in the same study area were estimated to have average home ranges with a diameter of 1392 feet and an area of 35.0 acres; the average female home range for this species was estimated to have a diameter of 1124 feet and an area of 22.7 acres. It should be noted that none of these estimated home range sizes includes the hibernaculum which was generally more than the diameter of the home range away.

Hirth et al. (1969) used radioactive tantalum wire tags and scintillometers to study the summer movements of Mormon racers (Coluber constrictor mormon), Great Basin rattlesnakes (Crotalus viridis lutosus), and striped whipsnakes (Masticophis t. taeniatus) from the hibernation den monitored by Woodbury et al. (1951) in Utah. In contrast to the evidence outlined above indicating definite home ranges, Hirth and his associates were seldom able to locate snakes in the same area on successive days and found, in general, that the distances between recaptures increased as intervals between recaptures increased. These workers concluded that the three species of snakes wander about extensively on their summer range. The snakes are actually on the summer range for less than four months of the year, and it appears unlikely that permanent home ranges are established for such a short duration. None of the snakes moved much during periods of ecdysis, but, otherwise, a few Mormon racers were the only snakes to center their activities in a particular area. These exceptional racers remained in a heavily vegetated gully throughout the summer. Thus, it appears that all three species behave like ecological opportunists. When, perhaps by chance, they crawled into a gully having an abundance of cover, prey, and basking sites, they tended to remain there. If they failed to find an optimal location, their movements appeared to be random and triggered by physiological needs. Hirth and his coworkers concluded that none of the three species exhibits stable home ranges in the cold desert and that their movements are best described by the concept of "total range;" total range, as applied here, includes the hibernaculum, the routes away from and back to the den, and the wandering of the snakes on the summer range. It seems relevant at this time to point out that Woodbury et al. (1951) found that individuals of all three species generally return to the same den year after year.

There is very little evidence to indicate territoriality in snakes. McCauley (1945) did observe what appears to be incipient territoriality in a large male racer in Maryland. This individual remained in a small area for several hours, moving about conspicuously with his head raised as if patrolling. When a larger male of the same species intruded, the former drove him away; neither of the racers paid any attention to a king snake (Lampropeltus getulus) that was within the same area. Fitch (1963a), however, was unable to find any evidence of territoriality in his observations of racers in Kansas. In fact, he noted that even in the breeding season several males may be present within a small area. The "courtship dances" of adders and other snakes, as mentioned earlier, have generally been shown to be combats between males of the same species during mating seasons. However, it has not been determined whether or not this aggressive behavior is involved in the defense of a territory.

The density of snake populations appears to vary widely among different species and conditions. Seibert (1950) estimated the density of a population of plains garter snakes (*Thamnophis radix*) in the vicinity of Chicago, based on mark and recapture data, to be between 342 and 360 snakes per acre while the density of a population of smooth green snakes (*Opheodrys vernalis*) in the same area was estimated at between 44 and 74 snakes per acre. Both of these figures are extremely high in comparison with estimates for other populations. Fitch (1963a) estimated the population of racers in Kansas at one to three adults per acre in early summer in areas of favorable habitat and noted that the population would be approximately tripled in late summer at its annual maximum. He (1963b) estimated that the population of black rat snakes in the same area fluctuated between 0.84 snakes per acre and 1.71 snakes per acre annually. Fitch (1960) estimated that the summer Kansas population of copperheads had a density ranging between five and seven snakes per acre, being somewhat higher than this in the brushy fields that are the snakes' preferred habitat and somewhat lower in woodlands. In various localities where habitat conditions approach the optimum, Fitch concluded populations of 10 to 20 copperheads per acre occur. In his study of the common garter snake, Fitch (1965) noted constant change in the size of the population and estimated that the density ranged from less than three snakes per acre at the time of the annual low in the summer of 1963 to a maximum of about 18 snakes per acre at the time of the annual population peak in early autumn of 1961.

While aggressive behavior appears to be a major factor regulating the density of lizard populations, snake populations seem to be regulated more by conditions of cover, food, and basking sites; areas having optimal conditions will support much higher densities of snakes than less favorable areas.

References

Agassiz, L. 1857. Contributions to the natural history of the United States of America. *1*:1–452. Boston, Little, Brown.

Alcala, A. C. 1955. Observation on the life history and ecology of *Rana erythraea* Schlegel, on Negros Island, Philippines. Silliman J. *2(3)*:175–192.

Altland, P. D. 1951. Observations on the structure of reproductive organs of the box turtle. J. Morphol. 89:599–616.

Anderson, J. D. 1967. A comparison of the life histories of coastal and montane populations of *Ambystoma macrodactylum* in California. Am. Midland Naturalist 77:323–355.

Anderson, P. K. 1954. Studies in the ecology of the narrow-mouthed toad, *Microhyla carolinensis carolinensis*. Tulane Stud. Zool. 2:15–46.

—— 1960. Ecology and evolution in island populations of salamanders in the San Francisco Bay region. Ecol. Monographs 30:359–385.

Bannikov, A. G. 1950. Age distribution of a population and its dynamics in *Bombina bombina* L. Tr. Akad. Nauk SSSR 70:101–103.

Barbour, R. W., M. J. Harvey, and J. W. Hardin. 1969. Home range, movements, and activity of the eastern worm snake, *Carphophis amoenus amoenus*. Ecology 50:470–476.

Barwick, R. E. 1959. Life history of the common New Zealand skink *Leiolopisma zelandica*. Trans. Roy. Soc. New Zealand 86:331–380.

Bishop, S. C., and H. P. Chrisp. 1933. The nests and young of the Allegheny salamander *Desmognathus fuscus ochrophaeus* (Cope). Copeia 1933:194–198.

Blair, A. P. 1943. Population structure in toads. Am. Naturalist 77:563–568.

Blair, W. F. 1953. Growth, dispersal, and age at sexual maturity of the Mexican toad (*Bufo valliceps* Wiegman). Copeia 1953:208–212.

—— 1960. The Rusty Lizard. A Population Study. Austin, Univ. Texas Press. 185 pp.

Blanchard, F. N. 1928. Topics from the life history and habits of the red-backed salamander in southern Michigan. Am. Naturalist 62:156–164.

—— 1937a. Eggs and natural nests of the eastern ring-neck snake, *Diadophis punctatus edwardsii*. Pap. Michigan Acad. Sci., Arts and Letters 22: 521–532, pls. 53–57.

—— 1937b. Data on the natural history of the red-bellied snake, *Storeria occipito-maculata* (Storer), in northern Michigan. Copeia 1937:151–162.

——, and F. C. Blanchard. 1940. Factors determining time of birth in the garter snake, *Thamnophis sirtalis sirtalis* (Linnaeus). Pap. Michigan Acad. Sci., Arts and Letters 26:161–176.

—— and —— 1942. Mating of the garter snake *Thamnophis sirtalis sirtalis* (Linnaeus). Pap. Michigan Acad. Sci., Arts and Letters 27:215–234.

——, and E. B. Finster. 1933. A method of marking living snakes for future recognition, with a discussion of some problems and results. Ecology 14: 334–347.

Bragg, A. N. 1940. Observations on the ecology and natural history of Anura. I. Habits, habitat and breeding of *Bufo cognatus* Say. Am. Naturalist 74:424–438.

―――― 1965. Gnomes of the Night; the Spadefoot Toads. Philadelphia, Univ. Pennsylvania Press. 127 pp.

――――, and J. Bresler. 1951. Viability of the eggs of *Bufo cognatus*. Proc. Oklahoma Acad. Sci. 32:13–14.

Brown, W. C., and A. C. Alcala. 1970. Population ecology of the frog *Rana erythraea* in southern Negros, Philippines. Copeia 1970:611–622.

Bustard, H. R. 1965. Observations on the life history and behavior of *Chameleo hohneli* Steindachner. Copeia 1965:401–409.

―――― 1970. The population ecology of the Australian gekkonid lizard *Heteronotia binoei* in an exploited forest. J. Zool., London 162:31–42.

Cagle, F. R. 1944. Home range, homing behavior, and migration in turtles. Univ. Michigan Mus. Zool. Misc. Publ. 61:1–34.

―――― 1946. The growth of the slider turtle, *Pseudemys scripta elegans*. Am. Midland Naturalist 36:685–729.

―――― 1948. Observations on a population of the salamander, *Amphiuma tridactylum*, Cuvier. Ecology 29:479–491.

―――― 1950. The life history of the slider turtle, *Pseudemys scripta troostii* (Holbrook). Ecol. Monographs 20:31–54.

―――― 1954. Observation on the life cycles of painted turtles (Genus *Chrysemys*). Am. Midland Naturalist 52:225–235.

Carpenter, C. C. 1952. Comparative ecology of the common garter snake (*Thamnophis s. sirtalis*), the ribbon snake (*Thamnophis s. sauritus*), and Butler's garter snake (*Thamnophis butleri*) in mixed populations. Ecol. Monographs 22:235–258.

―― 1954. A study of amphibian movement in the Jackson Hole Wildlife Park. Copeia 1954:197–200.

―― 1959. A population of Sceloporus undulatus consobrinus in southcentral Oklahoma. Southwest. Naturalist 4:110–111.

Carr, A., and L. H. Ogreen. 1960. The ecology and migrations of sea turtles. 4. The green turtle in the Caribbean Sea. Bull. American Mus. Nat. Hist. 121:1–48.

Cott, H. B. 1961. Scientific results of an enquiry into the ecology and economic status of the Nile Crocodile (*Crocodilus niloticus*) in Uganda and Northern Rhodesia. Trans. Zool. Soc. London 29:211–356.

Crenshaw, J. W. 1955. The life history of the southern spiny lizard *Sceloporus undulatus undulatus* Latreille. Am. Midland Naturalist 54:257–298.

Cunningham, J. D. 1960. Aspects of the ecology of the Pacific slender salamander, *Batrachoseps pacificus*, in southern California. Ecology 41:88–99.

Deraniyagala, P. E. P. 1939. The tetrapod reptiles of Ceylon. Ceylon J. Science 1:1–412.

Duellman, W. E. 1966. Aggressive behavior in Dendrobatid frogs. Herpetologica 22:217–221.

Einem, G. E. 1956. Certain aspects of the natural history of the mudturtle, *Kinosternon bauri*. Copeia 1956:186–188.

Emlen, S. T. 1968. Territoriality in the bullfrog, *Rana catesbeiana*. Copeia 1968:240–243.

Ferguson, D. E. 1954. An interesting factor influencing *Bufo boreas* reproduction at high elevations. Herpetologica 10:199.

Fitch, H. S. 1949. Study of snake populations in central California. Am. Midland Naturalist 41:513–579.

―――― 1954. Life history and ecology of the five-lined skink, *Eumeces fasciatus*. Univ. Kansas Publ. Mus. Nat. Hist. 8:1–156.

―――― 1955. Habits and adaptations of the Great Plains skink *Eumeces obsoletus*. Ecol. Monographs 25:59–83.

―――― 1956a. Early sexual maturity and longevity under natural conditions in the Great Plains narrow-mouthed frog. Herpetologica 12:281–282.

―――― 1956b. An ecological study of the collared lizard (*Crotaphytus collaris*). Univ. Kansas Publ. Mus. Nat. Hist. 8:213–274.

―――― 1958. Natural history of the six-lined racerunner (*Cnemidophorus sexlineatus*). Univ. Kansas Publ. Mus. Nat. Hist. 11:11–62.

―――― 1960. Autecology of the Copperhead. Univ. Kansas Publ. Mus. Nat. Hist. 13:85–288.

―――― 1963a. Natural history of the Racer *Coluber constrictor*. Univ. Kansas Publ. Mus. Nat. Hist. 15:351–468.

―――― 1963b. Natural history of the Black Rat Snake (*Elaphe o. obsoleta*) in Kansas. Copeia 1963:649–658.

―――― 1965. An ecological study of the Garter snake, *Thamnophis sirtalis*. Univ. Kansas Publ. Mus. Nat. Hist. 15:493–564.

Flower, S. S. 1925. Contributions to our knowledge of the duration of life in vertebrate animals. III. Reptiles. Proc. Zool. Soc. London 1925:911–981.

Gibbons, J. W. 1967. Variation in growth rates in three populations of the painted turtle, *Chrysemys picta*. Herpetologica 23:296–303.

―――― 1968. Reproductive potential, activity, and cycles in the Painted Turtle, *Chrysemys picta*. Ecology 49:399–409.

―――― and D. W. Tinkle. 1969. Reproductive variation between turtle populations in a single geographic area. Ecology 50:340–341.

Goin, C. J., and O. B. Goin. 1962. Introduction to Herpetology. San Francisco, W. H. Freeman. 341 pp.

Gordon, R. E. 1952. A contribution to the life history and ecology of the plethodontid salamander, *Aneides aeneus* (Cope and Packard). Am. Midland Naturalist 47:666–701.

―――― J. A. MacMahon, and D. B. Wake. 1962. Relative abundance, microhabitat and behavior of some Southern Appalachian salamanders. Zoologica 47:9–14.

Grant, W. C., Jr. 1955. Territorialism in two species of salamanders. Science 121 (3135):137.

Green, N. B. 1957. A study of the life history of *Pseudacris brachyphona* (Cope) in West Virginia with special reference to behavior and growth of marked individuals. Dissertation Abstr. 17:23692.

Hamilton, W. J., Jr. 1955. Notes on the ecology of the Oak Toad in Florida. Herpetologica 11:205–210.

Harris, V. A. 1964. The Life of the Rainbow Lizard. London, Hutchinson Tropical Monographs; New York, Hillary House.

Harrison, J. R. 1967. Observations on the life history, ecology and distribution of *Desmognathus aeneus aeneus* Brown and Bishop. Am. Midland Naturalist 77:356–370.

Harrison, L. 1922. On the breeding habits of some Australian frogs. Australian Zool. 3:17–34.

Harrisson, T. 1956. Tagging green turtles, 1951–1956. Nature 178:1479.

Hayne, D. 1949. Two methods for estimating populations from trapping records. J. Mammalogy 30:399–411.

Heatwole, H. 1962. Environmental factors influencing local distribution and activity of the salamander *Plethodon cinereus*. Ecology 43:460–472.

Hendrickson, J. R. 1954. Ecology and systematics of salamanders of the genus *Batrachoseps*. Univ. California Publ. Zool. 54:1–46.

Hendrickson, J. R. 1958. The green turtle, *Chelonia mydas* (Linn.) in Malaya and Sarawak. Proc. Zool. Soc. London 130:455–535.

Herreid, C. F., Jr., and S. Kinney. 1966. Survival of Alaskan woodfrog (*Rana sylvatica*) larvae. Ecology 47:1039–1041.

Highton, R. 1956. The life history of the slimy salamander, *Plethodon glutinosus*, in Florida. Copeia 1956:75–93.

_____ 1962. Geographic variation in the life history of the slimy salamander. Copeia 1962:597–613.

Hildebrand, S. F. 1929. Review of experiments on artificial culture of diamond–back terrapin. Bull. United States. Bur. Fish. 45:25–70.

_____ 1932. Growth of diamond–back terrapins. Zoologica 9(15):551–563.

Hirth, H. F. 1963. The ecology of two lizards on a tropical beach. Ecol. Monographs 33:83–112.

_____ , R. C. Pendleton, A. C. King, and T. R. Downard. 1969. Dispersal of snakes from a hibernaculum in northwestern Utah. Ecology 50:332–339.

Humphrey, R. R. 1922. The multiple testis in urodeles. Biol. Bull. (Mar. Biol. Lab.) 43:45–67.

Imler, R. H. 1945. Bullsnakes and their control on a Nebraska wildlife refuge. J. Wildlife Mgt. 9:265–273.

Inger, R. F., and B. Greenberg. 1963. The annual reproductive pattern of the frog *Rana erythraea* in Sarawak. Physiol. Zool. 36:21–33.

_____ 1966. Ecological and competitive relations among three species of frog (genus *Rana*). Ecology 47(5):746–759.

Jameson, D. L. 1955. The population dynamics of the Cliff Frog, *Syrrhophus marnocki*. Am. Midland Naturalist 54:342–381.

_____ 1956a. Growth, dispersal and survival of the Pacific Tree Frog. Copeia 1956:25–29.

_____1956b. Survival of some Central Texas frogs under natural conditions. Copeia 1956:55–57.

Jorgonson, C. O., and W. W. Tanner. 1963. The application of the density probability function to determine the home ranges of *Uta stansburiana* and *Cnemidophorus tigris*. Herpetologica 19:105–115.

Klauber, L. M. 1956. Rattlesnakes Their Habits, Life Histories, and Influence on Mankind. Berkeley, 2 vols, Univ. California Press. 1476 pp.

Littlejohn, M. J. 1956. Frogs of the Melbourne area. Victorian Naturalist 79(10):296–304.

Lutz, B. 1960. Fighting and an incipient notion of territory in male tree frogs. Copeia 1960:61–63.

Marshall, A. J. and R. Hook. 1960. The breeding biology of equatorial vertebrates: reproduction of the lizard *Agama agama lionotus* Boulenger at 1at 0° 01′ N. Proc. Zool. Soc London 134:197–205.

Martof, B. 1953a. Home range and movements of the green frog, *Rana clamitans*. Ecology 34:529–543.

_____ 1953b. Territoriality in the green frog, *Rana clamitans*. Ecology 34:165–174.

_____ 1956. Factors influencing the size and composition of populations of *Rana clamitans*. Am. Midland Naturalist 56:224–245.

_____ 1962. Some aspects of the life history and ecology of the salamander *Leurognathus*. Am. Midland Naturalist 67:1–35.

_____, and F. L. Rose. 1963. Geographic variation in southern populations of *Desmognathus ochrophaeus*. Am. Midland Naturalist 69:376–425.

McCauley, R. H. 1945. The reptiles of Maryland and the District of Columbia. Hagerstown, Maryland. Published by the author. 194 pp.

McIlhenny, E. A. 1934. Notes on incubation and growth of alligators. Copeia 1934:80–88.

Miller, M. R. 1951. Some aspects of the life history of the yucca night lizard, *Xantusia vigilis*. Copeia 1951:114–120.

Mitsukuri, K. 1895. How many times does the snapping turtle lay eggs in one season? Zool. Magaz. Tokyo 7(85):143–147.

_____ 1905. The cultivation of marine and fresh water animals in Japan. Bull. United States Bur. Fish. 24:257–289.

Nichols, J. T. 1939. Range and homing of individual box turtles. Copeia 1939:125–127.

Oliver, J. A. 1955. North American Amphibians and Reptiles. Princeton, N. J., Van Nostrand-Reinhold. 359 pp.

Organ, J. A. 1961. Studies of the local distribution, life history, and population dynamics of the salamander genus *Desmognathus* in Virginia. Ecol. Monographs 31:189–220.

Pearson, O. 1960. A mechanical model for the study of population dynamics. Ecology 41:494–508.

Pearson, P. G. 1955. Population ecology of the spadefoot toad, *Scaphiopus h. holbrooki* (Harlan). Ecol. Monographs 25:233–267.

Pope, C. H., and S. H. Pope. 1949. Notes on growth and reproduction in the slimy salamander *Plethodon glutinosus*. Fieldiana 31:251–261.

Rahn, H. 1942. The reproductive cycle of the Prairie Rattler. Copeia 1942:233–240.

Risley, P. L. 1933. Observations of the natural history of the common musk turtle, *Sternothaerus odoratus* (Latreille). Pap. Michigan Acad. Arts, Sci. and Letters 17:685–711.

Ryan, R. S. 1953. Growth rates of some ranids under natural conditions. Copeia 1953:73–80.

Saint Girons, H. 1966. Le cycle sexuel des serpents venimeux. Mem. Inst. Butantan. Simp. Internac. 33(1):105–114.

Savage, R. M. 1952. Ecological, physiological and anatomical observations on some species of anuran tadpoles. Proc. Zool. Soc. London 122:467–514.

Sayler, A. 1966. The reproductive ecology of the Redbacked salamander, *Plethodon cinereus*, in Maryland. Copeia 1966:183–193.

Seibert, H. C. 1950. Population density of snakes in an area near Chicago. Copeia 1950:229–230.

Sexton, O. J. 1960. Some aspects of the behavior and of the territory of a dendrobatid frog, *Prostherapis trinitatis*. Ecology 41:107–115.

_____ 1962. Apparent territorialism in *Leptodactylus insularum* Barbour. Herpetologica 18:212–214.

Spight, T. M. 1967. Population structure and biomass production by a stream salamander. Am. Midland Naturalist 78:437–447.

Stebbins, R. C. 1944. Field notes on a lizard, the mountain swift, with special reference to territorial behavior. Ecology 25:233–245.

_____ 1954. Natural history of the salamanders of the plethodontid genus *Ensatina*. Univ. California Publ. Zool. 54:47–124.

Stickel, L. F. 1950. Population and home range relationships of the box turtle, *Terrapene c. carolina* (Linnaeus). Ecol. Monographs 29:353–360.

Stickel, W. H. and J. B. Cope. 1947. The home ranges and wanderings of snakes. Copeia 1947:127–136.

Stille, W. T. 1952. The nocturnal amphibian fauna of the southern Lake Michigan beach. Ecology 33:149–162.

Test, F. H. 1954. Social aggressiveness in an amphibian. Science 120:140–141.

——, and B. A. Bingham. 1948. Census of a population of the red-backed salamander (*Plethodon cinereus*). Am. Midland Naturalist 39:362–372.

Tester, J. R., and W. J., Breckenridge. 1964. Population dynamics of the Manitoba toad, *Bufo hemiophrys*, in northwestern Minnesota. Ecology 45:592–601.

Thornton, W. A. 1960. Population dynamics in *Bufo woodhousei* and *B. valliceps*. Texas J. Sci. 12:176–200.

Tilley, S. G. 1968. Size-fecundity relationships and their evolutionary implications in five Desmognathine salamanders. Evolution 22:806–816.

——, and D. W. Tinkle. 1968. A reinterpretation of the reproductive cycle and demography of the salamander *Desmognathus ochrophaeus*. Copeia 1968:299–303.

Tinkle, D. W. 1958. The systematics and ecology of the *Sternothaerus carinatus* complex (Testudinata: Chelydridae). Tulane Stud. Zool. 6:4–56.

—— 1961. Geographic variation in reproduction, size, sex ratio and maturity of *Sternothaerus odoratus* (Testudinata: Chelydridae). Ecology 42:68–76.

—— 1967. The life and demography of the side-blotched lizard. *Uta stansburiana*. Misc. Publ. Mus. Zool. Univ. Michigan 132:1–182.

—— 1969. Evolutionary implications of comparative population studies in the lizard *Uta stansburiana. In:* Systematic Biology, Washington, D.C., National Academy of Sciences. Publ. No. 1692:133–160.

——, H. M. Wilbur, and S. G. Tilley. 1970. Evolutionary strategies in lizard reproduction. Evolution 24:55–74.

——, and D. W. Woodward. 1967. Relative movements of lizards in natural populations as determined from recapture radii. Ecology 48:166–168.

Turner, F. B. 1959. Some features of the ecology of *Bufo punctatus* in Death Valley, California. Ecology 40:175–181.

—— 1960. Population structure and dynamics of the Western Spotted Frog, *Rana p. pretiosa* Baird and Girard, in Yellowstone Park, Wyoming. Ecol. Monographs 30:251–278.

—— 1962. The demography of frogs and toads. Quart. Rev. Biol. 37:303–314.

Wiewandt, T. A. 1969. Vocalization, aggressive behavior, and territoriality in the Bullfrog, *Rana catesbeiana*. Copeia 1969:276–285.

Wilhoft, D. 1963. Reproduction in the tropical Australian skink, *Leiolopisma rhomboidalis*. Am. Midland Naturalist 70:442–461.

Williams, G. C. 1966. Natural selection, the costs of reproduction, and a refinement of Lack's principle. Am. Naturalist 100:687–692.

Wood, J. T., and W. E. Duellman. 1951. Ovarian egg complements in the salamander *Eurycea bislineata rivicola* Mittleman. Copeia 1951:181.

Woodbury, A. M. and R. Hardy. 1948. Studies of the desert tortoise, *Gopherus agassizii*. Ecol. Monographs 18:145–200.

——, B. Vetas, G. Julian, H. R. Glissmeyer, F. L. Heyrend, A. Call, E. W. Smart, and R. T. Sanders. 1951. Symposium: A snake den in Tooele County, Utah. Herpetologica 7:1–52.

Zweifel, R. G. 1969. Frogs of the genus *Platymantis* (Ranidae) in New Guinea, with the description of a new species. American Mus. Nov. 2374:1–19.

——, and C. H. Lowe. 1966. The ecology of a population of *Xantusia vigilis*, the desert night lizard. American Mus. Nov. 2247:1–57.

CHAPTER 15

MAN'S INTERACTIONS WITH AMPHIBIANS AND REPTILES

AMPHIBIANS AND REPTILES HARMFUL TO MAN

The class Amphibia is the only class of vertebrates which does not include pests or species which are dangerous to man. As noted by Cochran (1961), no amphibian has a poisonous bite; amphibians do not compete with man for his harvests of grain and fruit; no amphibian is ferocious or destructive; amphibians do not have poisoned barbs or stingers; unlike such invertebrates as insects or protozoans, no amphibian has inflicted a major disease upon man. It is true, of course, that nearly all amphibians have venom glands in their skin and, if taken internally, amphibian venoms could cause serious illness or death; some people are allergic to spadefoot toxins, for example, and will suffer a hay-fever type of response if they handle the animals or, in severe cases, are merely in the same vicinity as the amphibian. There are many superstitions regarding the danger of handling amphibians. The danger of amphibian skin toxins is really minimal, however, if one practices normal hygienics; the superstition that handling toads will give one warts has no basis in fact. Despite their innocuous relationship to man, humans have had, and continue to have, a far-reaching and lasting influence on the distribution and abundance of amphibians.

Although no species of reptile can be properly described as a pest, the class Reptilia does include species which are noxious to man. Some of the larger crocodilians, varanid lizards, and snakes may injure or kill man or prey upon his domestic animals such as dogs, pigs, goats, and poultry. In addition, of course, a variety of venomous snakes and the two species of poisonous beaded lizards (*Heloderma horridum and H. suspectum*) are capable of killing or seriously injuring persons. Folklore has tended to exaggerate the real menace of harmful species and, perhaps as a result, people tend to look upon all reptiles with disdain.

CROCODILIANS

The only two demonstrably dangerous species of crocodilians are the Nile crocodile (*Crocodylus niloticus*) and the saltwater crocodile (*Crocodylus porosus*), both of which are Old World species. The Nile crocodile is found wherever there is suitable habitat throughout the interior of tropical and southern Africa, on Madagascar, and on some of the offshore islands; it apparently occurred until recent times along the lower Nile and portions of the southern Mediterranean coast. The saltwater crocodile is primarily a marine and coastal species, ranging from Ceylon and the east coast of India throughout the Malay Archipelago eastward to the Philippines, Fiji, and the Solomon Islands, and southward to northern Australia; it may penetrate hundreds of miles inland up large rivers such as the Sepik in New Guinea. A third species, the mugger (*Croc-*

odylus palustris), which occurs in fresh-water drainages of India and Ceylon, has a sinister literary reputation exemplified by the following quotation from Kipling's *The Undertakers:*

Is there a green branch and an iron ring hanging over a doorway? The Old Mugger knows that a boy has been born in that house, and must some day come down to the Ghaut to play. Is a maiden married? The Old Mugger knows, for he sees the men carry gifts back and forth; and she, too, comes down to the Ghaut to bathe before her wedding, and—he is there.

The mugger may be dangerous to man in Ceylon (Deraniyagala, 1939), but most of its reputation seems to stem from its habit of eating corpses washed from burning ghats along the Indian rivers.

All crocodile attacks on people appear to have been attempts to capture food, and differences in feeding habits cause some species or individual crocodiles to be more dangerous than others. As noted by Schmidt and Inger (1957), crocodilians feed on a variety of animals ranging from shrimp and crabs through fish, turtles, snakes, aquatic birds, to larger mammals, including those the size of cattle. Those which feed on larger prey also tend to be the more aggressive, and whether or not a particular crocodile is a threat to man depends upon how aggressive it is, rather than how large it is. The Nile crocodile and the salt-water crocodile are both noted for their aggressiveness and their willingness to stalk and attack people, but neither is particularly large, averaging about 12 feet in length. Some of the larger species, such as the Orinoco crocodile (*Crocodylus intermedius*) and the sharp-nosed crocodile (*Crocodylus acutus*), which both reach lengths of 22 feet, have no records as man-eaters. Schmidt and Inger (1957) report that the disposition of the Nile crocodile varies from place to place in Africa. It is very dangerous in some localities and permanent stockades have been built at the water's edge to protect people filling water jars. At other localities no precautions are taken and few injuries sustained. It seems likely that these local differences in aggressiveness represent local variations in feeding habits and may reflect the abundance of alternative kinds of prey.

All crocodile attacks on people seem to have been made from the water, and victims have been swimming, wading, trailing their hands in the water, people who fell out of boats, or those who were working at the water's edge filling water jars, washing clothes, bathing, and so forth. Stories of crocodiles wandering into villages in search of victims probably result from the tendency of certain species, such as the mugger, to wander overland during droughts in search of a suitable habitat. After a victim has been seized and pulled into the water, the crocodile uses a twisting-dismembering technique to detach limbs or twist or tear the victim into pieces. While holding the victim tightly with its jaws, the crocodile throws its body into vigorous twisting or thrashing motions; if the victim does not escape, he is eventually drowned and dismemberment continues until he is reduced to manageable portions and swallowed.

It is impossible to determine accurately the human mortality which results from crocodile attacks. Neess (1970) feels that estimates, such as the suggestion by Earl (1954) that 1000 lives are lost annually on the lower Zambesi from crocodile attacks, are probably exaggerations; he has concluded that the significance of crocodile injury to human welfare in general must be very close to nil. This conclusion is supported by the observation that native populations exposed to crocodile attack do not seem to have responded by making any systematic attempts to control crocodile populations; in fact, crocodiles are often venerated as ancestors (Neess, 1970).

PREDATORY LIZARDS

Dragon lizards (*Varanus komodoensis*) are the largest of all living reptiles, reaching a length of 10 feet and a weight of 300 pounds. They occur on the Indonesian islands of Flores, Komodo, Padar, and Rindja, where they are the formidable carnivore taking the place of the tiger of Asia and the lion of Africa. Surprisingly speedy, keen of eye and nose, they are the fiercest lizards known; having long yellow-orange tongues, they may have inspired the mythical Chinese dragon. Despite the potential danger which they present to human populations on these islands, there don't appear to be any records of dragon lizard attacks on man. Both observations of feeding animals and evidence from stomach contents indicate that deer and wild boar form their principal diet, but it seems likely that given the opportunity they would also prey upon man's domestic animals. There is one record of a large dragon lizard, which had been chained to a tree on the outskirts of a village jumping upon an old pony that strayed within reach and so severely lacerating it that it had to be shot.

Depending upon the species, other monitors feed upon small mammals, lizards, snakes, fish, frogs, crabs, and insects. Most seem to have the protection of public opinion because they eat carrion and thus control the blowfly. Occasionally a large monitor will rob a chicken coop, and in some places the Malays claim that of every seven chickens lost to vermin, four are taken by "biawaks" (as monitors are called in Malay). Malayan natives also claim that monitor lizards enter their cultivated fields and eat melons, cucumbers, and rice.

All in all, monitor lizards may occasionally be pests, but they do not appear to be a serious menace to man; in many places, they are even held in high esteem. For example, the Nile monitor (*Varanus niloticus*) has long been given great respect and sometimes treated with veneration because of its habit of feeding upon the eggs and young of crocodiles, thus putting a check upon the populations of these reptiles. The ancient Egyptians must also have held this lizard in high esteem, for it is often seen depicted upon their monuments.

A few other large lizards, such as the New World common iguana (*Iguana iguana*) and teju (*Tupinambis nigropunctatus*, family Teiidae), are sometimes a nuisance around chicken coops, destroying eggs or killing young poultry, but it seems doubtful if they are ever abundant enough to cause serious losses.

SNAKES

Venomous snakes occur in the families Elapidae, Hydrophidae, Viperidae, and Colubridae, but out of over 2000 species of snakes, only about 300 are poisonous and about 50 of these are in the marine family Hydrophidae and represent no real danger to man. Despite the fact that some species have a very strong venom, large numbers of hydrophids are caught and handled alive by fishermen on the Indian, Malayan, and Philippine coasts. These fishermen apparently differentiate between sea snake species that generally do not attempt to bite and those that cannot be handled with impunity. The former apparently use their venom primarily to capture prey fish rather than for defense.

Except for the African boomslang (*Dispholidus typus*), which is dangerous because it is large, and the bird snake (*Thelotornis kirtlandii*), another dangerous African colubrid, the venomous colubrid snakes are considered harmless to man because they are all rear-fanged (opisthoglyphous) and can only inject venom into small objects which are in the back of their mouth.

Elapids range throughout Australia, Africa, southeast Asia, India, and the New World from southern United States to Paraguay. Viperids occur in most of Europe, Asia, Africa, Ceylon, and the New World except at high altitudes and latitudes. Because of their abundance and widespread distributions, these two families of venomous snakes represent a hazard to man. However, it must be emphasized that they constitute only a small proportion of the total number of snakes and cause only minor mortality anywhere in the world.

Estimates of the mortality rates from snakebites and from all causes in various countries around the world are given in the following table. The worldwide mortality from venomous snakes is probably between 30,000 and 40,000 per year, or about one per 100,000 of population per year (Swaroop and Grab, 1954; Christy, 1967). The greatest number of fatal snake bites is reported in Southeast Asia, about 25,000 per year, with the heaviest mortality in Ceylon, India, Pakistan, and Burma (especially in the Irrawaddy and Chindwin Basins). (See table on the following page.)

Christy (1967) reports that the most frequent cause of fatal bites in Thailand is the Russell's viper (*Vipera russeli*), while in India the cobra (*Naja naja*) and kraits (*Bungarus*) are foremost. Several venomous snakes occur in South Viet Nam. Interestingly, there are few deaths due to snakebite in Australia, a continent unique in having more poisonous than nonpoisonous species (Buckley and Porges, 1956). In Africa, Christy indicates that there are between 400 and 1000 deaths per year, pri-

MORTALITY RATES FROM SNAKEBITE AND FROM
ALL CAUSES, COMPILED BY NEESS (1970)

	Mortality from Snakebite: Deaths per 100,000 of Population per Year[a]	Total Mortality: Deaths per 100,000 of Population per Year[b]
Australia	0.07	910.0
Burma	15.40	810.0[c]
Canada	0.02[d]	860.0
British Guiana	0.80[d]	990.0[e]
Ceylon	4.20	1090.0
Colombia	1.56[d]	1160.0[f]
Costa Rica	1.93[d]	1170.0
Egypt	0.20	1480.0[g]
England and Wales	0.02[d]	1140.0
France	0.06[d]	1300.0
India	5.40	1500.0
Italy	0.04[d]	990.0
Japan	0.13	890.0
Mexico	0.94	1560.0
Spain	0.02[d]	970.0
Thailand	1.30[d]	810.0[c]
Venezuela	3.10	990.0

[a]Data from Swaroop and Grab (1954).
[b]Data from Encyclopedia Britannica Yearbooks, 1956 and 1967. Values given are for 1953 unless otherwise noted.
[c]Malaya, 1964.
[d]Includes bites or stings of all venomous animals.
[e]Venezuela, 1953.
[f]Peru, 1953.
[g]UAR, 1965.

marily from bites of puff adders (*Bitis arietans*) and black mambas (*Dendroaspis polylepis*), but acknowledges that these figures are crude estimates. Between 3000 and 4000 people are thought to die annually from snakebite in South America; the majority of these occur in Brazil and are due to bites of the rattleless pit vipers. Parrish (1959) estimated that about 6700 persons per year are treated for snakebite in the United States and that about 14 or 15 deaths result, giving a mortality of between 0.01 and 0.03 per 100,000 of population per year. Klauber (1956) estimated that about 1000 people are bitten per year by rattlesnakes and about 30 deaths per year result.

Despite the greater fear generally expressed of venomous snakes, Parrish found that during the period for which he had data (1950–1954) there were more deaths in the United States due to the stings of hymenopterous insects than due to snakebite (86 versus 71). Klauber (1956) believed that more people are killed in the

United States by scorpions than by snakes. By far the majority of deaths from snakebite occur in parts of the world having hot climates, agricultural economies, and many people who habitually go barefoot. These circumstances lead more readily to unpleasant encounters between snakes and humans (Christy, 1967).

Some large tropical snakes, particularly the boids, prey upon such domestic animals as chickens, ducks, rabbits, dogs, and pigs, but their contribution to rodent control around human habitations more than balances the damage they cause. Contrary to folklore, the large constrictors are not a serious hazard to man for they instinctively attempt to escape whenever it is possible to do so. If cornered, they may strike, and a large snake can cause severe lacerations. However, the only "authentic" record Neess (1970) could find of a person being swallowed by a large constrictor is that reported in Schmidt and Inger (1957) of a 14-year-old Malay boy in the Talaud Islands who was swallowed by a reticulate python (*Python reticulatus*).

Except for the examples noted above, the dangerous or destructive reputations of amphibians and reptiles are mostly unfounded; it is a shame that their beneficial characteristics are not as widely appreciated as are the harmful qualities of a small minority of their kind.

AMPHIBIANS AND REPTILES AS RESERVOIR HOSTS FOR PATHOGENS

The role of amphibians and reptiles as natural reservoirs for pathogenic organisms is essentially unstudied. Because environment greatly affects an animal's exposure and susceptibility to infections, the role of aquatic species and larval stages presumably differs from that of terrestrial forms. *Aeromonas hydrophila* is an example of an organism which is sometimes pathogenic for man and is commonly found in amphibians. In frogs and salamanders, this kind of bacterium produces a condition known as "redleg" because of the hemorrhagic erosion found in the skin and between toes and limb attachments. Redleg infections are commonplace in laboratory stocks of amphibians, and occasionally natural populations are found which are heavily infected.

Several species of parasitic bacteria in the genus *Salmonella* exist which are pathogenic for man, causing inflammatory reactions in the intestinal tract called salmonellosis and manifested in fever, stomach cramps, diarrhea, and vomiting. Studies conducted by the United States Public Health Service and various state agencies indicate that pet store turtles, generally hatchlings of the red-eared slider (*Pseudemys scripta elegans*), are carriers of *Salmonella*. As a result, a legislative bill banning the sale of these turtles has been proposed in Pennsylvania, and similar actions are being considered elsewhere.

Amphibians and reptiles may be intermediate or reservoir hosts for a variety of parasitic metazoans as well. For example, in Indochina, China, and Japan, frogs and snakes are commonly the intermediate transmitter of infective plerocercoids of the tapeworm *Diphyllobothrium erinacei* which infects man. Similarly, the ascaroid parasitic worm *Gnathostoma spinigerum*, found in Southeast Asia, China, and Japan, has a life cycle which commonly includes an amphibian or snake as an intermediate host.

AMPHIBIANS AND REPTILES EXPLOITED BY MAN

Man preys upon many large species of amphibians and reptiles because he has found their flesh to be a delicacy. The demand for frog legs is so great that the North American bullfrog (*Rana catesbeiana*) has been introduced into many parts of the world for breeding purposes. The commercial importance of this species is reflected in the fact that in 1948 there were 766,262 pounds of bullfrog legs exported from Cuba, a country where the species was introduced by man, into the United States, where the species is native (Mertens, 1960). Legs from the edible frog (*Rana esculenta*) have long been considered a gourmet's delight in Europe, and man has probably preyed upon this species since prehistoric time. In southern Asia, the legs of the Indian bullfrog (*Rana tigrina*) are equally cherished. The giant leptodactylid frogs in the genera *Batrachophrynus* and *Telmatobius*, native to the high lakes of the South American Andes, are roasted whole.

Actually, the flesh of all frogs and toads is edible, and the extent to which smaller species are hunted depends upon how hungry people are for meat. Koreans, which have very limited meat supplies, even use the rice-paddy frog (*Rana limnocharis*), a two-inch-long species, for food. Some of the largest species of frogs are preyed upon very little by man because they are difficult to catch. In the United States, for example, the pig frog (*Rana grylio*), nearly as large a species as the bullfrog, is seldom eaten because it is so wary and lives among floating vegetation where it is almost impossible to catch. The largest frog of all, the goliath frog (*Rana goliath*) of the African Belgian Congo which reaches a length of over ten inches, lives in deep pools of swift mountain streams and is only occasionally captured by natives with long-handled scoop nets. Few salamanders are utilized for food by people, but in Japan the giant salamander (*Andrias japonicus*, family Cryptobranchidae), which grows to a length of over five feet, is captured by fishermen with hook and line and its flesh, being held in high esteem, is eaten.

The meat of many large reptiles is also considered a delicacy. Among snakes, the large boas and pythons provide a favorite meat for people inhabiting the regions where they occur. In the United States, canned rattlesnake meat is an expensive delicacy. Thousands of sea snakes (family Hydrophidae) are caught annually by "fisheries" off the coasts of India, the Malay Peninsula, and the Philippines and shipped to various countries, especially Japan, where the meat is eaten fried or smoked. Large lizards provide a favorite kind of meat in many tropical countries. One New World species, *Iguana delicatissima*, even bears a scientific name reflecting its culinary value; this species has a restricted distribution in the West Indies. The common Iguana (*Iguana iguana*) is hunted commercially in South America and sold regularly in markets. Turtle soup, of course, is served nearly everywhere that it is available, the marine green turtle (*Chelonia mydas*) being much in demand for this purpose. In the United States, snapping turtles (*Chelydra serpentina*) and diamondback terrapins (*Malaclemys terrapin*) are prized for their flesh. Diamondback terrapins were greatly reduced in numbers by market hunters but are making a comeback in many areas. During the great seafaring days, the giant tortoises of the Galapagos, Aldabra, and other tropical islands were heavily depended upon to provide fresh meat for the sailors. Turtle eggs, especially those of the green turtle, are a favorite food nearly everywhere, and in the past nests were preyed upon to such a degree that many countries now have laws controlling the numbers of eggs which may be taken. Finally, crocodile meat, particularly that from the tail, is another popular food item in some tropical countries.

The skins of nearly all large reptiles and some of the anuran amphibians are valued for leather goods. The American alligator (*Alligator mississippiensis*) has been particularly exploited for this purpose and, without protection provided in recent years, might have become extinct because of the hunting pressure. Despite a continual increase in demand and hunting pressure, the number of alligator skins marketed dropped from an estimated 200,000 annually at the turn of the century to 190,000 in 1929, 80,000 in 1939, 18,000 in 1942, and 6800 in 1943. The price paid by dealers in Florida rose from $4.00 per hide in 1937 to $19.75 for a seven-foot hide in 1943. In 1944, Florida passed a law protecting the alligator during breeding seasons and prohibiting the capture of specimens less than four feet in length. Alligators are also currently granted protection in Georgia, in national parks, and in several refuge areas. Except for caimans (*Caiman*, *Melanosuchus*, and *Paleosuchus*), which have dermal armour on their ventral surface (the belly skin is commercially the most valuable portion on any species), other crocodilians are also hunted to various degrees for their hides. The skins of large lizards and snakes are tanned and used for a variety of purposes, from hatbands to shoe leather. Comparable exploitation of hawksbill turtles (*Eretmochelys imbricata*) has occurred, this species being hunted commercially for tortoiseshell.

The natives of many tropical countries kill countless numbers of young crocodilians and adults of smaller reptiles and amphibians so that they can be made into tourist souvenirs. The skins of these animals are removed, stuffed with padding of some kind, fitted with glass eyes, dried, and then varnished. It is appalling to view the number and variety of animals wasted in this manner around tropical tourist meccas. Crocodilians, lizards, snakes, toads, and frogs are all sold by vendors.

Countless amphibians and reptiles are captured alive and sold as pets or exhibited in "gardens" and zoos. Mertens (1960) reports that thousands upon thousands of tortoises, lizards, snakes, and tree frogs are imported every spring from Italy and the Balkan countries into central Europe for this purpose. In the United States, all pet stores sell vast numbers of juvenile turtles, lizards, and snakes, many of which are imported from tropical regions. Boas, iguanid lizards, and red-eared sliders are particularly popular with pet fanciers. Reptiles of all kinds, particularly the venomous and larger species, are exhibited throughout the world in "reptile gardens" of various descriptions. The exhibition of vipers and cobras by "snake-charmers" is commonplace in Mohammedan countries. Unfortunately, despite whatever good intentions the buyer of such animals might have about taking care of them, very few individuals survive very long in captivity unless cared for by an expert who is aware of their temperature, moisture, and food requirements and gives them a proper environment in which "normal" activity and rest are both possible.

Anuran amphibians are widely used in biological and medical teaching and in research laboratories around the world. European laboratories exploit the edible frogs (*Rana esculenta* and *Rana ridibunda*) as laboratory animals, Americans use leopard frogs (*Rana pipiens*) most commonly, and Asians the Indian bullfrog (*Rana tigrina*). Clawed frogs (*Xenopus laevis*) have the honor of providing the first reliable early test for pregnancy in humans. If urine from a pregnant woman is injected into the body cavity of an unfertilized female clawed frog, hormones present in the woman's urine will cause the frog to begin laying eggs in a few hours. After discovery of this test in the 1940's, gynecologists from all around the world requested female clawed frogs and they were raised in great numbers for this purpose. Subsequent studies have shown that a variety of anurans can be used for the same test, so laboratories now tend to use whatever species is most economically and easily obtained.

Throughout man's history, amphibians and reptiles have been used for medicinal and therapeutic purposes. Dried toads and frogs are sold in China for medicinal purposes, and in Japan salamanders are used as anthelminthics to rid people of intestinal worms. The medicinal skink (*Scincus scincus*) received its name because it was formerly used by Europeans as a general remedy for a diverse assortment of ailments. This lizard is still used for medicinal purposes by some Mohammedans. The Chinese have long used the dried venom of toads (probably *Bufo bufo gargarizans*) as a medicine. According to the Pên Ts'ao Kang Mu, the preparation (Ch'an Su) can be "employed by external application in the treatment of canker sores, sinusitis, and many local inflammatory conditions." When administered in the form of a compound pill, it is said to be able to break colds. Toad venoms are known to contain a number of biologically active substances, including cholesterol, epinephrine, and norepinephrine, and they are potentially useful as a source of digitalis. Consequently, there may be more basis in fact than superstition in their medicinal use by the Orientals. Any vermifuge value that the Japanese salamanders have must also stem from the venoms contained in cutaneous glands. Nearly every species of venomous snake is kept in captivity somewhere in the world for the production of antivenins (antisera or antivenoms) and anavenoms (anatoxins or toxoids). Antivenins are obtained by inoculating animals with sublethal doses of venom until immunity is achieved. Blood from such immune animals injected into another individual will provide the latter with immunity against the venom. Anavenoms are venoms which are partially detoxified by heat but which still retain their immunizing power. At the present time, there is at least one antivenin available for the most important venomous snakes of each continent (see Oliver and Goss, 1952; Klauber, 1956).

The natives of South America have long exploited the venom of poison frogs (subfamily Dendrobatinae, family Ranidae) to obtain food by using it to poison their arrows. They kill the frog by piercing it with a sharp stick and then hold it over a fire. The heat causes the cutaneous glands to secrete droplets of venom which are scraped into a container and allowed to ferment. Arrows are then dipped into the poison and allowed to dry. Although the venom on these missiles is not sufficient to be effective against large animals such as man, it will instantly paralyze a small bird or a mammal such as a monkey. Thus, the venom is used almost entirely in hunting rather than as a weapon against other people or large predators.

Finally, an attribute of both amphibians and reptiles which is little appreciated in "civilized" countries is that they include among their kind vast numbers of carnivorous and insectivorous forms which destroy pests and disease-carrying species. Tree frogs, toads, geckos, and other forms live in human habitations, particularly in the tropics, and destroy literally countless mosquitoes, flies, beetles, cockroaches, and other pests. This service is recognized by the natives of such areas and they, contrary to North Americans, include such animals among their welcomed guests. Amphibians and reptiles both play important roles in the control of garden and agricultural pests. Toads are particularly effective as a biological control for insect pests, and the giant toad (*Bufo marinus*) has been introduced into many different countries where sugar cane is grown for this purpose. Now that it is recognized that chemical controls are only temporarily effective against insects and are hazardous to the environment, perhaps more attention will be given to the great potential which toads have as insect controllers. There are many snakes which feed almost exclusively upon mice and rats, and there are even harmless snakes which feed upon venomous snakes. The value of snakes as controllers of vermin is recognized in such countries as Guatemala, where the government posts signs advising the rural people to protect the large boas rather than killing every snake they encounter. It is most unfortunate that, throughout the world, ignorant people needlessly slaughter every snake that they see because of the reputations of the relatively small number of venomous species.

MAN'S EFFECT UPON AMPHIBIAN AND REPTILIAN POPULATIONS

There is a complex relationship between the abundance and diversity of amphibian and reptilian faunas and the density of human populations. In both the African tropics and the New World tropics, there are relatively low-density human populations and great numbers and kinds of amphibians and reptiles. In the tropics of Southeast Asia, India, and Ceylon, rich amphibian and reptilian faunas occur together with high-density human populations. The North Temperate Zone contains high-density human populations but relatively low numbers and few kinds of amphibians and reptiles. The South Temperate Zone includes even fewer amphibians and reptiles together with relatively low-density human populations. Thus, as noted by Neess (1970), there are places where human populations are in contact with numerous species of amphibians and reptiles and have been for a long time. There are other places where large numbers of people have been contact with only a small number of amphibian and reptilian species. And there are just a few places, such as the Arctic, where humans are not in contact with amphibians and reptiles. The consequence of the above relationships is that man has had a varied effect upon the distribution and abundance of different species. Human populations and their alterations of natural environments have had a significant and long-lasting effect upon many species. Although man has not attempted to use pesticides to eradicate or control any species of amphibian or reptile, his technology now directly or indirectly influences all species of amphibians and reptiles whether or not they are influenced by ever-increasing human densities.

Some species of amphibians and reptiles have greatly benefited from their association with man and now depend to varying degrees upon symbiotic relations with humans. Agricultural development of land has had a positive effect upon many populations of amphibians and reptiles. The conversion of arid land to rice paddy fields, such as occurred in parts of Mexico (Fig. 15–1), provides an almost perfect habitat for a variety of anuran amphibians which could not otherwise exist in the region. Similarly, the construction of stock reservoirs through damming of gullies has provided breeding ponds for many arid land amphibians in western North America. Irrigation ditches have also been important in allowing certain amphibians to extend their ranges. In addition to the amphibian populations which these man-made aquatic habitats support, there is frequently an increase in the number of aquatic snakes and turtles present in such a region. In general, the more arid the region is, the more important will be artificial ponds to the aquatic species.

A variety of lacertid lizards in the Old World and iguanid lizards in the New World find agricultural land to their liking because rock walls, fences, rubbish dumps, ruins, and so forth provide them with ideal basking sites and cover in which to hide from their enemies. Snakes also tend to be attracted to farmlands because of the increased populations of rodents which are frequently present. Grass snakes (*Natrix*) are notorious for seeking manure heaps in which to deposit their eggs.

Human dwellings attract a variety of amphibians and reptiles because they provide an abundance of food, including insects, mice, and rats, and they also provide the amphibians and reptiles with considerable protection from their normal predators. In the tropics, there are many toads, tree frogs, and other anurans in nearly every house and building. Similarly, a variety of snakes, including *Lycodon* in southern Asia, *Boaedon* in Africa, *Ptyas* in India, and *Leptodeira* in the New World, commonly inhabit houses and other buildings (Mertens, 1960). Of all amphibians and reptiles, some lizards are the most closely associated with human dwellings. In the tropical areas of the world, geckos of various kinds are as closely associated with man as are house mice and, like the mice, are seldom found far from human habitations. Some species, like the red-spotted tokay (*Gekko gecko*) are almost impossible to find outside of houses; others, such as the Turkish gecko (*Hemidactylus turcicus*) occur away from humans but are most abundant in buildings. Other lizards exploiting man's buildings include the common agama (*Agama agama*) and an assortment of smooth lizards (*Mabuya*) in

Figure 15–1 A rice paddy in Guerrero, Mexico, exemplifying how man can alter an environment and, in this case, make it more favorable for amphibians and less favorable for xeric-adapted species formerly present. Notice the chaparral type of vegetation which was characteristic of the area now covered by water. (Photograph by Kenneth R. Porter.)

Africa, the Kabara Goya monitor (*Varanus salvator*) in Southeast Asia, keel-tail iguanas (*Tropidurus*) in Brazil, and anoles (*Anolis*) in tropical and subtropical North America. The close association of many of these species with man has resulted in a wide distribution by human agencies similar, again, to that of mice and rats.

While some species have derived benefit from their association with man, many more amphibians and reptiles have been detrimentally affected by human activities. As already mentioned, many species have been decimated because of their commercial value, danger to man, or misunderstanding and wanton killing. Many populations have been exterminated through destruction of natural habitats. Obviously, the creation of rice paddy fields out of arid land is as detrimental to the xeric-adapted species as it is advantageous to the aquatic forms. The current widespread clearing of tropical rainforests must be having a disastrous effect upon arboreal species formerly present. The draining and filling of swampland virtually eliminates those populations dependent upon that habitat.

Like life in general, all amphibians and reptiles are greatly endangered at the mo-

ment because of the awesome contamination of the biosphere by chemical pesticides. There is little information available on the susceptibility of different species to specific pesticides, but some data are available on the general effects of agricultural spraying on amphibians and reptiles. There appears to be a severe mortality of amphibians when DDT is sprayed at a rate of one pound per acre, and it is catastrophic when this compound is applied at a rate of two pounds or more per acre; mortality of lizards and snakes seems to begin at about two pounds per acre and is probably severe at about four pounds per acre (Rudd, 1964). When heptachlor was applied at a rate of one to two pounds per acre for fire ant control in southeastern United States, there was a significant mortality of salamanders, anurans, turtles, lizards, and snakes (DeWitt and George, 1960; Ferguson, 1963a); dieldrin applications also caused mortality in these groups. Ferguson (1963a) found an especially heavy mortality of reptiles following heptachlor applications and concluded that these were particularly susceptible to this chemical. Some mortality of diamond-backed watersnakes (*Natrix rhombifera*) occurred following applications of 0.25 pounds per acre of heptachlor; two treatments of 0.25 pounds per acre each resulted in mortality of common snapping turtles (*Chelydra serpentina*). The effect of any pesticide on amphibian and reptilian populations undoubtedly varies with the season of the year that it is applied, since this will influence where the animals are, what they are eating, and which age groups will be exposed to the poison. Boyd *et al.* (1963) demonstrated, for example, that recently metamorphosed cricket frogs (*Acris crepitans* and *A. gryllus*) are especially susceptible to DDT. These workers concluded that the insecticide is probably absorbed through the skin and the susceptibility of newly-metamorphosed frogs is probably related to their greater surface-to-volume ratio as compared with that of larger individuals. Since DDT is often present in an oil film on the surface of water, cutaneous absorption of the chemical could occur as the animals emerge through this film; it may also occur through the skin which contacts the substratum after emergence from the water. Consequently, the application of such pesticides at the time that metamorphosis is occurring could have a much more serious effect than if they were applied at another time.

It is known that amphibians and reptiles, like other animals, accumulate various pesticides contained in their food (Hunt and Bischoff, 1960; Pillmore, 1961; Rudd, 1964), but it is not known what delayed effects such accumulations may have on them. Cricket frogs (*Acris crepitans* and *A. gryllus*) are among a very small group of vertebrates which have shown an ability to evolve resistance to pesticides. Some populations of these frogs live in the vicinity of cotton fields which are heavily treated with DDT and occasionally sprayed with toxaphene, methyl parathion, and endrin; other populations live in habitats where they have little or no exposure to such chemicals. When individuals from these various populations are exposed to water contaminated with the insecticides, those from populations chronically exposed to pesticides survive in significantly greater numbers than those from unexposed populations (Boyd *et al.*, 1963; Ferguson, 1963b). Although the cricket frog appears to be gaining a natural resistance to pesticides, all species of amphibians and reptiles may not be so fortunate. Rapid evolution requires that the favored phenotype (resistance to pesticides, in this case) already exist in the population, because there may not be time for it to be produced by chance alone. Rapid evolution also requires short generation times, and in order to survive the population must have the reproductive potential necessary to tolerate the loss of unfit (pesticide-susceptible) individuals while changes in gene frequencies are occurring. Some species or populations of amphibians and reptiles may meet these requirements, others will not. Thus, one can expect that there will be reductions in the variety of amphibians and reptiles which will be present in areas where pesticides are continually being applied or where they are continuously accumulating.

When all kinds are considered, many more amphibians and reptiles have suffered from their association with man than have profited. The number of endangered species increases constantly; fortunately, a few have been granted legal protection. The tuatara, the last surviving member of a formerly very abundant order, has been provided with very strict legal protection by the New Zealand government. The African Convention for Conservation of Nature and Natural Resources was formed by 38 heads of state or their representatives in September, 1968, and is now in effect. Among the "totally protected" animals legally protected by this group are two species of toads (*Bufo supercilliaris* and *Nectophrynoides vivipara*), all marine turtles, the giant, angulated, and radiated tortoises, three genera of boas (*Bolyeria, Casarea,* and *Acranthophis*), the leaf-tailed gecko (*Uropeltis fimbriatus*), and the skink *Macroscincus*. Crocodiles are listed as "totally protected," but they may be hunted by special permit in some countries. Papua and New Guinea passed the Crocodile Trade (Protection) Ordinance in 1966 which is designed to prevent excessive killing of crocodiles by licensed traders and hunters and restricting the sale of skins from breeding crocodiles. Licenses which are issued specify the number of crocodiles which may be taken and the area within which they may be purchased or taken. Regulations are included which may prohibit the taking of crocodiles that are of particular species, below a minimum size, or above a maximum size. The salt-water crocodile (*Crocodylus porosus*), once heavily exploited for its skin, has been granted total protection for a period of 10 years by the Western Australian Cabinet. This body also approved intensified conservation efforts in regard to the fresh-water crocodile (*Crocodylus johnsoni*).

The United States government enacted Public Law 91–135, the "Endangered Species Act," in December, 1969; this provides protection to a variety of organisms including the crocodilians. The New York State Legislature passed the Agriculture and Markets Law, Section 358-a, in February, 1970, prohibiting the importation or sale of the animals or their skins which are "deemed to be near extinction." The animals listed include alligators, caimans, and crocodiles. The New York City Council passed the "Low Bill" (No. 943–1124) in March, 1969, prohibiting the sale of American alligator products in the City of New York. The combination of federal, New York State, and New York City laws provides full protection to crocodilians in this particular state and is particularly significant since New York is a major center for style, manufacturing, and sale of "alligator" products. Unfortunately, other such centers, including Paris, Rome, and Tokyo, have not yet passed comparable laws; if they do not take appropriate steps very soon it is likely that many crocodilian species will be extinct before the year 2000.

The Gila monster (*Heloderma suspectum*) has been granted legal protection in Arizona since 1952; this probably has saved it from extinction.

Despite the above-mentioned and a few other similar protective acts, I am afraid that man, through his ignorance and portentous technology, is going to continue to exterminate the very forms of life which he could be using to rid his villages of mice and rats or his gardens and fields of insects.

References

Boyd, C. E., S. B. Vinson, and D. E. Ferguson. 1963. Possible DDT resistance in two species of frogs. Copeia 1963:426–429.

Buckley, E. E., and N. Porges. 1956. Venoms. Washington, D.C., American Association for the Advancement of Science. Publ. No. 44.

Christy, N. P. (Ed.) 1967. Poisoning by venomous animals. Amer. J. Medicine 42:107–128.

Cochran, D. M. 1961. Living Amphibians of the World. Garden City, N.Y., Doubleday. 199 pp.

Deraniyagala, P. E. P. 1939. The tetrapod reptiles of Ceylon. Ceylon J. Science 1:1–412.

DeWitt, J. B., and J. L. George. 1960. Pesticide-wildlife review, 1959. Washington, D.C., Bur. Sports Fish. Wildlife, United States Fish and Wildlife Service. Circular 84, rev., 36 pp.

Earl, L. 1954. Crocodile fever. London, Collins. 255 pp.

Ferguson, D. E. 1963a. Notes concerning the effects of heptachlor on certain poikilotherms. Copeia 1963:441–443.

——— 1963b. Mississippi Delta wildlife developing resistance to pesticides. Agric. Chem. Sept. 1963, 3 pp.

Hunt, E. G., and A. I. Bischoff. 1960. Inimical effects on wildlife of periodic DDT applications to Clear Lake. California Fish and Game 46(1):91–106.

Klauber, L. M. 1956. Rattlesnakes; Their Habits, Life Histories, and Influence on Mankind. Berkeley, Univ. California Press. 2 vols., 1476 pp.

Mertens, R. 1960. The World of Amphibians and Reptiles. Translated by H. W. Parker. London, George G. Harrap. 207 pp.

Neess, J. C. 1970. Amphibians and reptiles as pests. In: Principles of Plant and Animal Pest Control, Vol. 5: Vertebrate Pests; Problems and Control, Washington, D.C., National Academy of Sciences, Subcommittee on Vertebrate Pests, Committee on Plant and Animal Pests, Agricultural Board, National Research Council. Pp. 42–57.

Oliver, J. A., and L. J. Goss. 1952. Antivenin available for treatment of snakebite. Copeia 1952:270–272.

Parrish, H. M. 1959. Deaths from bites and stings of venomous animals and insects in the United States. Arch. Intern. Med. 104:198–207.

Pillmore, R. E. 1961. Pesticide investigations of the 1960 mortality of fish-eating birds on Klamath Basin Wildlife Refuges. Washington, D.C., United States Fish and Wildlife Service, Wildlife Research Lab. 12 pp.

Rudd, R. L. 1964. Pesticides and the Living Landscape. Madison, Univ. Wisconsin Press. 320 pp.

Schmidt, K. P., and R. F. Inger. 1957. Living Reptiles of the World. Garden City, N.Y., Hanover House. 287 pp.

Swaroop, S., and B. Grab. 1954. Snakebite mortality in the world. Bull. World Health Organization, United Nations 10(1):35–76.

INDEX TO SCIENTIFIC NAMES

Page numbers in *italics* refer to illustrations.

503

Typhlopidae, 250, 406
Typhlopoidea. See Scolecophidia.
Typhlops, 184, 211, 249, 252
Typhlops avakubae, 406
Typhlops bibroni, 406
Typhlops braminus, 132, 406
Typhlops diardi, 406
Typhlops punctatus, 406
Typhlops schlegeli, 406
Typhlotriton, 84
Tyrannosauridae (Deinodontidae), 224
Tyrannosaurus, 224, 229

Ulemosaurus, 237
Uma, 187
Uma inornata, 401, 434
Uma notata, 401
Uma notata rufopunctata, 303
Uma scoparia, 299
Uranascodon superciliosa, 401
Uranocentrodontidae, 97
Urocordylidae, 100
Urocordylus, 100
Urodela, 89, 93, 107–111, 113, 245, 254, 275, 439
Uromastix, 178, 206
Uromastix acanthinurus, 280, 321, 402
Uromastix aegyptius, 282
Uropeltidae, 212, 250, 326
Uropeltis, 212
Uropeltis fimbriatus, 500
Uropeltis ocellatus, 406
Urosaurus graciosus, 303, 401
Urosaurus ornatus, 269, 297, 303, 402
Urostrophus torquatus, 402
Uta, 187, 310, 311
Uta mearnsi, 402
Uta stansburiana, 272, 297, 306, 310, 311, 389, 402, 434, 477, 478
Uta stansburiana stejnegeri, 400

Varanidae, 208, 251, 288, 297, 404
Varanoidea (Platynota), 208–209
Varanopsidae, 233
Varanosaurus, 100, 233
Varanus, 137, 145, 146, 161, 168, 205, 207, 208, 251, 330, 388
Varanus bengalensis, 387, 398, 403
Varanus brevicauda, 208
Varanus komodoensis, 208, 403, 492
Varanus niloticus, 398, 400, 403, 476, 493
Varanus salvator, 474, 499
Varanus varius, 398

Vaughniellidae (*Vaughniella urodeloides*), 109
Venyukoviamorpha, 237
Vipera, 140, 189, 214
Vipera ammodytes, 441
Viper aspis, 389, 408
Vipera berus, 250, 309, 408, 441, 482
Vipera russelli, 120, 408, 493
Viperidae, 119, 120, 156, 214, 250, 255, 295, 296, 493
Viperinae, 214
Virginia striatula, 408

Waggoneria, 98
Waggoneriidae, 98
Weigeltisauridae, 230
Weigeltisaurus, 230
Whaitsiidae, 236
Wolterstorffiella, 109

Xantusia, 178, 207, 310, 311
Xantusia vigilis, 178, 310, 379, 399, 402, 434, 478
Xantusidae, 207, 251, 297, 397, 477
Xenochrophis piscator, 408
Xenochrophis vittata, 408
Xenodon, 138
Xenodon merremii, 176, 177
Xenoderminae, 214
Xenodermus, 214
Xenopeltidae, 250
Xenopeltus unicolor, 212
Xenopholis, 214
Xenopus, 28, 75, *103*
Xenopus laevis, 22, 75, 79, 85, 104, 277, 335, 352, 367, 375, 496
Xenopus tropicalis, 34
Xenosauridae, 208, 251
Xenosaurus, 208
Xenosaurus grandis, 208, 404
Xenosaurus newmanorum, 208
Xenosaurus platyceps, 208
Xestops, 208

Younginiidae, 204

Zaisanurus, 109
Zatracheidae, 96
Zonosaurus, 207

SUBJECT INDEX

Page numbers in *italics* refer to illustrations.